T0399172

The Institutions of Extraterrestrial Liberty

The Institutions of Extraterrestrial Liberty

Edited by

CHARLES S. COCKELL

School of Physics and Astronomy, University of Edinburgh, UK

OXFORD
UNIVERSITY PRESS

Great Clarendon Street, Oxford, OX2 6DP,
United Kingdom

Oxford University Press is a department of the University of Oxford.
It furthers the University's objective of excellence in research, scholarship,
and education by publishing worldwide. Oxford is a registered trade mark of
Oxford University Press in the UK and in certain other countries

Published in the United States of America by Oxford University Press
198 Madison Avenue, New York, NY 10016, United States of America

British Library Cataloguing in Publication Data
Data available

Library of Congress Control Number: 2022945200

ISBN 978–0–19–289798–5

DOI: 10.1093/oso/9780192897985.001.0001

Printed and bound by
CPI Group (UK) Ltd, Croydon, CR0 4YY

Links to third party websites are provided by Oxford in good faith and
for information only. Oxford disclaims any responsibility for the materials
contained in any third party website referenced in this work.

Contents

List of Contributors

Zarinah Agnew, Social Science Observatory, Communes Research Commune, District Commons (with Eldridge Cruse, Kevin Bruce, Engelbert Wilfred Perlas, and Joseph Krauter).

Stephen Baxter, c/o Christopher Schelling, Selectric Artists, 9 Union Square #123, Southbury, CT 06488, USA.

Annalea Beattie, Amity University, Mumbai, RMIT University, Melbourne, Mars Society Australia, National Space Society of Australia.

Mukesh Bhatt, Birkbeck, Birkbeck College, University of London, London, UK.

Octavio Alfonso Chon Torres, Universidad de Lima, Lima, Perú.

Elena Cirkovic, Ecosystems and Environment Research Programme, Helsinki Institute of Sustainability Science (HELSUS), University of Helsinki, Finland.

Charles Cockell, School of Physics and Astronomy, The University of Edinburgh, Edinburgh, UK.

Raphaël Costa, Paris–Saclay University, Paris, France.

Ian A. Crawford, Department of Earth and Planetary Sciences, Birkbeck College, Malet Street, London, UK.

Janet de Vigne, Moray House School of Education and Sport, The University of Edinburgh, Edinburgh, UK.

Martin Elvis, Harvard–Smithsonian Center for Astrophysics, 60 Garden Street, Cambridge MA, USA.

Simon Malpas, English Literature, The University of Edinburgh, Edinburgh, UK.

Allan McKenna, School of Law, University of Glasgow, Glasgow, UK.

Tony Milligan, Cosmological Visionaries project, King's College London, London, UK.

Lucas Mix, Durham University, Durham, UK.

Ethan Morales, University of California — Berkeley, Berkeley, CA, USA.

Simon Morden, 13 Egremont Drive, Gateshead, UK.

Thomas Moynihan, St Benet's College, The University of Oxford, Oxford, UK.

Chris Newman, Northumbria Law School, Northumbria University, Newcastle, UK (with William Ralston).

Anthony Pagden, University of California, Los Angeles, CA, USA.

Stefania Paladini, Birmingham City University, Birmingham, UK (with Ignazio Castellucci).

Burkhard Schafer, Edinburgh Law School, The University of Edinburgh, Edinburgh, UK.

Jim Schwartz, Department of Philosophy, Wichita State University, Kansas, USA.

Michael Shermer, The Skeptics Society, USA.

Saskia Vermeylen, Law School, University of Strathclyde, Lord Hope Building, 141 St James Road, Glasgow, UK.

Matjaz Vidmar, The University of Edinburgh, Edinburgh, UK.

Frans von der Dunk, University of Nebraska-Lincoln, Lincoln, NE, USA.

Sheri Wells-Jensen, Bowling Green State University, Bowling Green, OH, USA.

Joanne Wheeler, Alden Legal, London, UK.

Robert Zubrin, Mars Society, 11111 W. 8th Ave. unit A, Lakewood, CO, USA.

Introduction

Charles S. Cockell

There is a view that when we establish permanent settlements beyond the Earth we should start again, that we should not take with us the mistakes and conflicts of our past and repeat them in space, that we should strive for new social arrangements that transcend the disagreements of the past. Yet humans cannot escape themselves and there are certain arguments and disagreements that will continue in space whether we wish them to or not. One of those is the eternal discussion on liberty and what liberties people should or should not have. As Thomas Hobbes recognized over three centuries ago (Hobbes 1651), this is a matter that has no obvious resolution:

> For in a way beset with those that contend on one side for too great Liberty, and on the other side for too much Authority, 'tis hard to passe between the points of both unwounded.

If we were to ask ourselves when people first began to consider the question of liberty in the extraterrestrial environment, we would realize that although people may not have explicitly used the word 'liberty' in connection with space exploration and settlement until quite recently, the matter is implicit in everything to do with these activities. When Konstantin Tsiolkovsky memorably wrote: 'Earth is the cradle of humanity, but one cannot remain in the cradle forever', he was making a statement about the choice to remain shackled to Earth or seek the liberty to move out beyond it (Tsiolkovsky 1911). This was in essence a statement about human liberty of a sort.

The first tentative steps to formulate ideas about 'planetary protection' in the 1950s and 1960s, regulations that govern the acceptable contamination of other planetary bodies with biota from Earth and vice versa, were always going to impinge directly on mission costs, planning, and even the locations where we could send space missions. Planetary protection is a discussion on the types and expressions of scientific and technical liberty in the space environment. Needless to say, all discussions about the weaponization of space are overt discussions about the liberties that people have in space and the way in which the conduct of people in space might influence others. As on Earth, no action in space, or decision about the use of space, whether scientific, technical, or legal, is devoid of consequences for the way in which others can operate. Therefore, arguments about liberty in space had begun as soon as humans imagined moving into that domain.

Nowadays, people unsurprisingly take much of their lead in these discussions from existing instruments of space law that touch on matters of freedom to operate in space such as the United Nations Outer Space Treaty or Moon Treaty. In this volume, these

Charles S. Cockell, *Introduction*. In: *The Institutions of Extraterrestrial Liberty*.
Edited by Charles S. Cockell, Oxford University Press. © Charles S. Cockell (2022).
DOI: 10.1093/oso/9780192897985.003.0001

legal instruments, and others, are cited often. It is altogether sensible that existing legal frameworks are developed, refined, and improved in an evolutionary way since a great deal of time has already been spent on developing and debating them. However, we should not underestimate the extent to which space may offer environments, and give rise to populations, that could lead to completely novel constitutions, legal structures, and societies that wish to part with earlier Earth-bound legal frameworks. A degree of the revolutionary is possible, particularly if organizations, mandates, agreements, accords, and other legal structures are perceived to be coercive or counter to the wishes of those in space. In this volume, we explore some of these possibilities.

We should not be under any illusion that the matter of liberty in space will ever come to an end or be resolved. It may seem paradoxical to state but we should not even wish for discussion on extraterrestrial liberty to be complete. A society in which there is no disagreement about liberty, and in which all people are in complete union on the matter of how much freedom and authority there should be, is a society in which freedom of thought is extinguished. Passionate disagreements on liberty are the symptom of liberty itself. Hobbes' conundrum will never be solved and nor should it be. It should follow us to the stars.

Nevertheless, it is also true to say that disagreements about the level of control that one person, or group of people, should be able to exert over others can lead to bloody conflict, and it does not require a long list here to convince the reader of this truth. A cursory examination of the twentieth century alone is sufficient. Thus, although we do not wish to stymie debate on freedom, it is also the case that we should encourage its peaceful and continuous resolution to prevent discord or war in space. That characteristic of disunion on the matter of freedom we should try to avoid at all costs. How is this to be done? Trivially, again, the most effective way to achieve the peaceful resolution of dispute is to encourage a vibrant and vigorous debate on the forms of freedom and their realization in space.

The Institutions of Extraterrestrial Liberty brings together a diversity of thinkers with a wide scope of expertise in matters of space science, ethics, law, philosophy, and other fields. Our interest was in considering the institutions that would be likely to secure liberty in space. How should early governance structures be assembled? What are the ideal forms of institutions, from science academies to schools and governments? Needless to say, although liberty encompasses ideas, and often ideals, the way in which we experience freedoms is necessarily shaped by human institutions and the way in which these institutions operate and treat human beings in an everyday sense. In this book, authors consider many dimensions of freedom in space and what our ends might be in securing different forms of human liberty. We especially wanted to investigate the practical implications of these goals for the way in which we go about building organizations, rules, and laws on Earth and in space.

Although liberty within future space settlements was of special interest, we also wanted to consider more immediate matters that are a concern in the present time— the laws that influence the freedom that organizations have to choose satellite orbits and the challenges of preventing war in space, for instance. Thus, the chapters in this volume consider both the near-term and long-term prospects for different types of freedom.

Our discussions underline the fact that space is not a homogeneous environment. The freedom one has to use a particular satellite orbit, and the arguments that might exist in who can claim it, and how an orbit is to be claimed, is a quite different matter to how coercive a station commander should be in encouraging a group of space settlers to increase oxygen supply on Mars. Just as the forms of liberty have varied across Earth, the space environment would seem to offer an almost infinite variety of permutations of different forms of liberty in different locations.

The types of liberty that can exist in the space environment will change not just over spatial dimensions but also over time. The liberties experienced by a small group of settlers on the Moon may be different from those experienced by a later hypothetical settlement containing many thousands of people. As on Earth, no agreement on the form of liberty will be static and changing circumstances, technologies, and scales of activity are likely to bring forth different combinations and requirements for social control. In much the same way that the infinite spatial scales of space might seem to make possible a vast manifestation of social arrangements, so we might also speculate that the great expanses of time over which human civilization might explore and settle this frontier will allow for a never-ending discourse on the changing conditions for liberty and how it is to be realized. Political philosophers have their work cut out for them.

Despite this vision of the infinite possibilities for new discussions on liberty, there may be, running through them, universal forms of freedom that we hope to encourage. For example, on Earth, freedom of conscience, the capacity to write or speak one's opinions without imprisonment, is regarded by many people as a freedom that is independent of geography, history, or culture. Similarly, in space, there are likely to be such forms of freedom that have no relevance to whether one is on the Moon, Mars, in Earth orbits, or in interplanetary space. We can only find out what these freedoms are through enlightened discussion about the forms of freedom beyond Earth.

The chapters we present here follow a four-day online conference that we held 8–11 June 2021, hosted by the UK Centre for Astrobiology. Originally planned as an in-person meeting in Edinburgh, ironically a virus confined us to our homes, just as the lack of oxygen will confine future space settlers to their habitats and spacesuits for much of their time, influencing the liberties they experience. The four days of discussion were open to the public and they provided a means for the authors to refine and consider the chapter contents contained herein.

There was the matter of how to order the chapters. I did consider arranging the chapters in spatial order from matters of liberty near Earth (satellite orbits) to the furthest reaches of liberty (interstellar worldships), but the contents of many chapters can be applied at many scales, making this a false separation. They might also be grouped according to content (e.g. legal, scientific, philosophical). However, this too does an injustice to the multidisciplinary nature of many of the problems we address. In the end, I opted for arranging the chapters alphabetically by author. This is the least invidious and provides the reader with the option of reading chapters in an order that might reflect the reader's own interests or priorities.

The work presented here should by no means be considered exhaustive and we hope that it is, albeit written by Earth-bound scholars, a useful contribution to a discussion on the depth and breadth of liberty in space, a discussion that will be as endless in space as it has been since the emergence of human societies on Earth. We also hope that this book may be of value as a snapshot of ideas in 2021. The exploration and settlement of space is changing at a remarkable rate and, as will be clear in these chapters, there are many unresolved questions about how humans should operate there. Future scholars may find these chapters useful to compare to their own perspectives as an historical record of how ideas and thoughts have changed since we wrote this book.

References

Hobbes, T. 'Letter to Mr. Francis Godolphin, from Paris' in *Leviathan* (first published 1651, Penguin Books, 2017).

Tsiolkovsky, K. 'A letter' (1911) <http://www.uranos.eu.org/biogr/ciolke.html>.

1

Are we ready for new liberties?

Stewarding mutually assured autonomy through place-based experiments

Zarinah Karim Agnew, Eldridge Cruse, Kevin Bruce, Engelbert Wilfred Perlas, and Joseph Krauter

1. Towards Posthumanism

'We must act as if the future is today.'

—Howard J. Ehrlich

The term 'human' comes from the Latin term humus, meaning soil. The settling and migration away from planet Earth, then, represents linguistically, symbolically, and materially (until soil is created or found on other planets), stepping into posthuman times. In relation to geologic time required to produce soil, our collective histories as a species have been both relatively brief and extremely dynamic. Human society has grown and shrunk, evolved and flourished, many times over. Presently, we have an international world forged on and now dominated by colonial and territorial frameworks and Westphalian state sovereignty. Yet for many the settling of space is the 'utopian impulse' (Jameson 2005), that is, both a response to and critique of existing social conditions, alongside a vision that goes beyond these, an attempt to transcend or transform those conditions to achieve a new society. As we bear witness to corporate interests and ultra-elite tourism forging their way to the forefront of exploration (Shammas and Holen 2019), we must cast a critical eye over whether our technological achievements are sufficient to promise space futures that are any different at all to our earthly pasts. It seems our social technologies are woefully lacking behind our technical achievements. In Murray Bookchin's 1978 speech (Bookchin 1978), he clarifies the distinction between futurism and utopianism (or what Bookchin refers to as 'ecology'); 'Futurism is the present as it exists today, projected, one hundred years from now'. What we need, he says, is utopianism. He asserts that we need to change the present in order to enable a future that differs from what we have today in order that the future can reveal itself to be different from the present. Much has been written about how our collective imaginations have in some ways stunted the futures that are available to us. Today our societies are organized around capitalism, and the system is entrenched to the point that, as Frederic Jameson famously noted, it is easier to envision the end of the world than the end of capitalism (Fisher 2009). Prominent in this field of inquiry is Mark Fisher's writing on capitalist realism (Fisher 2009) in

Zarinah Karim Agnew et al., *Are we ready for new liberties?*. In: *The Institutions of Extraterrestrial Liberty.*
Edited by Charles S. Cockell, Oxford University Press. © Zarinah Karim Agnew et al. (2022).
DOI: 10.1093/oso/9780192897985.003.0002

which, it is argued, our collective inability to imagine alternative futures renders the future cancelled.

What is to be done in this moment where the expansive futures we seek are out of reach? Not because of our technical inability but perhaps because we are blinded by a naïve optimism that new technologies forged in present day socioeconomics are sufficient to create new futures; a techno optimism that neglects to recognize that we are a deeply and fundamental social species, with subjectivities forged in societal pressures that in many ways keep us anchored in the past, stunting our ability to envision wide enough horizons for humanity. Understanding that the possibility of present-day change is perhaps a (necessary but not sufficient) prerequisite to realizing future extra-terrestrial liberties, it is necessary to define the who: that is, who we must be and who we must become in order to be agents capable of transforming our social structures in order to realize our desired futures. The term 'NewSpace' denotes the arrival of capitalism in space, and as Shammas and Holen describe, only benefits a select portion of Earth-based society (Shammas and Holen 2019). Yet if all humans are to participate and be agents in extra-terrestrial liberties, then all humans should at least be able to experience and access liberties now / in the present / in the here and now. On Earth, we are far from this reality. Defining what constitutes the category of human, and who qualifies for the protections that this category confers, has historically been a topic of debate and prone to dire errors. Defining, reflecting on, and expanding what and who constitutes the category of human is vital if we are to avoid repeating the mistakes of history as we send our first communities into extra-terrestrial domiciles. In their work on posthumanism, Braidotti writes: 'Not all of us can say, with any degree of certainty, that we have always been human, or that we are only that. Some of us are not even considered fully human now, let alone at previous moments of Western social, political, and scientific history' (Braidotti 2017). In a similar vein, Bratton's piece on terraforming reminds us that whilst terraforming usually refers to the transformation of landmasses of other planets to render them acceptable to life, we urgently need to do the same here on Earth for this planet to remain viable to life as we currently know it (Bratton 2019).

These lines of thinking beg the questions of who are the peoples of the future and how do we get from here to there? Here we posit that questions of extra-terrestrial governance will never be sufficient to achieve utopian societies without also addressing how to prepare humans raised in current social modes for lives of liberty that are not centred around the libertarian individualisms but instead focus on collective and mutually assured autonomies. In addressing the types of institutions that might be created in space to maximize liberty, we ask related questions: Are we as actors ready for sovereignty? Do we have the skills, relationships, knowledge, and experience to forge new liberties, and, if not, how do we prepare for the kinds of liberties, and challenges to liberty, that space migration will afford us? We argue that our work to prepare to be the species that we need to be—the actors, co-operators, and participants—must begin here and now. In so doing, we aim to adopt an approach outlined by Ferrando (Ferrando 2016) in which they adopt an onto-epistemological in which 'space will be thus accessed as "a way of revealing", allowing for an original understanding of the notion of space'.

As a case study for this type of experimentation, we describe two prototype spaces, alongside their lessons, theory, and praxis, that have explored the collective freedoms for perhaps some of the most liberty restricted and disenfranchised: the formerly incarcerated (henceforth 'returning citizens'). In these spaces, half of the residents have all served indefinite life sentences, and have collectively served many centuries incarcerated, and have not only been raised in the absence of access to effective government institutions but have also arguably internalized their constraints.

Small community domiciles are an ideal format for learning about and iterating on how to obtain freedom and be free, to explore new liberties and kinship. Those who comprise the first communities of 'space migration' (Ferrando 2016) (a term used to denote moving away from colonialist language) will also be small groups, sharing space, community, and vital lifelines; these will be small groups for whom the ability to cooperate, be cohesive, and steward each other's liberty will be a personal matter of life and death and also a collective matter of existence and persistence. These small, stable communities, in some ways then, parallel the dynamics of what are likely to be early 'space migrations'. Moreover, we argue, the social logics, norms, and principles that seed these early extra-terrestrial communities will be formative in the evolution and social reproduction of cultural norms as new generations join them in space. Thus, learning how to build new liberties, mutually assured freedoms, and kinship in small domestic groups is an essential part of the social preparation that the future of humanity in space, and on Earth, deserves.

2. Mutually Assured Stewarded Autonomy

I am most glad to have my personal liberty, but I only have it to the extent that there is a sphere of freedom in which I can operate. That sphere is coproduced by people who live together or who have agreed to live in a world in which the relations between them make possible their individual sense of being free.

> So perhaps we might regard personal liberty as a cipher of social freedom. And social freedom cannot be understood apart from what arises between people, what happens when they make something in common or when, in fact, they seek to make or remake the world in common. The world is given to me because you are also there as one to whom it is given. The world is never given to me alone but always in your company. Without you, the world does not give itself. We are worldless without one another.
>
> Butler and Berbec 2017

Hannah Arendt (2018) wrote that freedom 'does not come from me or from you; it can and does happen as a relation between us, or, indeed, among us'. Building on these ideas in an interview, Butler speaks to the idea of personal liberty as a 'cipher of social freedom' (Butler and Berbec 2017), and in doing so, speaks directly to the issue we focus on here. Liberty, both on planet Earth and elsewhere, is a socially held, mutually assured phenomenon. As such, the types of liberty and freedom that we can hope for will be determined by our sociality. Our social logics affect our development as individuals, the development of our relationships with each other, the kinds of kinship and communities that we forge, all the way to the institutions that shape us and the governance that empowers us.

We argue that liberty requires individual development and collective nurturing if we are to encounter new forms of it in extra-terrestrial societies. As individuals and as groups (collectives, teams, kin, corporations), we need to cultivate the ability to know what one wants and needs (autonomy) as well as collectively ensure that each individual has the material resources needed to enact those needs (agency). Mutually assured autonomy dictates an understanding of and relationship with the autonomy of others, an embodied sense that we are, as Butler puts it, making a sphere of freedom in common. Like all commons, mutually assured autonomy needs tending to if it is not to be degraded. Freedom is a commons, a commonwealth whereby 'Collective freedom and collective happiness can exist only as the sum of the freedom and the happiness of the individuals' (Bakunin). And thus when an individual freedom is degraded, taken from or used to benefit another (extracted), collective freedom is weakened. We posit that a form of 'anti individualistic individuality' (Gee 2003) is required to create the conditions for flourishing. By that we mean a move from individualism to mutually assured autonomy. As agents of the future, it is essential that we collectively develop and cultivate the skills for autonomy (moral, political, and personal) as well as agency and self-determination without leaning into individualism or collectivism in the extreme.

3. Place-based Experimental Spaces in Future Crafting and Social Innovation

> The shared impulse of all versions and understandings of social innovation is the effort to design initiatives in a particular part of society—an organisation, a practice or an area of activity—that signal a promising path of wider social change even as they meet a pressing need. The innovations that the movement seeks to advance convert experiments designed to solve social problems into transformative ambition: the effort to change some part of the established arrangements and assumptions of society. The focus of the movement falls on problems that have not been solved by either the state or the market.
>
> Unger, 2015

Domiciles, residential living environments and *home* are crucial sites of lasting and stable experimentation. As such, these are vital test beds for social technologies of the future that we suggest have much to offer in preparing communities of practice for extra-terrestrial liberties. We have witnessed a surge in test beds and living labs for the exploration of sociotechnical avenues. Engels, Wentland, and Pfotenhauer (2019) argue that such *in situ* sites of experimentation constitute a marked approach to advance sociotechnical innovation across geography and technical disciplines. These approaches serve to move beyond, or have arisen in reaction to, the increasingly clear observation that policy shifts to 'grand societal challenges' are lacking. 'Test beds and living labs represent an experimental, co-creative approach to innovation policy that aims to test, demonstrate, and advance new sociotechnical arrangements and associated modes of governance in a model environment under real-world conditions' (Engels, Wentland, and Pfotenhauer 2019).

Physical spaces are one locus for prototyping altered social logics. They are by no means the only site of collective transformation but these sites of domesticity are long-lasting, shared social enclosures that allow the development and exploration of decision-making, governance, conflict resolution, and resource allocation and distribution. Many of these collective behaviours, by definition, require collectivity to experiment with, and therefore experimenting with home is a powerful site for social innovation. Home and the economics of home are also widely scalable as a set of local behaviours that have historically been adopted in widespread ways, as we have seen, for example, with the emerging dominance of the nuclear family.

The home, as a major site of human social reproduction, is also a point of leverage for the altering of 'biopolitical production', a term that builds on Foucault's conception of biopower. For Foucault, the self is produced by regulatory power structures that surround it, and the subject, therefore, is the realization of both history and power. Biopower refers to the forces that administer and regulate human life, lives, and populations. Biopolitical power then describes the impact of these regulatory mechanisms as extended to 'quantitatively to social reactions and qualitatively to consciousness, intimacy etc' (Ruivenkamp and Hilton 2017).

Here we refer to commons-based places—that is, physical spaces that constitute a commons, where common*ing* is the mode of being together. In his essay 'Commoning as a Transformative Social Paradigm', David Bollier writes:

> In facing up to the many profound crises of our time, we face a conundrum that has no easy resolution: how are we to imagine and build a radically different system while living within the constraints of an incumbent system that aggressively resists transformational change? Our challenge is not just articulating attractive alternatives, but identifying credible strategies for actualizing them. (Bollier 2020)

This is precisely what these experimental spaces attempt to do.

We present the notion that place-based stewarded autotomy builds collective social logic for 'being otherwise' and social future crafting (Ratti and Claudel 2016). In his essay on social innovation, Unger (2015) writes that 'social innovations must point beyond themselves' and he warn us that these efforts are likely to be most impactful in the arenas of social life that are not encompassed by either the state or the market, that lie between the economic and political worlds. The home is precisely such a social sphere. Home, then, for these case studies are sites of prefiguration (Boggs 1977), a site to embody the modes of organization and social relationships that strive to reflect the future society that we seek. Prefigurative politics combines five processes: 'collective experimentation, the imagining, production and circulation of political meanings, the creation of new and future-oriented social norms or "conduct", their consolidation in movement infrastructure, and the diffusion and contamination of ideas, messages and goals to wider networks and constituencies' (Yates 2015).

You cannot advance liberation or support social transformation if you have not transformed your own practices and the ways your organization does things. Change begins with the organization, and the people within it, embodying what change looks like. That is a requirement to be meaningful contributors to the type of change that justice movements are envisioning and building every day.

—Solome Lemma (Van Deven 2021)

In this mode of exploration, the intervention of creating and prefiguring home economics is itself the thing that transforms each of the participants. In a similar vein, Howard Ehrlich describes a concept known as transfer culture (Ehrlich, 1996), an attempt at future worlds right here in amongst the *status quo*, the old world, an experiment in the future, manifested in the present. Sites of transfer culture support the manifestation of the ideas, processes, behaviours, skills, and activities that are needed to help humans transition from the current social formation to the incoming one. We posit here that home, both as sites of prefiguration and transfer culture, serves as a vital test bed for social technologies of the future.

4. Wisdoms of the Formerly Incarcerated

In the early 20th century California prisons followed a norm system called the convict code ('don't snitch' and some other things). By the end of the century, self-governance via the code had collapsed and the system reorganized around race gangs. Some institutional economists approach prisons as laboratories for emergent self governance, because it is a tough environment for stable governance to emerge in, but it emerges anyway, and that's important.

(Skarbek 2014)

4.1 Social order emerges in altered places

Social order emerges wherever and however we arrange humans, from the young humans depicted in the real life *Lord of the Flies* scenario (Bregman 2021), to online communities, or to the emergence of sign language in Nicaraguan schools for deaf children (Kegl and Iwata 1989). The same is true of prisons. As Skarbek notes (Skarbek 2014), gangs use a community responsibility system where all members are responsible for the actions of any member. The groups govern property rights, regulate internal markets, and promote cooperation when the state fails to or is unable to intervene or govern. These social relations of control and regulation reach outside prisons to the outside world. Crucially, these are, according to Skarbek and Freire, not based on trust and are focused on peacekeeping in order to maintain profitability of a few individuals. How then, on re-entry to civilian life do we transform these dynamics into mutual care, solidarity, and collective freedoms? Durkheim tells us that social order arose out of the shared beliefs, values, norms, and practices of a given group of people, and so it is here that these spaces start their work. US incarceration rates are higher than anywhere else in the world. One in nine prisoners is serving a life sentence. California has the highest percentage of prisoners serving life sentences, around 35,000. These humans often enter the system as children and frequently serve in excess of 20 years. On their return, housing, home, and community—fundamental parts of survival—are lacking. This renders much of their existence to bare survival. Under these conditions, the idea of being free is a treacherous one, and begs the question, to what are they returning?

In an essay about surviving incarceration Steven Powers reflects:

It's funny to appreciate a city from a jail bus, because it's that very city that's sending you away. And I mean all of it is sending you away—from the disciplinarians that fancy themselves educators, to the bullshit jobs that treat you like a crook from the first day, to the police who menace you everywhere you go. The city taught me the survival mechanisms it would ultimately punish me for. There's no other way to understand it. (Powers 2018)

This points directly to the issue that this group face on re-entry to civilian life, that returning to the same conditions that led to harmful behaviour is likely to lead to similar issues. We ask instead: what if this group returned to a different set of social conditions? Can the experiences and wisdoms of this liberty-disenfranchised population teach us about the future of liberty and emancipation?

Cockell (2018) has piloted a programme in Scottish prisons that allows incarcerated individuals to design settlements on Mars as part of a multi-week educational and future-oriented social reform process. Part of the programme ('Life Beyond') includes considering governance, social arrangements, and civic duties for these environments—crucially bringing their own experience of social logics in prison to these issues. Cockell's paper concludes that 'Life Beyond has helped harness the often latent talents of people in prison, helped generate positive contributions to society and stimulated hope for rehabilitation and future plans for life beyond prison'.

In many walks of society then, formerly incarcerated individuals have been shown to be a guiding force not only in personal transformation but in societal change also. Carcerally impacted leaders play a key role as catalysts in shifting public narratives empowering community leadership, influencing and collaborating at multiple levels of civic responsibility and government (Sturm and Tae 2017). Nowhere is this more evident than in the movement to reduce incarceration, to cultivate restorative and transformative justice and the revitalization of communities that have been affected by mass incarceration. This is down to having what is referred to as 'experiential carceral knowledge' (Binnall 2021). James Binnall, an associate professor of law, criminology, and criminal justice, writes:

In the end, experiential knowledge derived from a carceral experience has value in all contexts—the jury included. Without it, efforts to change our carceral trajectory are incomplete. As one of my students is fond of saying: 'You scooped us up and locked us in. Now listen to what we have to say—you may learn something.'

We argue that this community not only has much to offer the fields of criminal justice reform, violence prevention, carceral processes—what might be called the nature of being unfree—but also the nature of being and getting free, as well as the future crafting of liberty. The individuals contributing to this chapter have served indefinite sentences ('life sentences'), have been deeply liberty-deprived individuals who each have decades of experiential carceral experience but who have managed to transform themselves, overcome the chronic negation of their autonomy, to return to society, free individuals. As such, their journey, collective experience, and wisdom provide a unique and valuable set of lessons about the future of new liberties. Here we consider the process of learning to be socially free and liberated after the social death and chronic elimination of liberty that incarceration imposes.

The spaces that we describe here were built with this precise problem in mind, and we ask whether we can ensure that those returning to society are not returning to the same societal structures that let them down in the first place, but are leapfrogging into the autonomous participatory commons. Can entering into these new (or perhaps old) social logics not only help people recover from a chronic absence of freedom and bypass old modes but actually develop and prototype new forms of collective autonomy and sovereignty through home? In building a domestic commons centred around the needs and wisdom of this group, these spaces and communities forged the co-creation of liberatory spaces that provide lessons about how groups can achieve collective freedoms, and learn methods of assuring our mutual autonomy. These lessons are crucial, not just for returning citizens emerging from the social order of prisons but also for the general population who must emerge from an increasingly individualized, zero-sum based society if they are to be qualified, prepared agents of new liberties, on this planet and others.

5. Principles of New Liberties—Lessons from Place-based Stewarded Mutual Autotomy

'We have to talk about liberating minds as well as liberating society.'
—Angela Davis

5.1 What can be learnt for extra-terrestrial liberties from incarceration?

Both prisons and the settlement of extra-terrestrial loci can be seen as projects that abandon their negative externalities. Prisons are where we hold what we view as our failed projects, however we might be wiser to observe that these individuals are those whom we have failed.

> 'Prisons do not disappear social problems, they disappear human beings. Homelessness, unemployment, drug addiction, mental illness, and illiteracy are only a few of the problems that disappear from public view when the human beings contending with them are relegated to cages.'
> —Angela Davis

Similarly, the settlement of space is often motivated as a version of Albert Hirschman's form of 'exit' or escape from the negative externalities of the life we have made for ourselves on Earth; in his treatise on Exit or Voice, Hirschman (1970) describes the ultimatum that confronts humans in the face of deteriorating quality of goods; actors can either can *exit* by withdrawing from the situation, or can remain but instead deploy *voice* as a tactic in pursuit of repair and transformation of the situation (see Dowding 2016). How then to ensure that space migration does not fall into either category of exit or voice but instead serves as a platform of something, new, transformative, truly explorative, that neither rejects our pasts nor permits the past to obfuscate the future?

Laws are something put in place to deal with conflict, and until there is conflict there is little need for laws or rules, nor the enforcing of them. Indeed, anarchy does not mean 'no laws', it means *no need for laws* (Gee 2003). And certainly not laws made by some 'other' to be enforced non-consensually and uncritically. And so, in the communities described here, that are centred around mutual emancipation and collective autonomy—where we recognize that until we are all free, we are none of us free—we discuss modes of being together, in evolution, in consensual conflict, in mutual freedom and survival. In sites of transfer culture such as these, self-governed communities are formulating modes of conflict resolution that both help heal from the past and prefigures a future social logic to come. This section outlines some principles that have guided learning and praxis from these prototyped spaces.

> I've been a caged animal all my life, you can't just open the door and let me walk out into the jungle, I will be eaten alive.
>
> —Bert Perlas, incarcerated for 22 years

5.2.1 Social freedom

As earlier described, a number of authors have written about the idea of personal liberty as a cover for social freedom; that liberty, autonomy, and freedom are phenomena that are meaningless without our mutual respect and collaboration. In these emancipatory spaces and homes built by formerly incarcerated individuals, there is an understanding that the autonomy of each individual is stewarded by the others. This includes being given space to make mistakes, to push over the edge of what is acceptable in order to know that edge better and navigate with more data in the future. This is premised on recognizing that getting free is a journey that we are all on together, and different people who move at different rates will find different elements distinctly troubling, but that the group works through this process together. It is the recognition that getting free is not a linear process but a complex journey involving dynamic interactions with each other. Byung-Chul Han's analysis of modern day freedom (or lack thereof) indicates that there is no freedom other than self-realization in collective formation: 'Freedom is synonymous with a working community (i.e., a successful one)' (Han 2017).

5.2.2 Anti-individualistic individuality and transindividuation

> The sovereign human knows that it is not the whole, but that it lives as a part of the whole and knows how to relate to it.
>
> —Teoman Gee

One of our goals is to create conditions and social logics for becoming sovereign, self-valorized, and self-determined. Anti-individualistic individuality goes beyond the prioritization of either the individual unit at the expense of society, nor does it prioritize collective society at the expense of the individual. In present day western individualism we perceive ourselves as the singular drivers of our lives, enacting our free will along our individual paths, both singularly responsible for that which we achieve, and also for that which we do not. A sense of collective individuality, or anti-individualistic individuality is central to collective living, to allow for a diversity of individuals in a collective shape that is neither homogeneous nor totalitarian. Bernard Stiegler (see Crogan 2010) writes on the concept of 'transindividuation', a state that is neither individuated or interindividuated but rather is constituted by the processes of co-individuation whereby the 'I' and the 'We' influence, evolve, and transform each other. Stiegler claims that transindividuation is the platform for social transformation. In the domestic spaces described here, the social structure used to work towards this is the support circle. The horizontal structure of these circles distributes care such that no person is left out, as well distributing the skills for care work. This structure then dismantles hierarchies of care that are common in society. These circles include a diverse set of individuals with a wide range of life experiences, which allows the linking of the personal to the political, inspired by the consciousness-raising circles of the late 1960s feminist movement. This peer-to-peer structure is the platform used for the transindividuation of the group.

In his writing about how to cultivate such a collective sense of individuality, Teoman Gee names some central principles that have served as guiding principles for these communities, which includes taking individual responsibility for our collective. Gee states that '[t]o understand oneself as oneself, yet at the same time as an integral part of a collective that shelters one's own existence, demands an individual responsibility for this collective'. This means that we don't rely on retreating to private territory with individual rights but instead collectively contribute to the survival of the group. A rigid or authoritarian society can function—and even thrive—on passive individuals, but a vital, adaptive, and flexible society, such as one engaging in space migration, needs active participants.

> We all know what needs to be done, because in space exploration, we all know what's at stake. The truth be told there's really no room for failure. Under these conditions, it becomes less about who is going to do this, but how we get it done.
> —Eldridge Cruse, 29 years incarcerated, two years free.

5.2.3 Self-determination and alternative justices
This means taking responsibility for the consequences of our individual and collective actions, understanding ourselves as agents in the world with social responsibility. Gee (2003) calls for a form of self-respect that constitutes balance, calm, and sovereignty (not needing to rule or dominate over others) as well as self-knowledge. Gee rejects the idea of conformity to the group in favour of creativity and ingenuity, stating that 'a people reaches a stand still with it loses individuality' and warning that allowing such individuality to flourish in a collective prohibits totalitarianism.

He advocates for 'imminent diversity', which is a call for a form of individuality that guarantees diversity of talents and skills that are required for the health of society.

> One of the greatest tools I have developed was the ability to just accept everyone for who they are and work as a team that enhances one another's strengths while at the same time being present for each other's weaknesses. Living in collective responsibility ensures that this is not a chore; we share values, interests, resources and support, no matter what that entails.
>
> —Kevin Bruce, 35 years incarcerated, three years free

New liberties must seek a balance between feeling responsible for our decisions and understanding that we are a product of our environment and circumstances. Evidence suggests that too much of the former (determinism) can lead to individuals being less willing to help others, increased aggression (Baumeister, Masicampo, and DeWall 2009), and cheating (Vohs and Schooler 2008). It seems that believing our actions are predetermined and not self-determined might then introduce antisocial behaviours in our social worlds. On the other hand, an understanding that we are all intertwined and that our actions are largely reflected by the environments that we have inhabited, some argue, will increase degrees of compassion and forgiveness in society.

In this community, this means an emphasis on transformative justice and violence prevention, as well as conflict resolution. Transformative justice is a political framework and approach for responding to violence, harm, and abuse that seeks to respond to violence without creating more violence and/or engaging in harm reduction to lessen the violence (Kim 2018). In these liberatory spaces, the values of transformative justice are part of daily inquiry and praxis and conflict resolution strategies are built for each group, by themselves, such that at any given time, the processes of harm prevention and response are adapted to and opted into by the Individuals affected by them. This is distinct from more traditional models of justice, in which external values—external to the stakeholders and the situation—may take priority; for example, there may be prescribed methods of dealing with particular types of conflict, determined in advance of any situation.

Long-duration interplanetary missions confer many significant differences from present day space activity, and will require new tools of conflict mediation and justice to those used in Earth-orbiting missions such as the International Space Station. Distance-related signal delays will affect the quality of communications with Earth-based mission control, outside support, or family and friends, as real-time interaction will be limited to crew members. Thus these groups must learn how to be stakeholders and agents in their own conflict resolution processes and form the skills for processes that can adapt to rapidly changing circumstances. The need for robust conflict resolution systems for long-duration interplanetary human spaceflight missions is clear then, if only as a fall back. Whilst pre-flight measures such as crew selection and psychological assessment can minimize the chances of conflict, it is not possible to avoid conflict and harm entirely. Studies that attempt to simulate the psychological challenges that humans might face on a journey to Mars, for example, are likely to underestimate adverse psychological pressures and conflict-inducing scenarios. Conflict on a long-duration spaceflight, where crew members from different backgrounds and cultures live in close quarters for an extended period of time, could reduce their effectiveness or even completely destabilize the mission. It is important, therefore, that measures are put in place to resolve conflicts where they arise. These forms of justice will need to be adaptable, be able to be deployed without external guidance, and need to be internally recognized and respected by all stakeholders involved. Therefore, any extra-terrestrial communities forming settlements in space, must have (i) worked considerably on their existing kinship relations; (ii) have established and agreed-upon approaches to alternative forms of justice; as well as (iii) the skills for developing them *in situ*. As Tiarks writes:

> restorative justice offers a method for resolving disputes that requires no external intervention, because the crew members are empowered to act autonomously. It encourages reintegration, which is important when no crew member is expendable and there is unlikely to be room on a spacecraft for confinement facilities. Even in the event that no disputes arise, the crew benefits from the security of knowing that an accessible system for resolving disputes is in place. (Tiarks 2020)

What is needed to build such skills is training and relationships. All individuals must develop their own self determination and a respect for 'imminent diversity', as described by Gee, as well as the practical skills for alternative forms of justice.

To date, criminality in space relies on existing state-based legal and regulatory regimes. Codes of conduct are used to maintain internal coherence, and it is assumed that risk of harmful behaviour is low as astronauts are well trained and carefully selected (see Newman 2016). However, this does not address that should humans manage to forge settlements in space, away from the Earth's orbit, these selection criteria may not be sufficient to ensure collective liberties. We will be forced to consider alternative solutions to prisons for socially harmful behaviour, as well as recognizing that we don't yet know what those behaviours might be as these missions will be encountering novel circumstances. Whether one believes that they function to reduce harm or not, prisons on Earth are costly and require considerable materials that may not be available to extra-terrestrial communities. Prisons require guarding and surveillance, as well as considerable effort to maintain a boundary between those incarcerated and the rest of the population, and the utility of how other sanctions or punitive measures, for example financial or social isolation, work remains unclear. Moreover, in extra-terrestrial communities, where each person is crucial to the mission, removing someone who has irreplaceable skills from the general population may put the entire community at risk. Thus, whether one believes that punitive measures are the appropriate response to harmful behaviour, extra-terrestrial communities will have to explore other forms of protecting their societies and maintaining liberty. Space migration is a pivotal moment for humans to develop the awareness, skills, and tools for alternative modes of ensuring freedom, such as restorative justice, transformative justice, harm prevention, and iterative adaptive conflict resolution. Crucially, these skills, processes, and social relations must be built, used, and practised ahead of time.

5.2.4 Constituent power (*potenza*)

In these spaces, it is constituent power (*potenza*) that is cultivated. Ruivenkamp and Hilton (Ruivenkamp and Hilton 2017) describe the distinction between constituent power (*potenza*) and constituted power, the former being 'the power of the multitude that aims at the realisation of its potential, which is always already there as a space within capitalist structures and relations, and through which these are altered'. The latter constituted power (*potere*) being the 'power of the authorities that appears when formal constitutions are formulated, which comes from above and is imposed upon the multitude'.

> In prison there was a currency, or maybe a better description would be leverage? Safety was our currency. The whole point in prison politics was to keep people safe and to maintain some order to the potential chaos. I imagine some similarities coming together on a moon settlement. People have to come together and agree on collective decisions for the development of their culture and societal norms. I think how people come together would be subjective because people are comfortable with different forms of leadership. I personally like when everyone is in charge of their lives and works together to not trample over other's needs and wants for the collective good.
>
> —Bert Perlas, 22 years incarcerated, six years free

The activities of commoners (commoning) have historically been a site of the production of autonomy and 'common rights'. These are rights that have been recognized by the state but are not granted by the state. De Angelis writes: 'Common rights ... originate in their being exercised and therefore the state can only, at most, acknowledge them and confirm them. This recognition is precisely what happened in the history of the freedom chargers discussed by Linebaugh—the 1215 Magna Carta and the 1225 Charter of the Forests' (De Angelis 2017). Therefore, the practices of commoning are inherently related to the practice of autonomy, and as commoning is a social practice and a set of social relations for collective, mutually assured autonomy. Earlier we discussed how liberty is a collectively held social structure, that is produced in common, that freedom is a commons. Accordingly, then, we suggest that an understanding of the history of commons logics will be valuable for any communities exploring and settling new frontiers, because they are forging future societies. In his chapter 'The Production of Autonomy, Boundaries and Sense' in his *Omnia Sunt Communia: On the Commons and the Transformation to Postcapitalism* (De Angelis 2017), the author speaks to the concept of commons autonomy, suggesting that whereas autonomy is often understood in tension with some other social order, the commons strives for internal unity and order, controlled by its own self-produced laws as opposed to being subject to the laws of any other entity. This is a fundamental concept to consider for liberty and self-rule in extra-terrestrial communities, which will be facing unique challenges and constraints that institutions forged here on Earth will necessarily be underprepared to govern. There will surely be new constraints encountered, communication delays, and other scenarios that will require autonomy and self-governance in these communities. We suggest that the experience of learning to be a commoner, then, plays a fundamental role in the preparation for extra-terrestrial liberties.

Interviews with returning citizens in these spaces reveal that they value (i) the prioritization of communal interactions; (ii) the ability and freedom to regulate self-isolation; (iii) leadership principles that guide towards self-autonomy; (iv) awareness and observation of the formation of non-democratic hierarchies, as well as tools for avoiding and transforming abuses of power. This points to subsidiarity—a principle of social organization that holds that social and political issues should be dealt with at the most immediate (or local) level that is consistent with their resolution. Lessons from these case studies have determined that this principle is a fundamental part of taking individual and social responsibility for collective freedom.

5.2.5 The evolution and social reproduction of liberty

We have approached freedom as an evolving process, consisting of personal tools, social skills, and environmental factors. In these spaces we have taken inspiration from the concept of the freedom quotient (FQ); like the intelligence quotient (IQ), the FQ should tell us how free we are, and how we can become even more free. The FQ comprises many factors, including the capacity to generate options (autonomy), the capacity to be able to choose between them in a meaningful way, and the capacity to be able carry out the choices we have made (agency). As we start to understand, and learn to measure, the capacities that underlie behavioural freedom, we can begin to put this natural free will on a scale. Such a scale should give us new insights into the factors that hinder or enhance our efforts to shape our lives. Communities centred around new liberties should attend to factors that are described in this (chapter). Part of any mutually assured autonomy centres these factors and holds them as a collective responsibility.

> 'If you are doing anything for the money, or the power or the fame, you are already screwed. You should be doing it for the love of the thing you are doing.'
> —Eldridge Cruse, 29 years incarcerated, two years free.

Byung-Chul Han writes: 'Freedom will prove to have been merely an interlude. Freedom is felt when passing from one way of living to another—until this too turns out to be a form of coercion. Then, liberation gives way to renewed subjugation. Such is the destiny of the subject; literally, the 'one who has been cast down' (Han 2017). Thus, freedom appears to both have some prerequisites and a developmental trajectory. Freedom should not be seen as a state but a series of phases, predicated on both having some combination of needs met and social relations. Han argues that in this moment in history, modern-day humans have become entrapped in our present-day form of freedom. We no longer feel compelled to do things but are entitled to have things, moving us from should to can, where 'should' has limits but 'can' has none. As a result, Han says, we are paradoxically imprisoned by the compulsion to do and have the things we can do and have.

Where once freedom was something that transpired in relationships ('Originally, being free meant being amongst friends. "Freedom" and "friendship" have the same root in Indo-European languages. Fundamentally, freedom signifies a relationship', Han 2017, Chapter 1), the modern-day neoliberal entrepreneur of the self has no ability, space for, or knowledge of relationships outside of purpose. Together, Byung-Chul Han (2017) tell us, this has become a totalizing force, such that in our isolation and alienation from each other, caged by our focus on what we can do and have, we are less free than ever.

Finally, we wish to discuss the importance of the social relations that extra-terrestrial communities hold over the social reproduction of liberty. Within a commons, the social relations are reproduced, instilling a resilience that we believe is essential to forming new liberties. Commoners reproduce themselves through the ways in which they maintain internal cohesion, and the circuits of operation are passed on to others who join these communities in times to come. This is key as it pertains not only to the first generation but to the values and forms of freedom that are conferred to subsequent generations that will be migrating from Earth to extra-terrestrial domiciles and institutions. De Angelis writes that autonomy is the 'property generated by the recursive interaction of components across a social network in such a way that the network that produced those interactions is regenerated and a boundary is defined. The network therefore is reproduced through a recursive loop ... and thereby constitutes a unity' (De Angelis 2017). The reproduction of cultural capital (Bourdieu 1999) then is a fundamental element in determining what liberties will be achieved in future, and we argue that these spaces, prison and the family, serve as places to explore and learn as well as sites to explore new modes of social reproduction of all elements of cultural capital, including liberty and collective freedoms. Our experience has shown that *in situ* lessons from residing in these spaces of freedom, does in fact scale to incoming generations. Our residents, having found themselves well-tooled, have set off to seed new spaces with recurrent, iterative social logics that allow for cultural norms to remain present in the collective environment but which are also subject to change and adaptation as the environment and individuals' needs shift. We believe these learnt experiences are an important aspect to ensuring that cultural reproduction in space is intentional, adaptive, and liberatory.

6. Conclusions and Speculative Futures

Here we have shared our guiding theory and praxis from stewarding autonomy as a collective practice, from incarceration to mutually assured autonomy and liberty. We demonstrate that voices, experiences, and wisdom from those affected by and working to change one of the most unjust social systems on this planet have much to offer us in thinking about new liberties in the future crafting of extra-terrestrial societies.

We have argued that the creation of stable residential domiciles are ideal sites for building the relationships, prerequisites, subjectivities, and cultural tools for new forms of liberty, and that any community responsible for seeding and socially reproducing new forms of liberty in space must be ready for such an endeavour, and that this work must begin now.

Our learning leads us to conclude that new liberties depend on developing the skills and relationships for mutually assured stewarded autonomy that include the following principles: (i) that personal liberty is a cipher of social freedom, a socially held commons in which the relations between and amongst individuals is key; (ii) the need to cultivate an anti- and transindividualistic individuality that involves the reciprocal dynamic influence between the 'we' and the 'I'; (iii) self-determination in a form that confers individual and collective responsibility for our actions, resulting in the skills for conflict mediation and alternative forms of justice that are essential for communities living outside of earth orbiting missions; (iv) the forging of constituent power for a collectively held and coproduced forms of liberty (a freedom commons); (v) a framework that views collective liberty as an evolving and dynamic process with material and social prerequisites to ensure agency and autonomy.

We suggest that educational facilities that prepare astronauts for space flight must incorporate some of these principles and lessons, to ensure that social freedom is a collectively held commons and that this work must begin now, on this planet, in order for there be a chance that liberty in space might be superior to that on Earth, as well as be adaptable to allow individuals to live successfully in extra-terrestrial settings. It is our hope that such a project would ensure that what is collectively learnt about extra-terrestrial liberties and would also serve those of us on Earth in our evolving pursuit of freedom, and that in this way, space migration can avoid being a simple for of exit for the elites but instead can be the frontier on which we all become free, no matter what the geography under our feet.

Just as prisons do not make social problems disappear, we suggest that settling new geographies will not solve our existing societal issues. Here we propose that any settlement, on Earth or elsewhere, that hopes for a different and more just future requires not only institutions and organizations that create the subjects needed for such utopian visions but also that we—as subjects and relational beings—must do the work to earn ourselves a place in that future.

We have discussed the problems that come from a society with huge disparities in who is free and who is not, and how the overall collective liberty suffers under these conditions, our experiences as collective freedom is an evolving process, and the need for tools, kinship, and environmental conditions for anti-individualist individuality. Finally, that building of the conditions for freedom (constituent power) is a core part of the process for any extraplanetary futures that represent us all. What do these conditions entail? In their writing on prison abolition, Moten and Harney (2004) stake the claim: 'Not so much the abolition of prisons but the abolition of a society that could have prisons, that could have slavery, that could have the wage, and therefore not abolition as the elimination of anything but abolition as the founding of a new society.' We support this notion that the conditions for our species to be collectively free, for our mutually assured autonomy, must be met in order that we are people ready for the worlds and societies that we seek, on this planet as well as others.

References

Arendt, H. *The Human Condition* (2nd ed.) (University of Chicago Press, 2018) 370.

Baumeister, R. F., Masicampo, E. J., and DeWall, C. N. 'Prosocial Benefits of Feeling Free: Disbelief in Free Will Increases Aggression and Reduces Helpfulness' (2009) 35(2) *Personality and Social Psychology Bulletin* 260–8.

Binnall, J. 'Carceral Wisdom' (15 October 2021) *Inquest* <https://inquest.org/carceral-wisdom/> accessed 24 July 2022.

Boggs, C. 'Marxism, Prefigurative Communism and the Problem of Workers' Control' (1977) 6(Winter) *Radical America* 99–122.

Bollier, D. 'Commoning as a Transformative Social Paradigm' in *The New Systems Reader* (London: Routledge, 2020).

Bookchin, M. 'Utopia, Not Futurism: Why Doing the Impossible is the Most Rational Thing We Can Do' (24 August 1978). WFCR Radio Broadcast Collection (MS 741). Special Collections and University Archives, University of Massachusetts, Amherst Libraries [Mums741-b237-i005]. Radio Broadcast Collection (MS 741).

Bourdieu, P. 'Cultural Reproduction and Social Reproduction' (1999) 2 *Modernity: Cultural Modernity* 351.

Braidotti, R. 'Posthuman Critical Theory' (2017) 1(1) *Journal of Posthuman Studies* 9–25 <https://doi.org/10.5325/jpoststud.1.1.0009>.

Bratton, B. *The Terraforming* (Moscow: Strelka Press, 2019).

Bregman, R. *Humankind: A Hopeful History* (Boston, MA: Little, Brown and Company, 2021).

Butler, J. and Berbec, S. 'We Are Worldless without One Another: An Interview with Judith Butler' (2017) 26 June *The Other Journal*.

Cockell, C. *Life Beyond: From Prison to Mars* (2018).

Crogan, P. 'Knowledge, Care, and Trans-Individuation: An Interview with Bernard Stiegler' (2010) 6(2) *Cultural Politics* 157–170.

De Angelis, M. *Omnia Sunt Communia: On the Commons and the Transformation to Postcapitalism* (London: Bloomsbury Publishing, 2017).

Dowding, K. 'Exit, Voice and Loyalty: Responses to Decline in Firms, Organizations, and States' in *The Oxford Handbook of Classics in Public Policy and Administration* (Oxford: Oxford Handbooks, 2016) 256–271.

Engels, F., Wentland, A., and Pfotenhauer, S. M. 'Testing Future Societies? Developing a Framework for Test Beds and Living Labs as Instruments of Innovation Governance' (2019) 48(9) *Research Policy* 103826.

Ferrando, F. 'Why Space Migration Must Be Posthuman' in Schwartz, J., Milligan, T. (eds.) *The Ethics of Space Exploration. Space and Society Vol. 8* (Amsterdam: Springer, 2016) 137–52 <https://doi.org/10.1007/978-3-319-39827-3_10> accessed 24 July 2022.

Fisher, M. *Capitalist Realism: Is There No Alternative?* (London: Zero Books, 2009).

Gee, T. *Antiindividualistic Individuality. A Concept Pamphlet* (Alpine Anarchist Productions, 2003).

Han, B.-C. *Psychopolitics: Neoliberalism and New Technologies of Power* (London: Verso, 2017).

Hirschman, A. O. *Exit, Voice, and Loyalty: Responses to Decline in Firms, Organizations, and States* (Harvard University Press, 1970) 176.

Jameson, F. *Archaeologies of the Future: The Desire Called Utopia and Other Science Fictions* (London: Verso, 2005).

Kegl, J. and Iwata, G. (1989). 'Lenguaje de Signos Nicaragüense: A Pidgin Sheds Light on the "Creole?"' *ASL. Proceedings of the Fourth Annual Meeting of the Pacific Linguistics Conference.*

Kim, M. E. 'From Carceral Feminism to Transformative Justice: Women-Of-Color Feminism and Alternatives to Incarceration' (2018) 27(3) *Journal of Ethnic & Cultural Diversity in Social Work* 219–33.

Fred Moten and Stefano Harney 'The University and the Undercommons: Seven Theses' (2004) 22 (2 (79)) *Social Text* 101–115.

Newman, C. 'Exploring the Problems of Criminal Justice in Space' (2016) 2(8) *ROOM Space Journal of Asgardia.*

Powers, S. 'Facing Time' (2018) 3 April *Aeon.*

Ratti, C. and Claudel, M. *The City of Tomorrow: Sensors, Networks, Hackers, and the Future of Urban Life* (New Haven, CT: Yale University Press, 2016).

Ruivenkamp, G. and Hilton, A. (eds). *Perspectives on Commoning: Autonomist Principles and Practices* (London: Bloomsbury Publishing, 2017).

Shammas, V. L. and Holen, T. B. 'One Giant Leap for Capitalistkind: Private Enterprise in Outer Space' (2019) 5(1) *Palgrave Communications* 10 <https://doi.org/10.1057/s41599-019-0218-9> accessed 24 July 2022.

Skarbek, D. *The Social Order of the Underworld: How Prison Gangs Govern the American Penal System* (Oxford: Oxford University Press, 2014).

Sturm, S. and Tae, H. (2017). Leading with Conviction: The Transformative Role of Formerly Incarcerated Leaders in Reducing Mass Incarceration (2017) *Social Science Research Network* n. pag.

Tiarks, E. 'Interplanetary Spaceflight and Restorative Justice' (2020) 2(24) *ROOM.*

Unger, R. M. 'Conclusion: The Task of the Social Innovation Movement' in A. Nicholls, J. Simon, and M. Gabriel (eds), *New Frontiers in Social Innovation Research* (London: Palgrave Macmillan, 2015) 233–51 <https://doi.org/10.1057/9781137506801_12> accessed 24 July 2022.

Van Deven, M. (2021). How Thousand Currents' Solomé Lemma Shifts Power through Solidarity Philanthropy [Interview]. https://www.insidephilanthropy.com/home/2021/9/16/how-thousand-currents-solom-lemma-shifts-power-through-solidarity-philanthropy

Vohs, K. D. and Schooler, J. W. 'The Value of Believing in Free Will: Encouraging a Belief in Determinism Increases Cheating' (2008) 19(1) *Psychological Science* 49–54. <https://doi.org/10.1111/j.1467-9280.2008.02045.x> accessed 24 July 2022.

Yates, L. 'Rethinking Prefiguration: Alternatives, Micropolitics and Goals in Social Movements' (2015) 14(1) *Social Movement Studies* 1–21. <https://doi.org/10.1080/14742837.2013.870883> accessed 24 July 2022.

2

The voyage of 600 years

The ethical governance of a worldship

Stephen Baxter

1. Introduction

- 'It would not be inaccurate to say simply that children born in space will be the first humans to be reared in cages.' (Cockell 2008)
- 'I tried to impress upon [the crew] that they were a chosen group, but this had little effect. It stuck in their minds that *they* had had no choice in the matter.' (Wilcox 1940, 162)

The worldship, a.k.a. generation starship or slowship, is a starship which may take centuries or more to reach its target, so that *generations of crew* spend their whole lives aboard the ship, with no prospect of escape. In general a 'worldship' is taken to imply a large crew/population (> 1000) (Hein 2012), but whatever the size it is hard to think of a more confining 'cage' in which to be born, to quote Cockell.

The worldship case may be a significant one regarding the ethics of extraterrestrial habitats, and one worthy of study.

On the one hand, a worldship is a starship we could, conceivably, build in the coming centuries. We already have precursor technologies in slow ships to the stars, beginning with the Pioneers and Voyagers, and we have experiments such as the International Space Station in long-duration human habitation in space. Granted great extensions of these technologies, from fusion drives to fully closed life-support systems, will be required for a worldship. But by comparison a *Star Trek* faster-than-light warp drive (Alcubierre 1994), requiring the construction of an object not dissimilar to a black hole with the mass of the Sun, would be likely to challenge the resources of a Galaxy-scale culture.

And on the other hand, as Regis remarked in 1985, '[t]he concept of a multigenerational flight [is] a test case. If it will be morally permissible, then so will any other form of space colony, whether within the Solar System or outside it' (Regis 1985, 251).

2. Worldships: Fact and Fiction

The concept of the 'worldship' emerged from the ferment of speculation about space futures that marked the early twentieth century, in both science and engineering forums and in fiction. Useful reviews of the technical literature include Martin

Stephen Baxter, *The voyage of 600 years*. In: *The Institutions of Extraterrestrial Liberty*.
Edited by Charles S. Cockell, Oxford University Press. © Stephen Baxter (2022).
DOI: 10.1093/oso/9780192897985.003.0003

(Martin 1984) and more recently Hein, Pak, Pütz and Bühler (Hein et al. 2012). See Caroti (Caroti 2011) for a study of the trope in fiction.

2.1 Worldships: Fact

Technical speculation on worldships goes back to such thinkers as the US rocketry pioneer Robert Goddard. In a manuscript written in 1918, Goddard considered 'The Ultimate Migration', in which mankind might escape the death of the Sun in hollowed-out asteroids driven by atomic propulsion (Goddard 1985). However this work was not published until 1970.

The first worldship proposal accessible to the general Western public was given by J. D. Bernal. In *The World, The Flesh and The Devil* (Bernal 1929), Bernal imagined an expansive human culture in space living off solar energy in 'celestial stations', spherical habitats 16 kilometres in diameter, each housing perhaps 20,000–30,000 inhabitants. In time, the gathering cloud of stations would compete for resources, and some would leave for other stars, becoming what we know as worldships.

With the advance of rocketry and other space technologies after the Second World War, the first reasonably technical study of worldships seems to have been published by L. R. Shepherd (Shepherd 1952) in the *Journal of the British Interplanetary Society*. His worldship would have been a 10 million-tonne planetoid, carrying a 300-head crew for 30 generations. Shepherd briefly noted the need for artificial ecologies, population control, and cultural continuity in order to fulfil the mission.

Sporadic studies followed. By the 1970s the work of Gerard K. O'Neill and others on near-future space settlements (O'Neill 1978) had delivered relatively plausible space habitat designs, whilst such studies as Project Daedalus (Martin 1978) had delivered comparably plausible fusion-engine drive systems capable of sending large masses into interstellar space. An early worldship proposal, putting these studies together, was made by Matloff (Matloff 1976), who suggested using a Daedalus fusion propulsion module to send an O'Neill Island One habitat to the stars at about 1% the speed of light.

A major workshop was held by the British Interplanetary Society (BIS) in 1984 (Martin 1984). In February 2002 an American Association for the Advancement of Science (AAAS) symposium was held on *Interstellar Travel and Multi-generation Space Ships* (Kondo 2003). Later work has followed (Hein 2012).

Compared to the literature on the engineering technology of worldships, the exploration of the concept's ethical, sociological, and demographic aspects—the human dimension—has been sparse, with the most significant works found to date by Holmes (Holmes 1984), Regis (Regis 1985), and Schwartz (Schwartz 2021).

Meanwhile, however, such aspects have been explored extensively in science fiction.

2.2 Worldships: Fiction

In genre science fiction, the worldship idea burst onto the scene in 1940, with a story called 'The Voyage of Six Hundred Years' by Don Wilcox (Wilcox 1940).

After the pioneering work of Verne and Wells, the modern science fiction genre was consolidated in (mostly) US 'pulp' magazines of the 1920s and thereafter, beginning with Hugo Gernsback's *Amazing Stories* from 1926. Full of wonder, lurid storytelling, with garish covers—and often with a dose of pensive speculation at the core—these works found a ready audience. And this 'scientifiction' evolved into science fiction (SF) with the appointment of John W. Campbell as editor of *Astounding Stories* in 1937. Wilcox, then, publishing in 1940, was at a point of transition in terms of the literature.

Wilcox (1905–2000) (Clute 2019) was born in Kansas, studied sociology, and later taught in Chicago. He was a prolific and popular 'pulp' writer of his times—but part time, as most writers were by necessity, given the pay rates. What may be his best-known story was clearly informed by Wilcox's own background.

Wilcox's near-future tale has its pulpish elements—such as the viewpoint character's fiancée happening to wander onto a starship just before its launch—but it reveals a depth of speculative thinking which perhaps exemplifies the value of SF to such studies. By telling a human story, Wilcox was trying to work out how it might be actually to *live through* a worldship journey. And as such, Wilcox can be seen retrospectively to have gained some remarkable insights into many of the key tropes later associated with the worldship idea—especially the sociological.

One such insight granted was into the purpose of such a project. Wilcox's *Flashaway* was sent to the stars explicitly to carry American values to a new world. What goals could motivate any society to take in the generations-spanning ethical challenges of a worldship—and invest in its enormous economic cost?

3. Worldships—Costs and Goals

3.1 Costs

The costs of a worldship project, at least in an order of magnitude study, are estimable, and the results are indicative of the possible timeline and social context of such projects.

As noted, one straw-man worldship design was given by Matloff (Matloff 1976), in which the author imagined sending out an O'Neill Island One habitat at 1% lightspeed. And its costs, at least in energy terms, are calculable. An Island One habitat massed some 3 million tonnes, most of it radiation shielding. The kinetic energy required is therefore $\sim 10^{22}$ Joules.

An economic cost may be hinted at by testing this against the Kardashev scale of hypothetical advanced civilizations (Kardashev 1964). Our modern industrial society produces some 20 Terawatts of industrial power; it would take us $\sim$ 20 years of all our civilization's output to provide Matloff's kinetic energy alone (leaving aside construction costs). But it would take a K-I culture (Kardashev Type I), utilizing all Earth's insolation (200,000 TW), less than a day to provide. And a K-II culture, controlling the output of its star, would need only seconds. This suggests that a worldship project is likely to take place some centuries from now, is likely to be the product of an interplanetary culture rather than a terrestrial one, and indeed may itself take centuries to assemble (and millennia to fly).

This very roughly matches estimates produced by Bond and Martin (Bond and Martin 1984). In 1964 the Apollo project consumed 5% of the US Federal budget (Bizony 2006). Bond guessed that if gross annual world product grew at a reasonable rate from present-day values, then by 2500–3000 AD a worldship might be regarded as affordable as Apollo, costing only ~1% of the world product at that date.

So, as suggested by Bernal (Bernal 1929), worldship projects will be a small step following a mastery of the resources of the Solar System rather than a giant leap by a planetary civilization like our own. (For a deeper analysis of this point see Ashworth 2012.) We will probably not build a worldship soon.

But why build one at all?

3.2 Goals

As noted, Wilcox's story features a political goal (Wilcox 1940). At launch, with an initial crew of around 30, consisting of carefully selected reproductive couples, and facing a 600-year trip to the 'Robinello planets', the Captain declares: 'We go forth into space to live—and to die. But our children's children … will carry forward our great purpose.' In the future others could set off into space 'knowing that you will find an American colony planted there' (Wilcox 1940, 154).

So, presciently, Wilcox's goal is political, just as in the case of the US Apollo programme, whose *stated* purpose (Bizony 2006) was to beat the Russians in visible space exploits, following the Soviets' own politically motivated drive into space.

What other goals might be served? Martin (Martin 1984) speculated on motivations to build worldships, including:

- To fulfil *economic* ambitions
- To fulfil *political, cultural,*or other social goals
- To escape *existential* threats to a group or to humanity as a whole.

Economic targets could emerge following the spread of space colonies within the Solar System in search of resources, as suggested by Bernal (Bernal 1929), O'Neill (O'Neill 1978), and Holmes (Holmes 1984). The worldship then would emerge as a natural extension of such colonies because exploration and exploitation seeped out into the Sun's Oort cloud and beyond. Certainly it would require a very compelling commercial goal to retrieve goods from the nearby stars: a lode of antimatter, perhaps.

Political and cultural goals, as Holmes (Holmes 1984) argued, could include the mission being a nationalistic stunt like Wilcox's *Flashaway*, or some totalitarian grand gesture—or, *Mayflower*-like, the fulfilment of a desire by some group for a political or other form of cultural independence from an oppressive regime. Kondo (Kondo 2003, 7) suggested we should go simply to 'keep our civilization alive'—to avoid a stagnation that might follow a stalling of exploration given the growth limitations of the Solar System.

The goal might be still more primal: to use worldships to survive the extinction of mankind, as Goddard had suggested in 1918 (Goddard 1985). This trope has been explored before a mass audience in BBC's *Doctor Who*. In the story 'The Ark' (1966)

(Erickson 1987), set 10 million years in the future, Earth appears to be plunging into the Sun, and the surviving population set off on a 700-year journey to another world aboard a giant spacecraft complete with a contained 'jungle'.

In 2018, however, Schwartz (Schwartz 2018) argued that worldship technology will most likely *not* be useful in an existential planetary crisis. Such craft will not be ready for near-term crises like all-out war, lethal climate collapse, or an asteroid strike, but will probably be long obsolete in the face of a disaster on long-term geological or astrophysical timescales, such as the death of the Sun.

Regarding goals, a fundamental distinction is that if a worldship journey is essentially 'voluntary'—not something we *had* to do—then the ethics of the treatment of its crew should be assessed accordingly. And at the heart of that ethical assessment is the fact that most of the crew will live and die for the cause without even seeing the mission completed.

4. Worldships—Ethical Challenges and Solutions

On a worldship, the crew's two principal functions are to service mission-critical systems and to survive to reach the destination. Thus a key part of the planning for any such mission, and central to the role of any governance system, is to ensure that a viable, reproductively able, technically competent population is maintained through the mission. And this in turn implies both population and cultural control, as will be seen.

Regarding the ethical dilemma, the founding crew (we presume) *choose* to live out the rest of their lives in this apparently very confined, compulsion-heavy regime. But so must their children, and grandchildren—the *caretaker generations* in the literature, who live and die in the long centuries between launch and landfall at the eventual destination. It can be argued that a number of the caretakers' fundamental human rights are violated in the process.

The most stark of these may be a lack of freedom of movement. You can't get off a worldship.

4.1 Freedom of movement

Perhaps the single most significant difference in the ethical situation of a worldship caretaker-generation crew member and the inhabitant of even the most remote of space habitats in the Solar System is that *the worldship crew member is unable to leave.* A colonist even far out in the Sun's Oort cloud might know that it is at least physically possible for them to leave and return to Earth within their own lifetime. This is impossible for the caretaker generations aboard a worldship.

This is a clear violation of human rights, under common Earth protocols. For example, Article 13 of the Universal Declaration of Human Rights (1948) includes the right to freedom of movement within one's own 'country' or beyond (Amnesty 2021). Furthermore, as argued, for example, by Schwartz in Chapter 23 of this volume, freedom of movement is itself a buffer against tyranny of other sorts. Institutions are pressured to persuade free individuals not to leave; the right to leave offers a fundamental ability to assent to or dissent from tolerating conditions within a community.

However, further restrictions are imposed on the caretaker generations.

4.2 Environmental richness; vocational choices

The worldship craft itself will be a remarkably fragile environment, no matter how advanced its technology, if only because of the long duration of its mission.

Consider the closed loops of air, water, and other essentials in the ship's closed-environment support system. No such system will be leak-proof. Biosphere 2, our largest (Earthbound) closed-habitat experiment, lost 10% of its essential materials in a year of closure (Nelson et al. 1993). Mahon (Mahon 2020) considered the mundane challenge of resource recycling in a worldship journey, concluding that a recycling rate of 99.9% would be required if the ship were to last even a century. Solutions suggested include advanced technologies such as nanotechnological repair and fabrication, and a highly managed biosphere.

In such an environment there would be a steady pressure against curiosity, initiative, or inventiveness. This would constrain the psychological development of generations of crew, and might perhaps exert an evolutionary selection pressure in the long term.

Meanwhile, the caretaker generations would be faced with very limited choices of meaningful occupations. In a ship inevitably dominated by automation there might be little direct human involvement in the essentials of the mission: a command crew perhaps, and maintenance teams. There could be a role for scientists, perhaps observing the interstellar medium beyond the walls of the ship, and indeed studying the evolution of the human society within. And there would be a variety of human-focused roles such as teachers, doctors, counsellors, even religious leaders. There would be children to raise.

Our human rights regimes do not generally recognize aspiration and its fulfilment as rights. Nevertheless the worldship environment would be one of stifling control, in which many normal human actions would be rendered impossible, or banned.

And a necessary control over the crew's reproduction would perhaps be the most ethically troubling issue of all.

5. Population Control

5.1 Population maxima and cultural minima

Aside from the basic inability to leave, it is perhaps the necessary control of human reproduction that imposes the key ethical difficulties associated with worldships, as opposed to confinement issues faced on deep-space habitats in the Solar System.

As Holmes (Holmes 1984, 298) pointed out, aside from at times of famine, war, or some other disaster, human populations have a tendency to grow—hence historically the rapid population growth of newly accessed territories like Australia. But on a worldship, even if the population were allowed to double over a 1,000-year mission, say, the annual growth rate could be no more than 0.07%—in a population of 100,000, an excess of births over deaths of just ~ 70 per year. Such low growth rates have not characterized any human population since the early days of farming ~ 10,000 years ago. The containment of growth is clearly essential; a 'bloom' of population could easily overwhelm a ship's life support systems. How could such low *maximum* growth rates be imposed, across generations?

Conversely, and still more controversially perhaps, what if the population *falls*? What is the limit to viability?

Very small human populations can be viable biologically. O'Rourke (O'Rourke 2003) outlined genetic considerations pertaining to the survival of a small, isolated population of perhaps 150–200 people for six to eight generations. Whilst natural selection was unlikely to operate on such short timescales, inbreeding could cause genetic drift, with an increased expression of recessive genes and various undesirable consequences. For example the mortality rate of progeny of second cousins has been observed to rise by ~ 2%. This would appear to recommend the genetic screening of the initial crew to ensure maximum genetic diversity at the start of the mission, and to mandate rules to avoid inbreeding in flight.

But, as a consequence, the smaller the crew, the stricter the rules, and the more limited the choice of reproductive partners for any individual.

In addition, as Hein (Hein 2012) points out, the smaller the size of the crew, the more danger it runs of losing cultural continuity, and perhaps even basic knowledge concerning the running of the ship. A small population runs the risk of knowledge loss through random losses of key individuals, or even a lack of sufficient students to train in their place. Of course, in the context of an advanced technology worldship, smart systems might be expected to codify technical knowledge at least, and even a time-lagged contact with Earth could replenish education and libraries. But softer cultural richness may be lost by a small group.

In short either a population crash or bloom would be disastrous for the ship's mission, threatening a loss of genetic diversity on the one hand and famine on the other.

As a result, there have been a number of studies on the stability of worldship populations over time, generally focusing on the *effectiveness* of various reproduction rule sets (as opposed to their morality, discussed below). Some rules have been tested with computer modelling.

5.2 Reproduction rules

Grant (Grant 1984) used then-current world population growth computer models to study a suite of 'passive' (voluntary) and 'active' (compulsory) worldship population control measures, essentially linking the birth rate to the death rate.

If the birth rate was *above* the death rate, there could be 'birth licensing' to reduce the births to match the deaths—or even, conceivably, and still more ethically challenging, 'euthanasia licensing' to raise the *death* rate to match the birth rate.

But if the birth rate fell *beneath* the death rate, then 'birth forcing', mandatory births, could be introduced: an 'active and socially intrusive measure', as Grant put it.

However, Grant found in his computer modelling that through feedback effects, the link between the size of a population and the food supply, and the randomness of free human choices, *passive* measures generally resulted in instability, leading to crashes or runaway overpopulation. *Active* measures, however, if too crudely applied, also tended to result in population crashes—all save 'birth forcing', reproduction control that is *mandatory at the level of the individual.*

As one solution to the ethical issues this poses, Moore (Moore 2003), focusing on genetic diversity, suggested a 'constitution' *guaranteeing* that each individual would have the chance to marry (for reproductive purposes), and that each would have at least 10 possible choices of spouse close to their own age—and such spouses no closer than second cousins. Moore suggested that a crew as small as ~ 180, with an equal male/female split, could be 'organized' to achieve this, *if* an initial cadre of crew, all of similar age, avoided relatively early conceptions. If there were a wave of births with all the parents ~ 35 years old, and if subsequent generations followed this pattern, the total population would remain unchanged, but there would always be two 'bulges' separated by ~ 30 years, offering a maximal choice of reproductive partner to any individual in each age group.

Moore's work was interestingly revisited by Marin and colleagues (Marin 2017; Marin and Beluffi 2018), who used Monte Carlo simulations to model how populations governed by such rules might actually evolve. A safety threshold of 90% of the ship's crew carrying capacity was set. If the population was below that threshold, women in the allowed-reproductive window (a few years around 35 years old) were 'encouraged' (not forced) to have up to three children each; above that threshold, procreation was banned.

Unfortunately this led to crashes, due to the narrowness of the reproductive window; at any one time there were simply not enough available potential mothers. If that window was widened, however, a numerically sustainable population could endure, although at the risk of increased consanguinity.

These exercises at least suggest that a rule set to ensure long term crew survival might be feasible, theoretically. But computer modelling often proves that 'common-sense' approaches to the problem, when applied over generations, often fail counterintuitively.

Note that the discussion here, following the referenced works, centres mostly on reproduction through heterosexual coupling. In any population one would expect a range of sexuality and gender, with, hopefully, any mission 'rules' able to accommodate such choices.

But the key question in all such regimes is about the adherence to, or enforcement of, any such rule set.

5.3 Population control: Applying the rules

The sharp edge of the ethical challenge of the worldship is the image of an individual woman being told that she cannot bear a child—or, perhaps worse, that she must bear several—for the rather abstract good of the mission. This is a key point in Robinson's novel *Aurora* (Robinson 2015): 'There's a lot of pressure on all the women in this ship, to have at least one child, and better two ... So if a woman declines to have two, some other woman is going to have to have three ... It causes a lot of stress' (Robinson 2015, ch. 2, 1225).

Could education enable such rule sets to work without coercion? Certainly in a small population there would need to be a strong and visible link between the overall population size, growth rate, and genetic health on the one hand, and the choices of individual couples on the other, so that such choices could at least be informed.

No rules of any sort, however, are any use without an authority to impose them.

6. Cultural Drift

A key danger for any worldship mission arises from the simple fact that the caretakers' lives will be devoid of intrinsic meaning and walled in by rule sets even over such fundamental choices as whether or not to give birth—and all for a goal set out by founders who may have died generations ago.

A totalitarian regime, as emerged aboard Wilcox's *Flashaway*, seems all too likely an outcome. At the heart of the work of Cockell (Cockell 2008) is a warning that totalitarian societies might actually be an *adaptive* form for any space colony, in as much as such control might be *necessary* for the crew to survive an emergency. Indeed, in a sufficient emergency an appeal to 'lifeboat rules' might be mandated by the colony's constitution (the difficulty being to get rid of the lifeboat 'captain' once the emergency was over).

But even with strong authoritarian 'support' for the mission goals, it seems likely that the crew, increasingly remote from the mission's founders, will simply turn away from the regime that seeks to control them—turn away from the web of rules and compulsions that ensnares them—and simply find something more interesting to do. Thus, over the centuries, the crew of Wilcox's *Flashaway* (Wilcox 1940) creates food empires and fights clan wars, rather than listen to lectures about an America they have never seen and will never see.

And in the author's own *Ark*, in mid-flight,

[t]he [younger generation] moved in swarms like exotic fish in a tank, ignoring the adults and eyeing each other with suspicion. They wore basic wraps that left their arms and legs bare, their flesh adorned with tattoos that matched graffiti on the walls, markings incomprehensible to any adult, that badged their allegiance to one tribe or another. Holle knew that few of these kids ever attended formal classes. It worried her that they were so disconnected from the ship and its mission (Baxter 2009, 362–3).

With time, the mission goals may be lost altogether. This can go much further. Possibly the most resonant of all worldship tropes was introduced by Heinlein in his 'Universe' (Heinlein 1941). As the generations pass, ultimately *the caretaker crew may even forget they are on a ship.*

7. Ethical Dilemmas and Solutions

7.1 Morality and context

A worldship mission *by design* confines whole generations of unborn humans to a restrictive artificial environment, without any possibility of their giving consent to having been so confined.

By the time such a ship is launched, most probably many generations will have lived and died in the confinement of space colonies, and so perhaps some of the resulting ethical dilemmas arising from those colonies (Cockell 2008) will have been solved. But what is qualitatively different about the moral dilemma of raising children on a worldship, compared to other examples of space habitats, is a *lack of choice*. Even the most remote Solar System habitat offers at least the possibility of escape. There is no escape for the caretaker generations in a worldship.

And it is this aspect that, as Regis says, makes the worldship case an extreme example of the ethics of space colonization. This feels intuitively wrong, says Regis, like a form of 'kidnapping, involuntary exile, or repeated child abuse' (Regis 1985, 251).

But Regis also carefully points out that our notions of the rights of the unborn are hazy. Once born, an infant has basic human rights, and we would condemn a society that did not at least attempt to fulfil those rights. But what rights are being violated by simply being born in a worldship? Does a new-born have some kind of right of access to Earth itself? Or even to a space colony within the Solar System?

It is quite possible that the life of a starship caretaker-generation passenger of the future would have a life far richer and more fulfilling than that of many alive on Earth today. It may be, indeed, as is mentioned in some fictions, that many of the final crew generation will *choose* to stay aboard the comforting confines of their ship rather than to become pioneers at the destination, on an unknown and utterly bewildering world. Thus in Brunner's 'Lungfish' (Brunner 1957), the crew reach their target world but refuse to end the mission.

And there are plenty of terrestrial examples of children being born more or less *purposefully* into relatively impoverished environments—impoverished compared to their parents' younger lives, at least—such as the children of Polynesian settlers, or even those of early American frontier settlers. Parents may wish for better lives than their own for their descendants, but it seems acceptable to 'inflict' lives of comparable toil, isolation, and confinement on the children of pioneers as a means to an end, as if those children were available as a useful resource to their parents.

Regis (Regis 1985, 258) concludes that '[d]espite all appearances a multigenerational expedition does not differ in any relevantly fundamental way', that is relevant from the point of view of human rights, 'from ordinary human life on Earth', or, we could add, from life on a Solar System space colony.

Regis does, however, seem implicitly to assume a responsibility that purposeful or negligent design should not impinge on the crew's rights, beyond the issue of living one's life confined to a ship. To quote Milligan (Milligan 2016, 16), they should have 'a life which the agents themselves [the caretakers] could readily accept as meaningful in spite of suffering and meaningful and worthwhile in its own right rather than being simply apart of someone's grand plan'.

But whose ethical responsibility is it to ensure that such rights of generations of crew are respected?

As Schwartz (Schwartz 2018) points out, it surely could not be claimed, in mid voyage, that the parents of caretaker generations (being caretakers themselves) were exploiting their children unfairly by conceiving them. After the first generation, the parents will have had no say in the broad conditions into which their children are born.

So if there is any exploitation, it must be on the shoulders of those who did have a say: the designers of the mission, and perhaps the first crew if they volunteered for the mission, thus committing their descendants to it.

And if there is an ethical duty to provide as rich lives as possible for the subsequent generations, that duty must be the designers'.

There may however be technological, even social means to approach the goal of ethical equity on a worldship.

7.2 Solutions: Technology

It may be that comparatively modest technical advancements could alleviate some problems of the classic worldship design:

- Though the mission is long in time, the journey from Earth is only ever a few light years. A time-lagged contact with Earth and the Solar System could certainly be possible, though cultural drift over centuries might mar this.
- The use of cryosleep for many or most of the passengers could avoid issues of the caretaker generations. This was dramatized in the movie *Passengers* (2016, directed by M. Tyldum), in which a caretaker is woken by accident in the middle of the mission, finding himself doomed to live out his life alone, until he wakes another passenger.
- The use of longevity drugs could reduce the population turnover and perhaps stabilize society and ensure cultural continuity. (Perhaps the original crew could survive the entire mission, eliminating the need for caretaker generations altogether—and the consequent ethical dilemma.)
- Advanced reproductive methods could be used. Artificial insemination and cloning could make up a shortfall in birth rates without the need for coercion into biological reproduction. The clones themselves would deserve equivalent rights to the rest of the crew, of course. Or, rather than use adult humans at all, a 'seedship' could be sent, equipped with stored human embryos, or perhaps technology capable of fertilization and nurturing humans from the stage of sperm and eggs. This solution would of course bring its own challenges, with an initial generation of colonists presumably being raised without direct adult human contact of any kind (Hein and Baxter 2019).

7.3 Solutions: Social forms

Meanwhile, as a means of sustaining cultural continuity, Moore (Moore 2003), drawing on fictional speculation by Heinlein (Heinlein 1957), suggested that small communities (~ 100) could be structured along *family hierarchy* lines, building on natural human relationships and ancient forms. Alternatively, Thomason (Thomason 2003) suggested drawing the crew from a single *religious* background, to provide a shared culture and perhaps reduce the risk of inter-faith conflict, but possibly at the expense of genetic diversity amongst the founder crew.

Holmes (Holmes 1984), meanwhile, considered larger populations in enduring forms. Egypt appears to have been our longest enduring state, lasting some 3,000 years before the Roman invasion. But cities may endure longer, with Jericho, for example, emerging from a pre-farming village and enduring to the present across some 10 millennia. Perhaps then, Holmes suggests, the worldship should be organized along the lines of a single city-state, to ensure coherent governance and uniformity and continuity of culture.

7.4 Solutions: Ship size

The most obvious stratagem would be to maximize crew size, by, in turn, maximizing ship size. A large enough ship—presumably somewhere between a large cruise liner and planet Earth—could facilitate expansive lifestyles with a multitude of choices.

Schwartz (Schwartz 2018) touches on this, conceiving of levels of richness of potential as crew size increases, passing a 'point of labour surplus' where the crew would have a choice of mission-essential careers and roles, a 'point of reproductive freedom' beyond which reproduction mandates should never (or rarely) apply, and then a 'point of vocational freedom', beyond which a crew member could follow whatever career path they chose, whether utilitarian to the mission or not. One might add 'points of diversity' beyond which, for example, gender choices were not compromised by the need to maintain population numbers.

Schwartz gives no estimate on the number of such crew. Certainly several thousand might be necessary. Diamond (Diamond 1997) argued that there is a correlation between group size and the complexity of human technology and material culture. An example he quotes is the native Tasmanians, a population of 4,000 isolated for 10,000 years, who Diamond claims had the simplest material culture of any modern humans. This suggests a minimum size of perhaps 4,000; there are many island nations on Earth with less than 10,000 inhabitants.

8. Conclusion

Wilcox's story, 80 years old (Wilcox 1940), is prescient of the study of worldships in many ways. And Wilcox cuts right to the heart of the key moral dilemma surrounding worldships: the plight of the 'caretaker' middle generations of crew.

Certainly the worldship trope has endured. But in parallel, a literature of the science, engineering, and (to some extent) sociology of worldships has developed. And we should take this literature seriously. For all its titanic cost, the worldship is one of a very few conceptual technologies we have (perhaps the only one) which might realistically allow us some day to reach the stars.

It should be noted that worldship technology might evolve logically from long-duration, far-ranging space habitats within the Solar System. Perhaps ships that probe the Oort cloud will be their precursors, as the Portuguese caravels that explored the African coast were the predecessors of the ships that took Columbus across the Atlantic.

It may be that perhaps future cultural sensibilities might *pre-adapt* crew for the conditions of a worldship. Perhaps the cultural challenges of a worldship will seem less when the problems of liberty in Solar System habitats have been resolved.

And even on Earth, there may be a metaphorical resonance of the plight of a worldship's 'caretaker generations' for modern generations, who similarly may feel doomed to the task of preserving and repairing the Earth, before expansive dreams of a future of growth and optimism can be addressed. This perception goes back at least as far as 1965, in a speech given by Adlai Stevenson, US ambassador to the United Nations, in which he compared the planet to a fragile spaceship in our care (Caroti 2011).

We are already caretaker generations. And, through the climate crisis, we too are learning there is a link between our own individual choices and the continued viability of Worldship Earth.

References

Alcubierre, M. 'The Warp Drive: Hyper-Fast Travel within General Relativity' (1994) 11–5 *Classical and Quantum Gravity* L73–L77.

Amnesty. Universal Declaration of Human Rights (2021) <http://www.amnesty.org>, accessed 9 June 2021.

Ashworth, S. 'The Emergence of the Worldship (I): The Shift from Planet-based to Space-based Civilisation. The Emergence of the Worldship (II): A Development Scenario' (2012) 65 *Journal of the British Interplanetary Society* 140–75.

Baxter, S. *Ark* (London: Gollancz, 2009).

Bernal, J. D. *The World, the Flesh and the Devil* (Scottsdale, AZ: Prism Key Press, 2010).

Bizony, P. *The Man Who Ran the Moon* (London: Icon Books, 2006).

Bond, A. and Martin, A. R. 'Worldships: An Assessment of the Engineering Feasibility' (1984) 37 *Journal of the British Interplanetary Society* 254–66.

Brunner, J. 'Lungfish' (December 1957) *Science Fantasy* (London: Nova Publications, 1957).

Caroti, S. *The Generation Starship in Science Fiction: A Cultural History* (Jefferson, OH: McFarland and Co., 2011).

Clute, J. (ed). <http://www.sf-encyclopedia.com> Entry on Don Wilcox, accessed 17 October 2019.

Cockell, C. 'An Essay on Extraterrestrial Liberty' (2008) 61 *Journal of the British Interplanetary Society* 255–75.

Diamond, J. *Guns, Germs and Steel* (London: W.W. Norton, 1997).

Erickson, P. *Doctor Who: The Ark* (London: Target Books, 1987).

Goddard, E. L. (ed). *The Papers of Robert L. Goddard*, vol 1 (New York, NY: McGraw Hill, 1985).

Grant, T. J. 'The Population Stability of Isolated Worldships and Worldship Fleets' (1984) 37 *Journal of the British Interplanetary Society* 267–84.

Hein, A., Pak, M., Pütz, D., Bühler, C., Reiss, P. 'Worldships: Architecture and Feasibility Revisited' (2012) 65 *Journal of the British Interplanetary Society* 225–31.

Hein, A. and Baxter, S. 'Artificial Intelligence for Interstellar Travel' (2019) 72 *Journal of the British Interplanetary Society* 125–43.

Heinlein, R. 'Universe' in *Astounding Science Fiction*, May and October 1941 (New York, NY: Street and Smith, 1941).

Heinlein, R. *Citizen of the Galaxy* (New York, NY: Ballantine, 1957).

Holmes, D. L. 'Worldships: A Sociological View' (1984) 37 *Journal of the British Interplanetary Society* 296–304.

Kardashev, N. S. 'Transmission of Information by Extraterrestrial Civilisations' (1964) 8 *Soviet Astronomy-AJ* 17.

Kondo, Y. (ed.). *Interstellar Travel and Multi-Generation Spaceships* (Burlington, VT: Apogee Books, 2003).

Mahon, P. J. 'Worldships—Some Ecological and Resource Constraints' (2020) 73 *Journal of the British Interplanetary Society* 21–5.

Marin, F. 'Heritage: A Monte Carlo Code to Evaluate the Viability of Interstellar Travel Using a Multi-generational Crew' (2017) 70 *Journal of the British Interplanetary Society* 184–95.

Marin, F. and Beluffi, C. 'Computing the Minimal Crew for a Multi-Generational Space Journey towards Proxima Centauri B' (2018) 71 *Journal of the British Interplanetary Society* 45–52.

Martin, A. R. (ed.). 'Project Daedalus—The Final Report on the BIS Starship Study' (1978) *Journal of the British Interplanetary Society* Supplement.

Martin, A. R. 'Worldships: Concept, Cause, Cost, Construction and Colonisation' (1984) 37 *Journal of the British Interplanetary Society* 243–53.

Matloff, G. L. 'Utilisation of O'Neill's Model 1 Lagrange Point Colony as an Interstellar Ark' (1976) 29 *Journal of the British Interplanetary Society* 775–85.

Milligan, T. 'Constrained Dissent and the Rights of Future Generations' in C. Cockell (ed.), *Dissent, Revolution and Liberty beyond Earth* (New York, NY: Springer, 2016) 7–20.

Moore, J. H. 'Kin-based Crews for Interstellar Multigenerational Space Travel' in Y. Kondo (ed.), *Interstellar Travel and Multi-Generation Spaceships* (Burlington, VT: Apogee Books, 2003) 80–8.

Nelson, M., Burgess, T.L., Alling, A., Alvarez-Romo, N., Dempster, W.E., Walford, R.L., and Allen, J. P. N. 'Using a Closed Ecological System to Study Earth's Biosphere' (1993) 43 *Bioscience* 225–36.

O'Neill, G. K. *The High Frontier* (London: Corgi Books, 1978).

O'Rourke, D. 'Genetic Considerations in Multi-generational Space Travel' in Y. Kondo (ed.), *Interstellar Travel and Multi-Generation Spaceships* (Burlington, VT: Apogee Books, 2003) 89–99.

Regis Jr., E. 'The Moral Status of Multigenerational Interstellar Exploration' in B. Finney (ed.), *Interstellar Migration and the Human Experience* (Berkeley, CA: University of California Press, 1985) 248–60.

Robinson, K. S. *Aurora* (London: Orbit, 2015).

Schwartz, J. 'Worldship Ethics: Obligations to the Crew' (2021) 71 *Journal of the British Interplanetary Society* 53–64.

Shepherd, L. R. 'Interstellar Flight' (1952) 11 *Journal of the British Interplanetary Society* 149–67.

Thomason, S. 'Language Change and Cultural Continuity on Multi-generational Space Ships' in Y. Kondo (ed.), *Interstellar Travel and Multi-Generation Spaceships* (Burlington, VT: Apogee Books, 2003) 100–103.

Wilcox, D. 'The Voyage of 600 Years' in G. Benford and G. Zebrowski (eds), *Skylife* (New York, NY: Harcourt Inc., 2000).

3

Art, institutions, and liberty in extraterrestrial communities

Annalea Beattie

Especially in the early days of our habitat on Mars, our small group will be confined by the lethal environment outside, a lack of physical space inside, restricted resources, and a utilitarian routine that keeps daily systems functioning, so we can stay alive. Because of the danger, there will be necessary surveillance and coercive controls on the way we live. Our new home in space will be a disciplinary site and an environment of enclosure.

What is social freedom in a place like this? How will we remain healthy and well? To encourage autonomy and agency, to love and care for others, and to promote all different kinds of vernaculars, what is needed in our extraterrestrial environment is art. Creative work keeps our imagination alive. It stimulates us to explore other realities and modes of resistance courageously and critically. Whilst making art offers us immersion, contemplation, and ideation in the material and sensory, it also inspires us to perform, to be curious, to improvise, and to take risks. Through art, we can fail or dissent—we can reinvent ourselves. In art we find our potential and capacity for change. Prompting us to share meaning and helping us to belong, art-making brings people together and creates important cultural norms. In our tiny community off-Earth, engagement with art will define and alter our experiences of life, mobilizing democratic principles to create new citizens and societies on Mars.

On Earth, the institutions of art are at the intersections of art and its public. Not everywhere but in many countries, we rely on these cultural and educational bodies to mediate between art and people. Our global art institutions emerged from Western European and North American traditions, from an imperial world of forced migration and stolen artefacts. In constant pursuit of the new, imperial conditions are reproduced through the dominance of western art. As they continually expand their reach into different territories, art institutions perpetuate specific modes of cultural constructions, writing and rewriting both the subjects and objects of art into systems of power and knowledge.

As artists, part of our work is to question the many contexts for making and distributing art, including the nature of art institutions and the genealogies of values they present. For example, we recognize that the strategies and practices of art institutions and their mixture of historical and cultural narratives give rise to particular kinds of communalities. Art institutions themselves suggest these communalities are, in some way, inherent in art objects. Staged within selective environments as points of public cultural narration, exhibitions and collections are presented in museums

Annalea Beattie, *Art, institutions, and liberty in extraterrestrial communities*. In: *The Institutions of Extraterrestrial Liberty*. Edited by Charles S. Cockell, Oxford University Press. © Annalea Beattie (2022). DOI: 10.1093/oso/9780192897985.003.0004 and galleries as if the artwork has qualities that intrinsically tie it to specific periods or categories. Museum director and curator Jessica Morgan writes of the authoritative nature of these categories of cultural production:

> The collection and display of certain objects and artefacts, according to chosen curatorial techniques, represent not only the writing of specific colonial and national histories but also the circulation of particular values and ideals.
>
> (Morgan 2013, 23)

Through engagement with the public, art institutions like the art museum generate borders that shape societal and cultural norms as they attempt to define what art is. In fact, what art does is infinitely more important. Whilst artists are directly responsible for everything they make, art institutions on Earth do not always share this responsibility. For instance, the museum makes a choice when it gives authority to the accumulation and exploitation of the past rather than to the ongoing reality of a shared world. It performs a democratic role as it categorizes, classifies, and catalogues collections and displays them, selecting the context within which objects can be 'read'. At the same time, it circumvents histories and discourses that it perceives hamper reception. At the heart of the politics of the museum, within the collection, classifications occur within artificially constructed, discursive fields, such as 'community' or 'nation', or perhaps within specific historical periods or eras, as categories found within Western art history. For example, in relation to the museum, writer-curators Irit Rogoff and Daniel J. Sherman suggest:

> The museological context, in other words, exists within a larger signifying process that invokes notions of community. This may be a trans-historical community of art lovers, a local community that has formed the collection, a national community that the objects represent, a politically aspirant community that seeks alternative forms of representation and alternative identities or some combination of all four.
>
> (Sherman and Rogoff 1994, xii)

We know that the representations of our art institutions are never neutral or silent and each mediation establishes a border or delimitation. Is this always the dominant position of our art institutions—to assume a public through cultural display and then to constitute it from within?

To avoid past mistakes and to enable artists, institutions, and the public to affect change, it is critical we address a speculative model for extraterrestrial institutions before we leave our planet to make a home somewhere else. Art and its institutions are active sites for understanding liberty. To realize how they might contribute to social freedom, self-determination, and self-governance off-Earth, we can learn from the empowering work of both artists and art institutions in their recent attempts to reimagine democracy in action.

Institutional critique of art institutions developed as part of ongoing, artist-led movements in the late 1960s and early 1970s, through the critical practices of artists who questioned lines of art production, testing traditional formats of galleries, museums, and their collections. By the mid-1980s, art institutions themselves took action 'against neo-liberal, populist, cultural policies and authoritarian, repressive, cultural policies' (Raunig in Raunig and Ray 2009, 3). They began to find ways to engage the public with temporary, 'nomadic' institutional models, aiming to create conditions for activism that investigated and provoked other kinds of institutions, as well as the institutions of art. It wasn't until the end of the twentieth century that 'New Institutionalism' emerged in the art world from the political left (Kolb and Flückiger 2013). Aligned with curators and galleries in Northern Europe, it focused on institutional structures and agency, questioning large-scale privatization and the pressure to develop market-orientated profiles. Crossing disciplines to involve artists, curators, theorists, and scholars, New Institutionalism linked art organizations to social activism rather than individual art practices. Its proponents took an oppositional stance to policies reified within the key institutions of art, theorizing how institutions shape actions through their representations and claiming that, within art institutions, rationality itself is viewed as 'constructed by the cultural environment' (Di Maggio 1998, 700).

New Institutionalism proposed that agency and action are not merely created by context but are embedded within it, and:

> [b]ehaviour within institutions is driven by elements other than utility calculations, which include internalised principles and values, cultural features, identity and habit.
>
> (Lecours 2005, 10)

In the past 20 years, revisionist scholarship has continued to search for ways to reinvigorate art institutions. Institutional critique has become an analytical tool, recognized as part of artistic practice and used by institutions and artists alike. As the institutional power structures of art engage with the legacies of imperialism, colonialism, genocide, human rights violation, and our racist past, statues that have long symbolized systemic inequality are falling. Today art institutions must combine self-critique with outside debate, accounting for the relationships between their gallery programme and its institutional reality. 'Characterised by the rhetoric of the temporary—transient encounters, states of flux and open-endedness' (Doherty 2006, 2), experimental public programs emphasize art projects that are process-orientated, dialogical, and participatory, for instance, through the practices of installation, debate, workshops, and performance. To reconceptualize their ideas of a cultural public sphere by highlighting multiple publics, common spaces, and democratic principles, art institutions provide the public with opportunities to negotiate and confront pressing social questions, both from inside and outside institutional frames. Like artists, institutions now attend to the social role of art and connect the political relevance of art to wider society, advocating for directions that centre upon consultation and co-production in the 'experiences' of art, rather than its objects. What this means in part is that art institutions have moved away from the primacy of the exhibition, from the traditional white cube, from the endless shuffling and reshuffling of the permanent collection, and the easy framing of autonomous art objects. In spite of our oppressive, institutionalized history, the lineage left to us is open to restitution and reinvention. The kind of radical changes that have occurred in art institutions in the past few decades provide the necessary background to our discussion about what kind of art institutions we might need on Mars.

To understand what art might be for artists living and working on other planets, in my art practice I have sought to situate the experiences of making art in extreme environments on Earth, in conditions that might be analogous to those that are found on Mars. Living and working in simulation, for instance, for three months in the Mars 160 interdisciplinary team in a remote habitat in a hostile desert environment, I have learnt that all art-making in such a place begins slowly. I imagine that in space, art will need to find a role for itself in places where it might have little or no part to play at all. It takes time and energy to explore different languages and share new models of political engagement through art. Considering this, and rather than invent the new, I suggest that the art archive and the art school are appropriate as practical starting points for thinking about art and community and cultural growth off-Earth.

In terms of the archive, its roles and functions are much theorized. The spatial and temporal history of the archive is violent and privileged. Writing about institutional thresholds to the imperial archival regime, writer and curator Ariella Aïsha Azoulay claims that engagement with the archive was made a form of discipline that 'provides proof of the imperial citizen's identity, that is, one's place in the imperial world' (Azoulay 2019, 166), and:

> [a]rchives were advocated as a form of salvation; an incarnation of the promise to tame the individual lust for power; to defeat negligence, weariness, arbitrariness: to overcome frailty and vulnerability of individual memory and oral and corporeal traditions—in sum to survive human mortality.
>
> (Azoulay 2019, 167)

When we consider the many different ways in which the contingent nature of the archive influences history and memory, and how it is itself influenced by political, social, and technological powers, the extraterrestrial future of the art archive should focus upon more than what counts as knowledge. Although the archive is often perceived as neutral technology and is used as a model across disciplines, as Jacques Derrida famously writes:

> There is no political power without control of the archive, if not memory. Effective democratisation can always be measured by this essential criterion: the participation in and access to the archive, its constitution, its interpretation.
>
> (Derrida 1995, 4, note 1)

In art, in its broadest sense, the use of the archive is as a literal, detailed store of documents. Art movements such as Land Art and conceptual art relied on documentation of art as their work might be site-specific, ephemeral, temporary, or durational. In the case of Land Art, the art might cease to be visible altogether. Or Performance Art, for instance, might depend entirely on archival institutions—the video or photograph or the text *becomes* the artwork. Since the 1960s, archival material has been widely employed as part of the artistic method and is used in exhibition practices in the field of art. Focusing on the agency and materiality of the archival document, artists, curators, museum, and biennale directors have used the archive as a theme, not only for institutional critique—to rewrite the past or to re-present or provoke particular discourses and practices within institutions of art—but to incorporate archival material as the material remains of performance, in order to make new works.

How might we experience the art archive in enclosed, monotonous environments where art or art-making is not necessarily a priority for those who live there? History has shown us that the archive is capable of framing and imposing meaning upon our bodies, objects, and cultures, changing the way we view art and ourselves. But even though 'archivists aspire to a democratic facilitation' (Breaknell 2012, 3), in that they aim to give each person a similar experience of what might be found in the archive, the art archive is not a place where actual historical or creative knowledge is produced. Archival elements are inconclusive. The archive is what we make of it—our singular, performative encounter with the art archive is what produces our experience. Whilst we attend to disconnected, partial, aural, and visual fragments in the art archive, we are not simply informed, we participate. Within its fluid space, we are affected and activated by imaginary worlds, by sensory moments of connection or comprehension or possibility:

> Archival collections consequently order things towards a future, are caught in the midst of a process of place-making, of a setting of things in an order: a distribution of points yet to be joined, to be acted out and realized as histories.
>
> (Clarke et al. 2018, 13)

Within the art archive, each gesture, mark, line, or level opens up as palpable and tangible, and small things lead us away. Perhaps we sense an atmosphere or mood evoked by a colour, or mourn the loss of a species in the sound of an extinct bird call. The art archive is a place for our memories but also for longing and appropriation. It is, as curator and archivist Carolyn Steedman writes, 'the space of otherness' (Steedman 2011, 324) and 'the place of dreams' (Steedman 1998, 67).

On Mars, our priority for the extraterrestrial art archive will be its degree of integration into our small family. To find institutions that can easily be situated within this new context and explore useful, democratic models for the production and presentation of art in our community off-Earth, this may require what theorist Gerald Raunig calls 'instituent practices'; those practices which in fact are able to resist the dominant power of institutionalising processes (Raunig in Raunig and Ray 2009, 3–12). In the art archive, instituent practices can be used as methodological tools to create

> a productive tension between a new articulation of critique and the attempt to arrive at a notion of 'instituting' after traditional understandings of institutions have begun to break down and mutate.
>
> (Raunig and Ray 2009, xvii)

What this really means for our extraterrestrial art archive is that we should form a clear plan, one that involves an institutional infrastructure which is both self-critical and emancipatory, beyond the boundaries of particular fields. Our art archive needs to be not only adaptable and responsive to social criticism and the collective imaginary, it must rely on the slow, long-term process of building an interface between an emerging archival framework and internal and external social dynamics, which evolve as our multi-disciplinary community off-Earth evolves. Fortunately, like all archives, this discursive domain has the advantage of being able to accommodate many voices. Participation, ownership, and access will be key to creating something which, over time, will belong to the social space in a small community or village or city on Mars. The extraterrestrial art archive can be constructed by a community through transdisciplinary and collaborative practices that emerge from the social realities of everyday. The rules of engagement can be negotiated and coproduced in an ongoing way, positioning the art archive as part of a plural society where the art object is not necessarily the only key focus for our encounter with art. How this assembly comes together is of most significance. It is here that the concept of the art school can assist in developing open, debatable models.

On Earth, art schools are messy, experimental, interdisciplinary, both improvised and rigorous, transparent, non-profitable, inclusive, non-hierarchical, and invested in community engagement. In today's art school, the artist or the art collective is decentred as author and art is no longer preoccupied with fashioning and amassing more and more unique art objects. To engage audiences directly and spontaneously, art students quickly learn to make their own pathways as cultural producers. They leave behind ideas of monuments and temples of art to work with overlapping, multiple temporalities, rather than single strands of history spoken in the name of institutional power.

On Mars, the art school studio must account for complex but not totally unfamiliar conditions, not the least being small, shared spaces, few materials, and constant surveillance. A working art school will eventually need workshops, libraries, exhibition, and performance spaces, outreach and organization. However, art schools can 'start with a table', as curator Charles Esche writes (Esche 2010, 311). In an extraterrestrial space habitat, whatever physical form it takes, the art school studio will be an exploratory place for self-directed *and* collaborative art practice, as it is on Earth. Significantly, in long-term space duration, the art school will offer rare opportunities for failure in a low-stake environment. In the studio, anomalies, mistakes, ineptitude, amateurism, and a lack of aptitude or skill hold as much critical promise as established sets of technical skills or traditional frames of reference for art-making, as these 'test the limits of what is proper, good and acceptable' (Beech 2010, 53). Recognizing art as a sphere where challenges takes place as a daily part of life, through exchange, failure, defeat, conflict, and problem solving, the art school will establish its own criteria for how art should be viewed.

Writer, curator, and art educator Irit Rogoff writes that art prompts us to reimagine ourselves in a way that is essential to re-imagining our society and:

> [i]f 'academy' was a space of experimentation and exploration, how could we extract these vital principles from it and apply them to the rest of our lives? And if to our lives, then, perhaps, also to our institutions?
>
> (Rogoff 2010, 35)

My own belief is that not subject or object but the *experiences* of making art—the embodied and sensate *practising* of art-making itself; its play, its struggle to make different and bring about change—that belief in and advocacy of art-making, is one of the most significant contributions the extraterrestrial art school can make to the difficulties faced by our lonely micro-society living off-Earth.

In terms of the art archive, relationships between the art school and the concept of the archive are already well-recognized. In the art school, the archive stands as an institutionalised device and functions as an established methodology. As scholar Jenny Sjöholm writes, 'There is an archival character to the modern art studio' (Sjöholm 2014, 503). To make art and to build an art practice, art students must self-organize; they gather, collect, select, index, document, and archive their own material exploration and their knowledge to contextualize their practice and direct and develop future artwork. In their archival methods, art students attend to the moment when art is made, and before and afterwards. Negotiating cultural and political dialogue with non-art constituencies takes place as a day-to-day part of that critical learning as art students enrich their material experiences of art-making with speculative propositions, connecting art to other communities and disciplines. By acknowledging shared rights and by making sure the art archive is not only limited to activities defined by certain areas of speciality, an interdisciplinary art school will insist that an art archive is a truly democratic space for interrogation and discussion.

As art institutions are central to our cultural identity and our shared political existence, their role in understanding the complexities of a healthy, extraterrestrial democracy is decisive. It will be the institutions of art that support artists to take up their own place in a young, developing culture on other planets, ensuring that art is not relegated to leisure or therapy or entertainment. However we choose to think about art institutions in future societies—however we frame the exchanges that occur between artist and exhibition, between the public and the art in the gallery or within the museum, with its educational and pedagogical formats; however we understand the transition from student to practitioner in the art school or engage with the critical role of the archive; whatever position we take on the demands of a marketplace and its collectors or the domain of public art, its shrines, memorials, and monuments— how we envision art will influence how we understand ourselves and our growing society on Mars.

On Mars we will need a common world to care for, one that is not simply a resource-driven outpost in terms of its peoples and its objects, and one that pertains to more than the division of rights in a differential body politic. In the future, on other planets, yes, we will need all the institutions of art. Before we travel to other celestial bodies and in the first few years of our new home off-Earth, we should consider ethical infrastructures that recognize and fund artists and distribute their work. Initially though, to find futurity in art and build upon interdisciplinary histories, methods, and frameworks, in tandem the art school and the art archive have the potential to lead. These institutions of art are self-sufficient enough to rehearse and implement their own terms of engagement and socially responsive enough to question and negotiate their conditions, promoting coproduction as well as resistance. Realizing that material practices are about our rights made present, the art school and the archive will help us narrate different ways of being together.

In our small society off-Earth, maintaining the health and liberty of a socially cohesive culture through art-making will be juxtaposed with our commitment to institutional inclusivity. Art and artists will create democratic conditions for us to question what it means to be a citizen on Mars. If we are committed to a sense of belonging, to building cultural democracy on Mars, rather than homage earthly values, we must rethink our habitual selves, our collective cultural heritage, and the institutions of art.

References

Azoulay, A. *Potential History: Unlearning Imperialism* (London: Verso, 2019) 166–7.

Beech, D. 'Weberian Lessons: Art, Pedagogy and Managerialism' in P. O'Neill and M. Wilson (eds), *Curating and the Educational Turn* (London: Open Editions/de Appel, 2010) 47–60.

Breaknell, S. Perspectives: Negotiating the Archive. https://www.tate.org.uk/research/tate-papers/09/perspectives-negotiating-the-archive accessed 13 April 2021.

Clarke, P, Jones, S, Kaye, N., et al. (eds). 'Introduction; Inside and Outside the Archive' in *Artists in the Archive: Creative and Curatorial Engagements with Documents of Art and Performance* (London: Routledge Press, 2018) 13.

Derrida, J. *Archive Fever: A Freudian Impression*, Eric Prenowitz trans. (Chicago, IL: University of Chicago Press, 1995).

Di Maggio, P. 'The New Institutionalisms: Avenues of Collaboration' (1998) 154(4) *Journal of Institutional and Theoretical Economics* 696–705.

Doherty, C. *Protections Reader* (Craz: Kunsthaus Graz Publications, 2006).

Esche, C. 'Start with a Table' in P O'Neill, and M Wilson (eds), *Curating and the Educational Turn* (London: Open Editions/de Appel, 2010) 310–19.

Kolb, L. and Flückiger, G. 'New Institutionalism Revisited' (2013) *On Curating* 21, accessed 4 April 2021 <https://www.on-curating.org/issue-21-reader/new-institutionalism> accessed 24 July 2022.

Lecours, A. *New Institutionalism: Theory and Analysis* (Toronto: Toronto Press, 2005).

Morgan, J. 'What Is a Curator?' in J. Hoffman (ed.), *Ten Fundamental Questions of Curating* (London: Mousse Publishing and Fiorucci Art Trust, 2013) 1–10.

Raunig, G. 'Instituent Practices: Fleeing, Instituting, Transforming' in G. Raunig and G. Ray (eds), *Art and Contemporary Critical Practice: Reinventing Institutional Critique* (London: Mayfly Books, 2009) 3–13.

Rogoff, I. 'Turning' in P. O'Neill and M. Wilson (eds), *Curating and the Educational Turn* (London: Open Editions/de Appel, 2010) 34–46.

Sherman, D. and Rogoff, I. (eds). *Museum Culture: Histories, Discourse, Spectacle* (London: Routledge Press, 1994).

Sjöholm, J. 'The Art Studio as Archive; Tracing the Geography of Artistic Potentiality Progress and Production' (2014) 21(3) *Cultural Geographies* 414–505.

Steedman, C. 'The Space of Memory in an Archive' (1998) 11(4) *History of the Human Sciences* 65–83.

Steedman, C. 'After the Archive' (2011) 8(2) *Comparative Critical Studies* 321–40.

4

Space for opportunity

Transcultural and transnational sources of extraterrestrial liberty

Mukesh Chiman Bhatt

1. Introduction

The rule of law is seen as a prohibiting environment when enforced and propagates a view of the Outer Space treaties and associated United Nations (UN) resolutions and declarations as similarly restrictive. However, law is also an enabling and structuring mechanism, allowing the engineering of specific outcomes and consequences, albeit predictable or unforeseen, desired or not. When not restricted to a jurisdiction off Earth, the Outer Space treaties and documents extend extraterrestrial jurisdiction and responsibility and may be treated as part of an existing international legislation applicable to all future anthropogenic activities. More opportunities become available, grounded in natural law, with a rights-based approach. Implemented and used in conjunction with domestic national and extra-territorial legislation, new freedoms in civil, political, social, economic, cultural (and including environment, development, gender, and race) domains open up beyond Earth. Modified through traditional customary or soft law, values from beyond the Western world provide a larger framework for human society in space, and on Earth, emphasizing the truism: Everything not forbidden is allowed, that not allowed is prohibited.

Early settlements on Earth which led to later urbanization were probably the result of the development of agriculture. Such urban centres would have required access to air, supplies of water, and, given the assumption of agriculture, the need for sufficient land on which to grow crops. Of necessity, these activities require effort: a cycle is established whereby energy is provided through the produce eaten by the populace in order to ensure their survival and to continue to be able to irrigate and harvest the land for present and future use through the generations. Any settlement in outer space will undergo various stages of development which for the moment we presume to be non-specific but can be assumed to be parallel or similar to those of early settlements on Earth. Any settlement in outer space or on celestial bodies other than Earth is going to be heavily dependent and reliant upon technology and engineering. It is these engineering technologies that will be at the core of the production of power, air, water, and nutrients that are essential and necessary for the survival of humans and the simulated biosphere which will provide the necessary atmosphere and ambience off-Earth. The four constituents required for urbanization correspond to the four primæval elements of air, water, earth, and fire presumed to be the central constituents of the universe by ancient civilizations. These also equate to the modern notions of the three states of matter and their transformative catalyst: namely gas, liquid, solid, and energy.

The Outer Space Treaty (1967) refers to the province of all mankind and of the common heritage of mankind. These enshrine a possible set of rights shared by humanity as a whole, or within which resources are shared by humanity as a whole, or those resources as a form of commons which are a necessity for evolution and survival but which are not and cannot be owned or controlled by anyone. In principle, that which is in the province of or in the common heritage of all mankind cannot be separated by session, enclosure, or by borders h6and therefore be subject to ownership, although they may be controlled or managed. These rights may be shared between groups: a river can run through many countries; a mountain range has no borders or boundaries except those agreed by humans; the oceans and the atmosphere allow property-free travel for currents, winds, and ships. Although the common heritage of mankind is often compared to international waters, it is as much about the experience of living under the night sky, observing the Moon, the Sun, or the Earth from some far distant interstellar craft. There is an evolutionary requirement for air, water, and food. Fertile soils are needed for food, in turn requiring access to air and water for growth. Animals and plants require access to air and water separate from that needed by the fertile soils. The notion of common heritage thus includes the co-evolution and coexistence of species within a shared ecosystem for habitation, reproduction, and life. The atmosphere is part of the biosphere, and the experience of the moon part of the human condition: these are not subject to human-imposed borders. A virtual, fictive border can be actualized through technology—fences, domes, or other constructions and human activity such as war or occupation divides the indivisible and severs teleconnections across the global Gaian system.

The settlement, whether in actual space or on a celestial body, is essentially an enclosed environment. It requires the production and supply of air, water, and nutrients necessary for the health and survival of not just human beings but of those parts of the terrestrial biosphere that the human settlers have elected or selected to accompany them. This biosphere will include not just plants but may include animals and those elements of any biome and eco-system which are classified beyond that of the plant and animal kingdoms. This brings in considerations of law, technology and property: technology is a requirement not just for survival but for all other activity (Tranter 2018). Questions of ownership, possession, use, utilization, distribution, marketing, commodification of air, water, food, power, and other resources in common essential for growth and survival. Should there be a general right to these resources and the necessary technologies? Are these general rights part of a common heritage through collective use in commons or complicated by the individual's right to economic activity?

Law, like technology, is a local social anthropogenic artefact or terrestrial global structuring ethic. Law structures and shapes society, institutions, and enterprise through their interactions and relationship. It balances negative and positive freedoms; encourages new structures and explorations within well-defined limits where constraints act to encourage creativity and collaborative standardization (Contreras

2017); and supports and restrains cooperation, collaboration, and competition. Law is essentially a structuring agency, and can be viewed as engineering (Tamanaha 2006; Howarth 2013). Law is not necessarily statutory or founded in a definitive absolute or stable and unchanging philosophy or in sets of morals and ethics. Through positive soft law it is primarily reactive and retrospective (Marboe 2012; von der Dunk 2017), changing regulations and standards. Law is an enabling mechanism: it restricts unfettered individualism, and encourages cooperation, feeding into institutional freedoms that allow non-linear progression. With respect to the law of outer space, the relevant treaties and national legislations project the freedoms (and restrictions) of Earth through the extension of national jurisdictions onto human activities in space (Article III, Outer Space Treaty 1967).

2. Context: Air, Water, Nutrients, Power

In the human past, land has been enclosed and ownership imposed, leading to social, economic, and political changes. The English have had a particular preference for removing indigenous populations from specific and desired lands, whether in England, North America, Australia, or the African countries. This removes a shared right or part to a commons. Does this impact on property rights in space? Rivers pass through several countries, nomadic peoples—the Maasai between Kenya and Tanganyika—exist across post-colonial or colonial national borders as the result of colonial land acquisition and the removal of peoples onto reservations. Usufruct becomes ownership. Ownership includes possession, control, and jurisdiction, but these latter do not commute into ownership. If we apply these notions to air, can air be self-enclosed?

There appear to be no generalized rights to air in terrestrial legislation. Through the provision of extraterritorial jurisdiction in Article III of the Outer Space Treaty of 1967 this lack of generalized rights on Earth extends to air in outer space and the celestial bodies with particular respect to activities involving humans in space. Can the right to clean air, or the effect of pollution and therefore of clean air, be extended to the right to air and oxygen? Is this right separate from the right to an environment that implicates and derives from the right to life? Coase (1960), in his consideration of social cost of economic activities, thinks that a specific right to air and a provision for its entrance and exit as fuel and exhaust in particular reference to economic activities, however, does exist. In addition certain locales, London in the United Kingdom and Tokyo in Japan appear to have limited legislation in place for a generalized access to sunlight. Given the provision for economic activities through the use and exploration of outer space in the 1967 Outer Space Treaty, these approaches, however specific, may perhaps provide a route through economic activity to the generalized provision of air and a right to air for settlements in outer space and on celestial bodies. It is through the particularities of this approach, and through the examination of the patent and property systems applicable to existing and innovative technologies on Earth, that the provision of air for settlers of Earth may be considered to be acceptably regulated, or indeed constraints put in place to ensure that such provision and supply be constant, uninterrupted, and not misused. As described below and by others (Cockell 2019), control of the supply of air that can lead to unacceptable restraints on freedom and on what would be considered normal democracy for the present time on Earth. It is noteworthy that this 'freedom engineering' approach to the supply of air and its control considers the use of engineering technologies to ensure freedom from tyranny and dictatorship in settlements beyond Earth. Additionally, unless a viable terraformed biosphere is present, any method of producing oxygen and essential resources is likely to require power in some form. Proposed and existing technologies for these situations, in whatever type of habitat—space station or platform, planetary settlement, or gravity well, and a travelling, transient craft—are many (see e.g. Allen, Tainter, and Hoekstra 2003; Shayler 2017; Cheston and Winter 2019; Dominoni 2020). Whatever the specific situation, a colonizing settlement necessitates a changed and balanced environment (Fischer-Kowalski and Haberl 1993). A full survey of these technologies is beyond the scope of this chapter.

Air quality and quantity will be subject to control and may be monitored on an individual basis, on the basis of supply to a dwelling, or on the basis of supply to an individual or a group of individuals including families or other gatherings. This is analogous to the supply of a resource or facility through a universal service on Earth (water, electricity, telecommunications, health) but which universal service is subject to enclosures unlike those on Earth. If we assume that the settlement is primarily biased towards economic activities of all kinds, the presumption is then that these essential supplies of air, water, and nutrients are provided either free of charge or at some level treated as paid-for commodities. The inequity of having to pay towards personal survival is emphasized by those unable to contribute to the political economy of the settlement: infants and children, those who are sick or disabled temporarily or permanently, those who through old age have become what one might consider a burden on society and their families, and the final indignity of having to fund the disposal of one's corpse. As discussed elsewhere (Diamond 2013; Bhatt 2021), the situation could be resolved by killing those who are unproductive, by exiling them, or by committing their children to a lifetime of debt, slavery, or indentured servitude, or requiring a similar provision of insufficient savings and investments to act as a buffer or pension to support their personal unproductive futures. Resources may be distributed more equitably and free of charge through welfare, but which would still have to be paid for through a general tax on the wider population.

In the matter of enclosing air, Indian philosophers have long held that the space inside a jar is not separate from the space outside a jar. The analogy here is obviously that of the air inside a jar—it is not separate from the air outside the jar with a closed lid. So the question here is whether the enclosure of air can be regarded as a matter similar to that of the enclosure of the commons. Of course, air can be separated from air in the sense that it becomes non-contiguous, however contiguity remains in the sense that as soon as the lid or the enclosure is breached, the air mixes to become inseparable from the rest of the air. In this sense the atoms and molecules of the air are not trackable or separable from each other, being identical to each other. Should a tracker or marker be somehow introduced it would allow the possibility of designating and tracking ownership of each atom or molecule. This can of course be extended to all atoms, molecules, elements, vitamins, enzymes, and the light which are all essential for human survival in any situation but which are particularly important in the inimical environment of non-terrestrial space. In space these may be considered to not be natural as they would need to be manufactured in some manner for the use of the inhabitants of these settlements. Of course, the use of such oxygen or elements may be monitored in the same way that other universal utilities are monitored and in the same way that the quantity of oxygen used in hospital environments is monitored. This however raises immediate issues of surveillance and privacy which are to some extent beyond the scope of this chapter but which impinge upon liberties and freedoms enjoyed by individuals.

Air (and the atmosphere) may be regarded as part of the common heritage of mankind. It is certainly the province of all mankind but this is a circumstance specific to the terrestrial biosphere. Air, as with water and other utilities or resources and facilities, may be seen as part of a commons: resources that were owned or limited to use by the owning family or aristocracy but which use was extended to the serfs and possibly the tenants of that family. These commons were not part of the common heritage of mankind but limited in use to specific dependents. The notion is particularly opposed to the idea of land or other resources which were open and accessible for nomadic use or settlement by anyone without regard to their feudal relationships.

A non-separable commodity can be generated by multiple companies using different processes (feedstock, fossil fuels, nuclear, wind, solar at different sites) and can then be fed into a common distribution system, which is then itself specialized into a separable and differentiated domain (house, company, building) and charged by metered cost that reflects the subscription paid to a specific generator. However this model may not apply to any system that distributes air into a common area such as a domed city for general use by all. It could be charged at a standing rate for all users, or it could be limited to top-up through 'air stations' which are equivalent to the metered universal service of water, gas, and electricity in Earth cities. The distribution network remains a service in common from and to which is fed into and taken out of by producers and users.

Each stage discussed above and in the following paragraphs is about technology that is patentable. Air as a utility, whether it be essential or commodity, free or paid, is dependent upon the available and ownership of patentable technology. In essence at first this concerns the research and development of the technology that will provide the essential resource, in this case air. After the research and development stage, there are further stages of marketing, of distribution, of use of the product, of quality control, and of maintenance of the technology and infrastructure required for this particular commodity. Given this is considered to be an essential resource necessary for survival and not just for economic or other activity, examples on Earth of non-essential commodities and resources may be noted as subject to use for geopolitical advantage and leverage. This leaves the production, distribution, and use of an essential resource necessary for survival hostage to the whims and vagaries of political ideologies as well as economic pressures amongst other constraints, thus limiting freedom and liberty and therefore the rights of the people within a settlement in space, and endangering the survival of the populace and the space settlement.

3. Patents and Power

The essential and central issue in terms of providing air, water and nutrients is the equitable generation, provision and distribution of power needed for survival. The central issue becomes the governance of and control over the machinery that generates and distributes this power. Governance and control over its design, construction, and maintenance assumes such importance that it can be used to control through the provision of the universal resource both the behaviour and the responses of the population at large and of individuals within it, which at its extreme may become authoritarian, dictatorial, or tyrannical.

3.1 Patents: Origins and ownership

Patents are a subgroup of the general award of intellectual property rights held by the creator and innovator. The general rights include design, branding, and copyright, although patenting is best known and applicable to the science and engineering disciplines. Patents exist at multiple levels from its smallest component to the overall design of the machinery and its operation. Patent rights are awarded on the basis of innovation and specificity and not for a universal service, nor for general concepts developed for general purposes, nor for what are deemed the result of natural processes. They are also restricted to the national sovereign territory that is governed and over which the authority of the patents awarding body is valid. This makes patents comparable to the extensive national sovereign jurisdiction over space objects alluded to in the various outer space treaties that constitute space law. Note that there is no general international patent that can be awarded such that it is valid in all countries. The notion of a launching state where more than one country may be involved in a space activity does not appear to extend the award of a patent to these several cooperating nations. It thus also limits the notion of ownership except by compromise and mutual agreement. The patent remains within the initial awarding national sovereign state and its territory and is awarded for a limited time. Modern patents are essentially monopolies over an innovation granted to the owner of the innovation, which ownership is itself a matter of discussion. Ownership is in perpetuity and may be transferred, the patents time-limited, but may be licensed for use by others during that period (Bently et al. 2018, Part II, Chapters 14–24).

An early example of such monopolistic knowledge is the use of hieroglyphics in ancient Egypt. These were a method of keeping secrets and ensuring payment from the general populace in order to maintain a priestly caste. Aboriginal songlines are similar: an oral memory map used as a navigational aid incorporating landmarks to indicate the placement of and routes to hunting grounds or water resources; these are of necessity kept secret from other groups and non-initiates, and also between men and women. This is a similar construct to a patent. It also mirrors the behaviour of mathematicians and other academicians in mediaeval Italy who kept secret their solutions to mathematical equations to earn money through gaming and gambling (Torretti 1999, 362).

Intellectual property legislation and rights may also apply to microbes and bacteria which are used for space mining. However, the US legislation of 2015 does not allow for living matter to be used in space. This begs the question of whether microbes and bacteria are to be considered living matter and whether the relevant legislation applies only to that living matter that is found in space and not to that taken from Earth or developed in space from terrestrial antecedents. It also begs the question of the definition of life in modern biology with its kingdoms and domains. Note that rights of intellectual property holders can apply to a patent on materials derived from DNA and similar antecedents of living matter which confuses the issue. The nations that have thus far passed property rights legislation for space resources treat living matter differently, with Japan possibly allowing for the use of living matter in space.

As the award of patent rights and the application of extraterritorial authority is restricted to national sovereign territories which have their own jurisprudential sources of law, an alternative and hopefully more transcultural approach to the jurisprudence of such law is needed in order to improve upon the possible self-determination and practice of rights in the provisions of essential items necessary to survival such as air, water, and nutrients. Further, the availability of patents specific to space technologies in the international public domain and the fact that a grant of patent is not valid on an international basis argues for the transnational extra-territoriality of such patents. Such transnational application for the use of space technologies is supported by precedent. Previously the European nations imposed extraterritorial jurisdiction by agreement, treaty or occupation, by reducing a population's humanity or religion to subhuman or animal inhabitation inhabited, or by possession due to non-inhabitation. Possibly not as morally ambivalent given its adoption by the United Nations, this extension of jurisdiction is also imposed in outer space by Article III in the Outer Space Treaty of 1967 and agreed by existing and new nation-states.

3.2 Social structure

The installation, repair, and maintenance of the oxygen-generating technologies as well as the power technologies is likely to be a specialized field of knowledge and training. As such, the knowledge may well be guarded and kept not so much secret as unavailable to the general public, initially on safety grounds as non-specialized persons may well interfere with essential survival equipment and cause risk or danger. This fear of damage may well lead to other restrictions in terms of knowledge which would be equivalent to the enclosures discussed above or to the application and granting of patent for such specialized knowledge and creative technologies. In any case, the specialized knowledge will need to be passed along to future generations in order to maintain safety. It is plausible that the children of the persons involved in such repair and maintenance are likely to be exposed to the ideas and concepts of said repair and maintenance in preference to those outside the family or the recommended educational system and thus would have an advantage over others in terms of this knowledge. Such knowledge and therefore advantage would therefore be inherited by the immediate descendants and possible future descendants of the original repair and maintenance persons. Control by a group of technicians and engineers may also lead to members of their families being more exposed to notions and concepts relevant to the production and maintenance of the machinery, as well as the personnel networks, equipment, and infrastructure surrounding the machinery and innovators. This then gives these family members an advantage over others who may wish to enter into the society and contribute to the well-being of the enclosed environment, but who are thereby less able not having the familiar and familial experiences enjoyed by others. Those advantaged would continue through the generations to increase their advantage as their progeny would also be more familiar with machinery than those outside the families. This therefore describes the growth and establishment of a specialized group of personnel and their families tasked with production and maintenance. Given the partial or fully closed nature of the families, this stratification would through the generations lead to a technocratic occupational or function-based caste system This may well result in a hierarchical caste system where knowledge and specialized knowledge is kept within certain accepted groups and in particular family groups (Polanyi and MacIver 1944; Graeber 2001; Wilkerson 2020) and which is likely to lead to a closed, authoritarian, and centralized system (Hayek and Caldwell 2014). Note the similarities with similar stratified or caste-based systems on Earth. An essential characteristic of such stratified populations is their guardianship as gatekeepers of what, over generations, becomes esoteric knowledge. Another characteristic of such systems is likely to be somewhat draconian measures to protect and keep secret the knowledge that leads to the family having advantage over others—this may manifest itself as the use or abuse of authoritarian power and control of others.

4. Alternatives to Patents

Patents are considered rewards for the producers and creators of innovation, and which thereby generate an income restricted to these innovators. Patents are also seen as drivers of innovation as this reward and restriction of income encourages innovators to create and produce novel solutions. If, however, controlled by a closed group, then this innovation is stifled or restricted. Where such innovation is restricted to a closed group, creativity and new ideas may be few and far between, leading to a degeneration of the system of production as it can no longer respond to changes in the environment or to novel situations. An open or public domain patent system to encourage cooperation and innovation within the space industry and to encourage space exploration and exploitation rather than closing it off by awarding patent or resource rights. Openness allows building on each other's work and encourages both cooperation through pooled resources for enterprise without conflict of ownership, or perhaps encouraging joint ownership, or investment and engagement in a project or programme, and making use of the vast volunteer and space-interested amateurs, and competition—this helps the economy to grow if that is the goal. It also underlines the provisions and freedoms embodied in the outer space legislation—that of cooperation in the use and exploration of space. Moral and ethical arguments for and against patents have been considered by Ang (2013). The subject is somewhat complex and beyond the scope of this work,

Should patents exist for essential, universal services? If patents exist, they should be open source. Derivations and drifts from open source should not be patentable, or they should remain open source. Essential service engineering technologies and infrastructure should remain open source and public domain, with resources to develop and maintain these services and technologies also being open source and public domain without restrictions. Some proposals that might be considered for settlements in outer space until resolution of these concerns are to do away with the patent system altogether; to declare any innovation as part of a global commons (such as the presumed public domain) and therefore available for use by all and any individuals; to waive the patent in specific circumstances (the production and provision of air might be such an instance); to ensure that governance and control of critical systems is subject to rule by a hierarchy-free, decentralized, autonomous organization in which, in principle, every person is a stakeholder with input and control, although this system would still be subject to factionalism and cartels; to nationalize such production and provision; or provide multiple and plural installations such that no-one can monopolize or control any singular installation (Cockell 2019); and of course to ensure that service is both universal and fully funded without exception through taxation. All alternatives require a source of consistent and coherent but flexible source of law applicable in novel situations such as settling in outer space so that the jurisprudence moves away from terrestrial and anthropocentric concerns based on Eurocentric ideals and philosophies. Any alteranatives also need to consider whether existing rights to self-determination, independence, to property and economic activity, to exploit natural resources which exist in various jurisdictions, need to be abrogated or continued within a wider circle of common heritage.

The jurisprudence of law that underlies statute and the application of law in the domain of patent rights and of the more general intellectual property rights is highly dependent upon the sources of law considered foundational in the sovereign national territory where those laws are applied. Jurisprudence is therefore rightly complex but limited to sovereign national territory and cannot be applied in a more general manner to the international domain. The jurisprudence underlying space law suggests that space law is essentially a part of international law under Article III of the Outer Space Treaty of 1967. The laws of outer space, however, attempt to compromise with national sovereign legislation through the extraterritoriality of jurisdiction extended over objects in space that are essentially the result of human activities: this jurisdiction does not extend to outer space or celestial bodies. Therefore when considering the jurisprudence of patent rights in relation to the jurisprudence of space, some compromise is necessary. There remain, however, contradictions which cannot be easily or readily resolved. In such situations it may be appropriate perhaps to introduce a more general jurisprudence that has a somewhat more global or universal application in the sense of being limited to terrestrial law as applied to outer space. Such a foundational approach is an attempt to try to resolve some of these contradictions and to propose that perhaps sovereign national and international law may not be quite as oppositional if the appropriate sources of law are agreed upon. Law for space settlements should perhaps be based on a science-based understanding of local environments, ecosystems, and needs.

5. Transcultural Sources of Law

Given the applicability of patent law to a specific situation depends upon territoriality and therefore the jurisdiction of the launching state, whether singular or multiple, it is apparent that this situation may not continue in the case of a settlement or set of settlements which are extraterrestrial and maybe of multiple origins in terms of the applicable legal jurisdiction. In such cases, of course, the singular settlement or multiple settlements, if they are cooperating or have come together to form a larger political entity, may secede or declare independence from their respective founding states or other jurisdictional entity. However, given the arbitrary nature of anthropogenic law on Earth and its extension to off-Earth premises, an alternative framework for legal responses and applicable patent law should perhaps be founded upon a different and more coherent and consistent basis.

Western law recognizes several foundational jurisprudences of law: natural law based on theology, natural law based on considerations of moral conduct, positive law which essentially feels reactive and arbitrary, and autopoietic law which is a result of self-organization of society. It is the last which is closest to law by analogy to the laws of nature rather than derived from the laws of nature. Western law also seems to exalt the Hobbesian view of nature—namely that nature being red in tooth and claw, the essence of man is purely competitive and destructive. In contrast to the goal of non-Western philosophies (including Spinoza) is the identification of the individual with the universe (Fung 1944 1). This provides a different and consistent, non-arbitrary foundation for ethics and law. As Fung states at the start of his opus, by nature, Man is good. No greater oppositional view need be reconciled in any transcultural approach to law. The Eastern approach tends to view cooperation and social harmony as essential, and as such it is more likely to be a reflection of the social interactions that arise in a more 'natural' manner in human society—it reflects the harmony in heaven. Leaving aside the difference in language, the Eastern approach appear closer to an emergent phenomenon in terms of autopoietic self-organization than its Western counterpart. Moral conduct is thus based on observed social behaviour than on arbitrary social values derived from theology or other presumed spiritual practices. The right to air, water, nutrients, to power in an artificial environment of outer space, to light and vitamin D in an environment distant from solar ultraviolet radiation in these cultures is more likely to be based on the heritage common to humankind. Such responses may therefore be considered evolutionary and provide a foundation for ethical action, even where such needs are supplemented or provided by artificial methods, and which foundation would be free of differing cultural biases and values.

Such a cosmology (Bahadur 1978) in common is shared by the Hindu, Jain, and Buddhist cultures of India and the hegemonic Buddhist expansion into the Far East and South-East Asia (Panikkar 2010). This posits an origin for the cosmos from a primordial emptiness constituting what we may in modern parlance describe as a substrate or field which enters a state of disequilibrium amongst its constituent properties to manifest the observed universe (Vetury 1987). In the original Sanskrit this substrate is denoted '*ākāśa*' (pronounced 'ah-kahsh' using English orthography for the phonemes). The earliest instantiation of this word is in the *Chāndogya Upanishad* (*c.* 800 BCE in its written form) according to Apte's *Practical Sanskrit-English Dictionary* (1957–59). Apte's *Dictionary* conveniently provides actual examples from extant texts whose approximate dates are known through other scholarship. An approximate chronology can therefore be constructed for the changes of meaning through the centuries. '*ākāśa*' later comes to denote the substrate through which the primordial sound of the universe propagates, and still later in the modern Indian languages denotes the upper sky and sometimes the space beyond it.

The 800 BCE meaning assigned to the word by Apte is 'void, vacuity'. Void and vacuity are clearly to be considered synonyms. 'Void' is a translation for the return to the source of *nirvāṇa* (or annihilation of experience), the final state of meditative practices. It indicates an immersion, assimilation, or identity with the source of the universe. When considered in a legal context, vacuity and void are seen to be even more closely related. Both words derive from the same Latin root, and 'vacuity' can be described in terms of 'void': a 'vacuity' is to have all legal obligations made void—in other words, to be free of all legal obligations as instantiated in its later fourteen- and fifteenth-century Italian use from ancient Roman law, and thence to the notion of freedom from all obligations if the economic basis of law is used to interpret the concept. Note that this notion of freedom returns us to the concept of liberation (*mokśa* or *mukti*) from all Earthly conceits and demands which is the central goal of all Indian (Hindu, Jain, Buddhist, and related systems) philosophies. The Eastern concept of liberation from the cycle of life is therefore to be understood as equivalent to the liberty of the individual in Western culture. It is here we note the similarity of the Eastern concept of the origin of the universe (metaphorically the 'ocean of creation') to the primordial quantum vacuum and its sea of virtual particles popping in and out of existence, thus disturbing or perturbing the equilibrium of the vacuum (Boi 2011). Liberation equates to freedom, and thus liberty is founded upon and derived from the primordial void due to the disequilibrium or imbalance between its three constituent properties. The coincidence of complex emergence from this primordial void with the quantum vacuum, and in both the associated notion of chaotic disequilibrium is suggestive of the possibility that the mechanisms of the universe may well provide a foundational jurisprudence for the behaviours and interactions of its incumbents that is, at least notionally, free of cultural difference.

The eighteenth century in Europe saw a conflation of universal ethics with modern science, starting with Newton and Kant and continued by the mechanical scientists and philosophers of the nineteenth century. Laplace considered the Hamilton–Jacobi formulation of the principle of least action in physics to be absolutely deterministic such that the end result determines the path taken. This interpretation also reflects the European Christian culture of the time which privileged control and constraint; it assumed dominion over nature through a theological teleology that entrenched previous biases and beliefs. This determinism resulted in the Enlightenment belief in the correctness of science and its methods and appeared to increase social conformity, fatalism, inequality, and inflexibility in social structures and *mores*. Buddhist *karma* is similar: it too is a delayed deterministic retribution irrespective of choice or circumstance. Such absolute determinism was paralleled by the deterministic historical materialism of Hegelian Marxism and its commentary on the political economy of the Industrial Revolution. The notion of society being formed of agential entities belonging to a community alienated from the environment and the products of their labour is time-bound and therefore ineffective and elusive when considered in changed circumstances. It therefore suggests the use of agential rights to realize the potential of one's self. In this perspective, it is the agency of the individual to rebel because of historical circumstance and to change such circumstance. Unlike the absolute determinism of science with its immutable laws, law was now viewed as instrumentality: law is a political instrument selective in its application, dependent upon material conditions, and determines economic relations through statute. In the present context, law would determine the ownership and distribution of property but would not allow either ownership or distribution of shelter or of oxygen to those in need. However, the new sciences of the twentieth century can modify the determinism previously accepted. Introducing probabilistic approaches as in quantum mechanics or through statistical mechanics, new formulations are found leading to new notions of deterministic chaos and complexity and the emergence of new phenomena through self-organization. In this context, and this is the current proposal, the Hamilton–Jacobi formulation can therefore be read as contingency or contingent determinism rather than absolute determinism. It becomes forward looking rather than backward looking. Teleology is determined by the choice made amongst a number of probable options. This leads to the idea that an action taken will lead to a certain result in the future but not that it will definitely lead to that result. It allows an optimal path to be chosen to attempt to reach that result and it implies that there is a teleonomic freedom of choice in that the action taken is a means to an end. What is apparent in this proposal is that it can act as an evolutionary foundation for ethical action and freedom which is very similar to the social norms of eastern societies and to the genealogy of liberty as outlined by Quentin Skinner (2016). Apart from physical freedom and psychological liberty, the individual as agent has the potential for self-actualization whilst being part of a network of interconnectedness which is rooted in the universal laws of nature. These latter laws are seen as mutable in the Eastern cosmologies described above, and yet which mutability allows these Eastern cosmologies to replicate the philosophies behind Western science. It is therefore apparent that law as *praxis* is an emergent anthropogenic agency resulting from evolutionary mechanisms. It is axiomatic in this discussion that Western science, like all anthropogenic artefacts, may be considered derived from the interaction between multiple cultures.

6. Discussion

Any space settlement will be the seed of a new and developing nation, despite its start under the jurisdiction of an existing terrestrial nation. Even considering its new situation in outer space, some consequences of this approach become apparent. The etymology of the word 'vacuity' implies a freedom from all and especially legal obligations. This suggests that there are no obligations of or to terrestrial or other laws from elsewhere. Therefore terrestrial laws may be considered not to apply in space, and a degree of freedom, individual and societal, and in particular a derogation from societal norms becomes possible. It allows free and independent withdrawal from the various international, bilateral, or multilateral treaties that might be supposed to impact upon the settlement in space and in fact it may be considered that these treaties have no applicability in space given that they were developed for terrestrial situations. Clearly consequent to the approach outlined in very short form above, any space settlement is therefore obliged to develop its own legal system independent of terrestrial norms. And given the multiplicity of ethnic groups or cultures which are likely to make up that space settlement or groups of space settlement, these legal systems are likely to be plural hybrid, novel, and fertilized from the different cultures within its make-up. Above all, derogation from terrestrial and societal norms that exist on Earth provide a very distinctive right to self-determination which will lead to independence from the hegemony of the founding nation and of Earth.

Despite the freeing from obligations, it should be noted that the approach as described is compatible with the Outer Space Treaty of 1967. In particular it encourages the principles of cooperation and community, and of benefits to all which are complementary to the core non-appropriation and non-militarization of the use and exploration of outer space.

As discussed above, humans cannot survive without air or water, although they take much longer to die of starvation or from nutrient deficiency. Unproductive groups (children, disabled, elderly) add to the burden of the various settlements requiring, as they do, care and essential sustenance, namely air, water, nutrients, and power within these inimical environments of space. Children require care, but may be seen as social reproduction and therefore a source of labour, labour which can act by itself, replace or complement automation, but which requires education and training. This constraint may lead to child labour or indentured servitude as a result of essential requirements during childhood, although children are seen to guarantee the future of the settlement and are also a source of society, innovation, and new creativity. The elderly may be seen to be a source of both tradition and caregiving, in both a general and a specific sense within both the social environment at large and within that of the family. These may of course add to the education, care, and training of children and adolescence and may perhaps be induced to provide care for those who are sick, disabled, and unable to look after themselves. This does not preclude the inclusion of those persons deemed disabled but different and therefore able to contribute. The presumption of care and essential provision by the parents or the family does not recognize the extra burdens placed upon those implicated in that care, and as such provision of care and support may contribute to family solidarity, it would need to be recognized as productive labour in the same sense that medical personnel are considered essential and productive. However, there appears to be an underlying assumption that each person, whatever their age and ability, must be able to contribute to the space settlement and its interactions within itself and with other similar settlements whether in outer space or on a celestial body. This approach is a modification of the Buddhist economics approach: in consideration of the political

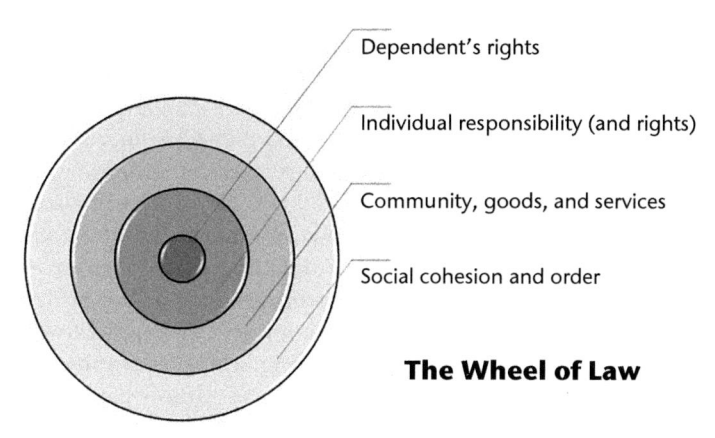

Dependent's rights

Individual responsibility (and rights)

Community, goods, and services

Social cohesion and order

The Wheel of Law

Figure 4.1 A Wheel of Law—protecting the vulnerable in outer space.

economy of a space settlement as a developing nation, Brown (2017)—basing her work on Sen's (1999) *Development as Freedom*—notes that the measurement of value depends on how well the economy delivers on the quality of life. This quality of life is not just material goods but includes consumption, the equivalent notions of equity, sustainability, and spirituality whilst protecting the environment (here meant as a general concept rather than a terrestrial one). The Buddha was also chary of pain and suffering: the proposed solution is illustrated in a modified Wheel of Law in Figure 4.1.

Freedom engineering is another solution to control and distribution of essential resources proposed by Charles Cockell (2019). It uses technological solutions to ensure that there are separable units such that the entirety of the settlement may not be controlled by any single individual or group or corporation. However, in engineering such individual units, these individual units may themselves become subject to control and misuse by other groups of corporate units, which then effectively transposes the problem from the higher entirety of the settlement to smaller groupings, which can then be used to create further division and a new form of technocracy and dictatorship at a lower level of group separation. This separation of groups into defensible units may also lead to a form of speciation in terms of social structure or of the redistribution of essential resources where different ideologies and norms may predominate, whilst effectively creating a rigid social structure within any one of these units or across the units. This would then also increase inequality within and between the various separate and separable units. Thus technological engineering as a solution has, in itself, the seeds to consolidate unacceptable structures within a social environment further limiting the freedom and liberties to which any individual or groups within these settlements are entitled.

If one arrives at a result, desired or not, and then looks backward in time in order to determine the path taken to arrive at this result, it will be seen that there is only one path that was taken and that the path was in some manner determined by the decision taken at each novel event or situation. History is, therefore, seen to be determined and considered deterministic through the decisions or events in the past which have manifested on the path taken and which are embued with a notional significance. To a greater or lesser extent, no significance is given to the constraints that might lead to a certain decision being taken and which might therefore determine the past at or after such an event or novel situation. History is therefore deterministic: all constraints are of equal significance and considered to have an equal and equitable effect on any decisions taken to determine the path. Hegel's notion of history was essentially a theological teleology in that he replaced the concept of an all-powerful transcendental deity with what may be considered an immanent and agential world spirit, but which, however, still leads to a specific goal indicative of purpose. Removing such a deity leads to the later equally deterministic and purposive notion of historical materialism as evolved by Karl Marx (Callinicos 2007). Like Marx, Laplace had no need of a theological basis or deity for his notion of history. He interpreted the Hamilton–Jacobi equations to be indicative of a deterministic history which, given the full knowledge of the constraints and laws of motion as devised by Newton to describe observed nature, constrains the path taken by a particle as determined by the final destination. In other words, the end result determines the path taken in the past, a retrospective decision that affects the past. In principle this allows for deterministic and inflexible prediction and planning of the future. A modern extension of determinism is sometimes considered to be super-determinism, the idea that the evolution of life and the appearance of humanity is the end result of a unique path starting from the Big Bang or the quantum vacuum, whichever came first. This latter modern version is however as restrictive and teleological as Laplacian determinism, and equally retrospective.

7. Conclusions

Is there a right to a viable ecosystem? Does this concern the availability of air, water, and a place to grow food as discussed in this chapter? Is there a corollary, a parallel, or some connection to the idea of the rights of nature applied to environmental protection? Should there be a recognition that one part of the environment, whether it is in a dome enclosed or otherwise, may have teleconnections through the land or through weather in climate and impact other parts of the ecosystem also enclosed underneath the same dome – perhaps a continent, a planet, an ocean or other similar enclosed structures?

The approach presented here decolonizes space, free of obligations to its terrestrial masters and the North Atlantic hegemony. As in the 1960s, the right to self-determination and to self-governance is recognized through the use of a diverse-valued human rights system specific to the new ecosystem and environment of outer space. It draws from the legal positivist conception of law in that it is a functional morality based on new and applicable norms rather than arbitrary cultural or religious values, thus showing variability and flexibility towards freedoms and liberties in outer space. This is because the legal system is constructed by analogy to physical systems and laws of nature hybridized with Indian philosophy and Roman law. It considers law to be an emergent anthropogenic artefact, consequent on evolutionary mechanisms. The system's autopoietic characteristics allow its internal elements to be reconfigured to attain newer stable states whilst being reactive to and communicative with external stimuli. Above all, it engineers, organizes, and maintains a diverse, valued system of liberties and freedoms.

Application of the above framework allows the use and consideration of new forms of political economy. A planned democratic structure established in its inflexible entirety from the outset and before the settlement has had time to adjust to a new environment inimical to humans and the terrestrial biosphere would likely cause the failure of such a settlemnt. Any planned structure for social organization and its institutions must have sufficient flexibility built in to respond to changed circumstances and to unknown constraints and situations. It is these very necessary and unpredictable answers that need to be catered for in responding to novel situations. Legal, political, and economic institutions and structures that are based on specific, perhaps national, human cultural norms, values, and institutional structures are likely to be a source of potential hazard detrimental to the settlement in space. To paraphrase a character from Robinson's *Red Mars* (2003), we are twenty-first-century scientists living according to nineteenth-century ideologies based on seventeenth-century philosophies. It is therefore perhaps necessary to examine and propose new conceptual frameworks for human societies and their institutions in space, that idea of outer space which is a completely new environment for humans, and able to throw up situations that could never exist on Earth.

References

Allen, T., Tainter, J., and Hoekstra, T. *Supply-side Sustainability* (New York, NY: Columbia University Press, 2003).

Ang, S. *The Moral Dimensions of Intellectual Property Rights* (Cheltenham: Edward Elgar Publishing 2013).

Apte, V. S. *The Practical Sanskrit-English Dictionary*, revised and enlarged edition Poona: Prasad Prakashan, 1957–1959).

Bahadur, K. P. *The Wisdom of Saankhya* (Sterling Publishers PVT Ltd, 1978).

Bently, L., Sherman, B., Gangjee, D., et al. *Intellectual Property Law* (Oxford: Oxford University Press, 2018).

Bhatt, M. C. Law and Political Economy in a Martian Settlement, presentation to the 21st virtual Mars Society convention, October 2021. Available at https://www.youtube.com/watch?v=b6S_OK2vlzI> accessed 20 December 2021.

Boi, L. *The Quantum Vacuum: A Scientific and Philosophical Concept, from Electrodynamics to String Theory and the Geometry of the Microscopic World* (Baltimore, MD: JHU Press, 2011).

Brown, C. *Buddhist Economics: An Enlightened Approach to the Dismal Science* (New York, USA: Bloomsbury Publishing USA, 2017).

Callinicos, A. *Social Theory: A Historical Introduction* (2nd edn, Wiley, 2007).

Cheston, T. S. and Winter, D. L. (eds). *Human Factors of Outer Space Production* (London: Routledge, 2019).

Coase, R. H. 'The Problem of Social Cost' in Gopalakrishnan, C. (ed.) *Classic Papers in Natural Resource Economics* (London: Palgrave Macmillan, 1960) 87–137.

Cockell, C. S. 'Freedom Engineering—Using Engineering to Mitigate Tyranny in Space' (2019) 49 *Space Policy* 101328.

Contreras, J. L. (ed.). *The Cambridge Handbook of Technical Standardization Law: Competition, Antitrust, and Patents* (Cambridge: Cambridge University Press, 2017).

Diamond, J. *The World until Yesterday: What Can We Learn from Traditional Societies?* (London: Penguin, 2013).

Dominoni, A. *Design of Supporting Systems for Life in Outer Space: A Design Perspective on Space Missions Near Earth and Beyond* (New York, NY: Springer Nature, 2020).

Fischer-Kowalski, M. and Haberl, H. 'Metabolism and Colonization. Modes of Production and the Physical Exchange between Societies and Nature' (1993) 6(4) *Innovation: The European Journal of Social Science Research* 415–42.

Fung Y.-L. *A Short History of Chinese Philosophy*, edited and with an Introduction by Derk Bodde (New York, NY: Macmillan, 1948).

Graeber, D. 'Value as the Importance of Actions' in Graeber, D. (ed.) *Toward an Anthropological Theory of Value* (London: Palgrave Macmillan, 2001) 49–89.

Hayek, F. A. and Caldwell, B. *The Road to Serfdom: Text and Documents: The Definitive Edition* (London: Routledge, 2014).

Howarth, D. *Law as Engineering: Thinking about What Lawyers Do* (Cheltenham: Edward Elgar Publishing, 2013).

Marboe, I. (ed.). *Soft Law in Outer Space: The Function of Non-binding Norms in International Space Law* (Vienna: Böhlau Verlag, 2012).

Panikkar, A. *Jainism: History, Society. Philosophy and Practice* (Delhi: Motilal Banarsidass, 2010).

Polanyi, K. and MacIver, R. M. *The Great Transformation* (Boston, MA: Beacon Press, 1944).

Robinson, K. S. *Red Mars* (London, UK: HarperCollins, 1992).

Sen, A. *Development as Freedom* (Oxford: Oxford University Press, 1999).

Shayler, D. J. *Assembling and Supplying the ISS: The Space Shuttle Fulfills Its Mission* (New York, NY: Springer, 2017).

Skinner, Q. Genealogy of Liberty. Lecture transcript. Presented 27 October 2016, Stanford Humanities Center as part of its Harry Camp Memorial Lectures. Available at <https://cluelesspoliticalscientist.wordpress.com/2017/01/27/a-genealogy-of-liberty-by-quentin-skinner-lecture-transcript/> accessed 20 December 2021.

Tamanaha, B. Z. *Law as a Means to an End: Threat to the Rule of Law* (Cambridge: Cambridge University Press, 2006).

Torretti, R. *The Philosophy of Physics* (Cambridge: Cambridge University Press, 1999).

Tranter, K. *Living in Technical Legality: Science Fiction and Law as Technology* (Edinburgh: Edinburgh University Press, 2018).

Treaty on Principles Governing the Activities of States in the Exploration and Use of Outer Space, including the Moon and Other Celestial Bodies [short name: *Outer Space Treaty 1967*] (United Nations, New York, 1967) as available on https://treaties.un.org/Pages/CTCs.aspx accessed 28 August 2022.

Vetury, R. R. *Selected Doctrines from the Indian Philosophies* (Delhi: Mittal Publications, 1987).

von der Dunk, F. 'Customary International Law and Outer Space' in B. D. Lepard (ed.), *Reexamining Customary International Law* (Cambridge: Cambridge University Press 2017) 346–73.

Wilkerson, I. *Caste: The Lies that Divide Us* (London: Penguin, 2020).

5

Expansion of humanity in space

Utopia or dystopia?

Octavio Alfonso Chon Torres

Humanity has long wondered whether we are alone in the universe and whether we will ever be able to reach the stars. So far, these questions remain unanswered, except for the second. An answer seems to be beginning to appear on the horizon, not so much whether we will reach the stars but if we are able to live in other planetary environments. Thus, a future awaits us in which human beings can form colonies far from Earth and these will be the guarantors of the future of humanity in the universe. Projecting into the future about the improvement of the human condition by expanding into the universe makes us think that perhaps a kind of modern myth is beginning to emerge: what once was the kingdom of Preste Juan but now revised in a new technological version of the myth (Chon-Torres 2021). It is almost as if thinking that thanks only—or mainly—to technological advances, humanity will be able to guarantee its relative perpetuity in time, without seriously considering ethical or social conditions. After all, it is not so much technological errors that can lead to death and desolation but mismanagement in the administration of resources and in decision-making which can condemn the lives of thousands and even millions of people.

However, why should we consider this situation if there is no guarantee that, in principle, our future could be off Earth? According to Baum et al. (2019), of all the possible futures that may lie ahead for humanity, the option consisting of the astronomical trajectory is a highly likely one. In their study they make a comparative analysis of four future trajectories for humanity: the first is the status quo trajectory, projecting that humanity will remain more or less the same as now, managed in the same way; the second option consists of the catastrophe trajectory, where humanity faces serious dangers that threaten its perpetuation as a species; the third is the trajectory of technological transformation, in which humanity makes great discoveries that make it change its current course radically; finally, the fourth trajectory is the astronomical trajectory, where humanity can migrate to other planetary environments to continue its existence. Of all the options, unless we become extinct, all the long-term trajectories involve our expansion in the universe; even if humanity must rebuild civilization, it will reach a point where it will have to explore other worlds. If we extrapolate the possibilities of survival to one of ethical and social coexistence, we could consider the following scenarios: humanity perpetuates its current moral behaviour (status quo), or humanity improves its moral condition and consolidates as a planetary and interplanetary species. Do we expect a future of humanity where modern forms of capitalism and democracy prevail in colonies on other planets, or a future where control predominates and freedoms are suppressed, such that they affect

Octavio Alfonso Chon Torres, *Expansion of humanity in space*. In: *The Institutions of Extraterrestrial Liberty*.
Edited by Charles S. Cockell, Oxford University Press. © Octavio Alfonso Chon Torres (2022).
DOI: 10.1093/oso/9780192897985.003.0006

the way we live in society? For this reason, this chapter aims to evaluate whether the future of humanity in space could be more akin to a utopia (morally desirable scenario) or a dystopia (morally undesirable scenario). In the following, I will outline different arguments in thinking about one or the other possibility.

1. Unpleasant Situations

Considering this situation, we are also presented with the astrobioethics challenge of considering humanity as a multiplanetary and interplanetary species. After all, the expansion of humanity into space poses quite specific moral challenges. We should not rush to claim victory if we already have the technological capacity to leave Earth. There are several problems that the natural sciences and engineering cannot solve, at least not as if these problems were mechanical. One of these problems is related to Mars and the possibility of putting at risk any extraterrestrial life that may exist, but that we are not able to identify and therefore do not take due care to preserve. After all, if there were a second origin of life other than terrestrial, we would be looking at a second genesis (McKay 1990, 2009) that does not necessarily depend on what we know about carbon-based life. Therefore, we must ascertain whether if extraterrestrial life exists before it is too late (Faire, et al. 2017).

The future of humanity in space must overcome this difficulty if it does not want to become a form of predatory species with respect to potential native life. And here we also see one of the fears that one might have with respect to humanity as a species expanding into the universe: the possibility of repeating the disasters connected with the colonization process here on Earth throughout history. How likely is it that humanity will repeat its predatory behaviour in other planetary environments? If we think about humanity's behaviour towards the past, we can see that we have grown and evolved in many aspects but we still insist on retaining behaviours that evoke our violent past. That is why I consider that as human beings, we are probably not mature enough to consider ourselves a multiplanetary species, properly speaking (maybe multiplanetary in the sense that we are technologically capable, but if technologically capable alone, this would be tantamount to being a shell with no contents inside). This situation is a shadow hanging over humanity, one that haunts us and reminds us that we are perfectly capable of repeating some of the same mistakes we have made on Earth. To what extent is it justified to think about this? The teams in charge of these missions will be skilled people and outstanding researchers, but that in itself is no guarantee of maintaining a stable society (in this case, a fully functional colony). It may take many centuries of moral development at least before we are truly worthy of being an interplanetary species. This is the dystopic scenario that some might outline as a powerful argument against our foray into the universe. But it is not the only one.

The other way of thinking that moving off-Earth is the fact that we might meet other, more advanced life forms, which could pose a great risk to us. After all, the possibility exists that we may become a subjugated species in the face of a more evolved one (Hawking 2015). We should not overlook a scenario in which (i) we are the ones who discover an extraterrestrial species, or (ii) we are contacted by an extraterrestrial species that is already mobile and has a lot of experience in being so. In the first case, much will depend on our own behaviour to determine whether we will make that extraterrestrial environment a dystopian one with respect to local life, or a situation of self- and other species preservation, or at least of care. In the first scenario, it is possible that fears of repeating an environmental disaster may arise due to the bad uses we make of a certain planetary environment, such as endangering microbial life on Mars.

In the second case, however, we have a major disadvantage, however we look at it. In what way? If in our expansion in the universe we encounter an extraterrestrial species more advance than us, their technology might be capable of putting us at risk. We might never have encountered them unless we had stayed in our own galactic neighbourhood, in the solar system, but leaving caused us to be in the situation that Stephen Hawking feared greatly. In this scenario, we are facing a case that moving off-Earth is far from offering salvation to humanity. We would be at the mercy of what our interplanetary companions decide.

Of course, we cannot guarantee either a favourable or unfavourable outcome for humanity in this second scenario. Nor could we guarantee that this species would do to us what, in parallel, was done to the Native Americans by the European colonizers. In the case of possible contact with other forms of intelligent life, there would be a great difference between the perception of humanity and how to manage the relationship and the environment ethically, between ourselves and the new species, in a way that cannot be compared to colonization efforts on Earth in the past. However, the unfavourable alternative for humanity is possible, and just as there is not enough reason to consider this first encounter will necessarily be hostile, there is no reason to think equally that it will be innocuous. Given that human beings can commit great acts of harm against each other without any help from different species, a possible looming dystopia may well stem from our own actions, acting against our own species.

2. A Silver Lining

However, I believe that a key concept to differentiate a future ideal humanity from a non-ideal one is that of a multiplanetary species. What does this mean? I am not using the word 'species' in the sense of the natural sciences. So, when there are arguments that we are beginning to be a multiplanetary species it does not mean that we are evolving into different genetic species, only that we are becoming able to travel to other planets, like Mars, for example, and that we can establish colonies and prolong the existence of humanity as such, that we are able to settle successfully on different planets.

The idea of moral community oriented to the context of living on other planets is interesting. To form authentic multiplanetary colonies, ideally we would first have a consolidated moral community. To become a multiplanetary humanity, each community will have differences that allow it to subsist according to the conditions and evolution in which it develops. Being a member of a planetary moral community would be the basis for the conformation of a concept of multiplanetarity.

This implies the consolidation of humanity at an ethical level, at a human level, and not only at a technological level. Why would we have to consolidate moral communities on a planetary scale to carry out space settlement? On a technological level, it is not necessary; we may well expand into the universe without having solved our most fundamental ethical problems. This could then be embodied in a legislation that would be characteristic of each set of colonies per planet.

According to Chon-Torres and Murga-Moreno (2021, 3):

> Can we be interplanetary without having resolved our situation as planetary citizens? Yes, because the interplanetary is more focused on one aspect of the technological possibility that can take us to other planets. Can you be multiplanetary without having solved the fundamental problems on Earth? Yes, and the future scenario is the fear of repeating the same mistakes of our current planet, taking with us corruption and human decay.

It may not be necessary to generate moral communities on different planets, the ethics originating from planet Earth is enough. Then, each planetary environment can be accommodated according to circumstances. However, to do so would mean that, for example, some forms of suppression of freedoms would be allowed on one planet and not on another, diminishing the idea of a constant human dignity. But this leads us to raise a situation that is striking. What in a planetary context may seem dystopian, such as living in conditions of suppression of freedom, like a prison, for other colonies on other planets may be the most ideal. To safeguard the survival of the colony, say on Mars, one may seek to restrict certain freedoms and operate a strong, top-down political system. For a colony with an Earth-like planetary environment, that would be a dystopia, but for Martian colonists, it would be the most effective way to survive because of the simple fact that you do not have much choice in the matter, particularly in the face of hostile weather conditions and certain death if you are beyond the safety of the settlement.

But continuing with the idea of being an interplanetary species, this does not refer to a quality of biological differentiation. As the name implies, 'inter' planetary would be the ability to be able to travel between planets, so its meaning may be more closely tied to that of a technological capability. Thus, we are currently not a multiplanetary species but a species in the beginnings of being interplanetary. Attempts to develop technology that will take us to Mars and, in the future, may take us beyond that remain dreams of making that kind of trip feasible. It could save us materially speaking, that is, in the hypothetical case that life on Earth cannot be tolerated by humans for some reason, but this would not guarantee that we would be saved from our own actions (as in a war). If we really want to save humanity, this must go beyond whether we can travel to and settle on other planets, otherwise we are merely romanticizing the idea of interplanetary travel, which is not wrong, but it rather loses sight of the perspective of safeguarding humanity.

Or do we want to save the version of humanity we know now, with its limitations and flaws, its betrayals and selfishness? Of course, if one is not too concerned by this, one can focus on what is relatively less complex, the technological part, but if we really want to make a change that improves the human condition, it will take more than just the material wherewithal to take us off the Earth. And this is an important aspect that astrobioethics also has to offer us: a perspective in which humanity sees itself in the universe from an ethical perspective linked to space. Just as we can use space technology to improve that used on Earth, astrobioethics must take it upon itself to evaluate our moral and human condition in a cosmic perspective.

Still, it remains to be defined whether this condition is fulfilled, considering that we do not yet behave as a true planetary species, dedicated to the ethical and moral survival of our entire species. We may live on the same planet but we do not act in a way that our morals give solid indications of being planetary citizens. There are some universal values such as those relating to human rights, for example, but even these are not fully respected by some countries, or even within some countries. In many nations, whole areas may appear liberal and liberated while others are under military rule or governed by dictators, the population's freedoms severely curtailed.

This compromises the idea of multiplanetary being understood as a moral community; to what extent can we define that a moral community exists on Mars, for example, if we cannot even define a planetary ethos on Earth? Will nationalities and religious and cultural differences be impediments to thinking of ourselves as a planetary species? Assuming a scenario where we already inhabit Mars in colonies of sufficient proportion to be able to identify a general administration on the red planet, does that administration have to be understood as administration of the red planet in general, or would it imply variations according to the different colonies established?

There is one issue that we must consider when we think about a particular planetary community: the climatic condition that exists on the planet in question. On Earth, we can breathe without difficulty and go wherever we want, whenever we want, and whenever we can. However, in a place as inhospitable as Mars, our possibilities are reduced and so are our ways of socializing with others. In other words, managing people and institutions in an other-planet context will demand limited ways of interacting with each other. An interesting proposal is suggested by Cockell (2018) when he indicates that a way of organizing on Mars would be like a prison, in the sense that no-one can do what they want with complete freedom. Again, this assumes that restrictions on freedom would not necessarily have to be seen as a bad thing. This issue is no different for us either, taking as an example the pandemic produced by Sars-Cov-2, which meant that we were practically forced, in many countries, to quarantine or avoid unventilated and crowded spaces.

Of course, despite being a pandemic with restrictions on individual freedom, we still could go wherever we wanted, with some limitations but few genuine prohibitions. This would not be possible on Mars, because it is simply not like going out to the park to walk the dogs or walk for a breath of fresh air. However, it gives us a perspective that by necessity and with the right conditions, our behaviour must change, and therefore the ethos that develops will characterize the moral community or ethical matrix shared by a set of human beings inhabiting it.

In this scenario then, will it be possible to think of divided territories like the ones we have on Earth? This is still a relevant question because here we have not been able to 'unify' as a planet. It would be unrealistic for us to unify in a planetary way on another planet, hoping that out there our future will be better than what we have achieved on Earth. There is no need for absolute condemnation. In fact, differences arise due to climatic, cultural, and general social conditions. Over time, each Martian colony will have its characteristics and differences, customs, and ways of understanding ethos. However, to succeed in such a hostile climate for humans, we should go one step further and thus approach the idea of being a truly planetary species. Establishing the concept of the human species as a planetary species should be our horizon of meaning, which, without necessarily reaching it completely, should always be in our North.

How are we to understand, then, being a planetary species? For us to ask about the future of humanity in the universe is also to ask ourselves about the human condition in the long term. Today, no matter how much technology we have managed to obtain, we are still afflicted by a lack of empathy and a lack of morality. To date, there is no scientific or technological way to improve the human condition as such. The social and economic differences between nations mean that we still perceive ourselves as a divided humanity, and an example of this can be seen in the hoarding of vaccines against COVID by rich countries, which has left developing countries at a disadvantage, which the World Health Organization (WHO) considers a moral failure (WHO 2021). And considering this, we could even raise the question that with the existing inequality between nations, who guarantees us that the opinion of the richer nations will weigh more when we must establish multiplanetary moral communities?

Based on what have been observe, we cannot consider the human species to be a true moral community yet, not at least as a community that engages in primarily moral activities. One could say that expanding into the universe will give us a different perspective on life and that just being able to travel to other planets could change us for the better. But the same could be said in reverse, since we have no real justification for thinking that leaving Earth will make us better morally as a species. We cannot think that the future of humanity in space will become a kind of utopia (as it could be represented) or a dystopia (a continuation of the ecological and moral disaster experienced on Earth).

Why would we think that humanity would improve if it became interplanetary, where does this desire come from? Travelling to other planets will save humanity, materially speaking and from a technical perspective, but it cannot necessarily save us from the effects of corruption leading to a crisis that would compromise our future. In any case, it is to be expected that the human species will continue with some of its vices and mistakes in its expansion in the universe, with successes and failures in the attempt to build a better interplanetary society.

3. Conclusion

The future of humanity is inevitably linked to its expansion in the universe. However, even if the technological difficulties that will allow us to become an interplanetary species can be resolved, the ethical difficulty remains as to whether that future will become something akin to a dystopia or utopia, although it is more concrete to speak of nuances. One of these nuances could be a version of humanity that does not manage to completely overcome the social and moral conflicts it currently faces, but that remains in its search and effort to be an authentically planetary species that brings it closer to the idea of moral community, and if it never completely achieves it, to consider that it would be more of a horizon of meaning for which to act. Possibly in this way, the future of humanity in the universe avoids ending up in a dystopian version, although we must emphasize that the concept, for example, of restrictions of freedom and ways of organizing oneself in society that are not necessarily completely democratic may be convenient in some planetary contexts, as it could be in future Martian colonies. However, the future of humanity in space is seen with a multiplicity of scenarios where the answer will depend very much on the reality of the habitability of each planet.

References

Baum, S. D., Armstrong, S., Ekenstedt, T., et al. 'Long-term trajectories of human civilization' (2019) *Foresight*, 21(1), 53–83.

Chon-Torres, O. A. 'Mars: A Free Planet?' (2021) 20(4) *International Journal of Astrobiology* 294–9.

Chon-Torres, O. A. and Murga-Moreno, C. A. 'Conceptual Discussion around the Notion of the Human Being as an Inter and Multiplanetary Species' (2021) 20(5) *International Journal of Astrobiology* 327–31.

Cockell, C. *Life Beyond: From Prison to Mars* (London: The British Interplanetary Society, 2018).

Faire, A., Parro, V., Schulze-Makuch, D., et al. 'Searching for Life on Mars Before It Is Too Late' (2017) 17(10) *Astrobiology* 962–70.

Hawking, S. 'Stephen Hawking: "I have learned not to look too far ahead, but to concentrate on the present"' *El Pais*, 25 September 2015 <https://english.elpais.com/elpais/2015/09/25/inenglish/1443171082_956639.html> accessed 29 June 2022.

McKay, C. P. *Does Mars Have Rights? An Approach to the Environmental Ethics of Planetary Engineering* (New York, NY: Routledge, 1990).

McKay, C. P. 'Biologically Reversible Exploration' (2009) 323(5915) *Science* 718.

WHO. Director-General's opening remarks at 148th session of the Executive Board (2021). <https://www.who.int/director-general/speeches/detail/who-director-general-s-opening-remarks-at-148th-session-of-the-executive-board> accessed 29 June 2022.

6

The cosmolegal approach to human activities in outer space

Elena Cirkovic

Without dimension; where length, breadth, and height,
And time, and place, are lost; where eldest Night
And Chaos, ancestors of Nature, hold
Eternal anarchy, amidst the noise
Of endless wars, and by confusion stand.
For Hot, Cold, Moist, and Dry, four champions fierce,
Strive here for mastery, and to battle bring
Their embryon atoms: they around the flag
Of each his faction, in their several clans,
Light-armed or heavy, sharp, smooth, swift, or slow,
Swarm populous, unnumbered as the sands
Of Barca or Cyrene's torrid soil,
Levied to side with warring winds, and poise
Their lighter wings. To whom these most adhere
He rules a moment: Chaos umpire sits,
And by decision more embroils the fray
By which he reigns: next him, high arbiter,
Chance governs all. Into this wild Abyss,
The womb of Nature, and perhaps her grave,
Of neither sea, nor shore, nor air, nor fire,
But all these in their pregnant causes mixed
Confusedly, and which thus must ever fight.

(Book II, *Paradise Lost*, John Milton)

The cosmolegal approach (or law for the cosmos, in the true sense and not as part of the more earth-bound philosophical tradition of cosmopolitanism) is meant to encompass the process of learning and law-making through which the law would recognize the unpredictability of human/non-human relations. As a normative proposal, it emerges from theories on post-human legalities that argue for a move beyond the centrality of the human subject that acts upon the world (cosmos), as its object. In other words, the cosmolegal gains its impact from non-human, human, and hybrid agencies. The hybrid here implies the interrelation of human bodies with their environments or artificial extensions of or insertions in the human body (e.g. brain machine interfaces (BMIs)). This proposal aims for a fundamental change

Elena Cirkovic, *The cosmolegal approach to human activities in outer space*. In: *The Institutions of Extraterrestrial Liberty.*
Edited by Charles S. Cockell, Oxford University Press. © Elena Cirkovic (2022).
DOI: 10.1093/oso/9780192897985.003.0007

in thought and practice of modern law. It also recognizes that, in the practice of politics, economies, law, and policy, this is not a likely outcome in the near future. The phenomena which it is trying address—climate change and outer space environmental degradation—are ongoing and unpredictable and require normative thinking for both near and mid-long-term range thinking and response.

What is the meaning of global norm production in the context beyond the Earth system? Do humans have a 'right' to exploit extraterrestrial resources and alter extraterrestrial environments? The question of a 'right' inevitably echoes the burden of history of rights stemming from natural law approaches (Garcia-Salmones 2017). Outer space is subject to the current state-extractive industry promotion of a 'rush' for resources (Feichtner and Ranganathan 2019), in the context where contemporary international regimes do not effectively and directly respond to the magnitude of the ongoing environmental degradation. While some of the 'earthly' discussions can be extended to the question of 'who owns the Moon?', there is also an important variable of the unknown in this context where humanity is still only developing its knowledge. The orbital environment is not limitless, and beyond it, humans do not have any finite knowledge, which can accommodate unchecked activities of the astral human, but can be as vulnerable as the Earth system. The orbital space also has capacity limits, which are not only determined by the number of anthropogenic space objects in a specific orbital neighbourhood but also the uncertainty in how these objects will behave in the future.

The cosmolegal proposal steps 'beyond the Earth system' with a proposal for the inclusion of non-human law on planetary and cosmic scales. This is crucial for outer space, a domain not inherently friendly or entirely known to humans. Scholarship on post-human law has been forbearing a nascent engagement with imagining a new legal future where non-human agents are intertwined with humans, and it provides the broader umbrella for the argument of this chapter (Philipopoulos-Mihalopoulos 2011, 2016a, 2016b; Grear 2020). More-than-human arguments have branched away from the inner critiques of the law, proposals for new types of 'governance', 'rights', or 'sovereignty of nature', all of which retain the current ontology of modern occidental law (Fitzpatrick 1992). Instead, cosmolegality argues for an overcoming of the fundamental correlation of human cognition to agency, personhood, and self-determination. While the focus is on outer space, the self-reflective theoretical and methodological approach situates the chapter within the pluri-epistemological context of the critique of modern law.

1. Cosmolegality and the Role of Transdisciplinary Thought and Practice

Outer space governance was established through a series of international treaties, enacted during the Cold War period. The Outer Space Treaty of 1967 (OST) is the most significant piece of legislation in this regard. It establishes an international legal framework for outer space, intended to preserve outer space as the *global commons*—that shall remain free for exploration by all states, and that should not be subject to national appropriation by any means (Article II).At the time of drafting of these treaties, outer space exploration was primarily an affair of national governments, which was principally undertaken for geopolitical purposes. Today, space activities are increasingly dominated by private actors, and corresponding state support. While states continue to be the principal subjects of international space law, and only states can be held responsible and liable (The terms 'liability' and 'responsibility' are here understood in accordance with the ILC, denoting that 'responsibility' only refers to State responsibility and 'liability' to State liability, both state and non-state actors are announcing the dawn of the 'commercial space age' (Feichtner 2019). While the geopolitical underpinnings cannot be ignored, companies are participating in the race for space exploration mostly for commercial reasons (Singh Sadcheva 2018). The dominant debates and narratives in international law as related to ongoing and future human activities in outer space have focused recently on the military and commercial uses of outer space, with international lawyers participating in the delineation of what the public- private, state-commerce nexus of relations should become (Bittencourt Nieto 2020). The classic doctrines of modern international law, with their focus on sovereignty, state consent, custom, and treaty, do not provide a satisfactory explanation for many of the practices and institutional structures that fill the global legal universe. Overlapping jurisdictions, regime interaction (Krisch, Kingsbury, Stewart 2005) inconsistent doctrinal interpretations, transnationalism and competing worldviews (ILC 2006), characterize the contemporary legal terrain. In this context, national laws on the utilization of space resources are examples of emergent national laws promoting exploration, exploitation, and utilization of space resources (Feichtner 2019).

The cosmolegal proposal moves towards the situatedness of international legal activities and related knowledge beyond social contexts (global or local). Anthropogenic and anthropocentric law is designed to regulate human behaviour. It is built on certain assumptions of how human beings behave and demonstrate agency. For instance, sociology and law focus on human interactions and give agency to humans or their fictitious creations: corporations, institutions, the state, etc. Thereby humans have a voice and a mode of expression, whereas non-human 'objects' are not recognized as having an agency. Non-human beings or physical and biological processes escape human control and react in unpredictable ways, which is why current legal assumptions of what has agency do not apply (Cirkovic 2021). However, cosmolegality does not seek to rebuild a different sort of ontology where whatever is considered as non-human gains rights, legal personhood, or sovereignty within current legal architectures. In the human context, rights claims imply that they are asked for, and granted by another (human) agent. In the situation where activities and impacts are not strictly human, the fundamental framework of reference needs to change.

In social sciences, tendency towards a totalization includes its approaches to physical laws, which are often discussed as if they were 'universal' in their reality or mathematical description. However, no 'law' is exact. Rather, all laws, recognized by a human mind, and defined as human and non-human, are constructed by humans, and as such, they are always laws by situated knowers; there is no 'outside' perspective. Physicists, lawyers, and cosmologists alike live in a 'case study' they have of the universe. Natural sciences cannot assume a 'God-like view' and need to recognize relevant social constructions (Jasanoff 2016).

In its various iterations, science and technology (STS) scholarship has pointed out that scientists are situated in social settings, which influence the knowledge production on both human bodies and 'nature' (Jasanoff 2016). References to physics in social sciences is just one example. Richard Feynman argued that some physical theories are difficult for humans to think and to accept because they are expected to comply with an understanding of 'common sense' and reasonableness. The theory of quantum electrodynamics, for instance, describes nature as absurd from the point of view of common sense (Feynman 1985). While the appropriation of quantum mechanics beyond physics has allowed for an understanding beyond ontologies of spheres or levels, challenging the nature–culture divide, we are still in the framework of how humans get to define 'things'. Rather, following Nancy Cartwright (Cartwright 1983) among others, Isabel Stengers suggests that the whole problem of indeterminacy and measurement in quantum mechanics is a false one (Stengers 1996). Quantum mechanics prevents humans *in principle* from having exact knowledge of every particle, as if it were independent of our interaction with the observed particles. Stengers argues that the limits of our knowledge in quantum mechanics are similar to the limits of knowledge of what another person is feeling and thinking (Stengers 1996, 21). She writes: 'cosmos refers to the unknown constituted by these multiple, divergent worlds, and to the articulations of which they could eventually be capable' rather than a particular territorial, geometric, or geographic, cosmos (Stengers 2009). The idealizing assumption of the world's total knowability and determinability in accordance with 'laws' or 'cosmograms' that humans perceive or create leads to a frustration with the limits imposed by them.

Hence, unlike the endless possibility of defined cosmograms (or complete images of the world), cosmolegality would need to be constantly open to contingencies in future coefficients of friction. The cosmolegal seeks to create a self-reflexive recognition that the human/non-human distinction, and its inherent hybridity, still have uncertain definition. The key working method is to facilitate cooperation and interaction among different disciplines and knowledge, including an opening for contingencies provided by the agency of the non-human, and other dimensions/spaces, where humans seek to extend their activities and life.

2. Law Beyond the Human Approach

Environmental degradation is a response of the non-human to human activities. A satellite is human-made. Nevertheless, its trajectory in an orbit can have unintended consequences resulting from non-human 'laws'. These are best studied and analysed by disciplines other than law, but they also have impact on the law. The ongoing reliance on the existing structures of international law often reinforces anthropocentric foundations and practice of international law. For instance, the principle of common heritage of humanity did not provide meaningful limitations on resource extraction, and prevention of environmental degradation (Feichtner and Ranagathan 2019).

How can international law, governance, and its scholars and practitioners extend themselves into outer space, or even spacetime? In trying to answer such questions, the cosmolegal proposal does not imply that 'planets' or 'asteroids' would act like humans. Instead, it challenges the appearance of distinctions and disparate attributes of the world. Agency does not require cognition (Bennett 2009). The interplanetary domain demonstrates that its apparent fracturing in human understanding, practice, and regulation does not stem from its own inherent multiplicity but from the human understanding thereof. The law, instead of being the mirror of permanently split human subjectivity would recognize the indeterminate nature of the world beyond it. And further, the material world and human bodies are not defined by strict binaries of subject–object and human–non-human. Non-human agency refers to all that is considered as non-human (sentient and non-sentient). It refers to, among others, theories posthumanism (Alaimo, 2016; Braidotti, 2013, Grear 2020), new materialist philosophy of science (Coole & Frost, 2010; Barad, 2007) and science and technology studies (STS) (Jasanoff 2012). As modern legal fiction branches from Cartesian thought, the traditional regulatory modalities (Friedman 2016; Pardo and Patterson 2015) seem bound to lose their traction. CelesLaw remains at the level of theory and philosophy in order to build on the hypothesis of profound discontinuity and uncertainty. This uncertainty is part of human interactions with outer space. As the very broad STS scholarship continues to point out, scientists are also situated in social settings, which influence the knowledge production on 'nature'(Jasanoff 2016). At the same time, with recognised 'complexity' it still has to account for the gap between current and institutionalised understanding of governance and law and how, even attempts to include 'scientific knowledge' do not succeed as they are still mitigated through those institutions and not the non-human agency itself (which is being impacted and impacting on the human). However, as indigenous scholars have argued, an alternative is that of a relational approach (Graham1999). Indigenous studies scholarship (Graham 1999, 2014l Coulthard 2014; Coulthard & Simpson 2016; Martin 2017; Simpson 2017), has already developed theories for extending subjectivities beyond the human species. Indigenous studies provide a basis for an ontology which moves beyond the location of 'where we live' to include the known universe (which has different narratives and imaginaries in various parts of the world). Hence, the ecology of a location would also need to extend out to the very boundaries of the totality of existence. Plural indigenous ontologies already assume a profound sameness, and therefore sense of recognition, between the abilities and sensibilities of objects and those of humans. For indigenous scholars, the struggle is to find a way to enable these ontologies to be recognised and reproduced in their academic work (Smith 2017). It then, becomes necessary to create a new vocabulary and to trouble the familiar language of empiricist or interpretivist social science to open up a space where objects can express their vitality—or, at least, where humans can experience and understand non-human agency. Predictability and causality (which are crucial for legal thinking) are specious when one ignores the underlying causes of 'disorder'- induced variability. Legal scholarship needs a capacity to see minute errors that increase over time with varied consequences (e.g., global warming and orbital debris). However, anthropogenic consequences are not entirely unpredictable if analysed with rigorous engagement with, and recognition of other expertise (climate science, engineering, astrophysics, etc.). The realization that complex systems are vulnerable to big (and sometimes catastrophic) shifts in behaviour stemming from small changes is a critical area of research (Nobel Prize Committee 2021). Technological innovation, for instance, has the capacity to trigger events and outcomes within the complex social and environmental systems that are hard to predict and manage. And finally, an understanding of responsibility (Graham 2014) in sciences, social sciences, and humanities, helps to challenge the misuse of 'systems' approaches in social science, as self-driven processes (Luhmann 1993). While specific non-human laws might govern the exponential growth of anthropogenic orbital debris, it is still, anthropogenic.

In addition, critical and post-human jurisprudence has offered extensive analyses of modern law's pretensions to secular origins, providing the background for critiquing the constitutive forces of race, racism, and colonialism in the structure and political, philosophical, and psychoanalytic imaginaries of modern law (Fitzpatrick 1992). What came to be considered as human by dominant global narratives of empire, commerce, and the international order shifted continuously. What it has yet to consider is that the 'border' of the human bodies is also not exact but part of transcorporeal and hybrid relations (Alaimo 2010) including the relationship with viruses and microorganisms, or pollutants,. Stacy Alaimo's seminal book *Bodily Natures: Science, Environment, and the Material Self* (2010) refers to transcorporeality, or movement across bodies and nature, which alters our sense of self. Beyond the scope of this chapter are two well-established problems that span across various disciplines: (i) the question of the extent to which the human body is somehow outside of its environment, and how it came to be represented as such; and (ii) the ontological differences and hierarchies among humans and human societies. What is in the interest of a human body and its survival is not necessarily represented in the ontology of the mainstream thought (e.g. is commercial gain more important than access to clean air and water? Would humans have to pay for air in outer space?). And finally, cognitive and other sciences have yet to understand fully the function of human bodies. The problem, then is not merely 'anthropocentric' (i.e. a human body acts to survive as much as any other organism); rather it is the established ontology of artificial mind–body and human–non-human binaries, which has not been shared by all human societies. The cosmolegal aims for a space for thinking beyond ontology where we do not (yet) fully know the entities in question (cosmos) even if we weave them into words and sentences of the law or science experiments (Jasanoff 2016). However, as Tsing et al. (2021) observe, the specific ecosystems where human action is already prevalent, there emerge new types of ecosystems, some potentially destructive to that same system (e.g. orbital debris). This does not mean that the category of the subjective experience is superfluous, or unnecessary, to the materialism displayed by matter.

3. Plural Ontologies in Cosmolegal Thinking

There never was an era where agents were exclusively humans, even if were self-define as living in the times of Anthropocene. By Anthropocene, we mean the present geological era, which is determined by human presence more than any other species. Humans continue to observe cognitive abilities, intentionality, and capacity to make autonomous decisions of what is considered to be non-human, as they redefine the 'natural right' of humanity to master nature. In philosophy, the ontological domain of the unknowable has reappeared within the various strands of continental materialism and realism that make up the so-called speculative turn (Hamzić 2019). However, speculation is on the opposite spectrum of legal determinism and, hence, combining scientific research on the potential unknowns that needed to be accounted for in outer space and the Earth system will require a different and transdisciplinary engagement.

Transdisciplinary collaboration cannot promote new knowledge without the recognition and elimination of discrimination in existing sites of knowledge production. In outer space law and policy, there are also problematic proposals to even greater exclusivity in future law-making due to unequal scientific and economic capacities of states.

In this context, Indigenous peoples' traditional knowledge has become a reference in various reports on the United Nations Sustainable Development Goals (SDGs) and the Earth system, and more recently, outer space exploration. However, these references often re-essentialize or romanticize Indigenous knowledges without an acknowledgment that 'Indigenous peoples' are not a monolithic constant (Johnston et.al. 2018). There is no one indigenous science, but a variety of processes depending on the locality of any given population and including various forms of migration. It is beyond the expertise and scope of this chapter to speak of specific Indigenous knowledges and traditions but the topic is especially relevant in what is now a discourse of 'outer space colonialism'. Co-production of knowledge implies a situation where pluriverses and other ontologies do have direct impact on lawmaking, and its conceptualization of 'what is owed' and to whom.

The 'colonizing' argument evokes frontier-focused history of international law, and the otherwise outdated 'openness' of empty spaces or *res nullius* (Cirkovic 2007). Imperial semantics remain in outer space law due to its outwardly nature where humans do not live. Some astronomers have warned against the use of colonial frameworks in outer space, in relation to other planets (Prescod-Weinstein et al. 2020). However, while collaborations between space agencies such as NASA aim to include Indigenous knowledge, the status of Indigenous peoples as non-state actors in international law has implications at the level of intergovernmental decision-making.

The rhetoric of right to space highlighted by some corners of the outer space sector reflects not only anthropocentric assumptions but also specific values of imperium and commercium. Outer space is imagined as a limitless resource—a space frontier. The metaphor of the frontier, with its associated images of pioneering, homesteading, claim-staking, and taming, has been persistent in the history of international law and colonialism. In the history of outer space law, this is what the 1979 Moon Agreement sought to address with only 18 states becoming parties to it. The Moon Agreement emerged from discussions over legal stewardship of the Moon. What has yet to be explored in scholarship and practice is the idea of recognizing outer space as possessing agencies in a similar manner to what is being discussed for the Earth system (including new materialisms and animal cognition). This does not refer to legal personality of celestial bodies because it would imply an ongoing ontology.

4. Cosmolegal and Complex Systems

The final consideration of this chapter is a reference to systemic thinking, and legal adaptations of complex systems, including the concept of design and adaptation. Beyond law, 'designing with nature' is not a new concept (McHarg 1969). Various disciplines, such as architecture, have argued in favour of planning adaptive to 'nature' for some decades, as an approach necessary to build a sustainable landscape (Marsh 2005). This architectural comparison is apt in a discussion on law and whether it can be 'designed for nature' as both disciplines share an emphasis for formalism and boundaries. Architecture is yet another discipline which is leading future ideas of outer space settlements. Law and architecture, in their modern Western formats, delineate and shape certain spaces and determine their form and function. How can law 'shape' air or water? Even architecture is more tangible in this respect as it directly shapes physical spaces, which can be either sustainable or detrimental to the environment (including human beings). Sustainable architecture emphasizes systems approaches and best management practices which include the relationship between the human and the Earth system. The interdisciplinary field of Earth System Science (ESS) studies the planet's oceans, lands, and atmosphere as an integrated and complex system. ESS integrates the principles of biology, chemistry, and physics to study problems involving processes occurring at the Earth's surface, such as climate change and global nutrient cycles, providing a foundation for problem-solving related to environmental sustainability and global environmental change. The complex systems approach studies how relationships between parts give rise to the collective behaviours of a system and how the system interacts and forms relationships with its environment. This means that a disruption in any of the subsystems affects the whole system. Conversely, architectural and engineering approaches that design with other priorities in mind, will seek to work 'against nature'. For instance, they may reorient water systems against their natural 'flow' and in favour to human imagination or design (e.g. commercial building) (Marsh 2005).

Social science has been borrowing from computing disciplines, physics, or mathematics to describe its own version of complex adaptive systems approaches. Previously, systems theorists have even borrowed from biology, while describing the so-called autopoietic systems (Luhmann 1993; Teubner 1997). However, danger of borrowing from other disciplines, be it biology or physics, is that it avoids structural complexities of each. While the recognition of complex systems is necessary, (i) the verdict is not complete even in other disciplines on what complex systems are, or how they function, and (ii) the component of responsibility via human agency and decision-making remains. International law is a highly specific system that forms just one part of Complex Adaptive Systems (CAS), and even more broadly as interrelating with Complex Physical Systems (CPS).

No one 'discipline' as a community of specialized knowledge is equipped with full answers to either the climatic changes in the Earth System, or the post-human arguments for unmanned vehicles in the outer space. Dietz et al. (2003) articulated the need for 'adaptive' governance of Social Ecological Systems (SES) arguing that our knowledge of any system is likely to be wrong or at least incomplete, and the required scale of governance may shift because of changes in the biophysical and social system components. However, complex systems of SES, still rely on traditional ontologies behind the global-local structures of law and governance, agency, and interaction. CPS are systems that have fixed components and rules of operation but can still have unexpected behaviours that in turn can create new rules of operation and even new systems (Holland 2014)., A climate system comprising natural elements and operating under laws of atmospheric chemistry and physics can produce unexpected weather patterns/systems. CAS, in contrast, comprise agents that have the capacity to learn and to adapt their behaviour to the behaviour or action of other agents in the system. In both CAS and CPS a rather small shift in initial conditions can have disproportionate outcomes (Cosens et al. 2021, B. Cosens, Craig, Hirsch, et al. 2021). While these are mere models because humans do not yet have a fundamental theory or understanding of the 'universe', ongoing research and experience points to a conclusion: that even when the underlying rules for a system are simple, the behaviour of the system can be arbitrarily rich and complex. For instance, ecological or societal shocks do not always lead to an emergence of new solutions. A dominant practice might become undermined because of societal or environmental changes (e.g. exponential growth of space debris; climate change), and the novel practice might develop an entirely novel thread of thinking and practice. Or, it might fail to adequately respond to the challenge, with uncertain outcomes. The realization that complex systems are vulnerable to big (and sometimes catastrophic) shifts in behaviour stemming from small changes is an ongoing critical area of research (Nobel Prize Committee 2021). The premise and an understanding so far is that no single prediction of anything can be taken as absolute truth and that without understanding the origins of variability we cannot understand the behaviour of any system. Only then, for example, could we understand that climate change or orbital environmental degradation is attributable to humans. The variability in the basic processes, from climate dynamics to orbital debris collisions, lead to the emergence of multiple length and timescales and is fundamental to interpretation of theory, experiment, and observation.

In the present context of space governance, the anthropogenic and anthropocentric novel, emergent, and potentially dominant practices, do not necessarily have a societal, ecological, even cosmic understanding. Nor do emergent practices consider all interests and possible outcomes. Meanwhile, phenomena such as space debris and/or climate change have impact on multiple societal sectors and production practices and consequently can gradually diminish the feasibility of current practices and production models, including international law.

The emergence of disorder from order, and with it, multiple scales in space and time, is a characteristic of complex systems and is not novel in human thinking. Some of the questions include: Are there limits to disorder? Does it 'choose' a particular spatial structure or many spatial structures? (Nobel Prize Committee 2021). This basic manner of thinking for any complex stochastic multiscale system, such as climate, it is essential to understand that disorder influences all systems and can determine the fate of some non-linear dynamical systems.

In the context of contemporary knowledge, the cosmolegal objective is a thinking from both the perspective of this indeterminacy of nonlinearity, and the normative undertaking where humans retain responsibility for their activities, and environmental degradation of spaces where they act.

The capacity to observe the Earth system from the vantage point of space has redefined not only our perspective of Earth as an interconnected system but also how we describe the dynamic processes that occur above, on, and beneath its surface. International law's anthropocentric concern with global environments has delineated inner and outer environments, where the 'outer' environments consist of the spaces beyond the atmosphere and beneath the lithosphere. At the same time, outer space becomes another wilderness to be tamed: "Our minds are taming it; our technologies are rendering it usable; our affluence is exploiting it; our power in general is transforming it." (Wapner 2010, 4). And further, "Wildness, as that dimension of nature that signifies genuine otherness, has been stamped out now that the human signature can be found everywhere" (Wapner 2010, 6). Outer space merely becomes yet another extension of the frontierist attitude.

5. Conclusion

The concept of 'chaos' or 'disorder' has always been part of human attempts to explain the universe (Lucretius, *De Rerum Natura* 5.436; Milton, *Paradise Lost* (1667)). Consider, for instance, John Milton's epic *Paradise Lost*, and the concept of atomic chaos as having 'responsibility', hence not existing only spontaneously but also in relation to a 'choice' (Leonard, 2000). This 'choice' Milton discusses in reference to both responsibility in human behaviour and it is neither systemic nor chaotic. John Milton was born in 1608, eight years after Giordano Bruno had been burnt at the stake for implying something similar, and two years before Galileo's *Sidereus Nuncius*.

Extending possibilities for rethinking of the world of law-making in a non-Anthropocentric direction is necessary because, as noted above, new activities (such as the mining of asteroids and celestial bodies), which continue to extend the borders of current international law into the cosmos, are becoming possible—along with the concomitant risks of unintended or unexpected reactions of various other-than-human environments. The objective of cosmolegality is to argue that human self-perception and relevant anthropogenic laws of societal organization will have to contend with more than just human interests. The space environment is always reacting, and this is already evident in the congestion and increased risk of activities in Earth's orbits. The question is what is overlooked when space is constructed as a passive empty receptacle of human agency and its products.

Acknowledgements

This chapter is part of the project 'Anthropocentrism and Sustainability of the Earth System and Outer Space (ANTARES)' (2017–) at the University of Helsinki Erik Castrén Institute for International Law and Human Rights and Helsinki Institute for Sustainability Studies (HELSUS).

This research was supported by the Arctic Avenue fund administered by the University of Helsinki and Stockholm University.

References

Alaimo, S. *Bodily Natures: Science, Environment, and the Material Self* (Indiana University Press, 2010).

Barad, K. *Meeting the Universe Halfway: Quantum Physics and the Entanglement of Matter and Meaning* (Duke University Press, 2007)

Bennett, J. *Vibrant Matter* (Duke University Press, 2009).

Bittencourt Neto, O, et.al (eds) *Building Blocks for the Development of an International Framework for the Governance of Space Resource Activities: A Commentary* (den Haag: Boomuitgevers, 2020)

Brooks, V. and Philippopoulos-Mihalopoulos, A. (eds). *Research Methods in Environmental Law: A Handbook* (Gloucester: Edward Elgar, 2017).

Calloway, K. 'Milton's Lucretian Anxiety Revisited' (2009) 32 *Renaissance and Reformation* 79–97.

Cartwright, N. *Nature's Capacities and Their Measurement* (Oxford: Oxford University Press, 1989).

Cirkovic, E. 'Self-Determination and Indigenous Peoples in International Law' [2008] *American Indian Law Review* 31.

Cirkovic, E. Working Papers: Space, Ice, and the Final Frontiers of International Law: Extreme conditions of climate change. Minerva Center for the Rule of Law under Extreme Conditions (Haifa University, 2018).

Cirkovic, E. 'The Next Generation of International Law: Space, Ice, and the Cosmolegal Proposal' (2021) 21 *German Law Journal* 2.

Cirkovic, E. 'The Earth System, the Orbit, and International Law' in T. Cadman, M. Hurlbert and A. Simonelli (eds) *Earth System Law: Standing on the Precipice of the Anthropocene* (Routledge, 2021).

Cirkovic, E. 'International Law beyond the Earth System: Orbital Debris and Interplanetary Pollution' (2022) 14 *Journal of Human Rights and the Environment* 1.

Coulthard G, *Red Skin, White Masks: Rejecting the Colonial Politics of Recognition* (University of Minnesota Press 2014)

Coulthard G and Simpson L. B. 'Grounded Normativity / Place-Based Solidarity' (2016) 68 *Amerian Quarterly* 2, 249–255

Coole D. and Frost, S. *New Materialisms: Ontology, Agency and Politics* (Duke University Press 2010)

Cosens. B. A., Craig, R. K. Hirsch, S., et al. 'The Role of Law in Adaptive Governance' (2021) 22 *Ecology and Society* 30.

Dietz, T, 'The Struggle to Govern the Commons' (2003) 302 *Science* 1907

Feichtner, I, 'Mining for humanity in the deep sea and outer space: The role of small states and international law in the extraterritorial expansion of extraction' (2019) 32 *Leiden Journal of International Law* 2, 255–274

Feichtner, I. and Ranganathan, S. 'International Law and Economic Exploitation in the Global Commons: Introduction' (2019) 30 *European Journal of International Law* 541.

Feynman, R. P. *Quantum Electrodynamics* (London: Penguin, 1985).

Fitzpatrick, P. *The Mythology of Modern Law* (London: Routledge, 1992).

Garcia-Salmones Rovíra, M. 'Natural Rights in Albert the Great Beyond Objective and Subjective Divides' in M. Koskenniemi, M. Garcia-Salmones Rovíra, and P. Amorosa (eds), *International Law and Religion: Historical and Contemporary Perspectives* (Oxford: Oxford University Press, 2017).

Graham, M. Some Thoughts about the Philosophical Underpinnings of Aboriginal Worldviews (1999) 3 *Worldviews: Global Religions, Culture, and Ecology* 2, 105–18

Graham, M. Aboriginal notions of relationality and positionalism: a reply to Weber (2014) 4 *Global Discourse* 1, 17–22

Grear, A. *Redirecting Human Rights: Facing the Challenge of Corporate Legal Humanity* (London: Palgrave Macmillan, 2010).

Grear, A. 'The Vulnerable Living Order: Human Rights and the Environment in a Critical and Philosophical Perspective' (2011) 2 *Journal of Human Rights and the Environment* 23–44.

Grear, A. 'Deconstructing Anthropos: A Critical Legal Reflection on "Anthropocentric" Law and Anthropocene "Humanity"' (2015) 26 *Law and Critique* 225–49.

Grear A. 'Introduction: Staying with the Trouble'—Environmental Justice for the Anthropocene–Capitalocene in Environmental Justice. International Library of Law and the Environment Series (Cheltenham: Edward Elgar Publishing, 2020).

Hamzić, V. 'What's Left of the Real?' in D. Fassin and B. E. Harcourt (eds), *A Time for Critique* (New York: Columbia University Press, 2019).

Holland, J. H. *Complexity: A Very Short Introduction* (Oxford: Oxford University Press, 2014).

ILC Conclusions 2006. ILC, Report of the Study Group, Fragmentation of International Law: Difficulties Arising from the Diversification and Expansion of International Law: Conclusions (A/CN.4/L.702) (18 July 2006).

Jasanoff, S. *The Ethics of Invention. Technology and Human Future* (WW Norton, 2016).

Jasanoff. S. *Image and Imagination: The Formation of Global Environmental Consciousness* (London: Routledge, 2012).

Johnston, R. et. al, *Indigenous Research: Theories, Practices, and Relationships* (Toronto, Canadian Scholars, 2018).

Krisch NB, Kingsbury B, Stewart, 'The Emergence of Global Administrative Law' (2005) 68 *Law and Contemporary Problems* 15

Latour, B. *Politics of Nature. How to Bring the Sciences into Democracy*, C. Porter trans. (Cambridge, MA: Harvard University Press, 2004).

Latour, B. *Reassembling the Social: An Introduction to the Actor-Network Theory* (Oxford: Oxford University Press, 2005).

Leonard, L. 'Milton, Lucretius, and "the Void Profound of Unessential Night"' in K. A. Pruitt and C. W. Durham (eds), *Living Texts: Interpreting Milton* (Selinsgrove, PA: Susquehanna University Press, 2000).

Luhmann, N. *Law as a Social System* (Oxford: Oxford University Press, 1993).

Martin, B. 'Methodology is content: Indigenous approaches to research and knowledge' (2017) *Educational Philosophy and Theory*, 49, 1392–1400

Marsh. W. *Landscape Planning. Environmental Implications* (5th edn, Wiley and Sons, 2010).

McHarg, I. L. *Designing with Nature* (Oxford: Wiley and Sons, 1969).

NASA. The Artemis Accords: Principles for Cooperation in The Civil Exploration and Use of the Moon, Mars, Comets, and Asteroids for Peaceful Purposes. 13 October 2020 <https://www.nasa.gov/specials/artemis-accords/img/Artemis-Accords-signed-13Oct2020.pdf> accessed 29 June 2022.

Nobel Prize Committee 2021, Press Release, October 5 2021 at https://www.nobelprize.org/prizes/physics/2021/press-release/

Otomo, Y. 'Law and the Question of the (Non-Human) Animal' (2011) 19 *Society and Animals* 383–91.

Philippopoulos-Mihalopoulos, A. 'Towards a Critical Environmental Law' in A. Philippopoulos-Mihalopoulos (ed), *Law and Ecology: New Environmental Foundations* (London: Routledge, 2011) 193–210.

Philippopoulos-Mihalopoulos, A. 'Lively Agency: Life and Law in the Anthropocene' in I. Braverman (ed), *Animals, Biopolitics, Law: Lively Legalities* (London: Routledge, 2016a) 193–210.

Philippopoulos-Mihalopoulos, A. 'Withdrawing from Atmosphere: An Ontology of Air Partitioning and Affective Engineering' (2016b) 34m *Environment and Planning D: Society and Space* 150–67.

Praet, I. and Salazar J. F. 'Introduction: Familiarizing the Extra-terrestrial/Making Our Planet Alien' (2017) 9 *Environmental Humanities* 309–24.

Prescod-Weinstein, C. et. al. 'Reframing astronomical research through an anticolonial lens—for TMT and beyond' (White Paper submitted to the Astro2020 Decadal Survey)

Ranganathan, R. 'Global Commons' (2016) 27 *European Journal of International Law* 693–717.

The Royal Swedish Academy of Sciences. Press release: 'The Nobel Prize in Physics 2021' <https://www.nobelprize.org/prizes/physics/2021/press-release/> accessed 29 June 2022.

Singh Sachdeva, G. 'Commercial Mining of Celestial Resources: A Case. 2. Study of US Space Laws' (2018) 13 *Astropolitics: The International Journal of Space Politics & Policy* 3

Simpson, L. B. *As we have always done: Indigenous freedom through radical resistance* (Minneapolis: University of Minnesota Press 2017)

Smith, J. L. 'I, River? New materialism, riparian non- human agency and the scale of democratic reform: I, River?' (2017) *Asia Pacific Viewpoint* 58, 99–111

Stengers, I. *Cosmopolitiques, vol. 1, La guerre des sciences* (Paris, France: La Découverte/Sciences humaines et sociales, 1996).

Stengers, I. *Au temps des catastrophes. Résister à la barbarie qui vient* (Paris, France: La Découverte, 2009).

Teubner, G. *Global Law Without a State* (Aldershot 1997).

Tsing, A., et al.. *Feral Atlas* (Stanford, CA: Stanford University Press, 2021) <https://feralatlas.org/>.

United Nations Office for Disarmament Affairs. Treaty on Principles Governing the Activities of States in the Exploration and Use of Outer Space, including the Moon and Other Celestial Bodies.

United Nations Treaty Series (online). Agreement Governing the Activities of States on the Moon and Other Celestial Bodies, Article 7, para 1, 5 December 1979, 1363 UNTS 3.

Wapner Wapner, *Living Through the End of Nature – The Future of American Environmentalism* (MIT Press, 2010).

Wilson, P. and Wilson, S. (2015). Indigeogy. Keynote presentation at Chiefs of Ontario, 'Charting Our Own Path Forward', Education Symposium, Thunder Bay, Ontario. <http://education.chiefs-of-ontario.org/es2015> accessed 29 June 2022.

7

Essay on the Scottish Islands, their lessons for extraterrestrial governance, and a sketch of the applications of this knowledge to settlements beyond Earth

Charles S. Cockell

1. Introduction

There is a tendency among researchers considering the future of human space settlement to attempt to predict the nature of institutions that might emerge by assuming these societies to be a blank canvas. Those studies that do make efforts to investigate the human factor and its influence on social conditions in real communities on Earth usually focus on the relatively short-term effects of isolation. For example, national stations in Antarctica, or deliberately designed simulated extraterrestrial habitats in remote locations, offer insights into the psychological and physiological effects of isolation (Harrison, Clearwater, and McKay 1991; Goemaere et al. 2019). However, these studies do not investigate the longer-term institutional arrangements that might emerge in an isolated group of humans because the communities established are generally of insufficient permanence.

Other studies that use terrestrial environments to understand extraterrestrial locations (so-called analogue studies of planetary exploration) are very numerous and diverse in their environments of study, but they tend to be focused on scientific studies, such as geology and biology (e.g. Preston and Dartnell 2014; Payler et al. 2016; Lim et al. 2019).

Although there are no environments on Earth that completely replicate conditions found anywhere in space, there are places where the long-term establishment of small populations over centuries, even millennia, might offer significant insights into how such populations can be sustained, and in cases where the populations have failed, what those factors were that led to discord or disintegration of the community. For example, in previous papers on the subject of extraterrestrial liberty, I have discussed some of the characteristics of communities of the High Arctic (the Inuit) and the Israeli kibbutzim (Cockell 2008, 2009, 2010). The limitation of these communities is that they do not offer a diverse set of communities of different sizes that are relatively easily accessible, whose governance histories are well documented, that have persisted for centuries or longer, and that might offer direct comparisons in the present day. It is in respect to these facts that I turned my attention to investigating the Scottish islands.

Charles S. Cockell, *Essay on the Scottish Islands, their lessons for extraterrestrial governance, and a sketch of the applications of this knowledge to settlements beyond Earth*. In: *The Institutions of Extraterrestrial Liberty*. Edited by Charles S. Cockell, Oxford University Press. © Charles S. Cockell (2022). DOI: 10.1093/oso/9780192897985.003.0008

In this chapter, I draw upon eight years of personal investigation on islands in the Hebrides (Table 7.1) as analogues for long-term human space settlements to propose a variety of lessons that might aid in the successfully governance of small populations in space, with a special emphasis on the conditions for liberty. I emphasize that my study is not exhaustive, nor is it intended to be a comprehensive historical and governance study of Scottish islands. I have spent these eight years attempting to glean general lessons of value to the specific problems of space and to notice what factors have led to successful settlement in their broadest scope. Scotland has many more islands than the Hebrides and I have little doubt that islands in the Orkney and Shetland Islands, for example, provide many lessons. My study focuses on islands in the Hebrides as they represent a cross-section of histories and population sizes, but the great quantity of Scottish islands that I do not consider should leave the reader in little doubt as to Scotland's potential to yield further information related to the line of enquiry I pursue here.

I underline that this study has no pretensions to be a quantitative study. It is extremely difficult to achieve such a goal when much information of relevance is anecdotal, or even gleaned from communities that no longer exist, such as observations on democracy on St Kilda. Nevertheless, there are many lessons to be gained from these islands.

I leave the completion of my Introduction and the start of the next section to John Morgan (1861) who gave his impression on arriving at the isolated island of St Kilda: 'There was a strange, indescribable look about all we saw, as though we had sailed into another planet or made a voyage to one of the little asteroids.'

2. From Muck to Mars: A Summary of the Potential of Scottish Islands

There are many island chains in the world with populations that could be of use to a study of the likely trajectory of isolated settlements beyond Earth, such as the Hawaiian island chain or the large number of islands across the Pacific Ocean.

Scotland is home to 168 islands of about 100 acres or greater (Haswell-Smith 2015). They vary enormously in their history, culture, and population. Some were inhabited but are now uninhabited (by a permanent population), such as St Kilda, Scarba, Pabbay. Of course, some have never been inhabited, such as Stuley and Flodday.

Scottish islands offer unique potential among island populations for the study presented here from a number of perspectives:

(1) There are a very large number of islands with different population sizes from less than 10 to over 10,000. This allows information to be gleaned not only on how population size influences the possible mode of governance but also the possibility of fashioning a theory of how an extraterrestrial settlement might successfully transition from initial settlers into a large-scale settlement. In that sense, Scottish islands offer a spatial analogue of the temporal development of a settlement in space.

(2) Their history is well documented. The best-known islands have attracted a number of books and publications specifically focused on their populations and their governance cultures.

(3) They have carried out diverse 'experiments in living' as John Stuart Mill would have put it (Mill 2004). They cover the range from coercive lairds to participatory democracies and between these two end-members, benevolent overseers of various shades.

(4) Although each island has unique environmental conditions and geology, at the broad planetary scale they are very much alike and thus one can consider their societies to be in rough approximation a controlled experiment in which environmental variables and living conditions are kept in general similarity, but the style of governance is the variable to be tested.

(5) Their present-day populations are well educated, making it possible to engage in impromptu conversations about how their lifestyles might inform Mars and lunar exploration without incurring disbelief. Thus, the populations are, in general, willing to discuss information for the extraterrestrial case.

(6) The islands are easily accessible and thus a considerable body of experience and information can be gathered in just a few years at relatively little cost (compared, for example, to a comprehensive study of islands in the Pacific Ocean).

In this chapter, I draw on several sources of information that I have investigated over the years:

(1) Informal discussions with selected individuals about space exploration and their own views on how island experiences might inform such efforts, for example with Lawrence MacEwen on the Isle of Muck (MacEwen 2018).

(2) Impromptu conversations with islanders in grocery shops and restaurants, and with owners of guest houses, museums, and other locations about how their lives might inform us about the settlement of the Moon and Mars.

(3) Books and peer-reviewed journal articles about the history of the islands.

(4) Information gained from museums and other institutions on the islands.

I split the islands into three broad categories of use for considering the extraterrestrial case. 'Small settlements' are those with populations less than 50. I consider these, in population size, to be analogous to the first settlements established in space. We might expect a scientific station initially established on the Moon, for example, to be no more than 10 people, potentially rising to a number still less than 50 in its initial years.

'Medium settlements' are analogous to societies where the initial station has grown to a considerable size—similar to many national Antarctic stations. The population is around 100 individuals or larger. Their dynamic is less intimate than small settlements, yet it is still small enough for people to know each other. However, the size is one that requires more coordinated governance structures. These islands provide useful insights into how such institutions can be formed.

Finally, 'Large settlements' I consider to be islands where the population approaches 1,000 or greater. These I consider to be analogous to large-scale towns in space and we do not expect these types of settlements to emerge in space for some time. At this scale, the governance structures again assume a more formal and expansive structure. The resources required for such settlements transition to more integrated and more demanding food and supply networks that often embody the characteristics of large towns with many more civic facilities.

In this chapter, I will focus on the small and medium settlements to discuss ideas that may be of value to initial populations in space and their subsequence successful permanence.

The reader may feel that this division is artificial and I accept that there is no obvious categorical change that occurs between these sizes. Indeed, one might consider that all islands, from the smallest to the largest populations, are spread across a continuum. However, over the last eight years, it has struck me that although there is a gradation, in terms of formality of governance, in very broad scape, islands can in general be fit into one of these three groupings.

Furthermore, for the purposes of my study, I also consider these three categories to usefully mirror three major stages of space settlement that we might contemplate: (i) Small initial landing population and its growth in the immediate years after establishment; (ii) larger matured settlement spreading into new spaces with a population approaching and exceeding 100; (iii) large space settlements in the further future housing many thousands of individuals.

In Table 7.1, I list the islands discussed in this chapter with a population number associated with a specific time (they are often fluctuating, so these numbers are a means to give the reader a general idea of the scale of society).

Table 7.1 Scottish islands discussed in this paper as analogues to space settlements. Population sizes are given for a specific year in which it was measured.

Size category	Island	Population Size
Small settlement	Soay	17 (1990)
	Canna	18 (2015)
	Rum	22 (2013)
	Muck	27 (2013)
Medium settlement	Colonsay	124 (2011)
	Eigg	105 (2017)
	St Kilda (Hirta)	Now 0 (but was medium, >100, before 1850).
Large settlement	Tiree	653 (2013)
	Islay	3228 (2011)
	Skye	~10,000

3. Lessons on the Establishment of Initial Settlements from Scottish Islands

In this section, I discuss a number of lessons learned about governance that are relevant to the extraterrestrial case. I have selected a number of lessons that I consider to be of singular use and importance, particularly with respect to ensuring a sense of liberty in small populations and in preventing the decay into despotism.

3.1 On the forms of community and liberty in small settlements

Many of the Scottish islands categorized as small settlements reflect the early stages of planetary settlement. Instructive in this regard is the Isle of Muck, which has been overseen by Lawrence MacEwen and his family since 1896. It has retained a stable population of about 25 to 50 for many years. The island has received new inhabitants over the years, some of whom leave after misidentifying the hard life on these islands as a romantic utopia. This is itself an important analogue to likely misperceptions of life beyond Earth, which I have addressed previously (Cockell 2002), and I will not consider further here other than to note that it is a trivial observation that those who travel into space need to be educated about the reality of the environment to which they travel. Unlike Muck, leaving extraterrestrial settlement to return to Earth once ensconced in these faraway places may not be easy.

In an interview with Lawrence MacEwen (MacEwen 2018) to try to establish the basis of the success of the island and its application to space, he summed up his overarching belief that the reason was that he 'helped people when they needed it, but otherwise let them be'. A similar culture seems to have pervaded the Isle of Soay (Cholawo 2016). MacEwen went on to speculate that this would be the case anywhere where a small population is established. The relevance of this point to the extraterrestrial case is that a community can hold together because of a common purpose in surviving, yet when there is no clear situation in which people must operate together, a philosophy of leaving people alone can be an effective counter-balance.

Muck is not without some formal structure. Major works are carried out through a community company with every adult a voting member. Works have included electrification and the development of fishing around the island. The company also oversaw the development of a community hall, which includes a meeting room, games area, and an area for parties. The whole community took part in raising the funds and building the hall.

The success of Muck is a template for an extraterrestrial settlement in which a community company could be used to raise support for new developments. The unchallenged position of MacEwen is largely because he has been a good 'father' to the community and he has family ties to the island that reach back to 1896 when they bought the island. To forestall the possibility of a one-man dictatorship becoming established in the extraterrestrial case, an equivalent position of laird could be rotated around the community by vote for two to three years at a time.

Crucial to the success of a small community under extreme circumstances is cohesion in the community. Instructive in this regard is the Isle of Rum. Whilst I spent some time on the island in 2018 to understand its governance, anonymous members of the community informed me that in recent years many of the community had turned to buying items online, and that the shops in Mallaig (the mainland ferry departure point) sent pre-ordered boxes of food, all of these items delivered by ferry. This had led to fragmentation. Individuals no longer needed the community shop to the same extent than in earlier years. The shop was a focal point of community discussion and generally oversaw orders of food and household items, and its reducing importance, combined with a growing sense of self-sufficiency, has eroded this lynchpin of the community. In the extraterrestrial case, food and other vital items are likely to be produced locally. Although some materials sent from Earth or other external locations might be ordered by individuals, these external supply networks are unlikely to allow people to act completely as individuals as the relatively benign environment of Rum makes possible. Nevertheless, it is instructive to observe that liberty can erode community. If the means exist for individuals to acquire an increasing number of their needs independently from others, antagonisms in the community that would otherwise be dealt with amicably can lead to an atomized community. In small populations, mutual interdependence, a limitation on individual self-sufficiency and by its nature a restriction on the freedom to act, is a necessary factor in the maintenance of the community.

In summary, we see two opposing possibilities. In contrast to overbearing control, the matter that MacEwan tried to avoid on Muck, a lack of cohesion and central focus can be a route to disembodiment of the community as suggested on Rum. In other words, perhaps unsurprisingly, there is a fine balance between individual liberty and central authority that small communities must achieve. A community must seek to create institutions that encourage group identity, cohesion and interaction, yet the overall management of the community must allow individuals to feel that they can escape the dictates of those in control.

3.2 On the forms of governance in medium settlements: Lairdship versus democracy

As a settlement grows in size to become an established group with a population of 100 or more, so daily deliberation and governance takes on a more formal structure. A fascinating example, and applicable to the extraterrestrial case, is the Isle of Eigg. In 1997, the islanders raised funds for a community buy-out. Fed up with decades of absentee lairds and their unpredictability, the community took ownership of their island by purchasing it through an organization they set up—the Isle of Eigg Heritage Trust (Dressler 2014).

Although many decisions are taken on an informal level, the number of people is within the capacity for direct participatory democracy. The community hall provides a venue for meetings to discuss matters of general importance to the whole community. Overall planning of the community is undertaken by the Trust. The Isle of Eigg Heritage Trust is a registered charity as well as company limited by guarantee. It was set up as a partnership between the residents of Eigg, the Highland Council, and the Scottish Wildlife Trust. Thus, the Trust has three members: the Isle of Eigg Residents' Association, the Highland Council, and the Scottish Wildlife Trust. Each of these members appoints directors to the board of the Trust. The Isle of Eigg Residents' Association has four directors who are elected by the community. The Highland Council and the Scottish Wildlife Trust appoint one director each. The Trust has an independent chairperson. The main board holds quarterly meetings, while the day-to-day management is entrusted to island directors, who meet on a monthly basis. The Trust is supported by a part-time Secretary.

Mapped onto the extraterrestrial case, the Resident's Association would be the people living in the settlement. The analogues of the Highland Council and the Scottish Wildlife Trust would be organizations on Earth who can provide additional care and influence on behalf of the isolated settlement (wildlife will not be relevant, but the general concept that the Trust has links to mainland organizations would be mirrored in directors from Earth being represented in the extraterrestrial settlement). As might also be replicated in space, the Isle of Eigg Heritage Trust also has subsidiary companies including Eigg Construction to oversee major building works and Eigg Trading that handles the shop and other amenities that cater to tourists, other visitors, and permanent residents.

In the 20 years that this structure has existed, it has proven highly successful. Part of its success is attributed by the islanders to their common sense of purpose. Having bought out the island, many of those who took part in this process feel duty bound to make it a success. The democratic nature of the island organization and the alternative—a return to lairdship—is probably a significant driver.

Despite the apparently utopian democratic structure on the surface, it is true, as with many human institutions, that there are specific people who provide much of the driving impetus. At the expense of violating anonymity (the revelation is a compliment) but in my own visits to the islands to understand the use of Eigg to extraterrestrial settlement, enquiries about governance frequently brought up the energetic personality of Maggie Fyffe, Secretary of the Trust and the lynchpin of the community buy-out of the island. One evening, when beer was flowing well, I was told that she was the 'unseen laird'. This latter epithet would probably appal her but it is worth mentioning because it underscores a general feature of the human personality. People often fixate their gaze on those who become prominent members of the community. There will always be specific individuals on whom the community seems to rely, whether that be mere perception or truth. If we accept this to be the case, then it follows that it is important that such individuals, realizing their influence, do not use it to acquire arbitrary power. This is as much about the self-control of the individuals who drive the community as it is about the robust structures of governance and leadership. In the case of Eigg, this mixture has been a remarkable success. However, there are examples of how such individuals have realized near-despotic regimes on some islands; the case of nineteenth-century St Kilda provides a lucid example to which I will return to later, and I will discuss ways to mitigate this tendency.

It is worth contrasting the Isle of Eigg to the Isle of Colonsay, an island of comparable size. Similar community structures have been assembled for the purposes of governance. Colonsay today is overseen by a Community Council which meets once a month and has up to 10 councillors. The Community Council is the most local administrative structure in Scotland and its purpose is to ascertain the views of the community. Like Eigg, the island also established a company, the Colonsay Community Development Company, to oversee island development. Established in 2000, any community member can attend its monthly meetings. Colonsay is a successful community and, in recent years, much of this may be attached to the development of these community democratic structures and development organizations.

An interesting feature of Colonsay is the documentation of recent times when the laird played a more prominent role in island life. Colonsay still has a laird, but the 1980s legislation giving crofters a right to buy their property reduced the power of the laird and was influential in driving the island community to its current community-led structure. McPhee's account of Colonsay (McPhee 2012) is significant because he documents the relationship between the community and the laird in these earlier days. It might best be described as a begrudging respect. If a laird has a paternalistic respect for those on his island, then a strong sense of community can prevail. An impression given by Alexander's account (2017) and my own conversations with islanders is that the balance between community and laird can sometimes be precarious and the felicity of the community at large is strongly dependent on the good will and benevolence of a single individual. As McPhee (2012, 101) observed about the laird of Colonsay: 'he is the enigmatic embodiment of good and evil, hope and fear, keeper of the gate of Heaven and Hell, fate's own fulcrum, overlord, landlord'. It is poetic, but this description raises shivers up the spine when one contemplates the potential for this to go wrong in an isolated extraterrestrial environment.

The problem with the lairdship model is that as McPhee's comment illustrates, one is beholden to their benevolence for a despotism not to emerge. When one depends upon the good will of a capricious or unpredictable leader for one's liberty, one is as good as enslaved. Whilst a lairdship, particularly in more recent years, is overseen by legal structures of the Scottish mainland then a collapse to abject despotism is avoided. An abusiveness would soon become common knowledge among the island communities and mainland council authorities. However, a lairdship on a remote extraterrestrial body with no oversight is inviting the possibility of disintegration to dictatorship.

Experiments conducted since the 1990s on community ownership, whose outcomes were uncertain at the time of their testing, have proven that participatory democracy, and local companies established under participatory democratic regimes to oversee commercial activity and development, have been notably successful. A benevolent lairdship might well be successfully established in the extraterrestrial environment but its unpredictable nature would seem to suggest that the modern models of Eigg and Colonsay, for example, are eminently applicable to the extraterrestrial case.

3.3 Larger settlements and their governance

In this chapter, I am less concerned with 'Large settlements'. I include in Table 7.1 some examples of large settlements for completeness and to provide a grasp of the full range of population sizes that can potentially be considered. Here, I will merely note that large settlements around or exceeding thousands of individuals tend towards municipal and governance structures similar to large conurbations. This is the case for the Isle of Skye, for instance, which is part of the Highlands Council. It has a capital (Portree), a police presence, many private companies that operate concerns from wind power enterprises to food provision, etc. It is represented in the Highland Council through representative democracy. Although the island retains a distinctly island culture, its governance structure resembles districts on the mainland of Scotland and in that sense, it offers fewer unique lessons on extraterrestrial management than small islands.

3.4 Conversation: An underappreciated mechanism of community cohesion

What other features of the social organization on Scottish islands are common to all the communities and play a role in their successful management? A characteristic of island life that seems common to all populations of small to medium size is intense conversation and gossip. In large societies, the idea that gossip plays some pivotal role in social organization may seem trivial, almost comical, but this is not the case for small, isolated populations. This has been described for the case of Colonsay by McPhee (2012, 99): 'when the population was two hundred and fifty, the gossip was extremely intense, but now that the population is half of that the intensity has doubled, for each person's turn comes up that much more often'.

One effect of the culture of conversation and gossip, although it may not be pursued deliberately with this end in mind, it that it acts as a pressure valve to give vent to concerns and allow individuals to feel that by expressing views about particular people or situations, perceived wrongs have been addressed. The extent of gossip can in some ways be regarded as a barometer of social liberty, since gossip can only thrive in a society where people feel comfortable expressing their point of view. Gossip has an ancillary benefit in allowing for the dissemination of useful information about events and situations. As Alexander (2017, 95) observed:

> Yet, so the passing weeks and months of a Colonsay year would reveal, the communal life of the islanders was essential to its sense of well-being, and it was woven together, not only by such focal points as the ferry arrival, a visit to the shop, the post office or even the pottery, but by an island-wide web of innumerable, interconnected conversations, and a hidden catalogue of organised gatherings that many bigger communities might envy.

Nevertheless, it is also apparent to the observer that gossip and local culture tend to impose a conformity of view and act as a strong deterrent to the slighting of perceived social mores. Alexander again on Colonsay (Alexander 2017, 104): 'The familiar goldfish bowl, which seems so friendly and comforting most of the time, can suddenly feel suffocatingly small ... Creeping peer pressure could come out of a clear blue sky.'

It may even degenerate into a tyranny of the majority (de Tocqueville 1835). As Mclean observes about individualism on St Kilda (McLean 2010, 171):

> Many people ... would regard life in St Kilda as anything but ideal ... because of the absence of opportunity for individual originality and creativeness, which, though to some extent dependent on the firm support of a culture, are easily smothered by a cultural pattern which is too strong ... Undeniably there was little room for individuality.

This reality is easily overlooked by an outsider. After his famous 17th century visit and account of St Kilda, Martin Martin concluded (Martin 1698): 'The inhabitants of St Kilda, are much happier than the generality of mankind, as being the only people in the world who feel the true sweetness of liberty', although it is significant to observe that in the very same volume, he recognized: 'the voice of one is the voice of all the rest ... their common interest uniting them firmly together'. Martin Martin had identified one form of liberty on the island—the escape from legal and other administrative structures associated with large-scale mainland societies—but he had not failed to notice, even if he did not recognize the apparent contradiction, that tyranny had emerged from within.

In extreme environments, this 'island-wide web' acts to reinforce social norms and may enhance the safety of the population, for example by influencing the next generation to accept ways of doing things that have been discovered to be beneficial by trial and error over generations. On the deleterious side of the balance sheet, gossip tends to stifle innovation (except where individuals take a pleasure in being the society maverick and violating social norms that will deliberately elicit discussion). On balance, gossip and small talk are likely to be vital and useful functions in small populations in extremes, provided that democratic means of governance at more formal levels are used to prevent them from becoming the primary means of social organization. On Eigg and Colonsay today, it seems to me that the informal lines of gossip continue to weave through the community, but the community council and development companies allow the society to make real progress and act as the true conduit of formal development above the capriciousness of human tempers and emotions. Thus, the most stultifying effects of conformity, encouraged by rigidly held social mores propagated through island gossip and conversations, may be mollified by providing the opportunity for individuals to express their capacities outside these structures through these island committees and companies.

3.5 On the need for common purpose and the control of faction

I begin this section with an amusing exchange with Lawrence MacEwen, Isle of Muck (MacEwan 2018). When I was discussing other islands and asked him how faction was avoided on the Isle of Canna with such a small population, he replied, 'well it does help if everyone has the same surname' (histories of Canna can be found in Perman 2016 and Campbell 2014). Although I have no doubt that this has a strong element of truth in it, it is difficult to apply this lesson to the extraterrestrial case unless one is proposing that a settlement be established with a single large family (an interesting idea that I have not seen suggested before, but I will not pursue it here, although one might point out that in the future such a settlement, for example on the Moon, is not inconceivable). Apart from familial allegiance, which has undeniably been a very powerful force in Scottish island politics in the past, what other means are used to control faction?

There is little hesitation in saying that a factor one does observe in all islands that seems to be immensely powerful in mollifying faction, whether lairdships or community run, is the common effort that is required to operate successfully in environments that can be extreme. Put colloquially, selfishness does not pay in environments where mistakes or a want of vital resources could be lethal or dangerous for an isolated individual. This observation is useful because exactly this state of affairs does apply in the extraterrestrial case.

In the modern age, conditions on the islands are less forbidding than they used to be. Regular ferry routes, tourism, and better shops and food have made the environments less potentially lethal than they once were. Nonetheless, winters are still extreme, unpredictable, and community ties remain an important part of securing cohesion. Even in the absence of extremes, the want of something as simple as a garden tool, rectified in large conurbations by driving to a store, must often be met with assistance from another islander. Isolation is a form of extremity and the unpredictable nature of what you might need, and from whom you might need it, makes people alert to maintaining good community relations.

There is another factor. An anonymous islander on Canna, when I asked them what motivates people to get on with one another, whilst discussing the role of extremities in social organization on the Moon, said: 'It's not just the extreme conditions that make us all get along. It's that you can't easily get off the island!' In the modern age, this is not strictly true since jobs and ferries do make possible an escape route to the mainland, depending on one's financial means. However, I found the comment of singular importance, partly because it is the identification of yet another factor that compels people to act with equanimity to their neighbours but also because this factor will undoubtedly be true in the extraterrestrial case where, depending on the extent and frequency of ships travelling between, say, the Moon or Mars and Earth, people will be relatively locked in for periods of many months to years with no easy escape. As on Scottish islands, particularly in the past (St Kilda being an extreme example; Seton 2012; Gannon and Geddes 2015) both the extremity of conditions and the long-term isolation act in unison to encourage the resolution of faction.

Nevertheless, despite the mechanisms to soften faction, the possibilities for it within small communities are ubiquitous and will be predictably present in all settlements beyond Earth. This is explained with clarity for the case of St Kilda by Fleming (2005, 113):

> It is hard, I think, to over-estimate the potential for social tension. In a population of 200, each individual has to handle up to 199 interpersonal relationships. The total number of such relationships, if one includes everyone from the youngest baby to the oldest inhabitant, is 19900. In terms of social harmony in a small island community, there is a lot to go wrong.

The need to encourage and instantiate powerful forces to prevent community dis-integration in small, isolated populations is a mathematical requirement. Of course, these calculations multiply for larger populations, but in small communes it is diffi-cult to avoid people and thus the mathematical permutations of possible interactions become more real and not mere theoretical abstractions.

It might be noted that, despite the mathematics just alluded to and the potential for 'a lot to go wrong', few of these islands see crime in any significant level requiring a police force. For example, on Colonsay, the slashing of tyres on a vehicle following a heated exchange in the island's hotel bar in 2013 was the first crime since 2006, when a visitor stole some money from an elderly resident's house. The strength of community cohesion, the networks of gossip and the sense of community responsibility seem capable of forestalling criminal activity. Quoting a MacDonald who lived in St Kilda, Donald Gillies reported (Gillies 2010, 124): 'There was never any fighting or crime on St Kilda. There was a community spirit, everybody used to help each other. There was no squabbling, and we all got on so well together. Life used to be just lovely on this island, despite the terrible winter storms.' These facts are observed in more recent texts (Fleming 2005). In the extraterrestrial case, the lesson to be learned is that the vision of a largely crime-free community is not utopian and can emerge naturally from a strong sense of community. Nevertheless, over long time periods criminality cannot be completely avoided. As on Colonsay, the appointment of a community member as 'special constable' can be used to provide deterrence and to ensure that there is a formal line of reporting should a crime occur.

Common purpose also emerges from the multiple jobs that many islanders do, which has the effect of binding people together in mutually reinforcing tasks. Alexan-der observed in Colonsay (Alexander 2017, 148): 'the islanders saw little amiss in the sometimes strange coupling, or even tripling or quadrupling, of different part-time jobs, for it was what lots of islanders had to do'. This is observed on many islands. For example on Muck (Pullar 2014, 186), '[e]veryone on the island has many roles'.

One might summarize by observing that although proximity might seem to create a wide field for interpersonal tensions, these disagreements really only manifest at the very local scale and that many islands in Scotland, for long periods of time, have retained cohesive communities where faction and crime has been minimized.

3.6 On the need for a sphere of self-governance

Even in the relatively open democratic regimes of Eigg and Colonsay, one cannot help feeling, as an outsider, the social intensity of the community structure. Within this collective environment, how is a sphere of personal liberty to be maintained, the form of freedom that Berlin (2002) referred to as 'negative liberty', or as had Hobbes expressed it: 'in those things which by his strength and wit he is able to do, is not hindered to do what he has a will to' (Hobbes 1651)?

In the case of Scottish islands, the solution is largely built into the very architec-ture of the settlement. Despite their small areas, many houses are separated and the open land available and the generally small population ensure that the population density remains low. Indeed, the everyday problem on the islands is not so much an overbearing sense of community but isolation. Many islanders are enthusiastic for fetes, summer festivals, dances, boating competitions, and other events at which the community can meet and socializes. Individual liberty can be so extensive as to be a challenge. As Alexander observed of Colonsay (Alexander 2017, 105): 'The apparent freedom offered by island life could prove hard to handle for even those born to it, let alone those who were discovering it. Most islanders had a rare degree of control over how they spent their time … .'

At once, we see an obvious probable difference from the extraterrestrial case. Habitats are likely to be closely packed on account of the need for oxygen, water, and other vital commodities, all of which will need to be manufactured or purified from extraterrestrial resources. One could attempt to replicate the Scottish island arrangement by deliberate separation of habitats to maximize the potential for a clear sphere of personal liberty outside community events, but logistics and safety may disallow or discourage this tendency. Thus, in the extraterrestrial case, it is important that the habitats themselves provide the means for private escape and seclusion. This is an ergonomics problem.

The more likely packing of people into smaller areas in the extraterrestrial case compared to Scottish islands would suggest that a greater emphasis be placed upon the cultural mechanisms to preventing an overbearing culture. Here we might return to the lesson learned from the Isle of Muck. Lawrence MacEwen's philosophy to help people when they needed it, but otherwise let them be, is a cultural solution to preventing an overbearing sense of community. It has a simplicity, but its implementation is far from simple, since it asks much of people to observe this behaviour as a matter of cultural priority. Nevertheless, this island culture of avoiding interference unless asked for, or unless absolutely necessary, may go some way to mitigating the social pressure resulting from the greater physical co-location of people in the extraterrestrial case compared to the islands.

3.7 On the tension between capitalistic or socialist economics

The question of self-governance and the forms of liberty it makes possible raise the economic question of whether the settlement is to be run along capitalist or socialist principles. That a small settlement can be run along a socialist path there is little doubt. The St Kildans shared most of their possessions communally. Wilson (Wilson 1842 24) explains: St Kilda 'may in many respects be regarded as a small republic in which the individual members share most of their worldly goods in common and, with the exception of the minister, no one seems to differ from his neighbour in rank, fortune, or condition.' Steel (1965, 38), notes that this system was 'partly induced by the conditions in which the St Kildans found themselves. The common survival was the prime concern and individual betterment was an idea scarcely conceived of.' In that the extraterrestrial environment also imposes extremes that require collective effort and coordination, one might cite the St Kildan experience as evidence that small communities, subjected to great environmental extremes, can operate a socialist economy.

The problem with the socialist economy at the small scale, as recognized by Wilson (Wilson 1842, 33), is that it encourages a 'tenacious adherence to uniformity', which makes the community less flexible to changing conditions. Wilson goes on to note (Wilson 1842, 24–5): 'Indeed, a peculiar jealousy is alleged to exist on this head, no man being encouraged to go in advance of those about him in any thing, which of course must be a drawback on improvement.' The sharing of items in short supply and the generally socialistic mentality of a strong sense of community can prevent a community-destroying selfishness from taking hold in extremes, as the community of St Kilda demonstrates. However, in places where private property is regarded with suspicion it is also generally the case that innovation is stifled and individuals are more likely to succumb to group peer pressure. The vigour of freedom of thought and the confidence to suggest and express new ideas is quickly eroded.

The community councils and companies established on islands such as Muck, Eigg, and Colonsay in more recent decades show that the culture need not become entrenched and unyielding. Business enterprise and personal advancement can flourish, still underpinned by a powerful sense of community. Indeed, the formation of these more democratic cultures and less overbearing community demands has attracted people back to islands and may have been largely responsible for the new-found 'optimism' (Byrne 2010) in the islands. Is seems that in the extraterrestrial case, survival under extreme conditions does not need to embody a slavish conformity or an ideological adherence to common ownership of all property.

3.8 On the role of information in diluting tyranny

An essential factor in the success of any settlement beyond Earth will be access to information and that this information is not under the control of a single source or person. The historical case of Scottish islands provides some exceptional examples of the truth of this matter. One such example was St Kilda. Before evacuation in 1930, the inhabitants were one of the most remote island populations in the world. In the nineteenth century, the island received irregular provision of victuals from the mainland but was generally self-sufficient as it had been for centuries prior. Its nineteenth-century state might be regarded as a useful analogy to the extraterrestrial case where external commodities arrive in the settlement, but it remains largely isolated. Clearly, the technological parallel does not apply, but the cultural parallels may.

St Kilda was run as a participatory democracy (Figure 7.1) as long as anyone can remember, long before this was recently implemented in places like Eigg and Colonsay. The population met in the Main Street in a small space between two houses (the reader should not misinterpret this to be something akin to a main street found in modern cities. This was the walkway running along the front of the houses). Each day, the Mòd or parliament was used to decide on plans for the day and resolve any community issues. Maclean summarizes (Maclean 2006, 37): 'Nor was the parliament above hearing the island's gossip ... But for all that the St Kilda Parliament was a serious institution.'

Despite the participatory democracy, the island did fall prey to despotic influences, mainly through the vehicle of religious representation (Hutchinson 2016). After 1843, when the Free Church split with the Church of Scotland, it took over the parish, which

Figure 7.1 The St Kilda Parliament. A photograph taken before evacuation in 1930 (date not precisely known) of the parliament in session.

included St Kilda. Steel observed (Steel 1965, 74): 'the stern faith of the Free Church of Scotland ...enslaved the people of Village Bay' (the bay of St Kilda).

The man the Free Church sent to look after the island, the Reverend John Mackay, was a tyrant and the description of the consequences is best left to Charles Maclean (2006, 126–9):

> Only one of them managed to fill the vacuum created by Neil Mackenzie [the previous minister]—not with enlightened leadership, but with blinkered despotism ... Shortly after his arrival ... he established a vibrantly harsh rule ... Services on Sunday at eleven, two and six o'clock were made to last for two to three hours each ... Mackay preached long, repetitious sermons in Gaelic, which invariably included the same message of hell-fire and eternal damnation to all sinners ... Children were brought to church at the age of two ... conversation between the islanders was forbidden from Saturday evening until Monday morning ... singing or whistling was a serious sin ... children were forbidden to play games and even made to carry bibles under their arms wherever they went.

There are two lessons that might be learned for the extraterrestrial case. First, even with a highly effective participatory democracy, small populations can come under the influence of zealots. In the case of St Kilda, the religious overtures were surely encouraged by the extreme climate and living conditions that made the population easy prey for a charismatic individual with pious and apparently good intentions. Faced with the lethality of the extraterrestrial environment, a convincing despot might easily usurp control over a community, even if their day-to-day management was achieved democratically. This is not a mysterious phenomenon. Remarking on it with respect to a discussion on the US Constitution and the efforts to avoid this problem, Morley (1972, 151) noted: 'This natural desire of the group for uniformity provides the opportunity of the demagogue, who is both the product and the menace of democracy.' Second, without regular contact with the mainland, there was no oversight of Mackay's regime and the islanders themselves had no access to information that might allow them to formulate some concern about their particular treatment. In a crude way, one might say that they lacked social calibration, and information, to be able to push back against the direction their society was moving in. In the extraterrestrial case, extreme isolation may encourage similar directions of social travel.

The St Kilda experience suggests the need for the regular arrival of new people and information from outside and the opportunity for departure from the settlement to reduce the chance for tyrannical control by single individuals. This flow of people and information is likely to dilute the possibility for an isolated community to develop an anomalous culture and to ensure calibration with the rest of human society by maximizing information flow in and out.

In short, isolation of people and information is likely to offer an open field to degeneration of a community. I should note that one could argue that the arrival of individuals such as Mackay was a consequence of opening up to the mainland in the nineteenth century. Ironically, these problems did not exist when the St Kildans were even more isolated in former centuries and had achieved a high degree of self-sufficiency. This observation has a large element of truth in it, but the same situation cannot be achieved in the extraterrestrial case (except by enforced exclusion) since settlements will need provision of commodities and people from the outside environment (such as from Earth). The point I wish to make is that assuming there is some contact with other parts of human society, it is better that it be made as comprehensive as possible rather than at a low level that allows the community to become infiltrated with outsiders who take over the settlement without challenge and drive it towards dictatorial ends.

4. A General Sketch of the Trajectory of a New Extraterrestrial Settlement Constructed from the Example of Scottish Islands

I finally turn to a brief survey and sketch of the general arc of the development of an extraterrestrial settlement suggested by the Scottish experience.

Upon establishment of a new settlement in a location such as the Moon, Mars, an asteroid, or the free plateau of space, the group can be run along informal lines of a direct participatory democracy with regular meetings to decide on all matters related to the running of the group. An extraterrestrial Isle of Muck should be established. At this stage of the development, it may be beneficial for the group to select among their number a director who will be responsible for the well-being of the group (there could be a separate director for the technical matters of the settlement). This individual is a benevolent extraterrestrial laird. The individual is always subject to accountability of abuse of power, and they might be changed at regular intervals.

When the settlement reaches a number of several tens, a new management structure is implemented. I suggest the following sketch based on the experiences of Scottish islands, and particularly drawing from the experience of Colonsay and Eigg.

The group shall form a community council responsible for the everyday affairs, well-being, safety, and group cohesion of the settlement. It shall meet at regular intervals (for example, every two Earth weeks or as demand requires). All settlement members are able to take part in the council. The council shall be run by six directors, four chosen by ballot within the community and two directors that shall represent Earth or other external organizations on which the settlement depends. These external directors provide the vital link between the settlement and the external resources, contact and communication on which its sustainability may depend. However, the number of external directors will never exceed a half of the total number of directors. The majority shall always be directors from within the settlement.

At the same time, the settlement shall form a company charged with the economic affairs and development projects of the settlement. Its exact form cannot be predicted, since it will depend on the corporate law, if any, that the settlement operates under. Nevertheless, regardless of the legal underpinnings, its purpose is to oversee development projects and consider approaches to the economic advancement of the settlement, from tourism to other economic operations. The settlement company will hold regular meetings at which all settlement members can attend. It shall be run by at least six directors and it shall work closely with the community council to ensure alignment of objectives with the well-being and safety of the community.

If the settlement should exceed ~250 individuals, then the structures described above will transform into organizations more in keeping with the large population. The council shall still hold meetings at which any member of the settlement can attend, but regions within the settlement will vote for councillors to represent their interests rather than voting in directors for the entire council. This modification will achieve a finer resolution of interests but it will also inculcate into the council the structures of representative democracy as the settlement transitions away from continuous direct participatory democracy, which becomes unwieldy at large scales.

The settlement development company will continue to grow in size and it will be run by those with an interest in its development or who are elected to its boards. Nevertheless, general meetings will still be held at regular intervals, which will not merely be open to shareholders or those who have some direct involvement in the company but to all members of the community. This process will mitigate the chance that the company becomes controlled by special interests that drive the settlement to economic tyranny. The company begins to take on the structure and culture of companies on Earth but retains a more open accountability because of the special conditions of isolation and potential for tyranny associated with the extraterrestrial environment.

It would be prudent to attempt to multiply these economic enterprises from an early stage to avoid monopoly. Although, as is the case with Eigg, we can imagine a single enterprise, dealing, for example, with construction, if it was to achieve the monopolization of oxygen supplies, this could be of grave danger. The special problems of monopolization of such vital resources recommend all attempts to diversify and create plurality in the companies and institutions created for the purposed of production. A community can attempt this at the earliest stages.

It must be noted that single community councils can operate by participatory democracy at populations greater than 250. For example, the Isle of Tiree has a community council that operates in this manner, as does Islay with an even larger population (Table 7.1) (a history of Islay is to be found in Storrie 2011). I suggest a threshold of ~250 as meriting a change to structures akin to representative democracy only because at these population scales it is rarely practical that the entire population can attend a meeting and even if they theoretically can, they rarely do. Thus, encouraging them to share their concerns with a representative who does make a point of attending such meetings can ensure that views are not lost.

If the settlement expands to beyond 1,000 individuals, then an additional layer of administration may be contemplated for community affairs. Each district or large grouping of people up to ~250 might form and operate their own community council in the manner of small settlements, representatives of which sit on a new body to present the whole settlement—a community parliament, which may even, at an early stage, take on a bicameral structure.

In this sketch of the progress of an extraterrestrial settlement, I do not discuss other measures that might be taken to enhance liberty and the evolution of institutions—in particular, to prevent the very mechanisms of democracy degrading into a tyranny of the majority. Elsewhere (Cockell 2022), I have suggested a rudimentary 'separation of powers', even at the earliest stages of a settlement, to instantiate the culture, habitats, and ideas of democracy in a settlement, whilst putting in place the checks and balances necessary to disseminate power and to prevent majorities or demagogues from wielding disproportionate influence—the perpetual bane of democratic republics. I previously have called this 'pre-emptive free governance'. Nothing in the scheme outlined above disallows this. The community council, for example, rather than merely being an organization run and managed by a set of directors, might be split internally into three bodies representing early forms of executive, legislature, and judiciary. My purpose in this chapter has been broadly to outline how the Scottish islands can inform the general shape of institutions, whose specific details are surely amenable to more detailed discussion.

Throughout the culture and social organization of the settlement, regardless of its scale, we should find three key ingredients that are to be found in the more successful Scottish islands. I might refer to these as the governance trinity of small settlements: gossip, democracy, and commerce. Gossip is the network of conversation, webs of interrelationships that meld a community together as one and keep information flowing freely and innocuously through a population. Democracy is the essential ingredient to mitigate the settlement falling under the control of a tyrannical laird and allowing the populace to determine the direction of the polity and its everyday needs. Commerce is the energizing ingredient that acts against an overbearing community culture to allow individualism and enterprise and encourage new solutions to the community's needs (Smith 1776). It dilutes the opportunity for an individual, even within the constraints of democratic deliberations and institutions, to seize control of a population that operates under strong peer-pressure.

Aside from its placating effects on social central control in small populaces, commerce introduces an extraterrestrial invisible hand, if you will, into the structure of a settlement that allows wants to be met and new opportunities for the betterment of the material conditions of the community to be realized. Like a well-baked cake, these are not the only three ingredients, as this chapter has made clear. However, I regard these three ingredients as the flour, eggs, and sugar of the governance cake. Without them, there is disintegration. With them, there is no guarantee of harmony and felicity, but there is every chance that with a sprinkling of forbearance, kindness, self-control, and a sense of purpose and dignity in the community, that an isolated island of free-thinking and happy individuals may come to be raised on the extraterrestrial frontier.

5. Conclusion

In an attempt to use the Scottish islands to consider the future of governance and liberty beyond Earth, one is instantly struck by a dichotomy. On the one hand, the diversity of governance structures, from lairds to participatory democracy, and every shade thereof, has been legion. Even within broad categories, every island has a unique 'feel' and culture that emanates from its means of organization. History bears out this enormous diversity and variety of governance. Yet shining through this, one is faced with universal facts. The success of all these experiments has depended on continuous means of deliberation and forbearance within the population and the structures to achieve this. These institutions have ranged from community companies and community buy-outs manifested in participatory democracies underpinned by more simple arrangements such as community shops and even the culture of gossip.

To be avoided are the extremes—a community in which individuals become so independent that the society becomes atomized and its integrity and sustainability is lost, or the opposite extreme, a population under the complete discretion of a tyrant. Both of these end-members have been seen on the Scottish islands. They can never be thwarted entirely, but if a community is focused on the design and implementation of continuous community involvement and active democratic deliberation among its members, it has every chance of avoiding them. What the Scottish islands show is that between these two extremes, there is no correct way to govern a community. Subtle permutations and combinations of individuals, institutions and particular local arrangements lead to a plethora of successful populations.

The extraterrestrial lesson to be learned from this analysis is that attempts should be made to establish open democratic means of deliberation from the outset and then the experiment should be left to run its course without micromanaging a perceived ideal endpoint. As the society develops, the objective is to keep it within the constrained boundaries of excessive atomization and excessive despotism, whilst allowing it to flourish and find its own equilibrium within those bounds. Once it is launched and on its way, the lessons and history of Scottish islands may well be a useful resource to future settlers who may find their social trajectory matching certain islands with higher fidelity than others.

The vast and multifarious details of the social history of Scottish islands is likely to be a useful well of information and of direct practical use in deliberating the management of extraterrestrial settlements.

Acknowledgements

I thank islanders who have indulged my interest in comparing them to space settlers. In particular, I have always been struck by the patience of islanders who in a howling gale have embarked on discussions with an academic from Edinburgh, who has appeared on their land without warning, about how they might help lunar and Martian settlers. Nevertheless, these spontaneous conversations have often revealed information of the greatest value. I was particularly privileged to share time with Lawrence MacEwen on the Isle of Muck to discuss the great success of that island and what it might tell us about our future on Mars. I hope that this paper may go some way to advancing the experience of Scottish islanders as a template to guide governance and liberty in space.

References

Alexander, D. *The Potter's Tale. A Colonsay Life* (Edinburgh, Birlinn, 2017).

Berlin, I. *Freedom and its Betrayal: Six Enemies of Human Liberty* (Princeton, NJ: Princeton University Press, 2022).

Byrne, K. *Lonely Colonsay. Island at the Edge* (Chippenham: CPI Antony Rowe, 2010).

Campbell, J. L. *Canna: The Story of a Hebridean Island* (Edinburgh: Birlinn, 2014).

Cholawo, A. *Island on the Edge. A Life on Soay* (Edinburgh: Birlinn, 2016).

Cockell C. S. 'Mars is an Awful Place to Live' (2002) 27 *Interdisciplinary Science Reviews* 32–8.

Cockell, C. S. 'An Essay on Extraterrestrial Liberty' (2008) 61 *Journal of the British Interplanetary Society* 255–75.

Cockell, C. S. 'Liberty and the Limits to the Extraterrestrial State' (2009) 62 *Journal of the British Interplanetary Society* 139–57.

Cockell, C. S. 'Essay on the Causes and Consequences of Extraterrestrial Tyranny' (2010) 63 *Journal of the British Interplanetary Society* 15–37.

Cockell, C.S. *Interplanetary Liberty* (Oxford: Oxford University Press, 2022).

Dressler, C. *Eigg. The Story of an Island* (Edinburgh: Birlinn, 2014).

Fleming, A. *St Kilda and the Wider World* (Oxford: Oxbow Books, 2005).

Gannon, A. and Geddes, G. *St Kilda. The Last and Outmost Isle* (Edinburgh: Historic Environment Scotland, 2015).

Gilles, D. J. *The Truth about St Kilda. An Islander's Memoir* (Edinburgh: Birlinn, 2010).

Goemaere, S., Brenning, K., Beyers, W., et al. 'Do Astronauts Benefit from Autonomy? Investigating Perceived Autonomy-Supportive Communication by Mission Support, Crew Motivation and Collaboration during HI-SEAS 1' (2019) 157 *Acta Astronautica* 9–16.

Harrison, A. A., Clearwater, Y. A., and McKay, C. P. *From Antarctica to Outer Space* (Berlin: Springer, 1991).

Haswell-Smith, H. *The Scottish Islands* (Edinburgh: Canongate, 2015).

Hobbes, T. *Leviathan* (London: Andrew Crooke, 1651).

Hutchinson, R. *St Kilda. A People's History* (Edinburgh: Birlinn, 2016).

Lim, D. S. S., Abercromby, A. F. J., Nawotniak, S. E. K., et al. 'The BASALT Research Program: Designing and Developing Mission Elements in Support of Human Scientific Exploration of Mars' (2019) 19 *Astrobiology* 245–59.

MacEwen, L. Discussion on the Isle of Muck and the exploration of Mars (personal communication, 2018).

Maclean C. *Island on the Edge of the World. The Story of St Kilda* (Edinburgh: Canongate, 2006).

Martin, M. *A Late Voyage to St Kilda. The Remotest of all the Hebrides, or Western Isles of Scotland* (London: D. Brown & T. Goodwin, 1698).

McPhee, J. *The Crofter and the Laird. Life on a Hebridean Island* (Exeter: House of Lochar, 2012).

Mill, J. S. *On Liberty and Other Writings* (Cambridge: Cambridge University Press, 2004).

Morgan, J. E. 'The Falcon among the Fulmars; or, Six Hours at St Kilda' [1861] *Macmillan's Magazine.*

Morley, F. *The Power in the People* (Los Angeles, CA: Nash Pub., 1972).

Payler, S. J., Biddle, J. F., Coates, A., et al. 'Planetary Science and Exploration in the Deep Subsurface: Results from the MINAR Program, Boulby Mine, UK' (2016) 15 *International Journal of Astrobiology* 333–44.

Perman, R. *The Man Who Gave Away His Island. A Life of John Lorne Campbell of Canna* (Edinburgh: Birlinn, 2016).

Preston, L. and Dartnell, L. R. 'Planetary Habitability: Lessons Learned from Terrestrial Analogues' (2014) 13 *International Journal of Astrobiology* 81–98.

Pullar, P. *A Drop in the Ocean. Lawrence MacEwen and the Isle of Muck* (Edinburgh: Birlinn, 2014).

Seton, G. *St Kilda* (Edinburgh: Birlinn, 2012).

Smith, A. *An Enquiry into the Nature and Causes of the Wealth of Nations* (London: W. Strahan and T. Cadell, 1776).

Steel, T. *The Life and Death of St Kilda* (National Trust of Scotland, Edinburgh: R & R Clark Ltd, 1965).

Storrie, M. *Islay: Biography of an Island* (Islay: Oa Press, 2011).

Tocqueville, de A. *De la Démocratie en Amérique* (Paris: Librairie de Charles Gosselin, 1835).

Wilson, J. *A Voyage Round the Coasts of Scotland and the Isles* (London: British Museum, 1842).

8

The law of Mars' colonization

Raphaël Costa

1. International Space Law to Regulate Mars' Colonization

In 1967, space became the final frontier of international law. In the geopolitical context of the Cold War, the international community felt the need to regulate future space activities in order to prevent an extension of the Cold War to outer space (for more details about the drafting history between the first UNGA resolution and the adoption of the Outer Space Treaty, see Hobe 2009). After a short drafting process, in October 1967 the United Nations adopted the Treaty on Principles Governing the Activities of States in the Exploration and Use of Outer Space, including the Moon and Other Celestial Bodies, called (for obvious reasons) the Outer Space Treaty (OST).

This Treaty is considered the 'Magna Carta' for human activities in outer space, including space colonization. It contains the fundamental principles of positive space law: the freedom of exploration and use of outer space carried out for the benefit of all mankind (Article I); the prohibition of state appropriation of outer space (Article II); the application of general international law to space activities (Article III); the total demilitarization of celestial bodies and the partial demilitarization of Earth orbits (Article IV); the assistance due to astronauts in distress (Article V); the precondition for private entities to obtain an authorization before operating their activities in outer space and to submit to their state supervision (Article VI); the liability of states for private and public activities in outer space (Article VII); the registration of space objects (Article VIII); and mutual respect and cooperation procedures (Articles IX–XIII).

As its title states, the OST is a *Principles* Treaty. That is why some of its provisions, broadly drafted, had to be detailed in subsequent treaties adopted between 1968 and 1979. Article V on the rescue of astronauts was detailed in the 'Rescue Agreement'. Article VII concerning the liability of states was detailed in the 'Liability Convention'. Article VIII on the registration of space objects was detailed in the 'Registration Convention', and finally, the 'Moon Agreement', establishing the bases for natural resources of celestial bodies exploitation, was adopted in 1979. This could be insignificant if all those treaties were ratified by the same number of states because a treaty only applies to the states that have ratified it. But the OST has been ratified by 104 states, including the spacefaring nations. The 'Rescue Agreement' and the 'Liability Convention' respectively received 94 and 92 ratifications. Ratification of the 'Registration Convention' was less successful, being ratified by only 62 states. And finally, as evidence of its failure (Mateesco-Matte 1993), the 'Moon Agreement' only received 16 ratifications.

Raphaël Costa, *The law of Mars' colonization*. In: *The Institutions of Extraterrestrial Liberty.*
Edited by Charles S. Cockell, Oxford University Press. © Raphaël Costa (2022).
DOI: 10.1093/oso/9780192897985.003.0009

According to Article 3 of the Moon Agreement: '1. The provisions of this Agreement relating to the Moon shall also apply to other celestial bodies within the solar system ... 2. For the purposes of this Agreement reference to the Moon shall include orbits around or other trajectories to or around it.' All provisions of space law also apply to life on Mars and life while being transported to Mars.

This demonstrates the importance of the OST but it also shows its weakness. Some states ratified the Treaty's major principles but not its contents. Space law is not unified. The OST is not self-sufficient, some would say it is 'unaccomplished' (Combacau and Sur 2008, 477) : it needed other treaties to complete it. We will use all those international agreement to express space law applying to Mars colonization, and all our conclusions will apply similarly to all states and to their nationals. We would like to outline here that even if colonization of Mars is undertaken by private companies, as SpaceX, we will be talking here about state colonization because according to the OST's Article VI, private entities conducting space activities need to obtain an authorization before conducting their activities and must submit to their state supervision. That is why no Martian colonization will occur without (at least passive) the will of a state. Moreover, all states bear international responsibility to ensure that their nationals respect the provisions of international space law bounding their state.

The Moon Agreement introduces the concept of the 'common heritage of mankind' applied to outer space and its natural resources. They are both held in trust for future generations and protected from self-serving exploitation by a state or a company. Unfortunately, as the Moon Agreement was only ratified by 16 states, and none of them is a spacefaring nation, we will not discuss this concept further as it is not admitted at all applying to outer space or celestial bodies.

To summarize, the status of international law according to space is that international law is applicable to space colonization, but not all treaties and concepts of international space law are applicable to space colonization as a space activity. Moreover, even though there is no relevant Treaty, international customary law remains useful. For space law, this concerns just a few principles which we will detail below.

1.1 The definition of Mars' colonization

Along with several other notions in space law, 'colonization' is not defined in law, nor is there a broad and internationally accepted definition (Martin 1992). We should therefore try to define it here. Colonization is usually deemed to mean a process by which an organized group of human beings extends its territory to new places, in our case, the Martian surface or subsoil. To colonize Mars means its final occupation. This human establishment must exceed and outrun its founders. When those die or return to Earth, the Martian colony will still exist. Thenceforth, colonization as an organized and planned process must be distinguished from immigration, which is an individual or familial act, and from occupation. The International Space Station (ISS) is for the moment the only permanent human occupation of outer space, yet it is not considered as part of a process of space colonization because the ISS programme will soon come to an end and the participating states do not intend to populate space by means of this station.

In our legal journey to the colonization of Mars we can distinguish two phases: the first is the installation of the colony: we called this the legal foundation of a Martian colony (in tribute to Isaac Asimov and our teenage days reading his stories) and this is discussed in section 2. Once this stage is completed, other problems will arise due to a long-duration Martian colony and this is discussed in section 3.

2. The Legal Foundation of a Martian Colony

In order to settle a human colony on Mars in law, we must first analyse the lawfulness of such a project before detailing its legal conditions.

2.1 The legality of Martian colonizing

Determining the legality, the international lawfulness of Mars colonization, requires two levels of assessment. First, we must establish the legality of colonization *stricto sensu*, which means the legality of a permanent and definitive occupation of a celestial body; note that space law does not define what constitutes a celestial body, so this notion includes both planets and asteroids (Pop 2001). This is the compelling function of a Martian colonization. Once this colony is installed, the settlers can adopt activities and goals beyond mere occupation: this means the legality of every optional function beyond occupation itself must be studied. For Mars colonization, we choose to focus on prospective mining and military colonization.

2.1.1 The legality of the definitive human occupation of Mars

OST Article I sets out the fundamental principle that states parties to the Treaty shall be free to explore and use outer space, which includes both outer space itself as the zone between celestial bodies and the celestial bodies themselves. Back in the 1960s, this principle was a consequence of the necessity to adopt a legal regime that distinguishes outer space from air space, over which each state is sovereign. The freedom of use recognized by the adoption of the OST in 1967 implies the freedom of Mars occupancy as a use of planet Mars. Indeed, outer space occupation is a legal use recognized, allowed, and sometimes encouraged among several provisions of the Treaty. At this point of our reasoning, Mars colonization is legal.

However, the freedom of use is granted to all states without any discrimination. To guarantee the proper use by each state the freedom of use cannot be absolute. The first limit is to be found in Article II of the OST: 'celestial bodies [are] not subject to national appropriation by claim of sovereignty, by means of use or occupation, or by any other means'. The first outcome of this provision of the Treaty is that the colonizing state cannot use the fact it is occupying some Martian ground to declare sovereignty over it or declare any area as entirely a part of its territory. Article II prevents states from constituting any property or sovereignty title on the occupied Martian lands. During the eighteenth century, unoccupied territories and sometimes occupied territories were considered as *Terra Nullius* by European empires in desire of territorial expansion in order to annex those newly discovered grounds. In our introduction, we gave a sociological and political definition of colonization, but classically, a legal definition of it would mean territorial appropriation of a land. This newly occupied place will become part of the national territory of the colonizing state, which will exercise exclusive sovereign rights over it and apply its law to people living there.

The OST's Article II removed any state's freedom to annex Mars. Outer space colonization cannot hold the same legal definition as classical colonization. Both common and science fiction uses of the word have resulted in its application to human expansion in outer space. But outer space colonization implies an alteration of the legal meaning of the term.

The meaning of Article II prevents sovereign appropriation of Mars by any means, but does the wording of the provisions imply that occupation always constitutes a national appropriation, therefore it is forbidden? Article II mentions several ways of forbidden national appropriation among which is use. Use cannot constantly be constitutive of national appropriation because its freedom is guaranteed by Article I of the OST. Thenceforth, the debate is open concerning definitive occupation. Some other provisions of the OST allow states to install bases, habitats on and below the surface of Mars. Therefore, we can argue that definitive occupation of Mars is internationally legal (under conditions) in order to guarantee the same possibilities of use to all states.

Even though we established the international legality of space colonization, Martian occupation can be perceived as form of appropriation that has no foundation or consent in the international community to proceed. But states are not bound by international community consent, as the US law on space resources exploitation has already illustrated. Some states conducting space colonization projects can find interpretations in the Treaty to justify colonization, however, the OST authorizes states physically to occupy celestial body surfaces. Colonization is then a form of appropriation that is authorized.

The prohibition of celestial bodies appropriation by commercial exploitation is more questionable.

2.1.2 The legality of Mars' mining

Space mining consists of the commercial use of space resources by some entity technically able to collect those resources *in situ*. This complicated task could be assigned to those settling on Mars. Once again, the liberty of use of outer space implies commercial use of it, as is the case for commercial satellite services. And here again, facing this liberty stands the non-appropriation principle. Concerning space resources, two interpretations of treaties can be found among the scholars. The one (Jenks 1965, 201; Lachs 1984, 217; De Man 2006, 305) considers resource exploitation as an appropriation, thereby totally forbidden by the OST. The second (Hobe and De Man 2017) is majoritarian and recognizes that the wording of the Article II is imprecise and does not expressly prohibit space resource exploitation.

The Moon Agreement, in Article 11.1, recognizes that space resources are the common heritage of mankind and can be exploited. Therefore, states parties to the treaty shall implement an international regime of exploitation guaranteeing in particular the rational exploitation of the resources and a distribution of the benefits to developing countries. This provision resulted in the failure of the Moon Agreement since the United States and Luxembourg have both adopted national laws allowing their citizens and companies to own space resources they have extracted and brought back on Earth.

In conclusion, Mars mining is theoretically allowed according to international space law by the Moon Agreement. If the colonizing state is a party to the Moon Agreement, it should adopt and respect the expected international regime. If the colonizing state has adopted a national legislation, its nationals may exploit Mars, although a debate is beginning concerning the international legality of such national laws regarding the provisions of the OST. If the colonizing state has not adopted a national space law, they remain is the grey zone emanating from the OST's Article II.

2.1.3 The legality of Mars' military colonization

Mars' military colonization is defined as the deployment of military personnel charged with occupying its ground with a defence-related mission. This hypothesis is quite realistic as astronauts are, for now, officials serving their states, as are military personnel (and indeed several hold a military rank). Furthermore, military objectives are at the basis of the development of space activities during Cold War (Nicui 1993; Achilleas 2015, 401). To prevent an extension of the Cold War to outer space, States parties to the treaty partially demilitarized it. According to the Article IV of the OST: 'celestial bodies shall be used by all states party to the Treaty exclusively for peaceful purposes'. This means states cannot send military personnel on Mars to conduct military activities even if such activities are non-aggressive. To conclude, despite the fact that the Treaty allows military forces to carry out scientific activities on the red planet, Mars' military colonization is forbidden.

2.2 The legal conditions of Mars' colonization: Respect of international space law provisions

States willing to colonize Mars are bound by all international space law provisions applicable to them and to any human activity conducted in outer space. States are only bound by the treaties they have ratified or are deemed law by international custom. This last element, regarding space law, concerns, for instance, the freedom of use and the non-appropriation principle. Those two rules apply to all states, even those who have not ratified the OST.

We would prefer that in the case of colonizing Mars, the colonizing state should follow the principle of the 'common heritage of mankind'. Unfortunately, none of the spacefaring nations have ratified the Moon Agreement. Not one of the states preparing projects of colonization are compelled to follow this desirable principle.

2.2.1 Exclusively use personal jurisdiction

Traditionally, jurisdiction is the tool by which a state applies its law. It is territorial and exercised at the exclusive will of the state. But a metropolis basing its jurisdiction over its Martian colony on territorial jurisdiction would violate the non-appropriation principle by recognizing sovereignty rights over the land itself. That is why the colonizing states must only use personal jurisdiction over their Martian colonies. Personal jurisdiction derives from the principle of space objects registration (OST Article VIII). State parties to the Treaty must register all space objects they intend to send to Mars. Space objects include every part of the launch, including the bases, the habitats, the rovers, and the astronauts themselves. Over those, the state of registry maintains its (personal only) jurisdiction and apply its law to them.

2.2.2 Protect the settlers

Article V of the OST recognizes that all astronauts are considered envoys of mankind and require a protective regime in case of situations of distress, but the provisions of the Treaty remain silent regarding any definition of the notion of astronaut. Certainly we can apply the astronaut regime to all humans in space for professional reasons, of non-commercial and non-private nature. This should be the case at the start of occupation of Mars, in order to install the colony, but once a settlement is ready and more and more settlers come to Mars, they will not be there for such activities. Activities will be private, maybe commercial, and not conducted for only professional reasons (Gibbon Wakefield 1849). Here again, two doctrinal interpretations exist. We will not detail them here because both conclude that considering all settlers as astronauts or not, all must benefit from the protective regime offered to them by the treaty in case of situations of distress, no matter when this occurs: during launch, their time on Mars, or their return to Earth (von der Dunk and Goh 2006; Achilleas 2007, 145).

2.2.3 The environmental obligations: The hypothesis of terraforming

In the preface of his *Mars Trilogy*, Kim Stanley Robinson defined Mars terraforming as the process of substantially modifying its surface in order to make it suitable for equipment-free human life (Robinson 2021, preface). Due to Article IX of the OST: 'States Parties to the Treaty shall pursue studies of ... celestial bodies and conduct exploration of them so as to avoid their harmful contamination.' Thus, states cannot proceed to Mars terraforming because it would constitute a harmful contamination of a celestial body, internationally prohibited. Moreover, as celestial bodies are not subject to appropriation and are free to use by all states, terraforming Mars would deprive other states of the natural state of the planet. Nevertheless, this obstacle can be overcome. The environmental protection established by Article IX aims to protect and preserve the correct conduct of scientific *studies* and exploration. This means that once the complete scientific study of Mars is complete, this article would not pose an obstruction anymore. Furthermore, a terraforming plan agreed at an international level, including all states or as many as possible, would be a good way to outshine a unilateral terraforming of Mars. But problems remain: there is no recognized authority that could declare Mars out of scientific interest, there is no international space agency leading international space programmes on behalf of all states.

2.2.4 The cooperation obligations

According to the provisions of the OST, cooperation, collaboration, and mutual assistance between states must lead all space activities. Whether it is about colonization or not, 'the use of ... celestial bodies shall be carried out for the benefit and the interests of all countries, irrespective of their degree of economic or scientific degree of development'. Moreover, when colonizing Mars, states are obliged by Article XII to inform the Secretary General of the United Nations, the scientific community, and people about the nature and the results of their activity and the location(s) of the colony. The principle and spirit of cooperation should form the bedrock of space activities and not be forgotten, particularly as space programmes conducted by private enterprise begin to loom on the horizon.

3. The Legal Problems of a Long-duration Martian Colony

There are two main categories of legal issues raised by a long-duration Martian colony : some concern the Martian base itself, others concern the relations with the earthian metropolis.

3.1 Legal issues within the Martian base itself

A lot of legal issues may arise, but we will concentrate on the main ones we can nowadays imagine will : the organization of the colony, its habitats, the use of resources, the progressive development of local institutions and the relationships with foreigners.

3.1.1 The legal organization of the colony

The law of each Martian colony depends on its instigators. If the entity proceeding to colonization is a single state or its nationals (e.g. NASA as representative of the US or SpaceX as its national), the organization of the colony and its applicable law will entirely belong to US law. This is the direct consequence of the personal jurisdiction granted by Article VIII to the state of registry of the colony, its equipment, and its people. The only limit to the organizational liberty of the colonizing state is that the legal arrangement of its colony must respect the OST provisions. This is also applicable if a private company is the colonizing body. For all these reasons, in the case of unilateral colonization, the organization of the colony only depends on the colonizing state. On the contrary, if the colonization of Mars is conducted by several states in international cooperation, the organization of the colony must be negotiated between those states in order to preserve each state's interests as well as an international interest, and hopefully the interest in space itself too.

This is the case concerning the ISS. In January 1998, the United States, Japan, Russia, Canada, and the European Space Agency (the last representing all its member states) adopted the International Space Station Intergovernmental Agreement (ISS IGA). From its first Article, the ISS IGA exposes the objectives of the ISS to develop an international pacific programme of permanent manned occupation of outer space. According to the provisions of the Agreement and the OST, each state conserves its jurisdiction and its property over the module it provides and registers. When sliding from module to module, an astronaut on board the ISS actually slides from one jurisdiction to another. The law of the state of registry of each module applies within that module. But the human element remains the most contentious. Article 11 of the

ISS IGA forecasts the Crew Code of Conduct. Once inside the ISS, all crew members, regardless of their nationality, must respect its provisions, agreed by all participating states. Crew members must take into account the multicultural and international nature of the entire crew in their personal conduct. The authority inside the ISS comes from the chief crew member, himself under the ISS ground authority.

In case of misconduct, various sanctions are prescribed: from an oral or written warning to the definitive expulsion from the ISS. The most interesting aspect concerns penal jurisdiction, dealt with creatively by the authors of the IGA against the mosaic of jurisdiction due to the registration tool. In case of a crime committed by one crew member against another of a different nationality, only the state of nationality of the alleged offender will retain jurisdiction to judge the affair. This was initially designed to prevent any Russian or US astronaut being judged by the rival state. However, in case of agreement or in case of lack of sufficient guarantees that the case will be correctly judged, the victim's state of nationality becomes the judge. The ISS could be a good starting point for the drafting of the legal organization of a Martian colony that would fit both the cooperation and common interest principles at the core of international space law.

3.1.2 The legal regimes of the habitats

Because of the inherent risk of national appropriation, the permanent occupation of celestial bodies is strictly constrained by the OST and the Moon Agreement provisions. According to the freedom of use (OST Article I) and Article 8 of the Moon Agreement, colonizing states are free to choose the site they want to install their colony above and under the surface of the red planet. Article 11.3 offers a large choice of architecture for the future colony in which the settlers can install themselves, their vehicles, their materials, their stations, their installations, or their equipment, including separate structures connected on or under the surface of Mars. Once several states colonize Mars, some won't be free to install their equipment wherever they want because of the existing presence of foreign colonies. Those prior installations must be arranged in order not to prevent the other state from accessing part of Mars and pursuing their own Martian activities. That is why each colony cannot be too extended and cannot isolate complete areas of the planet. To conclude, the habitats themselves must be accessible to foreign states. According to the Article XII of the OST, '[a]ll stations, installations, equipment and space vehicles on ... celestial bodies shall be open to representatives of other States Parties to the Treaty on a basis of reciprocity'.

3.1.3 The local use of Martian resources

What we mean by local use is direct use of Martian resources exclusively for the everyday needs of the settlers; we must distinguish the local use from the commercial exploitation of those materials. As long as the needed resources are renewable or endless, the total liberty of their utilization is protected by the freedom of use (OST Article I). This includes solar or cosmic rays, abundant minerals, or anything else without any scientific value. Article 9.2 of the Moon Agreement confirms that only the resources needed to sustain the colony can be used. The possibility to build bases under the surface of Mars outlines the fact that Martian resources can also be mined in order to protect the colony itself from cosmic rays.

3.1.4 The progressive development of institutions and of a complex legal system

At first, the colony must be materially and technically installed by a small and professional astronaut crew. The beginnings of the colony will be inorganic. This means no legal authority except that of the chef crew member will be established. Once the pre-installation is completed, the colonization process will take place via the regular arrival of individuals, groups, and families. They carry with them legal needs, as with any human developed society. Mankind is a source of conflicts, outrages, and disputes but humans also need organiation, order, law, and administration. But the colonizing state cannot decree that its earthly law will apply in the colony as it applies on Earth. This is because first, this would include property rights on the Martian ground or an implicit declaration of sovereignty over the Martian ground. The relevant state will have to introduce and develop political institutions elaborating a specific legal order inside the colony, as has already occurred in past terrestrial colonization activities.

These incremental institutional additions will progressively develop the legal consistency of the colony, just as cells grow into an embryo. The difficulty arises when it comes to predicting or prescribing the order in which each institution must be added. Several parameters such as distance, climate, and the uniqueness of this colonization prevent us from applying the same governance to metropolitan lands and to a Martian colony. In other words, the legal development of a colony is aleatory and opportunistic. Nevertheless, we can identity general steps in this construction. The first stage is the elaboration of a civil administration. Then, the setting up of a judge, courts, tribunals, etc. Each new institution removes some powers from the previous ones in order to restrict the concentration of such powers.

3.1.5 The legal relationship between Martian foreigners

As long as all Martians belong to the same state of registration, the legal relationships between them are unaffected by the lack of territorial jurisdiction. An eventual dispute will be regulated according to the law of their state of registration if it is a unilateral colonization or according to the rules decided by all states parties involved in case of international colonization. But once several states or entities colonize Mars, foreign settlers will interact. In case of dispute, which law will solve the conflict? When a legal conflict occurs between two foreign nationals on Earth, it is in the great majority of cases the law of the state where the facts occurred that will apply. But on Mars, there is no territorial jurisdiction. Positive space law does not envision this issue. At best, states can proceed to cooperative consultations provided by the OST. At worst, a personal and minor conflict can become an international dispute, involving not just the settlers but their respective states. For now, the only answer to this issue under positive space law would be a unique unilateral colonization or a unique international one.

3.2 The legal relationship between Mars and Earth

The following issues can be perceived as bold and prospective but they are based on observations made in colonial law and political history.

3.2.1 The legal status of the colony solely in the hands of the Earth State

We already explained, the state of registry retains its jurisdiction over the Martian colony. As long as terrestrial space law remains in force, this creates an asymmetrical link between Earth and the colony. This link gives to the Earth state the power to choose the nature of its relationship with the colony. There are several options concerning the regime linking a state and its colony over the past transposed to Mars colonization.

The first option is the subjugation regime, held in past colonization of France, Spain, and Portugal where representatives of the colonies were only admitted to their parliaments during the nineteenth century. The main characteristic of this regime is that the colonial populations are excluded from the elaboration of the law that will apply to them. On Mars, this regime in encouraged by the principle of the state's jurisdiction retained over the colony in eternity. Space law only prevents the subjugation of outer space itself and not humans living there.

The second option is the assimilation regime, by which the same legislation directs all parts of territory without any difference. Once a law applying to the state is adopted, this law fully applies automatically to the colony. We have already explained that this is impossible without recognition of sovereignty, even implicit sovereignty, over the Martian surface. However, it is still possible to proceed to this assimilation by including in each law a *mutatis mutandis* provision, for instance to exclude land rights on Mars. The final option is the self-governing regime. This regime was preferred by the United Kingdom. In this regime, colonies have their own representative and legislative institutions practising their own law themselves whereas they respect fundamental principles decided by the Metropolis. On Mars, these principles could be the OST provisions.

3.2.2 The hypothesis of a colony's Independence Day

Whatever the political regime chosen, the attitude of the Metropolis, the gap between the head and the body can grow inside the colony, raising a desire for legal independence. The risk of a unilateral declaration of independence is intrinsic to inegalitarian dyad colony metropolis. This issues are not only a matter of law. Turgot compared colonies to fruit: once maturity is reached, they do not remain on the tree. Generation after generation, the people of Mars will become the Martian people. What about this new people's right to self-determination? This right is guaranteed to all people by the first Article of the International Covenant on Civil and Political Rights: 'By virtue of that right they freely determine their political status and freely pursue their economic, social and cultural development.' But this principle does not enshrine the right to independence. It only protects one state from a foreign interference in domestic affairs, it protects the right of an already existing state to autonomy. Some resolutions of the United Nations General Assembly tend to create a right to independence in the case of colonization by a foreign state. But the settlers are not occupied by a foreign nation; they are within their state, they enrolled voluntarily. Moreover, resolutions from the United Nations General Assembly are non-binding. But as years and generations pass, this question will arise. De Tocqueville wrote that each generation is a new people.

3.2.3 The relevance of a maintained unequal link with the colony

The problematic question that develops is of the relevance of maintaining a link between Earth and a Martian colony. In our opinion, it is not appropriate to maintain unequal relations with colonies that have no natural bound with our primary planet. This naturalistic condition is based on the gravitational relationships between colonies and metropolises. It is logical that a space station orbiting around the Earth must be subject to Earth authority because this station depends on Earth for the position it is occupying, the frequencies it is using, and certainly for its supplies. On the other hand, a wandering station should not be subject to a state authority, especially if it drifts indefinitely out of our solar system. For a colonized solar system, authority should come from a central point. Mars should rule orbital stations around the red planet. And maybe, a common organization should govern all occupied planets of the solar system on an equal basis of participation to the elaboration of the common rules. The necessity of (and preference for) central authorities was already described by Shakespeare in *Troilus and Cressida*:

> And therefore is the glorious planet Sol
> In noble eminence enthroned and sphered
> Amidst the other; whose medicinable eye
> Corrects the ill aspects of planets evil,
> And posts, like the commandment of a king,
> Sans cheque to good and bad: but when the planets
> In evil mixture to disorder wander,
> What plagues and what portents! what mutiny!

3.2.4 The opportunity to create a state free of territory

If, in the very long term, subordination lead to the unilateral independence of the colony, there will arise the creation of a new state, a Martian state. Decolonization in the past often ended by the succession of states. At present, a state is composed of three elements: a people, a territory, and sovereignty over those two elements. If the non-appropriation principle in space law stays in force until that moment, will the Martian people be able to constitute a state without any owned territory? This new kind of state will be exclusively sovereign over its people by personal jurisdiction, compensating for the lack of territorial jurisdiction. Its law applies by the nationality criteria to people living on a territory. This state would have a territory to develop its operation but it would not own it. The jurisdiction of such a state would be total. The Martian state would have no territory, no borders, its national could never escape its jurisdiction unless they flee back to Earth inside the territory of a classical state. The only geographical border of the Martian state would be the borders of other classical states.

3.2.5 The end of actual space law?

The rupture of any subordination to Earth will probably sweep the application of positive space law, drafted on Earth, by Earthlings, assuring them the control of the Martian colony. However, the abandoned principles should leave a mark on the new ones: the spirit of theses law will remain Earth-centric, at least in its legal terms.

Let us dream bigger, when the entire solar system, or even the galaxy, will be colonized by humans: what will the legal scheme of the solar system or of the Milky Way be? Can we imagine, as it was the case on Earth, states acquiring their independence from the colonizing state, and then joining an international or interplanetary organization? At some point, will it be possible to see an interplanetary federalism and, maybe, an interstellar one? Despite its prospective and futuristic ambitions, positive space law will have to face a definitive and autonomous expansion of mankind. However, positive space law carries salutary principles useful to a peaceful and international colonization of the cosmos. Right now, this law is under attack by some authors (for instance, Gruner 2004), justifying the settlement on the ancient doctrine 'first come, first served'. According to them, the non-appropriation principle is an impediment to a commercial exponential growth of space activities.

4. Conclusion

Although these are new interpretations of space law, we must emphasize that by the past, colonization and decolonization, by renewing international law, contributed to progress in terms of international cooperation (Schmidt 1950). The legal heritage of outer space, humanist, pacifist, and internationalist, will perhaps survive. Space law also contributed to the development of international general law (Cheng 1986; 1997.

Carl Siger (Siger 1907) used to say that 'colonies could, at some point, be the safety valves of modern society. Even if this characteristic were the only one of colonization, it would be huge.' For now, space colonization has only been a dream but its impact is no less impressive.

References

Achilleas, P. 'L'astronaute et le droit international' in *L'adaptation du droit de l'espace à ses nouveaux défis—Mélanges en l'honneur de Simone Courteix* (Paris: Pedone, 2007).

Achilleas, P. 'La guerre des étoiles—de la science-fiction à la science juridique' in *Actes du colloque Les lois de la guerre* (Paris: Institut Universitaire Varenne, 2015) 57–80.

Cheng, B. 'The Contribution of Air and Space Law to the Development of International Law' (1986) 39(1) *Current Legal Problem* 181–6.

Cheng, B. *Studies in International Space Law* (Oxford: Clarendon Press, 1997).

Combacau, J. and Sur, S. *Droit international public* (8th edn, Paris: Montchrestien, 2008).

De Man, P. *Exclusive Use in an Inclusive Environment: The Meaning of the Non-Appropriation Principle for Space Resource Exploitation* (New York: Springer, 2016).

Gibbon Wakefield, E. *A View of the Art of Colonization* (John W. Parker, 1849).

Gruner, B. C. 'A New Hope for International Space Law: Incorporating Nineteenth Century First Possession Principles Into the 1967 Space Treaty for the Colonization of Outer Space in the Twenty-first Century' (2005) 35 *Seton Hall Law Review* 299–357.

Hobe, S. 'Historical Background' in S. Hobe, Schmidt-Tedd, and Schrogl (eds), *Cologne Commentary on Space Law. Volume I—Outer Space Treaty* (Cologne: Carl Heymanns Verlag, 2009) 12–17.

Hobe, S. and De Man, P. 'National Appropriation of Outer Space and State Jurisdiction to Regulate the Exploitation, Exploration and Utilization of Space Resources' (2017) 66(3) *Zeitschrift für Luft- und Weltraumrecht*, 462.

Jenks, W. *Space Law* (Stevens & Sons Ltd., 1965)

Lachs, M. in C. Q. Christol. 'Article 2 of the 1967 Principles Treaty Revisited' (1984) *Annals of Air and Space Law*.

Martin, P. 'Les définitions absentes du droit de l'espace' (1992) 182(2) *Revue française de droit aérien et spatial* 106.

Mateesco-Matte, N. 'The Moon Agreement: What Future?' in *Annuaire de droit maritime et aéro-spatial—Mélanges en l'honneur du Professeur* Mircea Mateesco-Matte (Paris: Pedone, 1993) 346–59.

Nicui, M. I. 'Considérations sur le droit international spatial' in *Annuaire de droit maritime et aéro-spatial—Etudes en homage au Prof. Mircea Mateesco-Matte* (Paris: Pedone, 1993).

Pop, V. 'A Celestial Body is a Celestial Body is a Celestial Body ...' (2001) *IISL Proceedings*, 100–10.

Robinson, K. S. *La trilogie martienne* (Paris: Omnibus, 2012).

Schmitt, C. *The Nomos of the Earth in the International Law of Jus Publicum Europaeum* (Cologn, 1950).

Siger, C. *Essai sur la colonisation* (Paris, 1907), quoted in A. Césaire, *Discours sur le colonialisme* (Présence africaine, 1955).

von der Dunk, F. and Goh, G. M. 'The Article V of the Outer Space Treaty' in S. Hobe, Schmidt-Tedd, and Schrogl (eds), *Cologne Commentary on Space Law. Volume I— Outer Space Treaty* (Cologne: Carl Heymanns Verlag, 2009).

9

Brightening the skies

Institutional solutions to the societal and geopolitical risks of space expansionism

Ian A. Crawford

There is only one rational way in which states coexisting with other states can emerge from the lawless condition of pure warfare ... they must renounce their savage and lawless freedom, adapt themselves to public coercive laws, and thus form an international state (civitas gentium), which would necessarily continue to grow until it embraced all the peoples of the earth.

(Immanuel Kant 1795)

In pursuing this inquiry, we must bear in mind that we are not to confine our view to the present period, but to look forward to remote futurity. Constitutions of civil government are not to be framed upon a calculation of existing exigencies, but upon a combination of these with the probable exigencies of the ages.

(Alexander Hamilton 1788a)

1. Introduction

In his important and thought-provoking book, *Dark Skies: Space Expansionism, Planetary Geopolitics, and the Ends of Humanity*, the international relations scholar Daniel Deudney (2020) provides a carefully argued critique of what he sees as overly optimistic visions of a human future in space. He coins the term 'space expansionism' for the popular view that an expansion of human activities in space is both desirable and inevitable, and he argues that some aspects of this agenda are so dangerous that they must be avoided by future generations. In this chapter, I will first briefly summarize Deudney's position as I understand it, and then go on to argue that at least some of his pessimism is misplaced and that appropriate institutional developments will be able to mitigate many of the risks that he foresees. In addition, and of particular relevance for the topic of this book, I will argue that some of these institutional developments will also help mitigate the risks of extraterrestrial tyranny identified by Cockell (2009, 2010).

2. The Three 'Space Expansionisms' of Daniel Deudney

Within the overall ideology of 'space expansionism', Deudney identifies three quasi-independent strands, which he terms 'military space expansionism', 'habitat space expansionism', and 'planetary security space expansionism' (Deudney 2020, 30). Fairly or otherwise, Deudney names these three strands after Wernher von Braun, Konstantin Tsiolkovsky, and Arthur C. Clarke and Carl Sagan, respectively. I will avoid that nomenclature here because I don't think the views and legacies of these very different advocates for a human future in space can be so easily pigeon-holed. However, I do agree with Deudney that von Braun's early involvement with Nazi Germany's war effort, and later advocacy of US military space projects, is deeply problematic and that this should not be overlooked, whereas Clarke and Sagan both had much more peaceable and inclusive visions that deserve to be celebrated.

Deudney's treatment of 'military space expansionism' is, at least to my mind, uncontroversial. We can all recognize the historical truth that the development of rockets and other space capabilities has been intimately connected with military activities, and that space technologies continue to have many military applications. The recent creation of a US 'Space Force' (USSF 2021) and the declaration that space is now viewed as a 'war-fighting domain' (e.g. AFSC 2019) clearly show that the dangers of space militarization are all too real. Deudney is right to draw attention to them and to call for their mitigation.

Similarly, Deudney's advocacy of 'planetary security space expansionism', calling as it does for the increasing use of space technologies to monitor compliance with international environmental and arms control agreements, and, if necessary, to protect Earth from asteroid impacts, appears sagacious. Moreover, although science isn't explicitly included in the names of any of the three 'space expansionisms', Deudney places space science in this category and argues that it should be greatly expanded. Finally, and this will become important later, Deudney places what he calls 'whole Earth identity formation' in this planetary security category (Deudney 2020, 241, 253–4). This refers to the thesis, proposed by multiple authors over the years (and discussed further below), that by increasing the ease of global communications, and by providing images of the Earth in its cosmic context, space activities may help trigger a greater sense of global identity which could help reduce international tensions and thus enhance prospects for global peace and security. All of these 'planetary security' applications of space technology are beneficial, and 'expansion' of these capabilities would seem to be positively desirable.

From the point of view of space advocates (and here I include myself), the most controversial aspect of Deudney's analysis is his negative stance towards what he calls 'habitat space expansionism'. Within this category Deudney includes schemes for human colonization of other planets, the mining of moons and asteroids for raw materials, and the construction of large-scale infrastructures in space (Deudney 2020, 186). He concludes that this will be as dangerous for the future of humanity as military space expansionism, and perhaps more dangerous in the long term. So dangerous, in fact, that humanity should refrain from undertaking these activities even if, as seems likely, we develop the technical capability to pursue them. Among the potential risks Deudney identifies are the possibility of conflict arising out of competition for space resources (exacerbating the risks of near-term military space expansionism); the possibility of armed conflict occurring between human colonies, and between these colonies and the Earth; the deliberate alteration of asteroid orbits (raising the spectre of asteroid impacts being used as weapons of mass destruction); and the distant possibility that human space colonists might evolve into post-human forms that would have little in common with humanity and might even cause our extinction.

It is noteworthy that Deudney's main reasons for opposing habitat space expansionism are based on geopolitical grounds and result from a lack of confidence that humanity will be able to manage space activities responsibly. As he notes, any expansion into new places, or development of new technologies, carries geopolitical risks, and outer space is no exception (Deudney 2020, 292):

> For geopolitical theory the question of whether more extensive space or more capable machines are desirable or disastrous depends on whether such enlargements are matched in their scope and powers by configurations of restraining institutions... But it provides no guarantee that humans will rise to the occasion to produce and sustain the political restraints vital to avoid disasters.

From this perspective, the question of whether humanity can safely expand into space essentially boils down to whether we can construct sufficiently robust *institutions* to govern human space activities. However, a moment's reflection will show that essentially the same considerations apply to ensuring human well-being on Earth, regardless of whether we expand into space or not. Thus, the question of institution building for human space activities can be seen as just one aspect of developing stronger institutions of human governance that will be required to manage multiple existential threats facing humanity in the twenty-first century and beyond.

3. Planetary Governance in the Twenty-First Century

The world currently faces a number of serious problems that can only be addressed effectively at a global level. These include (i) global environmental pollution, including, but not limited to, anthropogenic climate change; (ii) habitat destruction and biodiversity loss; (iii) the continuing risk of military conflict between major powers and the concomitant risk of a civilization-destroying nuclear war; (iv) threats arising from insufficiently regulated advanced technologies, including biotechnology, nanotechnology, and artificial intelligence; (v) global-scale natural threats such as pandemics and, on a longer timescale, mega-volcanoes and asteroid impacts; (vi) long-term development challenges, including the satisfaction of aspirations for higher living standards, for a growing world population; and (vii) inefficient, and often irresponsible, management of the global commons.

These problems, while widely recognized, are difficult to address in an anarchic international environment where independent nation-states act as judges in their own cause, and the perceived self-interests of these independent sovereignties are often in conflict. It seems likely that many of these global problems and associated risks will worsen as the twenty-first century progresses, and that much stronger institutions of global governance will be required to manage them effectively. Indeed, the creation of the United Nations (UN) in 1945 demonstrated that the international community recognized the desirability of limited supra-national governance, at least in the realm of global peace and security, in the immediate aftermath of the Second World War. Moreover, the adoption of the Universal Declaration of Human Rights by the UN General Assembly in 1948, and the expansion of the UN system to include numerous specialized agencies and programmes (e.g. UNICEF, UNESCO, UNDP, UNEP, UNHCR, WFP, etc.),[1] illustrates a broad consensus that coordinated global action is desirable in multiple areas to work for 'peace, dignity and equality on a healthy planet' (UN 2021).

Unfortunately, the UN, like the League of Nations before it, is predicated on the concept of nation-state sovereignty, and this greatly limits its effectiveness. Indeed, in practice, the UN is just one more forum where nation-states are free to exercise their own sovereignty in their own perceived self-interests. As Fremont Rider predicted just a year after its creation, the UN has, like its predecessor the League of Nations, been treated by national governments 'as merely another piece to be moved about on the international board in the game for national power—and as not a very important piece at that' (Rider 1946, 2). These considerations imply that much stronger systems of global governance will be required to deal effectively with global problems. Reves (1946, 279) put it well in the context of nuclear weapons when he wrote:

> It follows that the ultimate source of danger is not atomic energy but the sovereign nation-state. The problem is not technical, it is purely political.

This conclusion was reiterated by Derek Heater (1996, 205), who argued explicitly that the logic points in the direction of some form of planetary government:

> Individual states are at best powerless to prevent wars and environmental degradation, at worst they are the cause of these disasters. Only effective world government can protect mankind from these hazards.

Interestingly for the present discussion, in an earlier article Deudney (2018, 257) has himself considered humanity's current situation and argued that it:

> is marked by catastrophic and even existential threats stemming from ... nuclear war, climate change, and biotech pandemics, thus creating powerful *universal interests* that almost certainly require the erection of some version of substantial world government. (his emphasis)

There isn't room here to do justice to the extensive literature on the desirability, or otherwise, of establishing some form of world government, or the many different forms that such a government might take (for book-length treatments see Wynner and Lloyd 1944; Heater 1996; Baratta 2004; Leinen and Bummel 2018; Yunker 2018). My own view (e.g. Crawford 2015, 206–7) is that dealing effectively with planetary-scale problems will eventually require a federal system of planetary governance, constituted so as to implement the principle of subsidiarity on a global scale (where the federal world government would be responsible solely for global matters that cannot be addressed effectively at a local or national level).

I am aware, especially in a world where populist nationalism appears to be on the rise, that many people will instinctively object to the whole concept of a world government. Some of these objections may arise from an (in my view unhelpful and outdated) allegiance to the concept of nation-state sovereignty. Other objections, which need to be taken more seriously, will be based on legitimate concerns about preventing the rise of a global dictatorship. In principle, a federal system of world government could retain independent decision-making for nation-states at the national level, and so would be compatible with (limited) national identity, and, as in the case of the US Constitution, be based on balancing powers in such a way as to minimize the risk of totalitarianism (as discussed further in section 5 below). But, in any case, the sheer facts of increasing global interdependence, and worsening global existential risks, need to be addressed somehow. Those who object to the whole idea of a planetary government will need to explain how these risks might effectively be managed in some other way—bearing in mind that global institutional arrangements short of a world government, such as the League of Nations and the United Nations, have so far proven to be ineffective.

Of course, even if we agree that a suitably constituted world government would be desirable in principle, formidable political obstacles would need to be overcome to bring one into existence. An important reason for this is that despite our obvious common interests, the human race currently lacks a sufficiently strong sense of common identity to overcome allegiances to different nations, religions, and other forms of tribal identity that fall short of humanity as a whole. It seems likely that tribalism may be instinctive in *Homo sapiens*, possibly as a result of group selection during our evolutionary past (e.g. Wallace 1871, 313; Darwin 1874, 64; Wilson 1998; Wilson and Wilson 2007; Wilson 2012), and that this gets in the way of developing the kind of global institutions that the world increasingly needs. As Kwame Appiah (2006, xi) has put it:

> [t]he challenge, then, is to take minds and hearts formed over long millennia of living in local groups and equip them with ideas and institutions that allow us to live together as the global tribe we have become.

The importance of developing a common planetary perspective as a prerequisite for creating global institutions to deal with planetary scale problems has long been recognized in the international relations community (e.g. Morgenthau 1948; Herz 1962; Ward 1966). Significantly, Deudney (2019) has himself summarized the views of the 'realist' international relations scholar Hans Morgenthau as follows: 'humanity thus faces a tragic impasse: it needs a world state for security, but lacks a sufficiently thick sense of common identity both to make it possible and to prevent it from being threatening'. Because much of this discussion will use the US federal constitution as an example, it seems important to acknowledge Morgenthau's observation, made while arguing against the feasibility of a world government, that the American colonies had already developed a sense of common identity before the Constitutional Convention of 1787. As he put it, just as 'the community of the American people antedated the American State ... a world community must antedate a world state' (Morgen-thau 1948, 406). Morgenthau himself may have seen this as unrealistic but it is in this context that some aspects of 'space expansionism' may prove helpful.

Any society that is actively exploring the solar system can hardly fail to be aware that Earth is a very small and isolated planet when viewed in a cosmic context. Over the years, the social, cultural, psychological, and political importance of this perspec-tive has been noted by multiple authors (e.g. Clarke 1946, 1951; Hoyle 1950; Ward 1966; Sagan 1985, 1994; Poole 2008; White 2014; Deudney 2018b, 2020; Som 2019; Crawford 2021a). Sagan (1985, 280) articulated the core political implications in his science fiction novel *Contact*:

> Spaceflight, therefore, is subversive ... The nations that had instituted spaceflight had done so largely for nationalistic reasons; it was no small irony that almost everyone who entered space received a startling glimpse of a transnational perspective, of the Earth as one world.

Although Sagan's portrayal here is fictional, the psychological impact of seeing Earth from space appears real enough, and has been comprehensively documented by White (2014). In this context, it is especially notable that Deudney (2020, 241, 253–4) identifies 'whole Earth identity formation' as an important benefit of 'planetary secu-rity space expansionism'. Indeed, in an earlier work, Deudney (2018, 273–4) argued that the view of Earth from space has led to widening recognition of a 'practical geog-raphy of Planetary Earth' where 'the Earth as a whole is now a place' and that this 'type of Earth-place sensibility amounts to a kind of Earth nationalism'. It seems reason-able to suggest that if a sense of 'Earth nationalism', or what Barbara Ward (1966, 148) perhaps more felicitously termed 'a patriotism for the world itself', were to become suf-ficiently widespread it would imply a stronger sense of global identity. This, in turn, would help provide the psychological foundations on which the institutions of global governance might be built.[2]

Of course, even with a growing sense of planetary identity, stronger global politi-cal institutions will not emerge overnight, and this will be even more true of a federal world government. Rather, this must be an evolutionary process (e.g. Clark and Sohn 1960; Leinen and Bummel 2018; Yunker 2018; Bummel 2021), most likely including the gradual strengthening of the existing UN system (e.g. Weiss 2016; Lopez-Claros, Dahl, and Groff 2020). The point here is that because space activities can help pro-vide a supporting cosmic perspective on human affairs (e.g. White 2014; Crawford 2021a), they may act as a catalyst for the kind of 'Copernican revolution' in human-ity's self-image that Reves (1946, 26–9) argued would be a necessary precursor for the formation of a planetary government.

4. Interplanetary Governance

As discussed above, Deudney's (2020) objections to 'habitat space expansionism' rest largely on geopolitical concerns that conflict may arise if humanity expands out into the solar system, and that such conflict may pose existential risks to the survival of the human race. The risk of asteroids being used as weapons is of particular concern because this could indeed result in an extinction-level catastrophe (e.g. Sagan and Ostro 1994; Crawford and Baxter 2015). Moreover, there are multiple other reasons, ranging from the peaceable and efficient management of space resources to the implications of contact with extraterrestrial life, for wanting to ensure that a human expansion into the solar system is well-ordered and properly governed (e.g. Crawford 2021b). I therefore agree with Deudney that an anarchic expansion into space would be fraught with danger and must be avoided.

However, we already live in an anarchic geopolitical situation on Earth, facing existential risks of various kinds. We need to find ways to mitigate these risks, whether we venture out into space or not. The major threat facing humanity is therefore not space expansionism *per se* but the anarchic relationships between human societies, whether on Earth or in space. Deudney (2020, 368–9) recognizes this, when he writes:

Humanity's problem is not that it is stuck on Earth but that it is stuck in an inherited, fragmented, stratified, and violence-prone international system. If humanity is unable to overcome anarchy to establish mutual restraints and pursue mutually beneficial problem-solving on Earth, where so many factors are supportive, it is more unlikely to be able to do so in geopolitically malefic solar space … Humans and their institutions are not—and are not likely to become—capable enough to meet daunting solar space governance challenges.

The logic here is that if, *contra* Deudney, we are able to solve these political problems on Earth, which we will have to do in order to ensure our long-term survival, then we can also solve them in space. As discussed above, ultimately the solution to geopolitical anarchy on Earth will be global government, and it follows that the solution to interplanetary anarchy will be some kind of interplanetary government. To be fair, Deudney (2020, 352) does recognize this as a potential solution to the geopolitical problem of space expansionism, at least in principle, but then dismisses it as impractical (based on, as it seems to me, a rather forced analogy with failed attempts at federation within the British Empire). To my mind, the relationship between world and interplanetary government is the central aspect of this whole discussion and merits a much deeper analysis. For now, I'll just observe that the evolution of institutions for both world and off-world governance is likely to play out over a similar timeframe, say the remainder of the twenty-first century, raising the possibility of mutually supportive synergies between them.

Just as a planetary government for Earth will not emerge overnight but will be the result of a long evolutionary process, the same will be true of governance in space. Elsewhere (e.g. Crawford 1995, 2021b), I have identified a hierarchy of institutional proposals for the future governance of space activities. The nearest term of these proposals would involve strengthening UN oversight of space activities by building on the existing provisions of the 1967 Outer Space Treaty[3] and enhancing the roles of UN Office of Outer Space Affairs (UNOOSA)[4] and the UN General Assembly's Committee on the Peaceful Uses of Outer Space (UNCOPUOS).[5] Deudney (2020, 241–6, 372) favours this approach as part of what he calls a 'whole Earth security program', and he advocates expanding existing international agreements to include space traffic control and space debris mitigation. Although Deudney's discussion implies that he doesn't believe measures of this kind will be sufficiently strong to cope with the larger risks he associates with habitat space expansionism, it is possible to envisage stronger institutional responses evolving from the present UN system that may be helpful in this respect. One possibility, building on a suggestion by the 1995 *Report of the Commission on Global Governance* (Carlsson and Ramphal 1995, 251–2), would be to repurpose the now defunct UN Trusteeship Council to oversee human activities in space.[6] This would elevate oversight of human activities in space to one of the six principal organs in the UN system, placing them on a par, in principle if not initially in practice, with the deliberations of the Security Council.

In addition, and not incompatible with this suggestion, one could imagine the creation of a dedicated world space agency under UN auspices. To my knowledge, this was first suggested by the British rocket engineer Val Cleaver, a decade before the dawn of the space age, when he suggested the creation of an Interplanetary Agency (IPA)[7] to facilitate human missions to the Moon and planets (Cleaver 1948).[8] In this pioneering article, Cleaver wrote:

> One can visualise, therefore, that the IPA might be an international organisation ... modelled along the lines of the American proposals for an International Atomic Energy Authority, or on UNESCO ... the whole project being sponsored by UNO, or (better still) by a World Government.

Interestingly, proposals for an international space agency briefly received serious political attention once the space age became a reality and were discussed, although not acted on, in the context of the US National Aeronautics and Space Act which led to the creation of NASA (Shepard 1958). At about the same time, a suggestion for a 'United Nations Outer Space Agency' was advanced by Clark and Sohn (1960) in the second edition of their book on UN reform, *World Peace through World Law*. A few years later, a 'United Nations Space Administration (UNSA)' features in Arthur C. Clarke's short story *The Secret of the Men in the Moon* (Clarke 1963). More recent proposals for the creation of an international space agency/authority, with or without an explicit linkage to the UN, have been made by Brown and Fabian (1975); Crawford (1981, 1995); Tronchetti (2009); Pinault (2015); Koch (2018); Plattard and Smith (2018); and McKenna (2020).

In the present context, we may note that expanding the role of the UN to include responsibility for outer space affairs, and other transnational domains, would be fully consistent with its evolution in the direction of a federal world government. We have seen that Cleaver (1948) had already intuited such a connection. A more explicit linkage between outer space affairs and world government has been made by the international relations scholar James Yunker (2007, 60–1) where he notes that a future world government might need a 'Ministry of External Development' to coordinate human activities in space and suggests (Yunker 2007, 87) that a world government might one day be required to protect Earth from extraterrestrial threats.

Ultimately, however, if habitat space expansionism proceeds to the point where human (and perhaps eventually post-human) colonies become established throughout the solar system, reliance on an Earth-centric governance structure may cease to be workable or desirable. Yet, for all the reasons that Deudney carefully articulates, unrestrained political freedom for space colonies would amount to interplanetary anarchy and would potentially add yet another layer of existential risk for humanity. We need only recall Kant's (1795, 113) dictum that 'the separate existence of many independent adjoining states ... is essentially a state of war, unless there is a federal union to prevent hostilities' to realize that Deudney's fears are well-founded. In the present context, it is interesting to note that similar concerns were also very much on the minds of the founders of US federalism as they worried about competition between the states in the, as yet uncolonized, interior of North America. As Hamilton (1788b) put it:

> In the wide field of Western territory, therefore, we perceive an ample theatre for hostile pretensions, without any umpire or common judge to interpose between the contending parties. To reason from the past to the future, we shall have good ground to apprehend that the sword would sometimes be appealed to as the arbiter of their differences.

The possibility of 'swords' being replaced by space-borne nuclear weapons, or redirected asteroids, does not bear thinking about. But note that Kant explicitly identified the solution to the risk of such inter-state conflict, namely the creation of a 'federal union', and this was indeed the successful solution adopted by the US Constitution.[9] Inter-state anarchy did not develop in the interior of the North American continent and, with one significant exception, wars between the states have been prevented.[10] By analogy, I have argued elsewhere (Crawford 2015) that a federal form of government would be ideally, and perhaps uniquely, suited to the task of maintaining interplanetary peace, essentially for the same reasons that it will also the most appropriate form of a future world government on Earth.

The main point is that federal forms of government are inherently expandable. In his *The Spirit of the Laws*, Montesquieu (1748, 131) defined federalism as a form of government where 'many political bodies consent to become citizens of the larger state that they want to form. It is a *society of societies* that make a new one, which can be enlarged by new associates that unite with it' (my emphasis). This property of expandability was clearly recognized by the founding fathers of US federalism. For example, Hamilton (1788c) stressed that a key advantage of the federal constitution was 'the ENLARGEMENT of the ORBIT within which such systems are to revolve, either in respect to the dimensions of a single State, or to the consolidation of several smaller States into one great Confederacy' (capitals in the original), and Madison (1788a) noted that the 'practical sphere' of republican government 'may be carried to a very great extent by a judicious modification and mixture of the *federal principle*' (his emphasis). It was this expandability that enabled a federal government, designed initially to ensure cooperation between thirteen former colonies on the Atlantic coast of North America, to expand across the continent.

There is no reason in principle why such a form of government could not expand across Planet Earth (see, e.g. the discussion by Pistone 2013) or, in the fullness of time, the solar system. In the latter context, note that the islands of Hawaii, admitted as a state of the United States in 1959, might just as well be a colony on Mars as far as the federal institutions are concerned. Moreover, by employing the principle of subsidiarity, federal forms of government are compatible with significant local independence and diversity. This is because matters pertaining solely to individual constituent states are dealt with by each individual state government. This is an especially valuable aspect of federalism when the responsibility of the federal government extends over a large and diverse area, whether this be a continent, a planet, or a planetary system. In an interplanetary context, the component 'states' might be individual colonies, or perhaps entire planets (e.g. Mars) on which there are multiple colonies, themselves grouped together in a federal structure.

It might be argued that the solar system is in some sense 'too big' for even a federal form of government to function, but this is easily discounted. In *Dark Skies*, Deudney (2020, 266, 302–3) introduces the important concept of 'effective distance', where the effective distance between two points is defined by the time it takes to communicate between them. In terms of effective distance, the whole solar system is today far smaller than the area occupied by the original 13 American colonies that devised the US Constitution in 1787: it takes at most 20 minutes to get a message from Earth to Mars, and only about five hours to get one to Pluto, whereas getting a message from New Hampshire to Georgia in 1787 might have taken weeks. Thus, if a federal form of government could function in North America in the eighteenth century, it will be able to function throughout the solar system for any future interplanetary society utilizing electromagnetic means of communication.

5. Federalism as a Preserver of Liberty

In addition to what he sees as an unacceptable risk of interplanetary conflict, Deudney (2020) is concerned that space expansionism will stimulate the development of tyrannical forms of government, both on Earth and in space. There are multiple passages in *Dark Skies* where he recognizes that the answer to international/interplanetary anarchy would be a world/interplanetary government of some kind, but he shies away from advocating this as a solution, fearing that any such government would necessarily become 'hierarchical' and totalitarian. Indeed, in *Dark Skies* the phrase 'hierarchical world government' is used as an apparent synonym for totalitarian world government at least a dozen times. I was puzzled by this usage because, as it seems to me, *all* forms of government are hierarchical to some extent, including non-totalitarian forms. It is true that Deudney (2020, 284) draws a distinction between 'hierarchical' and 'republican' governments, where the latter are characterized by democratic control and checks and balances in a way that the former are not. However, federal forms of government are republican in Deudney's sense because individual citizens are represented at each level (e.g. Miller 1987, 151), and the checks and balances between the various levels of a federal system of government minimize the risk of one level usurping totalitarian control. It is therefore at first sight difficult to understand why Deudney never seriously considers world/interplanetary *republican* government as a non-totalitarian solution to world/interplanetary anarchy.

Digging a bit deeper, however, this appears (at least in part) to be because Deudney is concerned that habitat space expansionism may cause a non-totalitarian (republican) world government, which he has elsewhere advocated as a means of addressing pressing global problems (e.g. Deudney 2018), to become totalitarian in a way that would not happen otherwise. This is because external threats (real or perceived) may drive originally republican forms of government towards totalitarianism. This danger was also recognized by the founders of US federalism. For example, consider Hamilton (1788d):

> Safety from external danger is the most powerful director of national conduct. Even the ardent love of liberty will, after a time, give way to its dictates... the continual effort and alarm attendant on a state of continual danger will compel nations the most attached to liberty to resort for repose and security to institutions which have a tendency to destroy their civil and political rights. To be more safe, they at length become willing to run the risk of being less free.

Deudney's argument is basically the same: if humans do not expand into space, there will be no (human-caused) external threats to Earth, and therefore no pressures on a future world government to become totalitarian. As he writes, 'barring substantial space colonization or threatening aliens, a world polity would be alone and thus not need to mobilize, concentrate, or employ violence capacity against outside threats' (Deudney 2020, 308). On the other hand, if 'humanity expands into solar space, world government on Earth ceases to be a universal government and becomes one of many world governments' (Deudney 2020, 353) with attendant prospects for inter-world conflict and domestic repression. However, the same solution adopted in eighteenth-century America is open to us here: if new space colonies are embedded within an interplanetary federal government *from the start*, they will not pose a totalitarian-inducing threat to Earth.

It also seems to me that Deudney's argument ignores the fact that any future 'world polity' will need to mobilize and concentrate *some* power to overcome the common threats that face humanity on our home planet, quite irrespective of whether there are any external threats to worry about. This is why many people fear that even if desirable in principle, there is a risk that a world government would eventually devolve into a global tyranny. To the extent that we will need to build institutions of global government capable of solving global problems, this is a nettle that will need to be grasped regardless of whether humanity expands into space or not.

If, despite his entreaties, human colonization of the solar system begins to occur, Deudney follows Cockell (2009, 2010, 2015) in arguing that the nature of the space environment will favour totalitarian forms of government within individual space colonies. For Deudney, this is therefore another argument against space expansionism. However, as Converse (2010, 109) succinctly and correctly notes in his study of the lessons of US federalism, 'the liberty of any given society of people depends, to a great degree, upon the institutions that exist, or they create, to protect it'. I have argued elsewhere (Crawford 2015) that the simplest institutional way to minimize the risk of space colonies sliding into totalitarianism would be to ensure that all such colonies are, from their foundation, embedded in a larger political framework that guarantees individual rights and liberties in a manner that local administrators would find hard to overturn.

The evolution of American federalism again provides a valuable historical example of how federal systems of government are robust against the encroachment of tyranny. Anti-federalists at the time certainly shared some of Deudney's concerns; for example, consider George Mason speaking at the Virginia ratifying convention on 4 June 1788 (Kaminski et al. 1990, 937):

> Is it to be supposed that one National Government will suit so extensive a country, embracing so many climates, and containing inhabitants so very different in manners, habits, and customs? It is ascertained by history that there never was [such] a Government, over a very extensive country, without destroying the liberties of the people... Is there a single example, on the face of the earth, to support a contrary opinion?

And Patrick Henry, responding to Madison's advocacy of the federal constitution, was even more apocalyptic on 24 June 1788, using words that seem almost to anticipate Deudney's deep-seated concerns (Kaminski et al. 1993, 1506):

> He tells you of important blessings which he imagines will result to us and mankind in general from the adoption of this system—I see the awful immensity of the dangers with which it is pregnant—I see it—I feel it.

Yet, to respond to Mason and Henry with the benefit of hindsight, history has now shown that the checks and balances of the US Constitution, inherent to a federal system of government, have prevented the feared descent into tyranny.[11] Not only have the original 13 states retained republican constitutions but over the following two centuries 37 new states have been added and none of them have lapsed into totalitarianism either. It seems clear that an important reason for this success is that the new states were added to the existing federal structure, which guarantees to each a republican form of government, from their creation.

This, then, provides a possible model for preserving liberties within solar system outposts and colonies—space expansionism can safely proceed, but only within the framework of a pre-existing federal structure that guarantees the liberties of new colonies from the outset. This interplanetary federal structure will presumably be an extension of the federal government that, as I have argued above, we will need to develop on Earth in any case to tackle pressing global problems. I am aware, as Deudney (2020) also observes, that here is a strong libertarian wing among some space expansionists who view moving out into space as a way of freeing space colonies from all Earth-centred social and governmental restrictions. However, I agree with Deudney that this is a naïve (and, frankly, rather immature) vision. Not only will unrestrained interplanetary anarchy lead to the risk of civilization-destroying conflict, but it will also permit the evolution of the kinds of intra-colony tyranny feared by Cockell (2009, 2010, 2015).

This point can be illustrated by another example based on US history. In his talk, 'The Case for Space is Liberty', presented at the meeting on which this book is based, the libertarian-leaning space advocate Robert Zubrin (2021; see also his chapter in this volume) argued that totalitarianism is unlikely on space colonies because people will not freely emigrate to totalitarian colonies. That is, potential colonists will 'vote with their feet' and choose 'liberty'. However, even assuming that future colonists will have a choice,[12] this is not an argument for unrestrained freedom in space. Zubrin's main example was the millions of people who freely have emigrated to the United States over the last several centuries, attracted to a free society, whereas very few have voluntarily emigrated to totalitarian regimes elsewhere in the world. However, it is important to remember that, with the exception of the very first colonists such as the Pilgrim Fathers in the seventeenth century, these immigrants have not moved to a land without a government. Rather, they have been attracted to a 'New World' that was already effectively governed by a federal constitution that provided for a stable society and guaranteed certain political and other freedoms. North America would probably have been a much less attractive prospect for emigration had it consisted of numerous independent, and perhaps authoritarian, states constantly at war with each other, and where new immigrants risked ending up as cannon fodder in pointless wars between, say, New York and New Jersey.

It follows that avoiding government is not the answer to the problem of maintaining liberty in space. Rather, the answer is to get the form of government right from the beginning, and there are good reasons for believing that a federal interplanetary government is likely to be best suited to this task.

6. Conclusion

Daniel Deudney (2020) has argued that some aspects of what he calls 'habitat space expansionism', and especially the future colonization of the solar system, are so dangerous that they should be avoided.[13] Of particular concern are risks of civilization-destroying interplanetary conflicts and the rise of totalitarian forms of government, both on Earth and in space. I share some of these concerns, but I also think that forever relinquishing many of the activities that Deudney includes under 'habitat space expansionism' would severely, and unnecessarily, impoverish humanity's future (e.g. Lasser 1931; Clarke 1946; Sagan 1994; Crawford 2014; Smith 2016). Indeed, although Deudney is suspicious of this line of reasoning, there are deeper considerations: as we don't yet know how common life is as a cosmic phenomenon, self-quarantining humanity on Earth might also impoverish the whole future of life in the Universe (e.g. Dick 2012; Vidal 2014; Tegmark 2017; Rees 2018; Moynihan 2021). To be sure, we need to be clear about our reasons for expanding into space (Schwartz et al. 2021), but rather than throw away a potentially vast future it seems to me that we should identify, and implement, institutional innovations that will allow space expansion to proceed while minimizing the attendant risks.

In the short term, I agree with Deudney (2020, e.g. 241, 372, 376) that we should curtail military space programmes, strengthen the existing international institutions dealing with space activities (e.g. the UN and its associated treaty regimes), and implement something akin to his 'Whole Earth Security' space programme. But whereas Deudney implies that this may be sufficient if habitat space expansionism is curtailed, the logic points to stronger international measures being required if habitat space expansion is to proceed. Medium-term possibilities (say over the next several decades) might include the creation of a world space agency under UN auspices (e.g. Cleaver 1948; Clark and Sohn 1960; Brown and Fabian 1975), and/or the creation of a high-level UN organ at the level of the Security Council (perhaps by re-purposing the now redundant Trusteeship Council; e.g. Carlsson and Ramphal 1995; Crawford 2021b) to coordinate global space activities.

Unfortunately, the strength and efficacy of UN-based space institutions will depend on the strength and efficacy of the UN as a whole, which already appears insufficient for dealing adequately with a wide range of pressing global problems. As many others have argued over the years (e.g. Reves 1946; Heater 1996; Leinen and Bummel 2018; Yunker 2018; Weiss 2016; Bummel 2021), in order to deal effectively with global problems it may be necessary for the UN to evolve in the direction of a genuine world government, or for it to be superseded by some other form of planetary government. If we need to create a government for Planet Earth anyway, in order to address multiple common global problems, then it would also make sense for space activities to be placed under its jurisdiction. Importantly, the evolution of world government is likely to occur on the same timescale (e.g. the next century or so) as the initiation of habitat space expansionist activities, facilitating a co-evolution of planetary and interplanetary governments.

Although this would reduce the risk of anarchy in space, Deudney declines to advocate it as a solution to the geopolitical problems of space expansionism because he fears, as do many opponents of the idea of world government, that any such government may become totalitarian. However, not all proposals for world government are equally objectionable in this respect. Specifically, the multiple levels of political authority and representation, and associated checks and balances, which are inherent in federal forms of government minimizes the risks of totalitarianism. Furthermore, by applying the principle of subsidiarity, federal forms of government are able to maintain diversity and local autonomy among their members and are inherently expandable to ever larger spatial scales. For these reasons, I have argued here (see also Crawford 2015) that a federal world, and later federal interplanetary, government would be uniquely suited to minimizing the risks of both interplanetary anarchy and interplanetary tyranny. Of course, there can be no guarantee that such a federation will never fail—as Madison (1788b) pointed out in answer to similar concerns regarding the proposed US federation:

> It is a sufficient recommendation of the federal Constitution that it *diminishes the risk* of [calamities] for which no possible constitution can provide a cure... .
>
> (my emphasis)

It is true that humanity is still a long way from being able to construct a functioning federal world/interplanetary government, but we are also a long way from being able to colonize the solar system. My argument here is that both these activities may evolve over comparable timescales, and that mutually supportive synergies may therefore develop between them.

A key reason why a federal world government, even if acknowledged to be desirable, is seen as politically unrealistic is that humanity currently lacks a sufficiently strong sense of global community on which such an institution could be built. However, there seems little doubt that our long-term survival will depend on developing such a sense of community (e.g. Ward 1966; Appiah 2006; White 2014; Som 2019; Crawford 2021a) and, as Deudney (2020) himself recognizes with his concept of 'whole Earth identity formation', space activities have the potential to help create this perspective. Here, I am reminded of a passage in Arthur C. Clarke's short story *The Lion of Comarre* (Clarke 1949, 125), where the Council Chamber of a future world government is located in a high orbit about the Earth:

> When the members of the Council were in session it seemed as if there was nothing between them and the great globe spinning far below. The symbolism was profound. No narrow parochial viewpoint could long survive in such a setting.

Thus, the perspectives engendered by a human expansion into space may play an important role in laying the psychological foundations on which a federal world, and later interplanetary, government might be built. I am prepared to assert that no other form of political organization is likely to leave humanity in a better position to maximize the opportunities, and minimize the risks, associated with building an interplanetary civilization.

Acknowledgements

I thank Charles Cockell for the invitation to contribute this chapter, and for organizing the excellent and stimulating meeting at which it was presented. I thank Daniel Deudney for several interesting conversations (albeit carried out by email and Zoom), and especially for drawing my attention to the work of David Lasser (1931). I am also grateful to Allan McKenna for reminding me of Clark and Sohn's (1960) proposal for a United Nations Outer Space Agency.

Notes

1. UNICEF: United Nations Children's Fund; UNESCO: United Nations Educational, Scientific and Cultural Organisation; UNDP: United Nations Development Programme; UNEP: United Nations Environment Programme; UNHCR: United Nations High Commissioner for Refugees; WFP: World Food Programme; for a complete list of specialist agencies and programmes within the UN system, see UN (2021).
2. I should stress that I'm not claiming that space exploration is the only means of engendering unifying perspectives; elsewhere, I have argued that the common evolutionary perspectives engendered by the emerging discipline of 'big history' may also be helpful in this respect (e.g. Crawford 2021a; see also Dick 2018, 234–5).

3. Treaty on Principles Governing the Activities of States in the Exploration and Use of Outer Space, including the Moon and Other Celestial Bodies (herein the 'Outer Space Treaty', OST); <https://www.unoosa.org/oosa/en/ourwork/spacelaw/treaties/introouterspacetreaty.html>.

4. <https://www.unoosa.org>.

5. <https://www.unoosa.org/oosa/en/ourwork/copuos/>.

6. Carlsson et al. (1995) proposed bringing management of all the global commons, including the Earth's atmosphere and oceans as well as outer space, within the remit of a reformed Trusteeship Council. This suggestion seems to me to have considerable merit, but as human activities in space expand they will likely come to dominate and deserve a dedicated UN decision-making body.

7. The actual meaning of the acronym is not spelled out by Cleaver (1948), but the context (e.g. 25) implies that he had 'Interplanetary Agency' in mind.

8. As pointed out by Deudney (2020, 246), an even earlier suggestion for an 'International Interplanetary Commission' to organize future space exploration, with an eye to its possibly beneficial role in reducing international tensions, was made by David Lasser in his excellent and prescient book *The Conquest of Space* (Lasser 1931, 74). This long pre-dates the formation of the United Nations but it is interesting, and perhaps slightly disappointing, that Lasser does not suggest that the proposed Interplanetary Commission might be overseen by the then-extant League of Nations; indeed, it not clear that he had in mind any kind of intergovernmental organization, as elsewhere (Lasser 1931, 173–5) he drafted a proposed constitution for his Interplanetary Commission which indicates that he was thinking of a federation of national membership organizations engaged in spaceflight advocacy.

9. There is considerable academic debate on whether Kant had in mind what we would today call a 'confederal' structure or a 'federal' one (e.g. Barrata 2013, 254; Castaldi 2013, 242–3; Levi 2013, 37–8; Straus 2013, 145; and references cited by these authors), but this distinction doesn't really affect the essential point made here.

10. Of course, the United States did suffer a catastrophic civil war between 1861 and 1865, essentially a war of secession by the Southern states over the issue of slavery that was left unresolved by the constitutional convention of 1787. However, as discussed by Straus (2013), it is a mistake to see this as a failure of the federal constitution, not least because it would probably have been impossible to create a constitution that all 13 states could have agreed to in 1787 which would have prevented this conflict, and yet a failure to federate at that time would likely have resulted in many more wars between the states. Straus (2013, 122–3) argues that the North–South conflict over slavery was 'not entirely inevitable, but highly probable' following independence from the British Empire (within which the slave trade was abolished in 1807 and slavery itself was largely abolished in 1833). On the other hand, 'Federalism (the strengthening of the Union) averted many more potential inter-state wars and in several respects postponed the North–South one … What endures as an accomplishment of Federalism was to reduce this to only one more war before peace was restored among the colonies/states in perpetuity.' As discussed by Crawford (2015), a comparison of the number of wars fought in North and South America following independence from the colonial powers lends support to Straus' argument.

11. At least to date. US democratic institutions have recently been placed under considerable strain and, although the guardrails built into the system have so far held, the future emergence of an autocracy cannot be excluded. Nevertheless, the success of the federal constitution in maintaining intra-state liberties for over 200 years seems sufficient to show that the fears of the anti-federalists were misplaced.

12. There are darker possibilities. For example, as noted by Charles Cockell in the discussion following Zubrin's talk, some space colonies might be established by totalitarian Earth governments, in which case tyranny would presumably be locked in from the beginning. This is exactly the situation we would like to avoid, and why it would be desirable for space colonization to proceed under the auspices of a democratic federal world/interplanetary government and not be initiated by sovereign nation-states on an anarchic Earth.

13. Because of these perceived dangers, Deudney (2020, e.g. 139, 372) argues that all relevant enabling space capabilities should be 'relinquished'. This would seem to restrict humanity to Earth's surface until we become extinct owing to some natural or self-inflicted cause. Deudney (personal communication, 28 April 2022) has informed me that this was not his intention, pointing to his statements (364) that "[i]n the very long term, humanity must leave the Earth to survive" but that we should postpone expanding into space for "at least the next several centuries." However, by waiting this long, in the hope that we can use the time to solve all the other problems facing humanity, it is possible that we may miss an opening window of opportunity to gain a foothold in the solar system upon which, as Deudney himself acknowledges, our ultimate survival may depend.

References

AFSC. The Future of Space 2060 and Implications for U.S. Strategy: Report on the Space Futures Workshop, Air Force Space Command (2019) <https://aerospace.csis.org/wp-content/uploads/2019/09/Future-of-Space-2060-v2-5-Sep.pdf> accessed 4 July 2022.

Appiah, K. A. *Cosmopolitanism: Ethics in a World of Strangers* (London: Penguin, 2006).

Baratta, J. P. *The Politics of World Federation* (Praeger Publishers, 2004).

Baratta, J. P. 'The Complementarity of the Thinking of Kant and Hamilton in the United States' in R. Castaldi (ed.), *Immanuel Kant and Alexander Hamilton, The Founders of Federalism: A Political Theory for Our Time* (Brussels: P.I.E. Peter Lang, 2013) 253–69.

Brown, S. and Fabian, L. L. 'Toward Mutual Accountability in the Nonterrestrial Realms' (1975) 29 *International Organisation* 877–92.

Bummel, A. 'Towards a Planetary Polity: The Formation of Global Identity and State Structures' in I. A. Crawford (ed.), *Expanding Worldviews: Astrobiology, Big History and Cosmic Perspectives* (Cham: Springer, 2021) 325–40.

Carlsson, I. and Ramphal, S. *Our Global Neighbourhood: The Report of the Commission on Global Governance* (Oxford: Oxford University Press, 1995).

Castaldi, R. 'The Interaction between International and Domestic Politics' in R. Castaldi (ed.), *Immanuel Kant and Alexander Hamilton, The Founders of Federalism: A Political Theory for Our Time* (Brussels: P.I.E. Peter Lang, 2013) 237–51.

Clark, G. and Sohn, L. B. *World Peace through World Law* (2nd revised edn, Cambridge, MA: Harvard University Press, 1960).

Clarke, A. C. 'The Challenge of the Spaceship' (1946) 6 *Journal of the British Interplanetary Society* 66–78.

Clarke, A. C. 'The Lion of Comarre', reprinted in A. C. Clarke (ed.), *The Collected Stories* (London: Gollancz, 2000) 119–54.

Clarke, A. C. *The Exploration of Space* (London: Temple Press, 1951).

Clarke, A. C. 'The Secret of the Men in the Moon', *This Week*, 11 August 1963. Reprinted as 'The Secret' in A. C. Clarke (ed.), *The Collected Stories* (London: Gollancz, 2000) 817–21.

Cleaver, A. V. 'The Interplanetary Project' (1948) 7 *Journal of the British Interplanetary Society* 21–39.

Cockell, C. S. 'Liberty and the Limits to the Extraterrestrial State' (2009) 62 *Journal of the British Interplanetary Society* 139–57.

Cockell, C. S. 'Essay on the Causes and Consequences of Extraterrestrial Tyranny' (2010) 63 *Journal of the British Interplanetary Society* 15–37.

Cockell, C. S. 'Freedom in a Box: Paradoxes in the Structure of Extraterrestrial Liberty' in C. Cockell (ed.), *The Meaning of Liberty beyond Earth* (Cham: Springer, 2015) 47–68.

Converse, R. *World Government: Utopian Dream or Current Reality* (New York, NY: Algora Publishing, 2010).

Crawford, I. A. 'On the Formation of a Global Space Agency' (1981) 23 *Spaceflight* 316–17.

Crawford, I. A. 'Space Development: Social and Political Implications' (1995) 11 *Space Policy* 219–25.

Crawford, I. A. 'Avoiding Intellectual Stagnation: The Starship as an Expander of Minds' (2014) 67 *Journal of the British Interplanetary Society* 253–7.

Crawford, I. A. 'Interplanetary Federalism: Maximising the Chances of Extraterrestrial Peace, Diversity and Liberty' in C. Cockell (ed.), *The Meaning of Liberty beyond Earth* (Cham: Springer, 2015) 199–218.

Crawford, I. A. 'Widening Perspectives: The Intellectual and Social Benefits of Astrobiology, Big History, and the Exploration of Space' in I. A. Crawford (ed.), *Expanding Worldviews: Astrobiology, Big History and Cosmic Perspectives* (Cham: Springer, 2021a) 341–65.

Crawford, I. A. 'Who Speaks for Humanity? The Need for a Single Political Voice' in O. A. Chon Torres, T. Peters, J. Seckbach and R. Gordon (eds), *Astrobiology: Science, Ethics and Public Policy* (Beverly MA: Scrivener Publishing, 2021b).

Crawford, I. A. and Baxter, S. 'The Lethality of Interplanetary Warfare: A Fundamental Constraint on Extraterrestrial Liberty' in C. S. Cockell (ed.), *The Meaning of Liberty beyond Earth* (Cham: Springer, 2015) 187–98.

Darwin, C. *The Descent of Man* (2nd edn, London: John Murray, 1874).

Deudney, D. H. 'All Together Now: Geography, the Three Cosmopolitanisms, and Planetary Earth' in L. Cabrera (ed.), *Institutional Cosmopolitanism* (Oxford: Oxford University Press, 2018) 253–76.

Deudney, D. H. 'Going Critical: Toward a Modified Nuclear One Worldism' (2019) 15 *Journal of International Political Theory* 367–85.

Deudney, D. H. *Dark Skies: Space Expansionism, Planetary Geopolitics and the Ends of Humanity* (Oxford: Oxford University Press, 2020).

Dick, S. J. 'Cosmic Evolution: The Context for Astrobiology and its Cultural Implications' (2012) 11 *International Journal of Astrobiology* 203–16.

Dick, S. J. *Astrobiology, Discovery, and Societal Impact* (Cambridge: Cambridge University Press, 2018).

Hamilton, A. 'Federalist Paper No. 34' (1788a) in C. Rossiter (ed.), *The Federalist Papers* (New York, NY: Mentor Books, 1961) 205–11.

Hamilton, A. 'Federalist Paper No. 9' (1788b) in C. Rossiter (ed.), *The Federalist Papers* (New York, NY: Mentor Books, 1961) 71–6.

Hamilton, A. 'Federalist Paper No. 7' (1788c) in C. Rossiter (ed.), *The Federalist Papers* (New York, NY: Mentor Books, 1961) 60–6.

Hamilton, A. 'Federalist Paper No. 8' (1788d) in C. Rossiter (ed.), *The Federalist Papers* (New York, NY: Mentor Books, 1961) 66–71.

Heater, D. *World Citizenship and Government* (Basingstoke: Macmillan Press, 1996).

Herz, J. H. *International Politics in the Atomic Age* (New York, NY: Columbia University Press, 1962).

Hoyle, F. *The Nature of the Universe* (Oxford: Blackwell, 1950).

Kaminski, J. P., Saladino, G. J., Leffler, R., Schoenleber, C. H., et al. *Documentary History of the Ratification of the Constitution*, Vol. 9: Ratification of the Constitution by the states: Virginia (2). (Madison, WI: The State Historical Society of Wisconsin, 1990) <http://digital.library.wisc.edu/1711.dl/History.DHRCv9> accessed 4 July 2022.

Kaminski, J. P., Saladino, G. J., Leffler, R., Schoenleber, C. H., et al. *Documentary History of the Ratification of the Constitution*, Vol. 10: Ratification of the Constitution by the states: Virginia (3). (Madison, WI: The State Historical Society of Wisconsin, 1993) <http://digital.library.wisc.edu/1711.dl/History.DHRCv10> accessed 4 July 2022.

Kant, I. 'Perpetual Peace: A Philosophical Sketch' (1795) in H. S Reiss (ed.), *Kant: Political Writings* (2nd edn, Cambridge: Cambridge University Press, 1991) 93–130.

Koch, J. S. 'Institutional Framework for the Province of All Mankind: Lessons from the International Seabed Authority for the Governance of Commercial Space Mining' (2018) 16 *Astropolitics* 1–27.

Lasser, D. 'The Conquest of Space', reprinted in R. Godwin (ed.), *The Conquest of Space* (Burlington, VT: Apogee Books, 2002).

Leinen, J. and Bummel, A. *A World Parliament: Governance and Democracy in the 21st Century* (Berlin: Democracy Without Borders, 2018).

Levi, L. 'The Invention of Federalism' in R. Castaldi (ed.), *Immanuel Kant and Alexander Hamilton, The Founders of Federalism: A Political Theory for Our Time* (Brussels: P.I.E. Peter Lang, 2013) 15–42.

Lopez-Claros, A., Dahl, A., and Groff, M. *Global Governance and the Emergence of Global Institutions for the 21st Century* (Cambridge: Cambridge University Press, 2020).

Madison, J. 'Federalist Paper No. 51' (1788a) in C. Rossiter (ed.), *The Federalist Papers* (New York, NY: Mentor Books, 1961) 320–5.

Madison, J. 'Federalist Paper No. 43' (1788b) in C. Rossiter (ed.), *The Federalist Papers* (New York, NY: Mentor Books, 1961) 271–80.

McKenna, A. 'In Search of Global Security: Everett C. Dolman's *Astropolitik* and Daniel Deudney's *Dark Skies*' (2020) 18 *Astropolitics* 199–222.

Miller, D. (ed.). *The Blackwell Encyclopaedia of Political Thought* (Oxford: Blackwell, 1987).

Montesquieu, C. L. de S. 'The Spirit of the Laws' in A. M. Cohler, B. C. Miller, and H. S. Stone (trans. and eds) (Cambridge University Press, 1748[1989]).

Morgenthau, H.J. *Politics among Nations: The Struggle for Power and Peace* (New York, NY: Alfred Knopf, 1948).

Moynihan, T. 'The Summons of a Silent Universe: The Relationship between Existential Risk and Cosmic Silence' in I. A. Crawford (ed.), *Expanding Worldviews: Astrobiology, Big History and Cosmic Perspectives* (Cham: Springer, 2021) 65–90.

Pinault, L. 'Towards a World Space Agency: Operational Successes of the International Seabed Authority as Models for Commercial-National Partnering under an International Space Authority' in C. S. Cockell (ed.), *Human Governance beyond Earth* (Cham: Springer, 2015) 173–96.

Pistone, S. 'Peace as a Condition for Democracy' in R. Castaldi (ed.), *Immanuel Kant and Alexander Hamilton, The Founders of Federalism: A Political Theory for Our Time* (Brussels: P.I.E. Peter Lang, 2013) 91–111.

Plattard, S. and Smith, A. Optimising Human Space Exploration Policies and Strategies. IAC-18-E3.2.1, 69th International Astronautical Congress, 1–15 October 2018, Bremen, Germany. https://discovery.ucl.ac.uk/id/eprint/10079940/1/IAC-18-E3.2.1%20SP%20final.pdf accessed 23 August 2022.

Poole, R. *Earthrise: How Man First Saw the Earth* (New Haven, CT: Yale University Press, 2008).

Rees, M. J. *On the Future: Prospects for Humanity* (Princeton: Princeton University Press, 2018).

Reves, E. *The Anatomy of Peace* (New York, NY: Harper, 1946).

Rider, F. *The Great Dilemma of World Organisation* (New York, NY: Reynal and Hitchcock, 1946).

Sagan, C. *Contact: A Novel* (London: Arrow Books, 1985).

Sagan, C. *Pale Blue Dot: A Vision of the Human Future in Space* (New York, NY: Random House, 1994).

Sagan, C. and Ostro, S. J. 'Dangers of Asteroid Deflection' (1994) 368 *Nature* 501.

Schwartz, J. S. J, Wells-Jensen, S., Traphagan, J. W., et al. 'What Do We Need to Ask before Settling Space?' (2021) 74 *Journal of the British Interplanetary Society* 140–9.

Shepard, M. 'An International Outer Space Agency for Peaceful Purposes: A Brief Review of Various Proposals and an Analysis of Possibilities' in *National Aeronautics and Space Act: Hearings before the Special Committee on Space and Aeronautics* (United States Senate, 1958) 387–393.

Smith, K. C. 'Cultural Evolution and the Colonial Imperative' in C. S. Cockell (ed.), *Dissent, Revolution and Liberty beyond Earth* (Cham: Springer, 2016) 169–87.

Som, S. M. 'Common Identity as a Step to Civilizational Longevity' (2019) 106 *Futures* 37–43.

Straus, I. 'Democratic Peace and Federalist Peace Theories: Allies or Enemies' in R. Castaldi (ed.), *Immanuel Kant and Alexander Hamilton, The Founders of Federalism: A Political Theory for Our Time* (Brussels: P.I.E Peter Lang, 2013) 113–235.

Tegmark, M. *Life 3.0: Being Human in the Age of Artificial Intelligence* (London: Penguin, 2017).

Tronchetti, F. *The Exploitation of Natural Resources of the Moon and Other Celestial Bodies: A Proposal for a Legal Regime* (Leiden: Martinus Nijhoff Publishers, 2009).

UN United Nations: Peace, Dignity and Equality on a Healthy Planet <https://www.un.org/en/about-us/un-system> accessed 26 June 2021.

USSF United States Space Force <https://www.spaceforce.mil> accessed 27 June 2021.

Vidal, C. *The Beginning and the End: The Meaning of Life in a Cosmological Perspective* (Cham: Springer, 2014).

Wallace, A. R. *Contributions to the Theory of Natural Selection* (2nd edn, Macmillan, 1871). <http://www.gutenberg.org/files/22428/22428-h/22428-h.htm> accessed 27 June 2021.

Ward, B. *Spaceship Earth* (Columbia University Press, 1966).

Weiss, T.G. *What's Wrong with the United Nations and How to Fix it* (3rd edn, Cambridge: Polity Press, 2016).

White, F. *The Overview Effect: Space Exploration and Human Evolution* (Reston, VA: American Institute of Aeronautics and Astronautics, 2014).

Wilson, D. S. and Wilson, E. O. 'Rethinking the Theoretical Foundation of Sociobiology' (2007) 82 *Quarterly Review of Biology* 327–48.

Wilson, E. O. *Consilience: The Unity of Knowledge* (New York, NY: Little, Brown and Company, 1998).

Wilson, E. O. *The Social Conquest of Earth* (New York, NY: Liveright Publishing, 2012).

Wynner, E. and Lloyd, G. *Searchlight on Peace Plans: Choose Your Road to World Government* (New York, NY: Dutton and Company, 1944).

Yunker, J. A. *Political Globalization: A New Vision of Federal World Government* (Lanham: University Press of America, 2007).

Yunker, J. A. *Evolutionary World Government: A Pragmatic Approach to Global Federation* (Lantham: Hamilton Books, 2018).

Zubrin, R. 'The Case for Space is Liberty', Talk presented at the meeting on 'The Institutions of Extraterrestrial Liberty', UK Centre for Astrobiology, 8–11 June 2021. <https://www.astrobiology.ac.uk/events/extraterrestrial-liberty> accessed 4 July 2022.

10

Human conflict resolution in a non-Terran context

Janet de Vigne

> So that in the first place, I put for a generall inclination of all mankind, a perpetual and restless desire of Power after power ... because he cannot assure the power and means to live well, which he hath present, without the acquisition of more. (Hobbes 1651)

The words 'conflict' and 'humanity' might be considered inseparable. Our story, as humans, is marked by conflict—we grow through it, we learn how to overcome, step aside from, confront, and deny it at almost every step of our development process. Conflict is not necessarily bad, therefore, and it may be described as something that we need to think through, in all its varied dimensions, in order to understand how humans might develop and thrive in an off-world context. This chapter will examine different types of conflict from the point of view of the human story (the place of conflict in the narrative that informs and shapes our existence), explore ways of mitigating potentially damaging conflict as well as maximizing the mechanisms and benefits of positive resolution, and examine what we can plan for and what we cannot predict. It will make a plea for the development of a new language with which to consider human presence in places strange to us, by moving away from anthropocentric thinking (ego-centred perspectives) to more chthulucenic approaches (down in the soil from which we're made and its organisms (Haraway 2016)) in an attempt to address one of conflict's fundamental components, the affective issue of 'belonging' and its responsibilities, especially in an *ab origine* democratic context.

1. Humans and Conflict

Conflict is almost always described in negative terms, but for human beings learning to resolve conflict it can be a growth mechanism, particularly among adolescents. We rarely if ever equip people consciously to deal with and resolve conflict, not at school, not in human resources departments. Instead it is regulated to the farthest fringes of specialist practice, where experts are brought in to effect reconciliation between two parties, be they friends who disagree, divorcing parents, work colleagues in dispute, or warring governments. There is a verifiable need to address conflict in all these areas—it is something that at the individual level in the West we shy away from, possibly because of the divorce between head and heart that occurred during the age of

Janet de Vigne, *Human conflict resolution in a non-Terran context.* In: *The Institutions of Extraterrestrial Liberty.* Edited by Charles S. Cockell, Oxford University Press. © Janet de Vigne (2022). DOI: 10.1093/oso/9780192897985.003.0011 determinism, and the consequent fear of conflict's inevitable emotional roots. A lack of understanding of the power of affect also leaves us vulnerable to manipulation.

Addressing conflict requires an acknowledgement of human emotion that we still deny in the practice of science. Perhaps if we were to recognize it as a fundamental element of what it means to be human, as Haidt suggests, we might start out on a better foot:

> Conflict behaviour is ... a feature of our evolutionary design, not a bug or error that crept into minds that would otherwise be objective and rational.
>
> (Haidt 2013b, xx)

Accepting this statement requires us to recognize not just conflict behaviour but the possible irrationality of the factors contributing to it (conflict can of course have rational antecedents). So what is conflict? What danger does it present? To what extent can we plan for it and mitigate its potentially disastrous effects? We can train people to reason and we can encourage the practice of conflict resolution but we cannot, in any number of dimensions, either guarantee or breed for success.

At its most innocuous, conflict can mean incompatibility, variance, disagreement, argument. At its worst extrapolation, war, death, annihilation. Conflict immediately reduces its protagonists to binary positions—for or against. It is an essential 'othering' mechanism and herein lies its danger in an off-world context—as a species, humanity cannot afford to divide its loyalties. Perhaps we are unique in this particular capacity for self-destruction.

We have been here already in the imaginary. Neal Stephenson, in the novel *Seven Eves* (Stephenson 2015), outlines a scenario where a politician inveigles her way onto a spaceship designed to save the human race from impending environmental catastrophe. The scientists, in their perhaps naive focus on survival mechanisms, had never conceived of the potential division that would be fomented through the manipulation of discussion about individual rights and liberty, and therefore never thought through its possible consequences. Here, total destruction of the species is all too narrowly averted.

The capacity of story to imagine conflict scenarios and resolution might help us if we look to learn from them. We could consider conflict resolution in literature as the narrative of 'overcoming'—as such it performs several functions: examination of reasons for and the extent of the conflict; the thrill of facing an existential threat in the safe space of the imaginary; the feel-good factor of overcoming that threat and the consequent increase in trust of the human heart and mind. Episodes of the first iteration of 'Star Trek' are perfect examples of this, a very American ideology: 'if your intentions/heart are good, everything will be OK'. Decisions are led by the gut, not the head—instinct rules over considered judgement (think of Kirk's character traits versus Spock's as embodiments of this). In the twenty-first century we are beginning to complicate this type of story as the traditional boundaries of good and bad are muddied. This is happening for a variety of reasons, from post-human examination of the metaphysics to understanding the social construction of older women seen through the male gaze. This may help us in realizing that there is always more than one answer to complex problems. Consider the rehabilitation of the Wizard of Oz' Wicked Witch of the East in the Schwartz–Holzmann musical, 'Wicked' (based on Maguire's 1995 novel) and the so-called wicked fairy Maleficent, antagonist of the 1959 Disney film of Perrault's Sleeping Beauty, portrayed in a much more understanding and forgiving light by Angelina Jolie in 2014 and 2019. Conflict will always be there to feed the story, but we still need to know how we can overcome and understand the potential consequences of our choices in the way we construct society. Conflict in various forms shapes our identity. We will need to tell stories to remind us of who we think we are—perhaps even who we want to be:

> Stories can conquer fear, you know. They make the heart bigger. (Ben Okri)

A novelist might claim that:

> [c]onflict is what drives the narrative and without it a tale may exist, but there will be nothing exciting to tell—drama arises out of conflict. (*Writers' Digest*)

The sagas of old fulfil this remit. Writers today speak of certain universal plotlines: the hero's quest, the journey of love (un)requited, interaction with an Other (a monster?), overcoming challenges to achieve comfort, security, and happiness. So what types of conflict are there and how might they (i) be resolved and (ii) used to inform future resolution? The *Writers' Digest* suggests:

- Conflict with the self—where and what are your demons?
- Conflict with others—who are the 'others'? Your team? Others like or unlike you?
- Conflict with the environment—physical? Spatial? Ideological?
- Conflict with the supernatural—or things beyond our immediate understanding? (Monsters from the ID, if you remember 'Forbidden Planet').

What might be the causes of these conflicts? Here, 'you' might be a person, a thing, the space/place/land on which we find ourselves.

- I want something you've got (an object, perceived benefit, a talent, your respect/recognition?).
- I fear you because I think you're better/stronger than/have more than me (subconsciously or consciously).
- It is easier to remove you/your ideas (if they pose a threat) than understand you.
- I will gain relevance (power) by othering (subjugating/ oppressing) you.

We fight over who we are—over the space we occupy, the place we have/want to create in that space, the way we relate to each other, and our awareness of our environment—from micro to macro perspectives. Our identity is forged through these conflicts. Tim Hicks argues that our differences make us who we are, and that this process is dynamic and physical:

> We are biology and chemistry, but we are also our lived experience.
>
> (Hicks 2018, xiii)

Our interactions with self, other, and environment are in flux—constantly being negotiated in order to interpret the world, agree on action, engage with each other. Trust and knowing are essential components here. These processes are exhausting enough when the context is known—the unknown presents even more challenges. Consider the 2013 film 'Riddick' (the bad guy is in fact the good guy having fallen foul of societal processes) alongside the 1956 'Forbidden Planet', where monsters from the Id (the skeletons in the closet? Bad thoughts? Hidden fears?) take shape and, having caused the demise of the former, superior inhabitants now threaten the human visitors. Whom do you trust? In an altered context, an extreme environment— everything changes—you are not where you think you are. You are not who you think you are—nor are your colleagues.

The point of interest here is that our negotiations with our environments form and reform physical neural structures:

> constructed and established in response to our perceptual experience of the world up to that point in time.
>
> (Hicks 2018, 8)

If these negotiations, at their root interpretations, may be described as stories—and we live by stories—then these hermeneutics take form; they are physically present in the brain. Those structures are our stories—so the question remains, how do we understand and shape them? And in many dimensions, how do we do this to the benefit of the human race?

2. The Frontier Mentality and its Dangers

Much of the language around space travel is the discourse of the frontier (I suggest it would be difficult to find someone thinking about these issues who does not have 'the final frontier' imprinted somewhere in their neural processes):

> A frontier is created when a community occupies a territory. From then on the fron-tier is changed and shaped by the activity and growth of that community, or by the impact on it of another community.
>
> (Lattimore 1962, 469)

It is a boundary, a gatepost beyond which lie 'dragons'. There is also the sense that a frontier represents the farthest reach of that occupying community's society. If we consider 'belonging' as an issue to be faced with the establishment of the frontier, phrases such as 'no taxation without representation' come to mind. The frontier, then, delineates the boundary of a society and all its norms, ideological and practical, are enacted within that space, therefore a particular understanding of place is imposed, shared, agreed, believed by its inhabitants.

Is it possible to think about frontiers without thinking about military action, specif-ically? Do we still think that the best form of defence is attack? Lieutenant Colonel

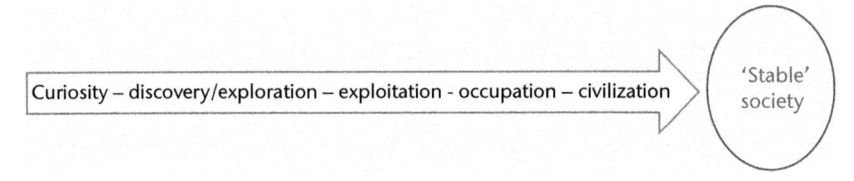

Figure 10.1 The process of the establishment of a stable society in space (Hyten 2000).

John Hyten, for whom conflict in space is inevitable, would define the settlement and establishment of a frontier thus (see Figure 10.1).

In the twenty-first century, we may have problems with the terms 'exploitation' and 'occupation', as we should by now have learned something about living with the consequences of the toxicity of these practices. Are they automatic steps on the road to civilization? That very much depends on your definition of 'civilization'. 'Imposition' might be a better term. It's hard to imagine any way that we could avoid 'imposition' of ourselves and our ways on a non-Terran place, but perhaps there might be ways in which this particular practice of conflict might be mitigated.

Who would decide when conflict becomes necessary? Teenage dystopian fiction is full of examples of this at the moment. They include the *Hunger Games* (a young girl volunteers in the place of her sister to take part in a televised fight to the death for the entertainment of the elite) (Collins 2008); *Divergent* (a young girl discovers that everything she believes about her world is wrong) (Roth 2008); the *Fifth Wave* (humans are set against each other by occupying aliens) (Yancey 2013). These are novels with conflict at their heart. Being managed by overlords using fear and othering to maintain the status quo, the internal conflict of the protagonists explodes into revolutionary action. Whom should we believe about so-called dragons (boundaries, others)? Or do we get caught out by the sudden appearance of the enemy at the gate? So many of our stories warn against both.

Where is leadership here and how might that lead to conflict? Do our stories of leadership (mostly military through computer games of the kill or be killed, establish and defend or die sort) feed us or perpetuate Machiavellian thinking? Warring communities seeking to gain control of trade routes? Requiring tribute to survive? Imposing their values, mores, rule on another situation? These model the 'hero's journey' image we have of ourselves, not to say the greed and destruction associated with perpetual expansion.

Do we need to be the strongest in order to survive? Would it matter if we weren't? What are alternative models? The film 'Avatar' might open up a different story—not so much the establishment of a Machiavellian principality ('He who has relied least on fortune is established the strongest' (Machiavelli 1532)) but instead a relationship of equals? For Machiavelli, fear trumps love:

> Because this is to be asserted in general of men, that they are ungrateful, fickle, false, cowardly, covetous, and as long as you succeed they are yours entirely; they will offer you their blood, property, life, and children, as is said above, when the need is

far distant; but when it approaches they turn against you... men have less scruple in offending one who is beloved than one who is feared, for love is preserved by the link of obligation which, owing to the baseness of men, is broken at every opportunity for their advantage; but fear preserves you by a dread of punishment which never fails. (Machiavelli 1532)

This thinking is at the base of poor choices and leads to tyrannical practices. Perhaps we can rise above the 'only the fittest survive' mentality (we seem to have misunderstood this in nature—there's a lot more collaboration that previously thought). Perhaps a different, less power-seeking idea may emerge. Could peaceful, harmonious co-existence, along the lines of the scientists' roles in the film Avatar emerge, rather than the 'kill, steal, destroy' mentality of the soldiers? We have imagined this. Where do we draw the line (if we are to survive) between respect for a new environment and, say, Starfleet's Primary Directive? (to summarize: no identification of self or mission; no interference with the social development of said planet; no references to space, other worlds, or advanced civilizations (Stenwedel 2015)). The problem here is that total non-interference is impossible—any human as alien presence will conflict with the natural surroundings of another planet. Also, the way in which we humans see ourselves as saviours—solving even those problems caused by us or problems the other never knew they had—interferes with a decentred perspective. We know this from our stories.

3. Inter-human Conflict

Conflict then is something constantly with us not only in reality but in our imagined futures—from an intrapersonal to interpersonal to broader but situated contexts. In examining these one by one and in relation to each other, our intention could help us form a 'knowing' that might enable positive resolution. Human capacities for moral reasoning and remembering must have a role here, as our memories will not just physically define us but influence our reasoning. In the twenty-first century, where much of our thinking around these processes has been affected by quantum mechanics, we now appreciate the dynamic relationships involved in mind–body constructs: 'memory seems to be altered by each effort to address it, not unlike the elementary particles of quantum physics' (Tucker 2007, 192). Our past informs our present, our present remakes our past:

Memory, it now appears, has not evolved to record things as they actually were, but to be able to predict things better in the future, should certain similarities in observed events occur, and to provide a script should such similarities arise.

(Lack 2012)

The stories that we experience today, before we carry them into space, need to be examined in the light of this—what might our collective power be if we aimed to provide our children with better scripts?

The stories we share help to bind groups together through the dynamic development of a group moral psychology. People form community through their beliefs, their beliefs form the community—this maps onto moral and religious practices:

> Shared gods, rituals, myths and values really do help people to trust each other and achieve together feats that they could not do on their own … .
>
> (Haidt 2013a)

Some of the ways in which we have imagined future societies include references to 'mash-ups' of belief systems, acknowledging the constant change in developing new ideas—the Orange Catholic Bible of the Bene Gesserit sisterhood in Frank Herbert's *Dune* comes to mind. Such systems act as internal and external controls, identity drivers, and conduits of power. If we want to protect the health—mental, emotional, physical—of space-going communities, perhaps we could explicitly address issues such as these. A sense of community might, therefore, help to mitigate the possible negative effects of external threat and ensuing conflict, but may also contribute to internal strife. Perhaps if community is seen as a dynamic, outward- and inward-facing system, this can be managed.

Managing community is another issue. One of the problems we are considering is the avoidance of tyranny—being in favour of democratic means of governance is a peculiarly recent Western idea that many suggest doesn't work in all human contexts. The possibility of a benign autocracy rather than a regime of violent oppression again feeds from memories of unsuccessful attempts at imposition of democracy into our present. Negotiators in these situations carry their own stories and beliefs—the listening quotient is short. Neither party can fully engage if they have not stood in the other's shoes. Where agendas are hidden, threat persists and the possibility of conflict is increased. Andrew White, the former Vicar of Baghdad and peace negotiator respected by all sides in the Middle East, states that understanding of is key:

> Established strategies … are mostly ineffectual. What is more productive, I have found, is to gain an understanding specifically of the people who are responsible for the violence and of their culture, religions, traditions and everything that shapes their expectations. These are the influences that propel people into conflict; these are the factors that complicate its resolution.
>
> (White 2009, 123)

Relationship, understanding, knowing ourselves and the other—these are at the heart of conflict resolution, at the heart of being human. In space, perhaps a shared understanding of expectations would be a good thing to work towards. But who is going to do this, and how can this work be accomplished? People in space will remain the same. Perhaps we could look at personality traits and the likelihood of conflict, mapping these onto behaviours. If we agree that our thinking forms us and that our choices also form our thinking, we might, through the power of story, change some important factors contributing to conflict situations.

In a 2007 study conducted among Korean teens, Jiang, Han, and Hur discovered that certain traits were more disposed to certain less helpful behaviours. A higher degree of neuroticism was associated with higher use of avoidance and attack strategies. Higher self-esteem and higher extroversion were associated with lower avoidance and attack (Jiang et al. 2007). (This may give us hope yet for the alpha Tybalt-type male in Shakespeare's *Romeo and Juliet*.) If we can influence the development of neurosis through a calm, nurturing, and collaborative environment, perhaps we could foster more collaborative strategies in solving conflict.

Along similar lines, it has been suggested that certain trajectories of education may produce undesirable effects, specifically the unusual number of engineering students among terrorists:

> This mindset is one that values established social structures, but not understanding, or accepting the ambiguity involved with real human problems and objectives, or the humanity of the people served by technology.
>
> (Frezza 2016)

Engineers are likely to 'spend most of their time with the same set of [positivist] epistemological tools', according to Donna Riley of Virginia Tech (Frezza 2016). Although this research is ongoing, complex, and by no means definitive, it might make us pause to consider the qualities we need in space-going and space-born individuals and how these might be influenced by their nurture. We may be able to encode, because we cannot breed for, reasonableness and collaboration. Could education then provide a move away from expectations of conflict in inevitable imposition scenarios? Could we think instead about finding another way to coexist?

If we can then consider conflict as a part of human nature that can be mitigated by reason and environment, we still need to plan for irrationality in conflict episodes—these will be the most dangerous. NASA's current human risk mitigation strategy mentions possible behavioural change due to the closed/hostile environment; lack of psychological connection to Earth; isolation, altered gravity, and radiation; and over 1,000 possible illnesses (Romero and Francisco 2020). The problem we don't have a solution for is the cascade effect—what is likely to happen should, for example, a urinary tract infection occur at the same time many other possible stressors. How might conflict in this situation be mitigated? It is an existential threat inflicted on us and caused by us, possibly without intention. Perhaps planning our environment might be a way forward. What kind of nest can we create for ourselves—what sensations will delight us?

Henri Lefevbre's discussion of in-between space may enable us to think through this. It's not just about the perceived space (emotional/behavioural) we inhabit, nor just about the conceived space (whose design is imposed on us, together with the designers' intention, ideology, etc.), but the space where all the spaces meet—subjective, objective, concrete, abstract, mental, physical (Saar and Palang 2009). How can we achieve harmony here for maximum benefit to the human?

The environment is not the only thing that might help—we also need to focus on relationship and that from the earliest stages of human development. Bowlby and Ainsworth's attachment theory is now recognized as one of the building blocks of healthy emotional growth (Bretherton 1992). Experiencing love and being able to trust mean secure relationships and better chances of being able to reason. Without, this, other strategies such as the arts, aiming to enable and enrich our humanness will only partially succeed.

Donna Haraway argues that the Anthropocene is not providing us with the tools we need to think through the future of humanity on this planet, let alone another. We can no longer afford to be human exceptionalists—we have already imagined the death of our planet through imposition of our will and ours alone on the environment ('Avatar' is just one of these iterations. We are the creatures that killed their mother.) 'The Anthropos myth system ends in double death, not ongoingness' (Haraway 2016, 149).

In Haraway's words again, the sciences humans have developed are restrictive systems. These and the evolutionary theories we have thought through (although important) are 'not able to think well about sympoeisis, symbiosis, symbiogenesis, development, webbed ecologies and microbes' (Haraway 2016, 150); in other words, pluri-dimensional relationships not just with the chthulucene but with each other and other beings, as well as the sacks of organisms we carry around in our human bodies—we are not alone. How would it be if we regarded each other not as what we appear to be, but what we might be in the future if all beneficial conditions were met? Just a thought. The so-called Four Freedoms (of President Roosevelt) must apply to them as well as us, surely?

In the present time, our talk of space is one of the conflict of imposition and greed, not really for survival. The universe is our enemy, we need to subdue it as we did the Earth ('we have tamed nature', someone said, not so long ago). It's a sobering idea, however, that a tiny virus, not even alive, has been able in recent times to reduce us to a dependency on each other that many thought impossible in these days of the neoliberal ascendancy. We need conflict to grow, for sure; in the words of the historian A. J. P. Taylor: 'If there had been no trouble-makers, no Dissenters, we should still be living in caves' but we don't need to take this kind of trouble with us ; that is the result of exploitation and injustice. That will destroy us.

Spaces become places through our possession of them. Our stories inform those spaces, internally and externally, from the physical development of our neural processes to our (fossilized) footprints on a mud plain. Everything we do, everything we are, leaks...

... any given place is never autonomous in its unity but forever bleeding and seeping into other places, both those of the past and those of the future ... . (Trigg 2012, 17).

We have the means to move beyond anthropocentric thinking, keeping the good and jettisoning the bad. Our concept of conflict needs to evolve—or we will repeat the history we seem presently unable to avoid, and ultimately destroy ourselves—if not in the immediate future, than as a swarm of Machiavellian, cannibalistic locusts devouring everything in our path.

References

Bretherton, I. 'The Origins of Attachment Theory' (1992) 28 *Developmental Psychology* 759–75.

Collins, S. *The Hunger Games* (New York, NY: Scholastic, 2008).

Frezza, S. 'Considering the Engineering Mindset—Does Engineering Education Breed Terrorists?' in J. W. Blake, et al. (eds), *Philosophical and Educational Perspectives on Engineering and Technological Literacy* (Bray: Reads, 2016) 1–7.

Haidt, J. 'Moral Psychology for the 21st Century' (2013a) 42 *Journal of Moral Education* 281–97.

Haidt, J. *The Righteous Mind: Why Good People Are Divided by Politics and Religion* (New York, NY: Vintage Books, 2013b) xx.

Haraway, D. *Staying with the Trouble: Making Kin in the Chthulucene* (New York, NY: Duke University Press, 2016).

Hicks, T. *Embodied Conflict: The Neural Basis of Conflict and Communication* (New York, NY: Routledge, 2018).

Hobbes, T. *Leviathan* (London, 1651).

Hyten, J. E. 'A Sea of Peace or a Theatre of War—Dealing with the Inevitable Conflict in Space'. ACDIS occasional paper, Program in Arms Control, Disarmament and International Security (ACDIS), University of Illinois, Urbana Champaign, 2000.

Jiang, H. S. Han, J. and Hur, G. 'The Effects of the Big 5 Personality Traits on Adolescents' Conflict Resolution Style' (2007) 27 *Korean Journal of Communication and Information* 418–51.

Lack, J. 'The Neurophysiology of ADR and Process Design: A New Approach to Conflict Prevention and Resolution' in Contemporary Issues in *International Arbitration and Mediation: The Fordham Papers* (2011)341–382. https://doi.org/10.1163/9789004231269_022

Lack, J. and Bogazc, F. 'The Neurophysiology of ADR and Process Design: A New Approach to Conflict Prevention and Resolution' (2012) 14 *Cardozo Journal of Conflict Resolution* 33–80.

Lattimore, O. *Studies in Frontier History: Collected Papers 1928–1958* (Oxford: Oxford University Press, 1962) 469.

Machiavelli, N. *The Prince* (1532) (trans. W. K. Marriott. A Blado d'Asola ch 5 Project Gutenberg, 1998, updated 2022).

Okri, Ben. *Songs of Enchantment / Ben Okri* (London: Cape, 1993). Print.

Romero, E. and Franciso, D. 'The NASA Human System Risk Mitigation Process for Space Exploration' (2020) 175 *Acta Astronautica* 606–15.

Roth, V. *Divergent* (New York, NY: Katharine Tegen Books, 2008).

Saar, M. and Palang, H. 'The Dimensions of Place Meanings' (2009) 3 *Living Reviews in Landscape Research* 5–19.

Stenwedel, J. D. Is the Prime Directive Ethical? 20 August 2015, *Forbes Magazine*.

Stephenson, N. *Seven Eves* (London: HarperCollins, 2015).

Trigg, D. *The Memory of Place: A Phenomenology of the Uncanny* (Athens, OH: Ohio University Press, 2012).

Tucker, D. *Mind from Body: Experience from Neural Structure* (Oxford: Oxford University Press, 2007) 192.

White, A. *The Vicar of Baghdad: Fighting for Peace in the Middle East* (Oxford: Monarch, Lion Hudson, 2009).

Yancey, R. *The Fifth Wave* (New York, NY: Penguin, 2013).

11

Scarcity in space

Challenges for liberty

Martin Elvis

1. Introduction

There is a vast wealth of resources in the asteroids. The Main Belt, located between the orbits of Mars and Jupiter is where, by far, the majority of these resources are to be found. The Main Belt has some 10 million times Earth's proven iron reserves (Elvis and Milligan 2019). To get a sense of scale, this is enough to build a Ringworld (Niven 1970),[1] a megastructure completely circling the Earth's orbit, that would make use of some of the Sun's 'waste energy' that does not shine onto a planet (see Chapter 18 of this volume). This material abundance has suggested to some, including Jeff Bezos, that human civilization could grow manyfold in size and resource use and could spread over at least the inner solar system (Powell 2019). Yet there are constraints even to a solar system-spanning civilization. In this chapter I look quantitatively at what physical constraints such a civilization would live under and at how those constraints may impose limits on our dreams of liberty in this future. Surprisingly, despite overall abundance, scarcity is a recurring theme.

Most of the resources will be contained in a relatively small number of the larger asteroids. This is because while there are many more small asteroids, there are not enough to make up for the large volume of the larger asteroids. Consider the biggest iron asteroid, 16 Psyche. If we want to make up its mass out of asteroids half its size, we would need to find eight of them, because volume grows as the cube of the radius. But there are fewer than eight such asteroids. (Technically, the log N–log D relation has a slope, b < 3, (Jedicke et al. 2015).) This concentration of resources will have implications for how we deal with the distribution of resources and hence for both justice and liberty.

We can expect three phases of human expansion into space:

(1) Near-term scarcity: With our current rockets we can plausibly only mine the most accessible asteroids (Elvis 2014). These are all near-Earth asteroids (NEAs). Even with the SpaceX Starship, there may be just a few hundred that have material worth a billion dollars or more that we will be able to mine in the 2030s.

(2) Medium-term abundance: If we can solve the propulsion problem, for example by developing fusion rockets (Paluszek et al. 2014; Adams and Cassibry 2014), then we can send large mining equipment to the Main Belt, opening

Martin Elvis, *Scarcity in space*. In: *The Institutions of Extraterrestrial Liberty*. Edited by Charles S. Cockell, Oxford University Press. © Martin Elvis (2022). DOI: 10.1093/oso/9780192897985.003.0012 up their enormous resources to human use. If population grows less quickly than our access to these resources, then we may enter a 'post-scarcity' era, where there is material plenty for everyone and all our basic needs (and more) may be provided for (Sadler 2010; Diamandis and Kotler 2012). (This assumes that machines, robots, do most of the work for us.)

(3) Long-term exhaustion: Vast is not infinite. Exponential growth has a way of surprising us. Population growth and CO_2 emission are well-known examples. The rate at which humans mine iron has doubled 10 times, once every 20 years, since the beginnings of the Industrial Revolution around 1800; at that rate another 20 doublings will use up all the iron of the Main Belt, which would happen in 400 years (Elvis and Milligan 2019). Then we have a big problem, with plenty of room for conflict. In order to keep a post-scarcity society running for more than a few centuries, limits to the exponential growth of resource use will be essential. Limits can affect liberty. But I will leave this third stage alone in this chapter.

How will space resources affect liberty in space? This is an exercise in environmental determinism, similar to Jared Diamond's *Guns, Germs & Steel* (Diamond 1997). It also resembles Marx's 'historical determinism' (see Chapter 4 in this volume) as well as political ecology which 'is the notion that systems of power in human societies shape and are shaped by physical environments' (Pluymers 2021). But environmental determinism is not fully determined. The environment only imposes boundaries, it does not require a particular form of government. Even worse, what we are attempting here is *predictive* environmental determinism. In that it is more like Hari Seldon's 'psychohistory' in Asimov's *Foundation* (Asimov 1951); in other words, impossible. But we try anyway because we must.

My method here is to take the numbers seriously. The actual values of the mass of iron, of planetary surface area, even of the number of medical specialties, will affect how justice and liberty thrive or not in our future in space.

I tackle the medium-term first and the near-term second because our near-term choices will greatly affect our medium-term prospects, so we should keep the endgame in mind as we think about space governance for the next decade or two. Our near-term choices are also of immediate significance, that is in the next five or maybe 10 years, as the space resource economy starts up. Choices we make now are likely to affect us for centuries.

2. Medium-term Plenty: A Few Centuries

First, then, let us think about the longer term "end state" once humanity has spread over the inner Solar System.

2.1 1 trillion humans

Amazon founder and space enthusiast Jeff Bezos has described his vision of '[o]ne trillion humans living and working in space' (Powell 2019). That sets up an interesting thought experiment. His dreamed-of population of 1,000 billion would be reached in only six to seven doublings from our current 8 billion people.

The first question this vision raises is where will they all live? Famously, Space is Big, but this is not really the case for liveable area. Only three planetary bodies other than Earth are even plausible for large-scale habitation: Mars, Venus, and the Moon. Mars and Venus might, in principle, be 'terraformed' to have breathable atmospheres and a temperature range hospitable to people. There are arguments against this colonizing view of the utility of other planets (e.g. Milligan 2015 145–51), but let's leave them aside for now.

Terraforming Mars and Venus, and throwing in the Moon too, gives only about four times the Earth's land area.[2] At the current population density of Earth 'only' 40 billion people can live on a planetary body. Many may feel that such a high density is missing the point of having whole new worlds to inhabit and so try to limit the number of Martians to a lower number.

As Bezos recognized, the remaining 960 billion people, in other words the vast majority, will have to live in large free-space habitats of the Gerard O'Neill kind, that will slowly rotate to provide simulated gravity. Nearly a million of these 'space cities' each with a million inhabitants will be needed!

Studies of O'Neill habitats, or space cities, with a million people were carried out almost 50 years ago (Bock, Lamro, and Simon 1977). Illustrations of these space cities show spacious environments with lakes and forests. However the Bock study allocated 157 square metres (m^2) per person, or a total of 157 square kilometres (km^2) for 1 million people. A cylindrical structure with a 1 km radius and 25 km long would have this surface area. In case this seems improbably large, it is noteworthy that the National Natural Science Foundation of China is funding studies of 1 km scale spacecraft (Gent 2021).

It would have a mass of ~50 million tons, 45 million of which are for radiation shielding. A million space cities would then need 45 trillion tonnes (t) (2.5×10^{13} tonnes total, 5×10^{12} tonnes of iron).[3] The Main Belt contains about 10^{18} tonnes of iron (Elvis and Milligan 2019, so the total resources required, at this population density, are only a millionth of the total available.

This area of 157 m^2 is not just living space but the total area necessary to support life, including agriculture, atmospheric processing, waste processing. From the Bock et al. report (Bock et al. 1977), it is unclear how much margin was allowed for, for example, to cope with crop failures.

The population density would be 6,470 people/km^2, a value only slightly lower than that of the most densely populated countries: Macau, Monaco, and Singapore have densities of ~21,000–~8000/km^2 (~48 to 125 m^2/person).[4] A space city at this density will more resemble Asimov's *Caves of Steel* (Asimov 1955) than the leafy suburbs seen in most illustrations of space cities. None of the most densely inhabited countries is self-supporting in food production. India is more or less self-supporting (Thomson 2012) at a population density of ~400/km^2, or 2500m^2/person, 16 times more than provided by an O'Neill space city. The Earth as a whole will soon have 10 billion people on a land area of 148 million km^2(Coble et al. 1987, 102). That gives 14,830 square metres (m^2/person), almost 100 times as much as an O'Neill space city would provide.

An area of 157 m²/person is not much. It is a total area just 12.5 × 12.5 m, about that of a spacious two-bedroom apartment, except that all your food is grown there too. It could be that food production won't eat into your space too much. Roughly, it takes at least 4000 m² of land, depending on diet, to feed one person (Farmland L.P. 2012). Vertical farming techniques may resolve this problem. Aeroponics consume ~300 times less land (Economist 2021), just above 10 m². With few ways to spend time away from high-density living, psychological problems may well arise among space city dwellers. What space cities provide is 'phenomenal cosmic power. Itty bitty living space'.[5]

Crowding in space cities will put limits on the behaviour of individuals, and so too on their liberties. Assuming no migration between space cities then, in a limited population, even if robots do all the basic work, there may be specialized jobs that have to be done (see section 2.4) but that no-one wants to fill. (The migration alternative is discussed below.) Some undesirable degree of persuasion, or even coercion, may be needed to keep the space city viable. The most controversial area will be limits on reproduction. Each space city will have a hard design limit on the population it can support that simply cannot be exceeded. Either you have no more than the number of children allotted to you, by whatever means, or you have to leave. In smaller space cities there may even be strong pressure for all women to have children to maintain the population.[6] Obviously, these are major intrusions into individual liberty.

Could we house people at low, Earth-like, densities in space cities? For one thing, Bezos may be wrong and we will not have so many people to accommodate. People in high-income countries are reproducing at well below replacement rates,[7] in which case maybe there will be fewer people to house and space cities can be more spacious, if cost is no object. To reach the average Earth densitiy a 10 km radius cylinder 250 km long would be needed for a million people. This size is close to the limit attainable with steel and aluminium structures (O'Neill 1977). Each of these large structures would have a mass of ~5 × 10⁸ tonnes (t). The total mass of the Main Belt is ~3 × 10¹⁸ t (Petit et al. 2002), enough to build 10 billion of these giant space cities. The maximum population that could be housed at Earth density with Main Belt resources would then be ~10 trillion. That is a surprisingly tight limit on a Bezos-like future. If some solution to radiation shielding other than cladding the space city in thick rock could be found (e.g. magnetic shielding, D'Onghia 2022), then this limit would be less strict.

The hope is that these space cities will create a wonderful diversity and flourishing of culture that will rival our favourite city states in their days of glory, Periclean Athens or the Florence of the Medici, for example. Robert Zubrin (Chapter 30 in this volume) believes we will have the opportunity to make new noble experiments. This could happen. New environments are one way that cultural and political flowering occurs. Ursula K. LeGuin's anarchist society in *The Dispossessed* is a well-realized example (LeGuin 1974). Or will they be less than idyllic? We have already seen that some personal liberties may be threatened. Will there be further Cockell-style threats to liberty (Cockell 2008)?

2.2 Necessary decentralization

The large size of space cities both allows and *requires* decentralization. Airtight modular construction is essential in case of a hull breach by an asteroid or by an uncontrolled space vehicle moving at speed. A design reminiscent of the watertight sections of the Titanic (only improved!) will be needed. Similarly, oxygen production will be needed in each airtight module, as will farming, as the time needed to repair a major breach could be months. For farming the isolation of modules should be particularly strong to avoid a single crop disease, like the potato blight that set off the Irish Famine, from destroying all the food of the entire space city. In addition, lifeboats capable of reaching another space city will be needed in case of a major unrepairable failure. All these safety requirements are also features of 'freedom engineering' (Cockell 2019) and so can be expected to inhibit autocracy by raising the cost of imposing it and to favour democracy.

Democracy may happen, but it is not inevitable. Feudal systems are decentralized but are locally autocratic. What systems could we imagine that will constrain a would-be autocrat or oligarch at the modular level? Will the existence of many other modules provide a moderating social force on the governments of others?

Perhaps it doesn't matter. If there are a million space cities some will surely find non-autocratic governance solutions among their myriad experiments in social forms.

Will seekers of liberty be able to immigrate into these freer space cities? Refugee crises are a continual source of friction on Earth. Yet there is far more space available on Earth, enough that, in the big picture, present restrictions on movement could already be considered a form of apartheid (De Blij 2009; Caplan and Weinersmith 2019) The friction may be heightened in space cities, and may even be unsurmountable, if they are sized for a given population and are fully occupied. Attitudes to migration will depend strongly on the margins for life support (oxygen, food) that these cities have. Liberty points us toward having generous margins, as does prudent engineering. Should space cities always preserve that margin by underfilling their living space to allow for immigration? How would rules like that be enforced? Margin may need to be a new feature of freedom engineering.

Perhaps there will always be new space cities being built and refugees could move into the newest one. That question would take us into the realm of post-scarcity economics, and pan-solar system governance. For now, let's avoid those considerations.

2.3 Autarky is impossible

A goal of some space settlement enthusiasts is that their space city would be fully independent of Earth and even of other space cities. The purpose of this pure autarky is that then their space city could pursue its own political and cultural agendas free of any outside interference. They would have the 'freedom to go where the rules have not yet been written' (Chapter 30 in this volume).

Full autarky is unattainable, however. One-hundred per cent efficient recycling is not possible. The International Space Station (ISS) does a remarkable job of recycling water and oxygen, at the cost of some comfort to the resident astronauts. The ISS though reaches roughly 90 per cent efficiency (Jones 2017), requiring a supply of only about 1 tonnes of water per astronaut per year (J. McDowell, private communication). At this rate 65 per cent of the water is gone after 10 years. That is far too short a timescale for a space city. For example, there is no significant growing of food on the ISS and the efficiency of an agricultural system is not yet known. On a longer timescale of 50 years, even if 10 times better efficiency is reached (99 per cent per year), only 40 per cent of the water will remain. Fifty years is a short time compared with the presumed space city design life. Further gains lead to diminishing returns: at 99.9 per cent efficiency half the water is still gone after 50 years. Space cities will need to replenish their raw material supplies on similar timescales. For a benchmark million-person space city that comes to at least 100,000 tonnes per year at ISS use and recycling rates; that's 274 tonnes/day. For a population of 1 trillion the total usage would be 100 billion tonnes per year. Sustaining such low usage rates for decades for so many people, rather than for a few hardy astronauts, will be socially difficult and will impose behavioural constraints. Such a regime would be far from the idyll many will have signed up for. A more plausible replenishment rate may be 10 times larger, 1 million tonnes per year. That's a lot of incoming traffic, equivalent to about 30 SpaceX Starship deliveries per day. Space cities could obtain water and other raw material supplies from another space city, but ultimately the supplies will come from freshly mined asteroids.

As noted in the introduction to this chapter, the resources of asteroids are concentrated in largest asteroids. Whoever controls those asteroids will have power over all space cities. Iron is a good example. It is abundant and is a strong building material in tension as well as compression. (Space cities spinning to create Earth-equivalent gravity on the inside of their surfaces will need considerable tensile strength.) A single Main Belt asteroid, 16 Psyche, which is 225 km across, has a mass of 2.4×10^{16} tonnes (Viikinkoski et al. 2018), and contains enough iron (~3.7×10^{15} tonnes)[8] for about 740 million O'Neill style space cities of a million people each. 16 Psyche also has enough rock for radiation shielding to build, though the NASA *Psyche* mission should pin them down far more accurately, starting in 2026. Even if they are wrong by an order of magnitude, though, it is clear that 16 Psyche has an abundance of raw material.

The next five Psyche-like asteroids in size will have a similar total mass, while the next 50 after them will have just 14 per cent of 16 Psyche's mass (Cellino, Zappalà, and Farinella 1991). These 50 will still each have enough iron and rock to build 2 million space cities so there is no shortage of raw materials. The complexity, and thus the cost, of constructing large numbers of space cities will be greater if the raw material has to be sourced from many asteroids, favouring an ongoing mining operation at a single site.

In an age of post-scarcity, perhaps cost will have become unimportant, but control of resources will surely remain an issue, only now on a solar system-wide scale.

It is easy to envisage a 'Big Six' cartel of suppliers for iron, for example, each controlling one of the largest iron-rich asteroids. Monopolies and oligopolies like this create power. The Big Six could throttle the supply and determine the rate of space city construction for everyone else.

A similar case can be made for water. The largest asteroid, Ceres, contains about 2×10^{17} tonnes of water (Prettyman et al. 2017, Konopliv et al. 2018; McCord and Rogez 2018), enough to supply a trillion people for 20,000 years at 10 tonnes/year, and could control the supply of this literally vital need.

The upshot is that space cities cannot be closed systems but must engage in trade simply to remain inhabitable.

2.4 Independence is hard

Elon Musk's call for humanity to become a 'multi-planet species' as a back-up for Earth (Musk 2017) demands full independence from Earth. Without that the back-up is incomplete. Even if imports are not constrained by oligopolies thanks to some solar system-wide agreement on the distribution of resources, full independence will still be tough to attain because of the way a highly technological society works.

This difficulty is due to the need for specialists. The number of specialties that our technological economy now makes use of is extremely large. The number is well beyond what any 1 million person city can support. Two examples give the concept some grounding.

First, the high-tech industry is extremely concentrated. The most advanced computer semiconductor chips that most of us have in our smart phones are made in advanced chip fabrication plants, or foundries. These foundries are so technologically demanding and so expensive to build and run that 8 billion of us depend almost entirely on just three manufacturers.[9]

Secondly, consider medical specialties. There are today at least 180 medical specialties.[10] Singapore, an advanced city-state with a population of around 6 million, has only 64 of them.[11] A doctor specializing in one of these fields must have enough patients coming to them to treat so that they stay current in what they do. For the more specialist areas that requirement will have to be satisfied by treating patients from many space cities. The need for imports of skills as well as water and other goods will make it harder to maintain a space city disease-free.

A full investigation of how many people it takes to support a complete technological economy such as we have today has not been undertaken, but some rough estimates exist. Charles Stross (2010), making the same point about space cities as I am making in this chapter, estimates between 100 million and 1 billion would be needed.[12] Settling in space, whether on Mars or in space cities, will need to be on this large scale if full independence as a back-up for Earth is the goal. Without a sufficient population the technological level may be hard to maintain. An oft-cited example is Tasmania, where many skills were lost after the island became detached from mainland Australia at the end of the last Ice Age (Henrich 2004).

For a space economy made up of space cities on scale of 1 million people imports of skills and of high-tech goods will be required, as well as imports of raw materials. That

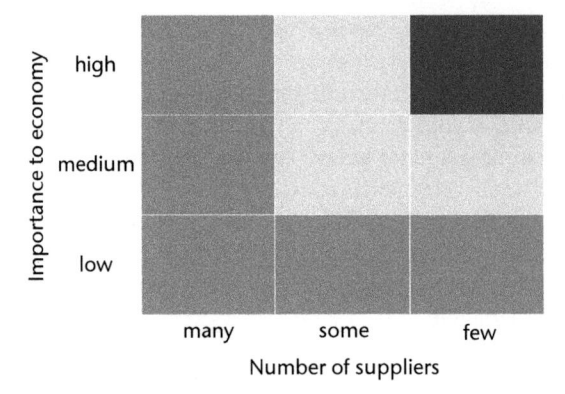

Figure 11.1 Importance to the economy—number of suppliers matrix. The threat to independence for any part of the economy depends on the number of suppliers and the importance of that part to the economy: medium gray = low threat, light gray = medium threat, dark gray = high threat.

leads to the possibility of controls on key imports, both internally by the space city government, and externally by oligopolistic suppliers. Situations like this are not rare on Earth. The price of oil is largely controlled by the Organization of the Petroleum Exporting Countries (OPEC) cartel (Smith 2005); rare earth elements are a strategic resource largely controlled by China, although its attempt to block exports in 2010–2015 was unsuccessful (Vincent 2018). Encouragingly, the necessary existence of large numbers of inter-space-city capable 'lifeboats' for safety could mean that trade is hard to control.

A simple way of envisaging the problem is a version of the 'likelihood—consequences' matrix used, for example, by NASA when considering potential failure modes. In this case it is an 'importance to the economy—number of suppliers' matrix (Figure 11.1). Only if there are few suppliers of high importance skills, products, or resources is the situation ripe for oligopolistic controls (the dark gray box in Figure 11.1).

What examples are there of highly important goods with few suppliers in space futures? It seems likely that all space cities will prioritize the independent maintenance of their air supply, food, and structural safety for their cities. As we do not know how these systems will work it is hard to know if they will need vital imports, but quite possibly not, or not in great quantities. The monitoring of these systems, though, will surely need vast numbers of sensors, computers, and controls. Specialized semiconductor chip production, as already noted, could well be an area where a quasi-monopoly situation could arise and the supply could be used to threaten the survival of a space city. Medical capabilities would also be a high priority but, as we saw, complete medical independence is not viable.

Are there ways to have a large enough population to sustain our full suite of technology? Only planets, and in practice this means Mars, seem likely to be able to support a large enough population. A single space city with 100 million people or more perhaps should not be ruled out. At O'Neill densities a population of one trillion would need 10,000 such giant space cities, each roughly 10 km in radius and 250 km long. For now, such a single giant space city seems to me to be a risky 'eggs in one basket' approach, but I may be too limited in my thinking.

Independence, it seems, is hard.

As an alternative to oligopoly, if the dependence on other space cities is mutual then that could be a countervailing force that ties them into a broader coalition. All space cities will need inputs from many other space cities, leading to a need for coalitions with well-established trading rules. Coalitions with many members will have to have some form of collegial decision-making, at least at the inter-city level. This could form the basis for a democratic society or may at least restrain individual cities from repressing their citizens' liberties.

2.5 Why monopolize in a post-scarcity economy?

The discussion so far assumes that the same ways of thinking we now find obvious will still apply in a post-scarcity economy; that the familiar human drives to domination will still operate.[13] But why should this be so? Imposing your will on millions of other people takes lots of effort. In this ideal post-scarcity economy there is nothing material to be gained from domination, as you can have whatever material goods you want. So, what's the point?

Hobbes, in his famous book *Leviathan*, gives three reasons for disputes (Ch. XIII):

> [I]in the nature of man, we find three principall causes of quarrel. First, Competition; Secondly, Diffidence; Thirdly, Glory.
>
> The first, maketh men invade for Gain; the second, for Safety; and the third, for Reputation. The first use Violence, to make themselves Masters of other mens persons, wives, children, and cattell; the second, to defend them; the third, for trifles, as a word, a smile, a different opinion, and any other signe of undervalue, either direct in their Persons, or by reflexion in their Kindred, their Friends, their Nation, their Profession, or their Name.

For Hobbes, competition is for gain. In a post-scarcity society almost everything would be automatically given to you, so this motivation will be much reduced. What if you wanted your own starship, though? The resources to build a spacecraft that could traverse interstellar distances may be too large for any one space city to manage. Assuming that faster-than-light travel is not possible, a starship would be a 'generation ship' that would take centuries to reach its destination. A generation ship would be quite like a space city; it 'just' needs to have propulsion added. (The moral dubiousness of committing generations to a voyage they did not choose (Regis 2015; Milligan 2015 145–51; Schwartz 2018) may seem less clear when most people are already living permanently in space cities.) In that case, building starships may not be out of reach for some groups who want it. Out of 1 trillion people, how many interstellar visionaries will there be? Few enough that they will not strain even a solar system-wide economy? Perhaps 'competition' really is removed as a motive for quarrels in a post-scarcity future?

Hobbes' 'diffidence' is about keeping your property and family.[14] Defending your property would not seem to be a big issue in a post-scarcity society. Your family, though, is another matter. Sexual jealousy can easily lead to long-lasting disputes, just as the abduction of Helen of Troy set off the Trojan War. We can easily imagine that this motive could arouse passions enough to cause an offended party, who had sufficient power, to restrict the lives of the citizens of some space city. We might blithely imagine that a revolution in human relations, for instance the widespread adoption of post-jealousy polyamory,[15] might have happened in this, already idealized, future. But there is no obvious reason to think any such social revolution would become more likely with the advent of space cities; though perhaps part of the hoped-for cultural flourishing would include such societies. Hobbes' 'diffidence' will then likely still operate, though moderated by the absence of property disputes.

Glory will surely still be with us in full force. The need for respect, relevance, and importance compared to one's peers will remain. Scientists already compete for these rewards rather than for money. As these are all positional goods, it is your ranking relative to your peers that matters, not the absolute value of them (Hirsch 1977). That being the case, the post-scarcity situation of having every material good you want does not help; abundance does not change the importance of our perceived ranking among others. There will inevitably be those who see no path to respect, relevance, and importance other than by physically dominating some group. Possessing the power to restrict the choices and actions of as many people as possible leads to respect of a (grudging) sort. Again, perhaps, in this ideal future such a mind-set will be seen as a sign of someone in need of psychological treatment. Maybe, but we can't assume that will be the case. Most likely 'glory' is not going away.

Reasons for seeking to exercise power over others will still be with us, if moderated, even in a post-scarcity future.

3. Near-term Scarcity: A Few Decades

The era of plenty will not be free of issues for liberty, as we have seen, but the near-term future will have strong, but different, issues too. The way we deal with those near-term issues will have influence well into the era of plenty.

3.1 Near-Earth asteroids

The near-term risks to liberty from space resources lie in the rarity of resource-rich asteroids. Rich deposits of materials are necessarily geographically concentrated, in space as well as on Earth. Otherwise, they wouldn't be rich. The concentration of resources leads to conflicts.

Only the near-Earth asteroids (NEAs) are accessible for mining for the next decade or two. There are far fewer NEAs than there are Main Belt asteroids, and they are relatively small. The largest NEA is Ganymed [16] at just 32 km across, with 0.1 per cent the volume of 16 Psyche. It is stony, and so of no immediate interest for resources. There are about 20,000–40,000 NEAs bigger than 100 m across (i.e. sports stadium-sized, Mainzer et al. 2012; Harris and D'Abramo 2015), but only a few percent are accessible with large mining equipment (multiple tonnes) using current rocketry. About half will probably become accessible with the new rockets now in development, notably the SpaceX Starship. As in the case of the Main Belt, the relatively low numbers of smaller NEAs (Jedicke et al. 2015) means that most of the resources will be in the largest few.

Initial space mining operations could well concentrate on very small (house-sized, ~5–10 m diameter) NEAs, in energetically favourable orbits (Jedicke et al. 2018). NEAs of this size number several millions and, for some, water will be 10 per cent of their mass. This approach is based on selling that asteroid water in orbit. The current cost of delivering a ton of water to LEO is ~$1.5 million with the SpaceX Falcon 9 Heavy.[17] This high price could allow asteroid water to undercut water from Earth. A single water-rich, 10 m sized NEA could have some 400 tonnes of water. Even if sold at half the cost of Falcon Heavy supplied water that would be worth $300 million. The current 'market' for water is space is just the few tonnes per year supplied to the ISS, so a rapid and major market expansion is required to get a return on investment within an investor's short time horizon. Perhaps surprisingly, I find that the prospects for a large growth in demand for water in LEO over the next decade are pretty good (Elvis 2021). This may well be the business strategy that begins commercial asteroid mining, and space resource utilization more generally.

To grow into a multi-billion-dollar enterprise, asteroid mining will have to graduate to larger NEAs. This is plausible to achieve by 2030. As we saw, the advent of new and more powerful rockets will allow a large tonnage of equipment to reach a much larger number of NEAs. In addition, we will know by then where almost all the larger NEAs are. Two surveys for NEAs are beginning in the next few years: the Vera Rubin Observatory Legacy Survey of Space and Time (LSST) will begin operations from a mountaintop in Chile in 2024. LSST will find three-quarters of all NEAs bigger than 100 m across (Jones 2018). This 10-year survey will operate in normal visible light, detecting the reflected sunlight off the NEAs. Many of the water-richest NEAs, though, are as dark as fresh asphalt and so will be missed by the Rubin Observatory survey. To overcome this problem NASA is planning a mission that will survey the sky in infrared light, where the asteroids glow simply because of their temperature. The Near-Earth Object Surveyor mission is due to launch around 2026 - 2028 and take five years to complete its survey.[18] It will find 90 per cent of all NEAs bigger than 140 m diameter. There will be considerable overlap of the two surveys. NEAs that are found in both will have their reflectivity (albedo) measured which will narrow down the search for water-rich asteroids. The results from these two US-funded surveys will be made public as they occur. That will not be the case for privately gathered prospecting data.

Water-rich NEAs make up only a few percent of the total population. Most (85 per cent) are stony, 5 per cent are metallic, and 10 per cent carbonaceous (Stuart and Binzel 2004); the carbonaceous ones can be water-rich, in which case they are called Ch-type (h is for hydrated.) However, based on meteorite statistics, of the carbonaceous and metallic NEAs only about a quarter are rich in water or platinum group metals (Elvis 2014), so about 1 per cent overall will be water-rich. This may be too pessimistic. Rivkin and DeMeo (2019) note that meteorites may underestimate the fraction of water-rich asteroids, and that these may comprise 6 per cent of all NEAs (with a factor of 2 uncertainty). With this more optimistic measurement, the total number of water-rich NEAs larger than 100 m diameter that can be reached by a rocket system with the energetic capability to land on and return from the Moon, is about 300 (also with a factor 2 uncertainty (Rivkin and DeMeo 2019). Most of the water will lie in the handful (~10 to 30) of the largest accessible NEAs, those bigger than 1 km across. These few NEAs will be the glittering prizes of the early asteroid-mining industry.[19]

Identifying the few, or few hundred, water-rich NEAs requires significant additional prospecting work. Even knowing from the two surveys that an asteroid is dark and so potentially rich in water still leaves only a one in four chance of actually being so (Elvis 2014). That success rate implies that to have a high chance (90 per cent to 99 per cent) of finding a water-rich NEA requires the investigation of 10 to 20 good candidates (Elvis and Esty 2014). Prospecting must cost much less than the revenue the asteroid could produce. A remote prospecting technique is to use a sensitive infrared spectroscope in Earth orbit to measure a telltale signature of water (the '3 micron' feature, Rivkin et al. 2015, 65–87). Such a telescope is likely to cost several hundred million dollars[20] but can study many asteroids. Significant savings may be possible. The Twinkle mission will be a test of this (Edwards et al 2019). Near-asteroid ('*in situ*') prospecting techniques include laser ablation spectroscopy and X-ray fluorescence, but one spacecraft per asteroid will be needed to keep the prospecting time short. That means that each one must cost far less than present asteroid missions like the roughly billion-dollar OSIRIS-REx mission, perhaps no more than $20 million each. That is a challenging target. The information gained will be the private intellectual property of the companies that carry out the prospecting.

3.2 Miners' rights?

The high cost of prospecting, let alone mining, NEAs, means that would-be mining companies will need incentives to undertake the enterprise. At a minimum some form of binding assurances that their work will give them some claim to the resources that they find. Present space law explicitly denies the possibility that any celestial body can be 'appropriated' by any means.[21]

Yet without a recognized legal regime of mineral rights, there will be great temptation to short-cut the process by using someone else's work to find a good asteroid to mine. This could take the form of claim-jumping, where you either steal the prospecting data and get to the asteroid first, or you simply arrive at the same asteroid as another outfit, set up a separate mining base and dig faster. More radically, if you find out what someone's target asteroid is, you could engage in 'asteroid rustling' by getting there first and changing its orbit so that the original prospectors can never find it, but you can.

To encourage space resource utilization, countries may start making broad interpretations of 'appropriation' that favour miners. The US-led 'Artemis Accords', signed by 21 nations to date, make this claim (NASA 2020). Already several countries recognize that material from a celestial body, including asteroids, belongs to whoever removed it from the body. Extensions of this legalization trend could include some form of mineral rights, at least for a period of time, to fend off competitors from claim-jumping or rustling 'your' asteroid.

To start up this risky industry these rules may need to be strong. If they are too strong, though, then the concentration of resources in a few best asteroids could lead to nascent monopolies held by those who establish their rights over those few. Their rights may become entrenched and counterproductive. What body will issue and control these mining rights? There is none at the moment.

If there is expensive mining equipment being used for years at a large NEA, then occasionally it will surely break down and need maintenance infeasibly far from Earth for teleoperation of repair equipment. People on site will be essential. This is because, for now, artificial intelligence (AI) depends on huge training sets. One place there will not be huge training sets is in the novel environment of space. People, instead, quickly learn not to touch a hot stove twice. That still uniquely human adaptability will be vital in space.

We can expect, then, that skilled workers will be going to asteroid mining operations. They may have minimal legal protections at first. The mine owners' interests may well prevail, as so they do often on Earth. For the initial six-month-long expeditions to NEA mines workers' rights, health, and safety won't seem like a top priority and so may not be legislated by the mining companies' state of registration. In addition, the rare expertise of the repair crews may be enough to protect them from exploitation.

As humanity begins to explore and use the resources of the Main Belt, though, repair trips could extend to several years and perhaps longer, if, for example, a permanent maintenance base is set up on the moons or surface of Mars. These bases will begin as fully owned company towns, with workers dependent on the good will of the company. Elon Musk has already made comments about working off loans on Mars that, to others, suggest indentured servitude (McKay 2020; Horn 2021), and about not accepting international law on Mars (Cuthbertson 2020; Horn 2021). These starting positions do not bode well.

The threats to individual liberty at these remote locations will be strong and hard to police.

4. The Founder Effect

The problem created by these near-term measures for the post-scarcity stage of human settlement of the solar system that follows it is that legal regimes are 'sticky' (Heller and Salzman 2021). That is, once set up they are hard to change. Remnants of law from the initial times of scarcity will persist into this later time.

Heller and Salzman (2021) give the example of oil rights on land in Texas. The law is based on the shallow well technology used at the time the legislation was passed. This technology created no interference with neighbouring properties. Decades later the technology had transformed. Deep drilling and pumps led to major interference between neighbouring claims. Texas law did not adapt to these changed circumstances, leading to a situation where whoever pumps fastest wins.

This history tells us that we can expect that regimes we set up in the initial era of scarcity of space resources will persist into a post-scarcity future. We need to be mindful of how we set up these regimes now.

More broadly, what the small group involved in the development of a space economy do now will affect the culture of the space economy beyond the strictly legal regime. As Wilbur Zelinsky (1973) notes:

> [E]ven a few score, initial colonizers can mean much more for the cultural geography of a place than the contributions of tens of thousands of new immigrants a few generations later.

An analysis of the persistence cultural influences in the United States over centuries by Woodard (2011) reinforces this conclusion. Woodard finds that, of the 11 cultures he identified in North America, only that of the western United States was formed more by the environment than by the cultural history. The US West is the most extreme environment he considered. Space is even more extreme. Perhaps it will be extreme enough that it will change the cultures people arrive with?

DeVigne (Chapter 10 in this volume) discusses how we might deliberately build a de-colonized culture that limits conflict as we begin to inhabit space, by getting away from the frontier paradigm. Doing so would fit with our long-term needs but would face strong resistance from would-be space miners for whom it would be at best irrelevant, and from those deliberately seeking the freedom of a frontier. Possibly we will first develop this culture in space communities in Earth orbit, where long-term living in space is likely to develop first.

How we act now will set the stage for decades or more to come.

5. Solving Common Pool Resource Problems

Liberty in space will depend on how we govern space resources. Space resources, which are dominated by those of the Main Belt asteroids, are an example of a 'common pool resource'. The problem is most famous as 'the tragedy of the commons'

(Hardin 1968), wherein it is in every individual's interest to graze as many cows as possible on the common land, while the group as a whole would benefit from everyone limiting their cows' grazing. The case mentioned above of Texan oil rights leading to overuse is another example. Free-for-all situations, where it is a case of everyone for themselves, are, somewhat paradoxically, destructive of liberty.

The good news is that people routinely solve the problem of common pool resources without guidance from Edmund Burke's 'divine architect' via bottom-up processes (Chapter 19, this volume). Space resources will almost certainly be managed the same way. It is no longer plausible to rely on the top-down architecture of the treaty-making process. Not least this is because the number of players involved has grown dramatically. In 1967 the only players that mattered were the United States and the Soviet Union, making negotiations easier even for these sworn antagonists. Now there are at least a dozen nations, international agencies, and private companies planning to land on the Moon within the decade (Krolikowski and Elvis 2021). Some form of 'legal pluralism' with partial solutions implanted *ad hoc* will be essential (as Saskia Vermeyelen notes in Chapter 25 of this volume). Eventually they may converge into a consistent system. Vermeyelen also talked of the view of people in the resource-poor Kalahari of 'property as generosity'. Learning to share scarce resources is at the heart of common pool resource questions.

We have particularly followed the work of Elinor Ostrom and her colleagues. Ostrom won the 2009 Nobel Prize in Economics 'for her analysis of economic governance, especially the commons'.[22] Her approach was highly empirical; that is, she looked at actual solutions worked out by many groups around the world on the sharing of resources.

Led by Alanna Krolikowski, we have distilled a few techniques that work out of the writings of Ostrom and her colleagues (Elvis, Krolikowski, and Milligan, 2020):

1. work locally, incrementally
2. identify shared interests
3. define the problems
4. lengthen the time horizon
5. design accommodating platforms
6. establish habits of cooperation
7. create withholdable carrots

In addition, as Tony Milligan stresses, timeliness and justice matter. Action is needed now, in the next five to ten years, before there are strong vested interests in space resources. As noted by Elvis, Milligan and Krolikowski (2016) and Elvis, Krolikowski, and Milligan, (2020) for a brief period we have a real 'veil of ignorance' (Rawls 1971).

Rawls proposed that a just system of laws is more likely if the designers of the system do not know the part they will play in that society, be it princess or pauper; they act behind a veil of ignorance. On Earth this veil is merely a thought experiment;[23] on the Moon, the asteroids, or Mars, it is the real situation. We do not yet know who among the many players will emerge with the greatest resources, or even where these resources are in detail. So long as there are no established winners and losers it will be easier to negotiate a just solution. Justice matters as a matter of principle, but also because just solutions are more likely to be stable solutions. Action now is timely. Once a few players have large investments in specific locations the veil will be ripped away.

6. Conclusions

The physical constraints on the use of space resources and the nature of space cities lead both to a number of challenges to extraterrestrial liberty, and to a number of liberty-enhancing features (Table 11.1). The key points are:

1. Most people in a much-expanded future population will not primarily live on planetary surfaces but in space cities that may well support a million people each. The living space per person in these space cities will probably be very limited, like Singapore today. This crowding will limit individual's behaviour.
2. Space cities will need to be designed for robustness in the case of large-scale failure of any life-support system. The resulting modularity will build-in important constraints on autocracy or oligarchy, favouring liberty, in line with 'freedom engineering' principles.
3. In the long-term space cities will be limited in their autonomy. Full autarky, or even full independence, is not attainable. Major imports of materials and skills are essential. This situation will lead to dependence on other space cities, and this could lead to their losing control over their own affairs. On the other hand, the need to trade with other space cities could be a countervailing force that may tie space cities into a broader coalition that may restrain individual departures into repressions of liberty.
4. In the near-term, rights given to asteroid mining companies may need to be generous to kick-start the industry; these rights may then encourage oligopolies soon afterwards.
5. Maintenance bases for expensive mining equipment will be staffed by humans; their conditions in these remote locations are likely to resemble those in company towns, with little way for limits to be imposed by distant authorities on the corporations that own and run them.
6. Legal regimes set up in the early times of space resource scarcity will be 'sticky' and the cultures arising from these conditions will be hard to change.

Mindfulness of these threats to liberty needs to be at the forefront of our discussions as we set up the law, policy, and culture that will determine the future of liberty in space for the next centuries. We need to act now, while the veil of ignorance is intact.

Table 11.1 Physical Constraints on Liberty in Space over Centuries and Decades.

Factor	Reason	Effect on Liberty	Section
Long-Term: A Few Centuries			
Hard population limits	Design capacity of air, food, power supply.	Promotes coercion in reproduction, migration; threatens liberty.	2.1
Modularity	Safety of air, food supply. Escape.	Raises cost of autocracy; promotes liberty.	2.2
Imperfect recycling	Closed ecologies are hard. ISS achieves 90%.	Need for substantial imports reduces autarky of cities. Oligarchy of suppliers readily achieved.	2.3
Limited skill set available.	Need >>1 million people to have all skills of a technological society.	Need high-tech imports and temporary, but ongoing, imported skilled personnel. Independence is impossible. Limits options for city governments. Essential imports with few suppliers make external control plausible. *Mutual* interdependence may promote liberty.	2.4
Hobbes' 'glory' still applies	Desire for positional goods not satisfied by abundance.	Some people will remain motivated to control others for prestige ('glory'), though 'competition' and 'diffidence' will be reduced.	2.5
Near Term: A Few Decades			
Resources concentrated in a few locations	Most resources in a few large asteroids.	Oligarchy relatively easy to set up.	3.1
Mining company rights	Needed to encourage mining.	Encourages oligarchy if too strong.	3.2
Remote human bases	Needed for repair, maintenance.	Exploitation of workers is easy and hard to police.	3.2
Legal regimes are 'sticky'	Law reform is slow, esp. with established interests.	Legal regimes appropriate for scarcity may be ill-matched to abundance; restricting actors' choices.	3.3
'Founder Effect'	Cultural patterns established early on can persist for centuries.	Patterns of exploitation originating in scarcity may continue into abundance.	3.4

Acknowledgements

I am extremely grateful to my long-term collaborators, Alanna Krolikowski and Tony Milligan for their wisdom. Eric Huff (JPL) provided the link to Charlie Stross' blog post; Ross Centers pointed me to the maximum space city size. Charles Cockell is a source of inspiration and the inventor of the whole field of liberty in space. Other participants in Charles' 'Institutions of Extraterrestrial Liberty' conference provided excellent insights, especially Janet DeVigne, Jarita Holbrook, Thomas Moynihan, Christopher Newman, Jessy Kate Schingler, and Saskia Vermeyelen.

Notes

1. Note that actually building a ringworld would be a really bad idea as the megastructure would not be stable. A collision with a comet or similar body could push the ringworld out of its location, sending one side crashing into the Sun.
2. Area of planetary bodies in million km^2: Earth (land area) 148, Moon 37, Mars 144, Venus 460, Total 838 (Cox 1999).
3. This is for 70 per cent of Earth's gravity (g); a 1g habitat would require a stronger and more massive structure.
4. Countries By Density 2021 <http://www.worldpopulationreview.com>.
5. Robin Williams as the genie in 1992 Disney *Aladdin* movie.
6. There will not be a problem with inbreeding as the space cities we are considering have far more than the roughly 5,000 individuals required to avoid this problem (Bradshaw 2018).
7. 'Fertility, Below-Replacement', *Encyclopaedia Britannica* <https://www.encyclopedia.com/social-sciences/encyclopedias-almanacs-transcripts-and-maps/fertility-below-replacement> accessed 1 September 2021.
8. Using the measured density for 16 Psyche of 3.99+/−0.26 (Kiinkokoski et al. 2018) and densities of 7.8 for iron meteorites and 3.2–3.4 for ordinary chondrite meteorites (Britt and Consolmagno 2003), gives a composition of 15 per cent iron, 85 per cent silicate rock. If the higher *16 Psyche* density of 4.2 +/−0.6 found by Ferrais et al. (2020) is used, then the Fe:rock ratio is 20:80. If 16 Psyche is a rubble pile with many interior voids, then the iron fraction will be larger.
9. Taiwan Semiconductor Manufacturing Company, Limited (TSMC), United Microelectronics Corporation (UMC, Taiwan), and Samsung Electronics Co., Ltd. (South Korea).
10. <https://www.sgu.edu/blog/medical/ultimate-list-of-medical-specialties/> accessed 4 July 2022.
11. Yap 2020.
12. Stross 2010.
13. Though maybe we are too resigned. See 'The real Lord of the Flies: what happened when six boys were shipwrecked for 15 months' <https://www.theguardian.com/books/2020/may/09/the-real-lord-of-the-flies-what-happened-when-six-boys-were-shipwrecked-for-15-months> accessed 2 September 2021.
14. Although this seems at odds with Ristroph (2012) who says that by diffidence Hobbes means the 'uneasiness or anxiety that all individuals … have about their own security and standing vis-à-vis one another'. That sounds like 'glory'.
15. <https://en.wikipedia.org/wiki/Polyamory> accessed 20 July 2022.
16. Not to be confused with Ganymede, the moon of Jupiter.
17. <https://www.spacex.com/media/Capabilities&Services.pdf> accessed 20 July 2022. This is the raw cost/tonne to LEO; the true price for water will be higher as one has to allow for the mass of the water-containing spacecraft, including the water tank, the transfer system to the customer, propulsion, and control systems to deliver the water to where it is needed, and profit.

18. The Near-Earth Object Surveyor mission: <https://neos.arizona.edu/> accessed 9 September 2021.
19. Mining the Moon does not solve the problem. Lunar resources are also mostly highly concentrated (Elvis, Krolikowski and Milligan 2020).
20. Cf NEO Surveyor at ~$500 million. <https://www.businessinsider.com/nasa-advances-asteroid-tracking-space-telescope-after-years-of-limbo-2021-6?op=1> accessed 2 September 2021.
21. In the 1967 'Outer Space Treaty', Article II. 'Treaty on principles governing the activities of states in the exploration and use of outer space, including the moon and other celestial bodies.' United Nations Office for Outer Space Affairs (UNOOSA) <https://unoosa.org/oosa/en/ourwork/spacelaw/treaties/outerspacetreaty.html> accessed 2 September 2021.
22. <https://www.nobelprize.org/prizes/economic-sciences/2009/ostrom/facts/> accessed 2 September 2021.
23. Although a less stringent 'veil of uncertainty' quite often applies in negotiations concerning a new field.

References

Adams, R. B. and Cassibry, J., 'Pulsed Fission Fusion (PuFF) Propulsion System' (2014) <https://www.nasa.gov/sites/default/files/files/Adams_PulsedFissionFusionPropulsionSystem.pdf> accessed 4 July 2022.

Asimov, I. *Foundation* (New York: Gnome Press, 1951).

Asimov, I. *Caves of Steel* (New York City: Signet Books, 1955).

Bock, E., Lamro, F., and Simon, M. 'Effect of Environmental Parameters on Habitat Structural Weight and Cost' in J. Billingham, W. Galbreath, and B. O'Leary (eds), *Space Resources and Settlements, NASA SP-428* (NASA, 1977) 33–60.

Bradshaw, C. 'Why Populations Can't Be Saved by a Single Breeding Pair' *The Conversation*, 22 March 2018 <https://phys.org/news/2018-03-populations-pair.html> accessed 2 September 2021.

Britt, D., and Consolmagno, G.J. "Stony meteorite porosities and densities: A review of the data through 2001", *Meteoritics and Planetary Science* (2003) 38: 1161.

Caplan, B., and Weinersmith, Z. *Open Borders: The Science and Ethics of Immigration* (New York: First Second, 2019) ISBN-13 978-1250316967.

Cellino, A., Zappalà, V., and Farinella, P. 'The Size Distribution of Main-Belt Asteroids from IRAS Data' (1991) 253 *Monthly Notices of the Royal Astronomical Society* 561–74.

Coble, C. R., Murray, E. G., and Rice, D. R. *Earth Science* (Englewood Cliffs, NJ: Prentice-Hall, 1987).

Cockell, C. S. 'An Essay on Extraterrestrial Liberty' (2008) 61 *Journal of the British Interplanetary Society* 255e275.

Cockell, C. S. 'Freedom Engineering—Using Engineering to Mitigate Tyranny in Space' (2019) 49 *Space Policy* 101328 <https://doi.org/10.1016/j.spacepol.2019.07.002> accessed 4 July 2022.

Cox, A. N. (ed). *Allen's Astrophysical Quantities* (Berlin: Springer, 1999).

Cuthbertson, A. 'Elon Musk's SpaceX will "Make its own Laws on Mars"', *Independent*, 28 October 2020 <https://www.independent.co.uk/life-style/gadgets-and-tech/elon-musk-spacex-mars-laws-starlink-b1396023.html> accessed 28 September 2021.

De Blij, H., (2009), "The Power of Place: Geography, Destiny, and Globalization's Rough Landscape", Oxford University Press, ISBN-13: 978-0195367706.

Diamond, J. *Guns, Germs, and Steel* (New York, NY: W.W. Norton, 1997).

Diamandis, P.H., and Kotler, S. *Abundance: The Future Is Better Than You Think* (Free Press, 2012) ISBN 9781451614213.

D'Onghia, E., 2022, "CREW HaT: Cosmic Radiation Extended Warding using the Halbach Torus", NASA NAIC, https://www.nasa.gov/directorates/spacetech/niac/2022/CREW_HaT/ (accessed 22 Ajugust 2022).

Economist 'Vertical Farms Are Growing More and More Vegetables in Urban Areas' <https://www.economist.com/technology-quarterly/2021/09/28/vertical-farms-are-growing-more-and-more-vegetables-in-urban-areas> accessed 4 October 2021.

Edwards B., Lindsay, S., Savini, G., et al. 'Small Bodies Science with the Twinkle Space Telescope' (2019) 5(03) *Journal of Astronomical Telescopes Instruments and Systems* 1. DOI: 10.13140/RG.2.2.34108.49287.

Elvis, M. 'How Many Ore-bearing Asteroids?' (2014) 91 *Planetary and Space Science* 20.

Elvis, M. *Asteroids: How Love, Fear, and Greed will Determine our Future in Space* (New Haven, CT: Yale University Press, 2021).

Elvis, M. and Esty, T. 'How Many Assay Probes to Find One Ore-bearing Asteroid?' (2014) 96 *Acta Astronautica* 227–31.

Elvis, M., Krolikowski, A., and Milligan, T. 'Concentrated Lunar Resources: Imminent Implications for Governance and Justice' *Philosophical Transactions of the Royal Society* doi: 10.1098/rsta.2019.0563.

Elvis, M. and Milligan, A. 'How Much of the Solar System Should We Leave as Wilderness?' (2019) 162 *Acta Astronautica* 574.

Elvis, M., Milligan, A., and Krolikowski, A. 'The Peaks of Eternal Light: A Near-Term Property Issue on the Moon' (2016) 38 *Space Policy* 30.

Farmland, LP, 'One Acre Feeds a Person' (2012) <https://www.farmlandlp.com/2012/01/one-acre-feeds-a-person/#.YVsX2n2SnPx> accessed 4 October 2021.

Ferrais, M., Vernazza, P., Jorda, L., et al. 'Asteroid (16) Psyche's Primordial Shape: A Possible Jacobi Ellipsoid' (2020) 638 *Astronomy & Astrophysics* L15.

Gent, E. 'China Wants to Build a Mega Spaceship That's Nearly a Mile Long', *Space.com*, 3 September 2021.

Hardin, G. 'The Tragedy of the Commons' (1968) 162 *Science* 1243–8.

Harris, A. W. and D'Abramo, G. 'The Population of Near-Earth Asteroids' (2015) 257 *Icarus* 302–12 <https://doi.org/10.1016/j.icarus.2015.05.004> accessed 4 July 2022.

Heller, M. and Salzman, J. *Mine!* (New York, NY: Doubleday, 2021).

Henrich, J. 'Demography and Cultural Evolution: How Adaptive Cultural Processes Can Produce Maladaptive Losses: The Tasmanian Case' (2004) 69 *American Antiquity* 197.

Hirsch, F. *The Social Limits to Growth* (London: Routledge & Kegan Paul, 1977).

Horn, B. 'Criticism of Space Cowboys Isn't Enough' *The Space Review*, 27 September 2021 <https://www.thespacereview.com/article/4253/1> accessed 28 September 2021.

Jedicke R., Granvik M., Micheli M., et al. 'Surveys, Astrometric Follow-up, and Population Statistics' in P. Michel, F. E. DeMeo, and W. F. Bottke (eds), *Asteroids IV* (Tuscon, AZ:

University of Arizona Press, 2015) 795–813 doi: 10.2458/azu_uapress_9780816532131-ch040.

Jedicke, R., Sercel, J., Gillis-Davis, J., et al. 'Availability and Delta-V Requirements for Delivering Water Extracted from Near-Earth Objects to Cis-Lunar Space' (2018) 159 *Planetary and Space Science* 28–42 <https://doi.org/10.1016/j.pss.2018.04.005> accessed 4 July 2022.

Jones, H. W. 'Would Current International Space Station (ISS) Recycling Life Support Systems Save Mass on a Mars Transit?' 47th International Conference on Environmental Systems, 16–20 July 2017, Charleston, South Carolina, ICES-2017-85. https://ntrs.nasa.gov/citations/20170007268. (Accessed 22 August 2022).

Jones, R. L, Slayter, C.T., Moeyens, J., et al. 'The Large Synoptic Survey Telescope as a Near-Earth Object Discovery Machine.' (2018) 303 *Icarus* 181–202.

Konopliv, A.S., Park, R.S., Vaughan, AT., et al. 'The Ceres Gravity Field, Spin Pole, Rotation Period and Orbit from the Dawn Radiometric Tracking and Optical Data' (2018) 299 *Icarus* 411.

Krolikowski, A. and Elvis, M. 'Space Resources and Policy' in S. M. Pekkanen and P. J. Blount (eds), *Oxford Handbook of Space Security* (Oxford: Oxford University Press, 2021).

LeGuin, U. K. *The Dispossessed* (London: Harper and Row, 1974).

Mainzer, A., Masiero, J., Grav, T., et al. 'NEOWISE Studies of Asteroids with Sloan Photometry: Preliminary Results' (2012) 745(1) *Astrophysical Journal* 7 <https://doi.org/10.1088/0004-637X/745/1/7> accessed 4 July 2022.

McCord, T. B. and Castillo-Rogez, J. C. 'Ceres's Internal Evolution: The View after Dawn' (2018) 53(9) *Meteoritics & Planetary Science* 1778–92.

McKay, T. 'Elon Musk: A New Life Awaits You in the Off-World Colonies—for a Price', *Gizmodo*, 17 January 2020 <https://gizmodo.com/elon-musk-a-new-life-awaits-you-on-the-off-world-colon-1841071257> accessed 28 September 2021.

Milligan, T. *Nobody Owns the Moon* (Jefferson NC: McFarland, 2015): Ch. 10, 145–151.

Musk, E. 'Making Humans a Multi-Planetary Species' (2017) 5 *New Space* 46 <https://doi.org/10.1089/space.2017.29009.emu> accessed 4 July 2022.

NASA. 'The Artemis Accords: Principles for Cooperation in the Civil Exploration and Use of the Moon, Mars, Comets, and Asteroids for Peaceful Purposes'. U.S. National Aeronautics and Space Administration (2020) <https://www.nasa.gov/specials/artemis-accords/img/Artemis-Accords-signed-13Oct2020.pdf> accessed 4 July 2022.

Niven, L. *Ringworld* (London: Ballantine Books, 1970).

O'Neill, Gerard K. (1977). *The High Frontier: Human Colonies in Space*. New York: William Morrow & Company. ISBN 0-9622379-0-6.

Paluszek, M., Pajer, G., Razin, Y., et al. 'Direct Fusion Drive for a Human Mars Orbital Mission', *65th International Astronautical Congress*, (Toronto, Canada: International Astronautical Federation September 29–October 3, 2014), IAC-14, C4,6.2.

Petit, J.-M., Chambers, J., Franklin, F., et al. 'Primordial Excitation and Depletion of the Main Belt' in W. F. Bottke Jr., A. Cellino, P. Paolicchi, et al. (eds), *Asteroids III* (Tucson, AZ: University of Arizona and LPI, 2002) 711–23.

Pluymers, K. *No Wood, No Kingdom: Political Ecology in the English Atlantic* (Philadelphia, PA: University of Pennsylvania Press, 2021).

Powell, C. S. 'Jeff Bezos Foresees a Trillion People Living in Millions of Space Colonies. Here's What He's Doing to Get the Ball Rolling', *Mach NBC* <https://www.nbcnews.com/mach/science/jeff-bezos-foresees-trillion-people-living-millions-space-colonies-here-ncna1006036> 4 July 2022.

Prettyman, T. H., Yamashita, N, Toplis, M.J., et al. 'Extensive Water Ice within Ceres' Aqueously Altered Regolith: Evidence from Nuclear Spectroscopy' (2017) 355 *Science* 55.

Rawls, J. A. *A Theory of Justice* (Cambridge, MA: Harvard University Press, 1971).

Regis, E. 'The Moral Status of Multigenerational Interstellar Exploration' in B. Finney and E. Jones (eds), *Interstellar Migration and the Human Experience* (Berkeley, CA: University of California Press, 1985) 248–59.

Ristroph, A. 'Criminal Law for Humans' in D. Dyzenhaus and T. Poole (eds) *Hobbes and the Law* (Cambridge: Cambridge University Press, 2012) 97–117.

Rivkin, A. S., Campins, H., Emery, J. P. 'Astronomical Observations of Volatiles on Asteroids' in P. Michel, F. E. DeMeo, and W. F. Bottke (eds), *Asteroids IV* (Tuscon, AZ: University of Arizona Press, 2015) 65–87.

Rivkin, A. S. and DeMeo, F. E. 'How Many Hydrated NEOs Are There?' (2019) 124 *Journal of Geophysical Research: Planets* 128–42.

Sadler, P. *Sustainable Growth in a Post-Scarcity World: Consumption, Demand, and the Poverty Penalty* (Gower Applied Business Research, 2010).

Schwartz, S. J. 'Worldship Ethics: Obligations to the Crew' (2018) 71 *Journal of the British Interplanetary Society* 53–64 <https://bis-space.com/membership/jbis/2018/JBIS-v71-no02-February-2018-kf92bx.pdf> accessed 4 July 2022.

Smith, J. L. 'Inscrutable OPEC? Behavioral Tests of the Cartel Hypothesis' (2005) 26 *The Energy Journal* 51.

Stross, C., (2010), "Insufficient Data", Charlie's Diary, http://www.antipope.org/charlie/blog-static/2010/07/insufficient-data.html (Accessed 4 July 2022).

Stuart, J. S. and Binzel, R. P. 'Bias-corrected Population, Size Distribution, and Impact Hazard for the Near-Earth objects' (2004) 170(2) *Icarus* 295–311 <https://doi.org/10.1016/j.icarus.2004.03.018> accessed 4 July 2022.

Thomson, W. 'India's Food Security Problem' (2012) 2 April *The Diplomat* <https://thediplomat.com/2012/04/indias-food-security-problem/> accessed 4 July 2022.

Viikinkoski, M., Vernazza, P., Hanuš, J, et al., (2018), "(16) Psyche: A mesosiderite-like asteroid?", *Astronomy &Astrophysics*, 619, L3.

Vincent, J. 'China Can't Control the Market in Rare Earth Elements because They Aren't All That Rare' (2018) 17 April *The Verge*. https://www.theverge.com/2018/4/17/17246444/rare-earth-metals-discovery-japan-china-monopoly. (Accessed 22 August 2022).

Woodard, C. *American Nations: A History of the Eleven Rival Regional Cultures of North America* (London: Viking, 2011).

Yap F. 'Medical Specialist in Singapore 2020: Organized by Specialty, Hospital and Treatment' (2020) <https://onedaymd.aestheticsadvisor.com/2015/07/how-to-look-for-medical-specialist-in.html> accessed 4 July 2022.

Zelinsky, W. *The Cultural Geography of The United States* (Englewood Cliffs, NJ: Prentice-Hall, 1973) 13–14.

12

'We have come to Mars for good'

Science fiction, sovereignty, and the challenges of liberty

Simon Malpas

1. Introduction: Living on Mars

Rachel Greene's *Once Upon a Time I Lived on Mars* reflects upon her time as a member of NASA's first HI-SEAS mission in 2013, which simulated taking part in the colonization of Mars by living for four months with a small group of crewmates in a geodesic dome on the Hawaiian volcano of Mauna Loa. Towards the end of the book, she speculates on the potential for founding a real Martian colony in the coming decades, and concludes that '[a]ll things considered, and due to the funding structure of NASA, it looks like the private space industry is best poised to get people to actual Mars' (Greene 2021, 194). Without the nationalistic and ideological imperatives underlying the drive for technological supremacy that lay at the heart of the Cold War Space Race, neither the United States nor any other nation is likely in the near term to have the economic will to devote such a major part of its budget to a venture of this type, and the onus, she suggests, will more probably fall on entrepreneurs and private companies who are more able to rapidly monetise the research and development involved in order to recoup the outlays required (for other versions of this argument, see Persson 2015; Reinstein 1999; Velocci 2012). Greene identifies Elon Musk's SpaceX as at the forefront of the push for Mars colonization, and this provokes some far-reaching questions about the status of such a colony:

> It's dark, but imagine the scenario of a struggling SpaceX Mars colony. Reports of longing for fresh water, for fresh air, for blue skies. What kind of sympathy would they receive from people back home? Empathy might be even less likely. What would the effect of a commercial Mars colony have back on Earth, if any at all? Might it just feel like another spectacle? But also, what policies, action, art, or perspective shifts might it inspire? (Greene 2021, 170)

Without the immediate capacity for appeals to national identity, calls for empathy and assistance for colonists, ranging from requests for charitable donations of blue-sky mementos to the patriotic-style demand to 'bring our boys [and girls] back home' in the event of an emergency, may be much more difficult to make. On the other hand, the existence of a 'non-national non-terrestrial' community might also have the capacity to generate new questions and perspectives not just about off-world existence but also about governance and culture on Earth. Greene's questions stem from a more fundamental problem that a 'private' or a 'private–public' colony poses: what would

Simon Malpas, *'We have come to Mars for good'*. In: *The Institutions of Extraterrestrial Liberty*.
Edited by Charles S. Cockell, Oxford University Press. © Simon Malpas (2022).
DOI: 10.1093/oso/9780192897985.003.0013

be the sovereign status of such a colony, and how might its political relations with the nation states and corporations of Earth be regulated?

A straightforwardly 'nation verses corporate' model is overly simple, however: the exceptional financial, scientific, technological, and social costs of not just reaching our nearest planetary neighbour but establishing a permanent colony there will require significant cooperation between multiple bodies. Greene, quite sensibly, doesn't finally assert a simple choice between an entirely national or a wholly private colony; instead, she argues that the scale of resource involved will require extensive collaboration between multiple nations and numerous companies (Greene 2021, 191–206), in a relation she that she can only conclude must be 'perhaps best categorised as "complicated"' (Greene 2021, 194). The complications here, as well as being economic, are political, and the 'perspective shifts' Greene evokes will necessarily go to the heart of the identity, governance, and social structures of such a colony, as well as its relationship with nations back on Earth. In potentially resolving some of the financial and technological challenges, this sort of national–corporate collaboration (which is already becoming increasingly common in space science) will only add further levels of complexity to the political ones involved in establishing a long-term colony. At the heart of these will lie the problem of liberty and the fundamental question of what it might mean to be a part of such a community: what would it be in such a circumstance to be an extraterrestrial subject or a citizen of Mars?

Providing concrete answers to the questions raised so far is beyond the scope of this chapter. Even so, I hope I can begin to clarify some important aspects of them and point towards a potential source from which at least the beginnings of answers might arise. Work undertaken in psychology provides extensive data about the stresses of non-terrestrial survival, but the impact of moving from 'long duration' stays aboard, for example, orbital space stations to something approaching 'semi-permanent' settlement on another world raises challenges not immediately amenable to empirical testing. Supplementing this work with explorations of fictional projections of colonization, I want to argue, might contribute to extending our understanding of not just the psychology but also the political and social impacts of extraterrestrial colonisation.

2. 'For Good': Politics and Colonization in Kim Stanley Robinson's Mars Trilogy

Kim Stanley Robinson's Mars trilogy (*Red Mars* (1993), *Green Mars* (1994), and *Blue Mars* (1996)) is one of the most influential pieces of literary extraterrestrial colonization science fiction published in the past few decades. The trilogy charts the settlement and terraforming of the planet, and the establishment of a permanent Martian community. Criticism of the novels has tended to focus on the utopian politics of the depictions of colonization (Jameson 2000) and the ethics of terraforming (Pak 2014), but I want to argue that they, and *Red Mars* in particular, also pose important questions about the nature of liberty in an extraterrestrial community where commercial and national sovereignties collide.

Near the beginning of *Red Mars*, those about to become the first hundred settlers, having arrived in orbit around the planet and as they are about to touch down for the first time, gather to discuss the impending birth of their new colony. One, a Russian scientist named Arkady Bogdanov, presents his crewmates with the following challenge:

> 'We have come to Mars for good. We are going to make not only our homes and our food, but also our water and the very air we breathe—all on a planet that has none of these things. We can do this because we have the technology to manipulate matter right down to the molecular level. This is an extraordinary ability, think of it! And yet some of us here can accept transforming the entire physical reality of this planet, without doing a single thing to change ourselves, or the way we live. To be twenty-first century scientists on Mars, in fact, but at the same time living within nineteenth century social systems, based on seventeenth century ideologies. It's absurd, it's crazy, it's—it's' he seized his head in his hands, tugged his hair, roared 'It's *unscientific*! ... We must terraform not only Mars, but ourselves.'
>
> (Robinson 1993, 112–13)

The provocation here is one to which all three novels in the trilogy frequently return. It arises from a discrepancy between the cutting-edge science and technology and the much older social, ethical, and political systems, which the novels depict as struggling to keep up with and adapt to the challenges presented by the new settlement and the new science it generates. What makes this discrepancy so pressing is that, in contrast to previous missions of space exploration, the settlement of Mars will be 'for good': the aim of the mission is to settle, to live, rather than just to explore or visit, and this, Bogdanov asserts, will require not just the technological terraforming of the planet but 'must' also entail the transformation of their selves and the social systems and ideologies of the community and society in which those selves exist.

Bogdanov's 'for good' emphasizes an important distinction proposed by Adam Stevens (2015) between a temporary 'outpost' with short-term or frequently changing staff and a colony proper, which would be made up predominantly of longer-term or permanent inhabitants. This distinction has consequences for both the psychological challenges faced by inhabitants and the range of potential relationships between liberty and sovereignty that these challenges will make necessary. Put bluntly, a semi-permanent or permanent member of a colony will have quite different demands in terms of personal freedom and self-determination than someone who is merely suspending their normal liberties in what I will identify in the next section as a 'state of exception' while in pursuit of short-term personal or collective goals.

Extensive empirical research to model the tensions and strains of the longer-duration non-Earth survival is being undertaken by psychologists (and Greene's account can be considered as a contribution to that), but the key difference of 'semi-permanence' places limits on the efficacy and applicability of those models to colonization, and that research might benefit from interaction with the alternative modes of exploration made possible by artistic and literary representations. In the next sections, I should like to focus, first, on recent discussion of the psychological challenges produced by non-terrestrial environments; second, on the ways in which the intensification of these pressures in longer-term colonization requires an exploration of the fundamental philosophical problems of sovereignty and liberty; and, finally, on beginning to make a case about how literary science fiction might provide a means to investigate the ways in which these categories can be rethought for a potential non-Earth community.

3. A 'State of Exception': The Psychological Challenges of Extraterrestrial Existence

Physically and mentally, extraterrestrial living will be hard. Frederik Pohl and C. M. Kornbluth's 1952 novel *The Space Merchants* depicts the struggles of an advertising agency executive to persuade the population of the United States to sign up for a voyage to colonize Venus (admittedly, a planet with a considerably less hospitable environment than Mars). An initial pitch generates the following response from the head of the company's 'Industrial Anthropology' department:

> You *can't* make people want to live in a steam-heated sardine can. All our folkways are against it. Who's going to travel sixty million miles for a chance to spend the rest of his life cooped up in a tin shack—when he can stay right here on Earth and have corridors, elevators, streets, roofs, all the wide-open space a man could want? It's against human nature ... ..
>
> (Pohl and Kornbluth 2003, 33)

Although, as the narrative progresses, the novel's focus shifts from the mechanics of colonizing and terraforming Venus to become more interested in the politics of advertising, identifying extraterrestrial colonization as the quintessential 'hard sell' for any marketing campaign speaks directly to a question I should like to explore in this section: given the hostility of any non-Earth environment and the restrictions on day-to-day freedoms required to keep settlers alive, how might the psychological challenges faced by permanent (or, at the very least, long-term) colonists be defined, explored, and mitigated?

The technological difficulties involved in establishing an extraterrestrial colony are matched by the psychological ones (Kanas and Manzey 2008). The dangers to life presented by a non-Earth environment mean survival and liberty will end up at odds with one another as the physical restrictions and necessity of constant monitoring required to maintain a safe ecosystem impact negatively on an individual's sense of personal freedoms. The balance between these two factors is ultimately a matter for political determination: the physical survival of colonists in the hostile non-terrestrial environment will require forms of technological organization and surveillance that necessarily conflict with those modes of liberty that have traditionally been considered essential for individual psychological flourishing on Earth. Charles Cockell identifies what he calls an 'inherent tension' between them when he describes the 'friction between the collective efforts needed to survive in an extreme environment and prevent instantaneous death and the deeper human urge to individual liberty and an independent state of mind. The extraterrestrial environment has a tendency to centrifugally drive these two states apart to their utter extremes' (Cockell 2015, 47). The capacity for individual self-determination held sacrosanct by liberal politics that give rise to the liberties evoked in the passage from Pohl and Kornbluth as 'folkways' essential to 'human nature' will be tested to their limit in the hostile environment of an extraterrestrial colony. And, as a result, the familiar political models that underpin social cohesion on Earth will require, at the very least, revision and, potentially, radical transformation if a non-Earth community is to function.

Research into the psychology of survival in non-terrestrial environments is an established part of space science. The investigation of the psychological impact of long-duration space missions has developed in close relation to the extended orbital voyages undertaken by astronauts working on the space station programmes of the last 50 years (Salyut in the 1970s and 1980s; Mir between 1986 and 2001, and the International Space Station (ISS) from 2000). This research has focused particularly on questions about the personality types best able to adapt and cope with the stresses of space, the protection of behavioural health, the maintenance of complex cognitive and performance skills, and the organizational structures best suited for continuing productive interpersonal relationships (Kanas and Manzey 2008; Kanas et al. 2009). Among the general principles that have been established is a broad consensus that extending the duration of any space mission increases the potential for four key areas of conflict. These, according to Kanas, et al., are: '(1) decreased crew morale and compatibility, (2) withdrawal or territorial behaviour as crewmembers cease to interact with each other, (3) the scapegoating of an individual as a "solution" to group conflict, and (4) the formation of subgroups that compete with each other and destroy crew unity' (Kanas et al. 2009, 671). A good deal of empirical data has been generated by analysing astronauts (as well as volunteers such as Greene who have simulated the isolation of space in Earth-based environmental models), and detailed recommendations for the selection, training, and support of astronauts have been shared between research psychologists and those running such space programmes as ISS to minimize these sources of conflict. As I suggested at the end of the last section, however, in terms of analytical understanding of the range of psychological challenges posed by non-Earth colonization, these models have significant limits.

First, even landing humans on Mars will present psychological problems greater than any yet faced by astronauts. As researchers such as Dietrich Manzey (2004) and Douglas Vakoch (2011) argue, the durations and distances involved in a mission to Mars will intensify the pressures beyond anything currently modelled:

[T]here are also major differences which render missions to Mars a unique challenge from a psychological point of view ... Of course, the Russian space program has proven that a stay in low-Earth orbit of up to 438 days is possible, but this evidence is based on just one cosmonaut who never experienced a period of extreme social monotony that lasted longer than a few months (due to crew exchange and visiting crews), and who got a large amount of ground-based support. During a voyage to Mars and a stay on the martian surface, crew members are expected to endure extraordinary long periods of confinement and social isolation which amount up to 1000 days.

(Manzey 2004, 783)

As well as the extended duration of a mission to Mars, the added pressures of time delay in communications, the difficulties of resupply or rescue from Earth in the event of any problems, and currently unquantifiable challenges such as the 'Earth-out-of-view' phenomenon (Horneck et al. 2006, 756; Kanas and Manzey 2008, 228–9) will require increased levels of autonomy from astronauts that has implications and consequences empirical research currently finds difficult to specify (Manzey 2004, 785). This increase in autonomy from mission control, and the need for similarly enhanced self-reliance and team unity, will amplify the pressures on the four areas of conflict set out above and necessitate the development of what Horneck et al. refer to as 'specific psychological countermeasures' (Horneck et al. 2006, 757) to allow astronauts to adapt to the challenge of living and working in a non-Earth environment for extended periods. At present, as each of the writers cited here acknowledges, precise specification of what such measures might entail has not been possible and is the subject of ongoing research.

Establishing a permanent colony, however, will amplify the psychological challenges even further than this, and in a manner that brings the relationship between liberty and sovereignty into sharp relief. The psychological and physical pressures of space travel mean the norms of liberal freedom must be suspended for the duration of any voyage as the models and recommended mitigations noted above make clear. More than this, though, the distances, physical challenges and costs associated with travel between Earth and Mars will mean that most of the inhabitants of such a colony will need to commit to long-term or even semi-permanent settlement. One cannot simply commute to Mars. Devoting a limited period of one's life to the exploration of space is one thing but settling permanently on another planet will generate challenges of a different order. To define this difference, it is helpful to identify the former case, the suspension of many Earth-based liberties for a time-limited period to achieve the goals and projected positive outcomes of a space mission, in terms of the notion of a sovereign 'state of exception', whereas the latter will entail the emergence of a new sovereignty.

Broadly understood, a state of exception can be defined as 'a multitude of different crisis reaction mechanisms that exist in the context of differentiated statehood [and] have in common that they all enhance the government's power of action and decision-making, if the relevant conditions to call out a [state of exception] are met. The declared aim of these instruments is to return as quickly as possible to the pre-crisis situation' (Lemke 2018, 374). Although examples of this practice can be found going back to classical civilizations, such as the frequently cited suspension of democracy and appointment of a dictator to cope with crises in the Roman republics, the term's most influential modern formulation occurs in Karl Schmitt's 1922 text *Political Theology*, which defines a state of exception as the sovereign decision to suspend the rule of law (Schmitt, 2005). In the context I am suggesting here, the state of exception identifies the required suspension of terrestrial freedoms to survive the 'crisis' presented by the hostile environments of space: mission control assumes the sovereign powers necessary to monitor, determine, and regulate the actions of astronauts to enable them to survive, and this, for the time-limited duration of the mission, requires forms of surveillance and discipline that effectively suspend the astronauts' capacities to act with the freedom and independence normally enjoyed by citizens on Earth. The psychological pressures explored in this section are, in part, generated by the imposition of such a state of exception, and techniques suggested for their amelioration rely significantly on the astronauts' recognition that things will 'return as quickly as possible to the pre-crisis situation' on their return to Earth.

In contrast to this, it makes no sense to define the longer-term citizens of an extraterrestrial colony as facing a 'permanent suspension' of liberties as such 'permanence' would entail the tyrannical extirpation of liberty itself rather than any meaningful 'suspension'. A non-time-limited state of exception falls inevitably into tyranny (Agamben 2005), and this means that for the longer-term settlement of space to have any ethical foundation, an alternative notion of sovereignty that can generate different senses of liberty and respond otherwise to the psychological pressures of the hostile environment will be required.

4. 'A Common Power to Keep Them All in Awe': Extraterrestrial Sovereignty

While there is no specifically defined time limit beyond which a 'state of exception' becomes tyrannic, the move from a temporary suspension of liberties to what is experienced as a permanent redrawing of the relationship between individual freedom and the rules required for social cohesion marks a fundamental shift in the social structure. The point at which an outpost transforms into a colony, where the temporary becomes permanent, is a moment of change in the sovereign nature of that community and its relationship to the nations and/or corporations that founded it 'back home'. Historical research identifies numerous examples of the political and social consequences of this shift on Earth, but the additional challenges produced by the restrictions required to survive a hostile non-terrestrial environment and the psychological pressures associated with space travel will amplify the impact of such a change of status in ways that are impossible to empirically assess. The costs and technological challenges of setting up a model of such a colony to test things out in advance of actual settlement are matched by the ethical problems associated with asking astronauts to become life-long (or, at least, very long-term) guinea pigs. If concrete experimental modelling of such a circumstance is impractical, therefore, I want to argue that fictional explorations of the implications of a shift from outpost dwelling to colonial living might offer resources to contribute to the analysis of potential political structures and explore their implications and consequences. And this moment of transition is one that has frequently been depicted in science fiction.

Stephen Baxter's *Ark* (Baxter 2009) depicts just such a point of sovereign crisis. The novel tells the story of a generation ship launched from a rapidly flooding Earth to take a carefully selected group of young astronauts on a journey to build a colony on an Earth-like planet in another solar system. The chaos of the launch means that ship's crew is made up of only some of the officially chosen candidates who are mixed with various other refugees that make it onboard, and the gradual loss of contact with mission control as distances and communicational time lags increase and Earth society fragments further as its environment collapses, produce an extreme version of the isolation and need for autonomy that the space psychologists cited in the previous section describe. There comes a moment of sovereign crisis (the first of many in the novel) where the crew come together to reassess the way their society will work:

> Even while we were at Jupiter, we still had ... the Nimrod project executive as a chain of command above us. But *now* there is no higher chain of command outside the Ark. And we need to find a new way of running things.... . As for leadership—well, we need a leader, a focus for decisions and disputes.... . Furthermore, when it comes to the laws by which we order our lives, we have a manual, a law book drawn up by the social engineers back at Denver. But they aren't here—and neither are half the Candidates they were meant to apply to. We can use that as guidance, but I propose that instead we should develop what we already refer to as 'Ship's Law'. Iron rules regarding safety and maintenance of the ship and its systems, rules that we all accept can be the basis of a set of laws which will emerge as we need them, by precedent, on a case by case basis. A law we don't need is a bad law, in my book.
>
> (Baxter 2009, 303)

In short, cut off from home and those who had hitherto controlled and determined their day-to-day behaviours, the young crew are faced with designing their society from first principles. Three linked things are invoked here as requiring reassessment on the basis of their isolation and independence from Earth: a new social order, a new leadership, and new rules and laws to maintain the systems necessary for survival. The way in which these three aspects of sovereign power are presented in the speech recalls the framing of the 'seventeenth century ideologies' evoked by Bogdanov as the basis of contemporary political order in the passage from *Red Mars* cited in the second section of this chapter.

Perhaps the most foundational and influential account of sovereignty from the seventeenth century comes from the philosopher Thomas Hobbes in his account of the nation state in *Leviathan* (Hobbes 1996). Returning to a seventeenth-century philosopher when there are much more fully developed political, social, legal accounts of power, freedom, and statehood might seem odd at this point, but, as Isaiah Berlin argues in his influential work on liberty, the 'mechanical structure' of Hobbes' account of sovereignty marks a key 'turning-point in the history of political thought' (Berlin 2017, 287–8) and still functions as a ground for theories of state-hood today. It is not so much Hobbes' solutions I want to focus on here, though they are worthy of attention and still generating discussion and analysis today (we haven't simply 'moved beyond' him), but rather his framing of sovereignty as a problem that needs to be solved if liberty, and even social coexistence, is to be possible. For Hobbes, liberty is not simply 'an exemption from Lawes', but, rather, 'the Liberty of a Sub-ject, lyeth ... only in those things, which in regulating their actions, the Sovereign hath praetermitted' (Hobbes 1996, 147–8). In other words, while individual free actions may consist of those that are unrestricted by legal constraint, the possibil-ity of such liberty in and of itself is achievable only on the basis of the existence of a sovereign legal structure that defends and protects individuality. Without the neces-sary limitations imposed on absolute personal freedom by sovereign power, Hobbes asserts, real liberty is impossible as we would exist in what he famously defines as a state of 'Warre':

> [D]uring the time men live without a common Power to keep them all in awe, they are in that condition that is called Warre; and such a Warre, as is of every man, against every man ... Whatsoever therefore is consequent to a time of Warre, where every man is Enemy to every man; the same is consequent to the time, wherein men live without other security, than their own strength, and their own invention shall furnish them withal. In such condition, there is no place for Industry; because the fruit thereof is uncertain: and consequently no Culture of the Earth; no Navigation, nor use of commodities that may be imported by Sea; no commodious Building; no Instruments of moving, and removing such things as require much force; no Knowledge of the face of the Earth; no account of Time; no Arts; no Letters; no Society; and which is worst of all, continuall feare, and danger of violent death; And the life of man, solitary, poore, nasty, brutish, and short.
>
> (Hobbes 1996, 88–9)

Although writing more than 100 years before what is often considered the birth of modern psychology (which some date to the publication of William Battie's *A Treatise on Madness* in 1758 and others to William Wundt's founding of the Laboratory of Experimental Psychology in 1879), Hobbes bases his argument about the need for sovereign power on a recognisably psychological account of a form of essential aggression that lies at the heart of human nature:

> [I]n the nature of man, we find three principal causes of quarrel. First, competition; secondly, diffidence; thirdly, glory. The first maketh men invade for gain; the second for safety; and the third, for reputation. The first use violence to make themselves masters of other men's persons, wives, children, and cattle; the second, to defend them; the third, for trifles, as a word, a smile, a different opinion, and any other sign of undervalue … .
>
> (Hobbes 1996, 88)

On this account, human interaction is premised on a lack of natural balance between individuals who are driven by desire simultaneously to increase their power, defend what they have, and gain recognition from others in ways they do not immediately wish to return. For Hobbes, humans are intrinsically competitive, and, in a context of limited resource, this leads to conflict.

The hostile environments of non-terrestrial existence impose significant limits on the resources required for survival and call for extensive cooperation and coordination between colonists to maintain survivable conditions, and this will require the enforcement of rigid discipline on heavily surveilled individuals in a manner liable to significantly exacerbate such conflicting drives. In fact, in each of the psychological studies cited above, the starting point for their arguments is precisely the sort of conflict-generating psycho-social situation that Hobbes identifies.

The aim of framing the distinction between an outpost and colony in terms of a state of exception and a new sovereignty is to make explicit the degree of difference between current space exploration and the challenges of establishing a permanent non-Earth settlement. Carried over to the moment of decision about extraterrestrial sovereignty depicted in *Ark*, the complexities of the types of social, leadership, and legal structures that might maximize safety, liberty, and individual flourishing in such a circumstance, and their differences from current Earth-based states becomes a crucial subject for investigation.

5. Writing Liberty: Science Fiction's Explorations of Extraterrestrial Sovereignty

As I said above, it is Hobbes' formulation of the problem rather than his explicit solutions in which I am interested here. In fact, the account of sovereignty he gives in *Leviathan* is ambiguous about the political status of colonies and their relation to the 'home' state. For Hobbes, writing in the early days of European colonization and torn between classical precedents from the Roman Empire and emergent modern models, a colony is one of two contrasting things: either part of the 'Bodies Politique' of a nation that remains 'subordinate' within an overall sovereignty and incompetent to legislate on its own behalf with regard to anything other than trivial day-to-day matters without reference to the express will of the sovereign (Hobbes 1996, 158–60), or, alternatively, its citizens are fully 'discharged of their Soveraign that sent them' so they must establish as 'a Common-wealth of themselves' (Hobbes 1996, 175) and form a new, separate state. In each case, a distinct sense of unease pervades the discussion as it attempts to skirt the fundamental threat of a 'division of Soveraign power' that Hobbes, writing amid civil war, sees as 'against the essence of a Common-wealth' because 'Powers divided mutually destroy each other' (Hobbes 1996, 225). This disquiet arises because, without a unified sovereign authority to hold them in check, the psychological forms of aggression he identifies between unconstrained individuals will play out between branches of the state and result in a collapse into civil strife or even all out civil war. The challenge for *Leviathan* is to reconcile the separate and often distant settlements within the necessary unity of the sovereign state, and its struggle to do so is telling in this current context.

For any potential extraterrestrial colony, the division of powers required to provide the financial, technological, and general logistical resources to make the venture possible in the first place, and the reliance of any new colony on Earth for the supply of essential materials, is likely to necessitate precisely the sorts of sharing of decision-making and profit between different states and non-state agencies that entail the 'division of Soveraign power' that Hobbes warns against (Velocci 2012; Persson 2015). While an examination of the colonial and imperial power structures that shaped world history in the years that followed *Leviathan* and built on his accounts of sovereignty find ways to come to terms with this and will themselves provide crucial resources for modelling extraterrestrial colonization, their roots in early modern, and predominantly Hobbesian, notions of the supremacy of the sovereign nation-state (as well as a legitimate desire not to simply repeat the mistakes and injustices of the last few hundred years of imperialism) call for the investigation of alternative visions and possibilities. It is on this basis that Hobbes' anxiety about the sovereign status of the colony remains an important ground for investigation.

As well as exploring the history of colonial power on Earth, often by projecting it into space (Robert Heinlein's classic novel *The Moon is a Harsh Mistress* from 1966 has been plausibly read as imaginatively replaying the politics of the American Revolution, for example), science fiction writing has often taken the challenges of non-terrestrial survival and sovereignty as a central theme. Moreover, its fictional approaches to these matters provides material for exploring the psychological impacts of such situations. Each of the three novels introduced in the previous sections gives an account of colonization that considers alternative possibilities to the historical nation-state and empire model. And, by looking very briefly at these different visions, I hope I can begin to set out some ways in which engagement with literary writing might contribute to the analysis of extraterrestrial liberty.

Although *The Space Merchants* finishes just before the founding of a colony on Venus occurs, it nevertheless examines relationships between sovereignty, freedom and identity as the novel's protagonist is forced to confront the realities of the economic systems that underpin his work as an advertising executive. This is a world in which powerful companies have supplanted national governments as controllers of the world's economy. Caught up in a vicious conflict between rival firms, advertising executive Mitch Courtenay is kidnapped, stripped of his identity and elite social position, and forced to become an indentured labourer in a food-processing plant where the costs of goods required to sustain life that must be bought from the company store exceed the wages paid, with the result that, through their perpetual debt, the workers' subservience to the tyrannical employer/sovereign is enforced. The novel's account of space colonization and the excesses of corporate power presents, according to Roger Luckhurst, 'an overt critique of the model of corporate-led economic imperialism undertaken by the United States in Latin America in the 1940s and 50s', and ends with a reformed and now radical Courtenay heading to Venus with his band of revolutionaries 'like new Puritans, leaving the Fallen Earth behind' (Luckhurst 2005, 111). The novel thus presents a choice between colonization as a brutal process of exporting unfettered commercial imperatives to other worlds or a revolutionary rejection of those values that seeks to rediscover liberty on a new world.

Published 40 years later at a moment in the 1990s where the reach and power of global corporations in relation to nation-states had become a subject of intense political debate (Malpas 2022), *Red Mars* explores the conflicts that arise between the first colonists, those who follow them, and the national and corporate interests who have invested in the venture and seek to control the patents, commodities, and other profits that colonization has made possible. Rather than depicting a simple either-corporate-or-nationalist opposition, the novel questions how such an opposition might emerge and what its consequences for extraterrestrial survival might be: it simultaneously charts 'the political-economic clashes and failures that have shaped historical trajectories towards the twenty-first century' and examines 'how a human civilisation that is utterly dependent on technology for its survival [in the hostile non-terrestrial environment] confronts a "second" Enlightenment' (Knoespel 2012, 110) as those clashes and failures re-emerge. The Mars trilogy can be classified as 'social realist' fiction (as are *Ark* and *The Space Merchants*) in that its narrative presents the story by depicting '"typical" characters in a "total" context' (Edgar 1987, viii). The mode of storytelling associated with social realism aims to depict a recognizable world that, in the words of Friedrich Engels who first used this formulation, is made up of 'the truthful reproduction of typical characters under typical circumstances' (Engels 1975, 269). In other words, the narrative portrays characters not just as recognizable psychological individuals but also in such a way that they 'represent a significant element in an analysis of a concrete social situation' (Edgar 1987, viii). As such, the interactions between psychologically rounded but typical characters (the utopian political thinker, pragmatic diplomat, resourceful engineer, etc.) enable *Red Mars* to explore the question of a transformation (of both the planet and the self) in terms of the violence and dissension that emerges from conflicts between the worldviews of the different groups of settlers and the pressures placed on them by corporations and national governments with their own agendas who are trying to exert political control.

Ark provides perhaps the most extended and thoroughgoing exploration of the attempts to find suitable sovereign structures to enable liberty without jeopardising survival in the hostile environment of a dying Earth, interstellar flight, and on new worlds. It depicts the processes of conflict and cooperation between entrepreneurs, innovators, private companies, the military, and a national government to develop the project on Earth; the first fumbling attempts at forming a non-terrestrial sovereign community by the young astronauts; subsequent political and technological challenges that require the reworking of legal structures; splits between factions as moments of crises occur (should the new planet with its less-than-perfect environment be settled, should they return home or journey on to an alternative world?); and, eventually, decisions about who might survive and form a colony in the face of limited resources. In each case, recognizable political choices are presented, and the social and psychological effects of the decisions reached are detailed in the characters' responses. *Ark* ultimately poses crucial questions about the sacrifices necessary for extraterrestrial survival, and tests out a variety of political, social, and psychological alternatives.

The contribution analyses of these texts might make to understanding the psychosocial and political challenges of extraterrestrial settlement and colony building is not based on any sense that they might have 'got the future right'. Science fiction is not prophecy, and the scientific discoveries, technological developments, and social changes it has correctly anticipated are massively outweighed by things depicted that have not and never will come to pass. Rather, the importance of investigating texts such as these lies in their capacity to set up analogues to be explored where the forms of political, social, and psychological structure that might make possible the maintenance of liberty in an extraterrestrial milieu can be tested and examined in a 'safe' environment. This approach is perhaps best summed up by the writer and critic Samuel Delany, who argues that science fiction is:

> a tool to help you think about the present—a present that is always changing, a present in which change itself assures there is always a range of options for actions. It doesn't tell you what is going to happen tomorrow. It presents alternative possible images of futures, and presents them in a way that allows you to question them as you read along in an interesting, moving and exciting story... . Without an image

of tomorrow, one is trapped by blind history, economics, and politics beyond our control. One is tied up in a web, in a net, with no way to struggle free. Only by having clear and vital images of the *many* alternatives, good and bad, of where one *can* go, will we have any control over the way we may actually get there in a reality tomorrow will bring all too quickly.

(Delany 1984, 34–5)

By setting out some of the grounds for this approach to science fiction, I have tried to show how its analysis might contribute to wider research in the field of extraterrestrial liberty. Working in the context of current developments in space science, and bringing the questions posed by experts working in psychology, law, engineering, and the other disciplines immediately invested in rising to the challenges of long-term extraterrestrial survival, science fiction provides a space in which some of the issues not susceptible to experimental empirical testing can be explored. The careful analysis of science fiction's various and differentiated 'images of tomorrow', even perhaps those not based on technologically feasible tomorrows, has a role to play in contributing to our understanding of the changes required to get to a future that might be desirable.

References

Agamben, G. *State of Exception*, trans. K Attell (Chicago, IL: University of Chicago Press, 2005).

Baxter, S. *Ark* (London: Penguin, 2009).

Berlin, Isaiah. *Liberty*, ed. Henry Hardy (Oxford: Oxford University Press, 2017).

Cockell, C. 'Freedom in a Box: Paradoxes in the Structure of Extraterrestrial Liberty' in C. Cockell (ed.), *The Meaning of Liberty Beyond Earth* (London: Springer, 2015) 47–68.

Delany, S. 'The Necessity of Tomorrows' in S. Delaney, *Starboard Wine: More Notes on the Language of Science Fiction* (New York, NY: Dragon Press, 1984) 23–35.

Edgar, D. *Plays: One* (London: Methuen, 1987).

Engels, F. 'Letter to Margaret Harkness (April 1888)' in D. Craig (ed.), *Marxists on Literature: An Anthology* (Harmondsworth: Penguin, 1975) 269–71.

Greene, K. *Once Upon a Time I Lived on Mars: Space, Exploration and Life on Earth* (London: Icon Books, 2021).

Heinlein, R. *The Moon is a Harsh Mistress* (London: Hodder & Stoughton, 2015).

Hobbes, T. *Leviathan (1651)*, R. Tuck ed. (Cambridge: Cambridge University Press, 1996).

Horneck, G., Facius, R., Reichert, M., et al. 'HUMEX, a Study on the Survivability and Adaptation of Humans to Long-Duration Exploratory Missions, Part II: Missions to Mars' (2006) 38 *Advances in Space Research* 752–9.

Jameson, F. '"If I Find One Good City I Will Spare the Man": Realism and Utopia in Kim Stanley Robinson's *Mars* Trilogy' in P. Parrinder (ed.), *Learning from Other Worlds: Estrangement, Cognition and the Politics of Science Fiction and Utopia* (Liverpool: Liverpool University Press, 2000) 208–32.

Kanas, N. and Manzey, D. *Space Psychology and Psychiatry* (2nd edn, London: Springer, 2008).

Kanas, N., Sandal, G., Boyd, J. E., et al. 'Psychology and Culture during Long-Duration Space Missions' (2009) 64 *Acta Astronautica* 659–77.

Knoespel, K. 'Reading and Revolution on the Horizon of Myth and History: Kim Stanley Robinson's Mars Trilogy' (2012) 20 *Configurations* 109–36.

Lemke, M. 'What Does State of Exception Mean? A Definitional and Analytic Approach' (2018) *Zeitschrift für Politikwissenschaft* 373–83.

Luckhurst, R. *Science Fiction* (Cambridge: Polity Press, 2005).

Malpas, S. 'The New McWorld Order: Postmodernism and Corporate Globalisation' in J. Evans (ed.), *Globalisation and Literary Studies* (Cambridge University Press, 2022) 144–58.

Manzey, D. 'Human Missions to Mars: New Psychological Challenges and Research Issues' (2004) 55 *Acta Astronautica* 781–90.

Pak, C. '"All Energy Is Borrowed"—Terraforming: A Master Motif for Physical and Cultural Re(up)cycling in Kim Stanley Robinson's *Mars* Trilogy' (2014) 18 *Green Letters* 91–103.

Persson, E. 'Citizens of Mars Ltd.' in C Cockell (ed.), *Human Governance Beyond Earth* (London: Springer, 2015) 121–37.

Pohl, F. and Kornbluth, C. M. *The Space Merchants* (London: Gollancz, 2003).

Reinstein, E. J. 'Owning Outer Space' (1999) 20 *Northwestern Journal of International Law and Business* 59–97.

Robinson, K. S. *Red Mars* (London: Harper Collins, 1993).

Robinson, K. S. *Green Mars* (London: Harper Collins, 1994).

Robinson, K. S. *Blue Mars* (London: Harper Collins, 1996).

Schmitt, K. *Political Theology: Four Chapters on the Concept of Sovereignty*, trans. G. Schwab (Chicago, IL: University of Chicago Press, 2005).

Stevens, A. H. 'The Price of Air' in C Cockell (ed). *Human Governance Beyond Earth* (London: Springer, 2015) 51–61.

Vakoch, D. A. (ed.). *Psychology of Space Exploration: Contemporary Research in Historical Perspective* (Washington, DC: NASA, 2011).

Velocci, A. L., Jnr. 'Commercialisation in Space: Changing Boundaries and Future Promises' (2012) 4 *Harvard International Review* 49–53.

13

Securing the long-term peaceful use of space

Allan McKenna

> We could fill books with problems of fundamental importance to the
> human race which can only be solved by spaceflight, more easily by
> spaceflight, or more probably by spaceflight.
>
> (Cole 1965, 32)

1. Introduction

Space expansionists propose that solutions to global problems will be found by moving terrestrial activities to outer space.[1] For example, Jeff Bezos urges the relocation of heavy-polluting industry in order to avert ecological disaster (D'Onfro 2016). Space migration may be essential for species survival in the event of a large asteroid impact or global nuclear war. Space-based solar power could help to resolve the climate crisis. Martian settlements could provide a clean slate upon which to construct egalitarian societies. However, sceptics contend that space expansion will not bring greater security, freedom, or save the environment (e.g. Mumford 1970; Schwarz 2020; Klinger 2021). Daniel Deudney's recent monograph, *Dark Skies: Space Expansionism, Planetary Geopolitics and the Ends of Humanity*, argues that ambitious space expansion inevitably leads to interplanetary conflict, hierarchical world government, and the destruction of the environment. It calls for the abandonment of expansionist plans due to the risk of 'astrocide': human extinction caused directly or indirectly by space expansion (Deudney 2020, 362).

This chapter is divided into four main sections. The first section critically examines four scenarios projected to result from space expansionism in *Dark Skies*: (i) perpetual war; (ii) space dominance; (iii) degradation of the environment; and (iv) inadvertent escalation. The second critiques Deudney's proposal for a 'Whole Earth security' space programme in relation to the twenty-first century space race. The remaining sections address the potential contribution of space expansionism to global security. The third traces the historical connection between disarmament negotiations and the 1967 Treaty on Principles Governing the Activities of States in the Exploration and Use of Outer Space, including the Moon and Other Celestial Bodies.[2] It highlights the problems of treaty verification and the prevention of surprise attack raised by these negotiations. It reviews proposals by the first space lawyers for a World Space Agency or inspectorate to address these problems. The final section draws inspiration from these proposals to chart a brighter future for space

Allan McKenna, *Securing the long-term peaceful use of space*. In: *The Institutions of Extraterrestrial Liberty*.
Edited by Charles S. Cockell, Oxford University Press. © Allan McKenna (2022). DOI: 10.1093/oso/9780192897985.003.0014

expansion than the 'worst of all possible worlds' (Deudney 2020, 351) anticipated by *Dark Skies*.

2. Catastrophic Scenarios in *Dark Skies*

Dark Skies predicts catastrophic events will occur as a result of space expansion. These futuristic scenarios provide the justification for relinquishing specific expansionist projects. This section examines the main elements of each scenario and assesses their plausibility.

2.1 Perpetual war

Dark Skies argues that if the first space colonies are planted by a consortium of states 'terrestrial anarchy will be modestly modified', but if military rivals establish independent settlements, it will compound the threat of large-scale nuclear war (Deudney 2020, 352). The probability of existential risk reduction approaches zero once settlements become self-sustaining and politically independent from Earth. Until this point, a rebellious colony could be suppressed by the withdrawal of life-support systems. Instead, Earth will become a vulnerable 'island' in the 'Solar Archipelago' (Deudney 2020, 41) and the site of perpetual conflict over frontier access. Limited surveillance capacities, uncontrollable migration patterns, divergent interests, and speciation will diminish the prospects for interplanetary law. A hierarchical world government and planetary-wide fortification will be required to protect the Earth from bombardment with asteroids (Deudney 2020, 339–40). Rival military forces will exploit their vantage ground at the top of Earth's gravity well and conceal weapons in the vast reaches of outer space. As solar space becomes more populated the threats will multiply and amplify. Therefore, plans for space colonization ought to be relinquished (Deudney 2020, 374).

Dark Skies suggests that an irreversible line will be crossed once space settlements become 'independent' or 'self-sustaining' (Deudney 2020, 363–4). However, if colonies begin life as asteroid-mining operations, scientific observatories, or military bases, it seems unlikely that they will subsequently become 'untethered' from Earth. Isolation would leave settlements transfixed in historical time whilst technology, science, and medicine advance on Earth. The rapid depopulation of remote tribes following exposure to common infectious disease (Walker, Sattenspiel, and Hill 2015) suggests that it would not be in the interests of a colony to separate from the most densely populated planet in the galaxy. In *The High Frontier: Human Colonies in Space*, Gerard K. O'Neill imagines that workers will commute between space and Earth (O'Neill 2021, 11–13). It seems likely that space settlements will require a steady influx of people and materials from Earth, providing strong incentives to maintain harmonious relations. The weaponization of asteroids and their role in future warfare also seems highly speculative. The capabilities required to divert objects hurtling toward the Earth are not an inversion of precision targeting (Oberg 1998). Therefore, the scenario ought to be regarded as low-risk.

2.2 Space dominance

Dark Skies warns that orbital megastructures to harvest solar energy, facilitate space industrialization, terraform planets, and 'urbanize' orbital space will lead to a hierarchical world state. The monopolization of vast quantities of energy and material resources will guarantee its economic superiority. Orbital battle stations, powerful energy-beam weapons, and weaponized asteroids will ensure its military supremacy. The planetary panopticon of real-time surveillance enables totalitarian 'crushing of individual freedom and group identities' (Deudney 2020, 137). Therefore, the construction of orbital infrastructure ought to be relinquished (Deudney 2020, 373).

The fantasy of 'space dominance' entered the public imagination in the earliest years the space age. For example, in October 1960 Massachusetts Senator John F. Kennedy warned: 'If the Soviets control space they can control earth, as in past centuries the nation that controlled the seas dominated the continents' (Newlon and Howard 1960, 12–13). *Dark Skies* utilizes the concept to underscore the perils of orbital infrastructure: 'In simple terms, who controls Orbita controls the Earth ... its masters dominating the rest of the planet' (Deudney 2020, 325). The conceptual separation implicit in the notion of 'space dominance' belies the intrinsic unity of Earth–space activities (McDougal 1963, 621–2). As a place of human action, space is an extension of the terrestrial environment. Military aggression in or through space will be met with a proportionate response in other operational domains. The fragility of space objects and the wide variety of ways in which their activities can be disrupted suggests that they cannot be relied upon by a small group to disempower a large population. In highly asymmetric conflict, the degradation of orbits through intentional creation of debris could thwart attempts to establish 'space dominance'.

Dark Skies advances a strong case for considering the safety of orbital structures, including their potential weaponization (Mankins and Kaya 2011, 80–3). Risk assessment and the design of safeguards ought to be conducted on a multilateral basis through established international fora, such as the United Nations Committee on the Peaceful Uses of Outer Space (UNCOPUOS). However, the possibility that Orbita will be hijacked by rivalrous states ought to be regarded as an implausible worst-case scenario. According to *Dark Skies*, orbital infrastructure will be built only under an effective governance regime which allocates rights, safeguards property, and restrains military activity (Deudney 2020, 324). The level of cooperation required to build and maintain Orbita safely is more likely to pacify and unify rather than divide the Earth. Therefore, 'space dominance' should be evaluated as a low-risk outcome.

2.3 Degradation of the environment

Dark Skies predicts that opening the space frontier will increase environmental degradation and Earth orbits will be ruined by human activities if space debris is not properly managed (Deudney 2020, 305). Space debris includes inactive satellites and rocket fragments, flakes of paint, leaking fuel, and microparticulate matter orbiting the Earth in an uncontrolled manner. Debris originates through natural events such as space weather; unintentional events such as the collision between the defunct

Russian satellite Cosmos-2251 and active US satellite Iridium-33 in 2009; and intentional events, such as China's ground-based anti-satellite weapon (ASAT) test in 2007 which destroyed the Fengyun-1C satellite in Low Earth Orbit (LEO). By 1 July 2012, 90 per cent of the 5,500 catalogued debris generated by these two collisions constituted 36 per cent of all trackable LEO objects (NASA 2012, 2). Cascading collisions will exponentially increase the debris population (Kessler and Cour-Palais 1978). Space warfare would create debris which persists for decades, centuries, or millennia in some orbits (Jakhu and Pelton 2017, 281).

Debris affects all current users of space and impacts upon future generations. Satellite mega-constellations will produce a steep rise in the number of objects launched over the coming decades, and this increase is expected to generate more debris over time (Pardini and Anselmo 2021). Recent studies suggest projections of debris mass in LEO have been underestimated (Brown et al. 2021). Space Debris Mitigation Guidelines (UNOOSA 2010; IADC 2021) promote information sharing, safety practices, and spacecraft design to minimize the production of orbital debris. However, as these are not legally binding rules there are issues with widespread implementation (UN Doc. A/AC.105.C.1/2011/CRP.14). There are also concerns that the launch of unreliable, first-generation systems with limited operability will fail to comply with the guidelines (UN Doc. A/AC.105/C.1/111, p. 8). The limited manoeuvrability of nanosatellites increases the probability of collision events. These concerns are amplified by the risk of accidents involving nuclear power sources (UN Doc. A/AC.105/C.1/2017/CRP.12). Radioisotope thermoelectric generators pose a risk of contamination due to fuel leakage, collision, and launch failure (Taylor 2011, 268). For example, in 1978 the malfunctioning satellite Kosmos-954 dispersed 31 kilos of highly enriched uranium over 124,000 square kilometres of Canada (Galloway 1979). The impact of greatly increased launch activity upon the environment is largely unknown (Ross and Vedda 2018).

Dark Skies suggests an extension of the Outer Space Treaty (OST) to accommodate a 'space traffic control regime' (Deudney 2020, 246). Space traffic management would help to reduce the number of collisions but establishing a regime poses technical, operational, political, and legal complexities (Jakhu and Pelton 2017, 319). The registration of space objects in accordance with Guidelines for the Long-term Sustainability of Outer Space Activities (UN Doc. A/AC.105/2018/CRP.20) would assist spacecraft operators to plan and conduct missions with minimal risks of collision (Blount 2021a, 5). Active debris removal methods are under development but these systems also pose a risk of 'weaponization' (Pelton 2015, 68). Until the problems of debris mitigation and space traffic management are resolved, this scenario represents a growing risk.

2.4 Inadvertent escalation

Military space expansionists advocate the interception of ballistic missiles in boost phase with rows of satellite battle stations orbiting the Earth. *Dark Skies* claims that an 'Earth-net' of interceptors configured for automatic operation will increase the risk of accidents and 'general conflagration' (Deudney 2020, 173). *Dark Skies* holds that the advent of the space age increased the risk of human extinction because intercontinental ballistic missiles (ICBMs) utilize orbital trajectories to deliver nuclear warheads (Deudney 2020, 309). The 'Archduke Francis Ferdinand of World War III' may be a reconnaissance satellite hit by space debris during a crisis (Deudney 2020, 238). For these reasons, further weaponization and militarization of space ought to be restrained (Deudney 2020, 373).

In the twenty-first century, the number of states conducting ASAT tests has increased (Weeden and Samson, 2021). ASAT tests demonstrate the capabilities to destroy, damage, or interfere with critical military reconnaissance satellites, global navigation systems, and nuclear command, control, and communications architecture (NC3). Satellites function as information force-multipliers, enhancing the joint utilization of land, sea, and air forces (Shalikashvili 1995, 15). ASATs thus complicate scenarios of 'war by mutual apprehension', for example, through the hacking of ground-station controls by a third party which brings two other states into conflict (Kobe 1962; Johnson 2021). Interference with NC3 satellites may be perceived as an attempt to disable early warning capabilities prior to a surprise nuclear first strike (Acton 2019). Conventional attacks upon US or allied NC3 warning and attack assessment capabilities could lead to retaliation with nuclear weapons (US Department of Defense 2018, 21). In the coming years, the proliferation of hypersonic, highly manoeuvrable missiles may influence more states to adopt launch-on-warning strategies, lowering the nuclear threshold and increasing risk of accidental launch (Hill 2019, 68). Accidents can occur through collision, misfire, malfunction, and miscalculation. Commercial satellites could be hijacked by hostile non-state actors and converted to weapons of mass disruption (West 2019, 112). As weapons systems improve in speed and accuracy, the length of potential decision-making time in a crisis is reduced. For these reasons, inadvertent escalation constitutes a growing risk.

3. Cooperation or Competition?

We have considered the scenarios of environmental degradation and inadvertent escalation to be the most plausible, but other catastrophic outcomes may become more probable in the future. *Dark Skies* urges the abandonment of expansionist plans and an 'Earth-centred pro-space agenda focused on nuclear security and environmental protection' (Deudney 2020, 7). Let us now evaluate this proposal.

3.1 Whole Earth security

The 'Whole Earth security' space programme promotes planetary-scale information technology and international collaboration to study climate change and map near-earth objects (NEOs); remove orbital debris; explore the moon, Mars, and asteroids; and construct scientific observatories or lunar bases (Deudney 2020, 371–7). It explicitly relinquishes plans for orbital infrastructure, space colonies, and space-based interceptors. It would abolish the development, testing, and deployment of space weapons, including ASATs, ballistic missiles, and nuclear warheads. *Dark Skies* holds that international scientific cooperation will defuse terrestrial conflict, paving the way for an international consortium to defend against NEOs and exploit asteroid resources (Deudney 2020, 248–9).

The proposals for global security in *Dark Skies* aim to complement the system of sovereign territorial states rather than replacing it with authoritative world government (Deudney 2020, 242). In contrast, world federalists call for a supranational military organization to enforce the decisions of an International Court of Justice and World Parliament (e.g. Einstein 1939, 210; Russell 1958). *Dark Skies* suspects that a monopoly on the use of force in conjunction with 'the substantial disarmament of all states' (Deudney 2020, 232) would lead to totalitarianism. In terms of preventing extinction, is hierarchical world government less preferable to a world of rivalrous nations poised on the brink of nuclear war? According to *Dark Skies*, advocates of 'nuclear one-worldism' fail to show a plausible path of transition to a world sovereign (Deudney 2020, 231–2). Would the path to an international asteroid consortium be less problematic? Computer simulations indicate that nuclear weapons are required to deflect extinction-level NEOs (Syal, Dearborn, and Schultz 2013, 109). How will the consortium solve the problem of the ownership and control of nuclear weapons? In any case, solving the potentially intractable political problems of a planetary defence consortium appears to be of secondary concern due to the temporal proximity of resource extraction operations, their importance for the success of planned expansionist projects (e.g. Kornuta et al. 2019), and the apparent remoteness of civilization-ending asteroids.

Dark Skies regards the OST as the essential framework for future space cooperation (Deudney 2020, 376). It assumes that legal instruments and institutions constructed in the decades following the Second World War are still fit for purpose. Critics of the OST argue that the rules and principles are stated in such an ambiguous way that support for a contradictory set of activities may be derived from it (Viikari 2008, 203). The OST fails to monitor compliance beyond voluntary requests for consultation and is lacking enforcement mechanisms. Furthermore, since the creators of the treaty did not anticipate extensive private sector involvement in space, it cannot provide the legal certainty required for operations such as asteroid mining and lunar resource utilization (Gaspari and Oliva 2019, 189). The principles of free access to all areas of celestial bodies (Article I) and non-appropriation (Article II) will only be put to the test in the coming decades. Even 'scientific investigation in outer space' (Article I) raises complex problems unanswered by the provisions of the treaty (Elvis, Milligan, and Krolikowski 2016). Will operating 'Safety Zones' on the Moon (NASA 2020, 73) comply with the open access provisions in Article XII? Will nuclear explosions for excavation purposes be permitted given the obligation to protect celestial bodies from harmful contamination under Article IX? Will nuclear weapons for planetary defence be compatible with the prohibitions on weapons of mass destruction in Article IV?

Dark Skies suggests that the OST is 'preserved' and 'expanded' by phasing out ballistic missiles and banning ASAT tests (Deudney 2020, 244–6, 376). Disarmament as a verifiable reduction in the aggregate number of armaments such as 'zero ballistic missiles' (Deudney 2020, 235) is unlikely to succeed in the current space security environment. The 'dual-use problem' (Jakhu and Pelton 2017, 280)—virtually any space object could serve as a weapon—compels a moderation of expectations.

A broader definition of disarmament which embraces 'all the forms of military coop-eration between political enemies in the interest of reducing the likelihood of war, its scope and violence if it occurs, and the political and economic costs of being prepared for it' (Schelling and Halperin 1961, 2) possesses a greater chance of success. Mitigat-ing the risks of inadvertent escalation and environmental degradation may necessitate arms control agreements which involve more, less, or different types of arms; or com-bine an increase in one category with reductions in another (Schelling 1960, 130–1). In space arms control the ground- or air-based high-intensity laser could be regarded as less dangerous than a kinetic energy ASAT because satellite destruction will not produce debris (Harrison, Johnson, and Roberts 2018, 3). Nevertheless, a realistic arms control proposition must take into account the gulf in space capabilities between states and not have the effect of placing the national security of any participating state in an inferior position.

Let us now consider several other features of the twenty-first century security environment which vitiate against the 'Whole Earth security' space programme.

3.2 A new Space Age

We find ourselves on the cusp of a transformative new Space Age. More governments, companies, and individuals have access to space than ever before. Commercial oper-ators such as Virgin Galactic, Blue Origin, Axiom, and SpaceX now offer the wonders of spaceflight to paying customers. In the coming years, technological advances in miniaturization, artificial intelligence, robotics, manufacturing, and reusable launch vehicles will continue to lower the cost of access. 'New Space' entrepreneurs plan to build global Internet connectivity with satellite mega-constellations, extract natural resources from celestial bodies, and establish human settlements beyond the Earth. In our daily lives we increasingly depend upon space systems for defence, communica-tion, transportation, weather forecasting, agriculture, banking, energy distribution, and a wide variety of satellite-enabled data applications. However, our reliance on space appears to be a mixed blessing. Increased dependence entails greater vulner-ability in the event of the failure of space-derived services. Low-earth orbits are becoming increasingly 'congested, contested, and competitive' (US Department of Defense 2011). The New Space era is characterized by its diversity of actors but also resurgent 'great power competition'.

In 2019, the North Atlantic Council declared space an operational domain (NATO 2019). In recent years China, Russia, and the United States reorganized their armed services to consolidate space forces; India, France, Germany, and the United King-dom also created space commands (Barrie 2021, 10). In 2019, then US Vice-President Michael R. Pence stated that the United States is engaged in a new 'space race', advert-ing to Chinese plans to become the leading nation in space (Pence 2019). The US-led Artemis programme aims to land astronauts on the Moon for the first time since 1972. It will establish an outpost to facilitate long-term human presence and support crewed missions to Mars. In 2017, the Chinese Academy of Launch Vehicle Tech-nology unveiled plans for a docking station in Earth orbit, a permanent lunar base, space solar power, and nuclear-propulsion spacecraft to mine asteroids and colonize the solar system (Goswami 2019; Goswami and Garretson 2020, 187–206). In 2021, China and Russia signed a Memorandum of Understanding to establish a lunar base for human occupation by 2036 (CNSA 2021).

The emerging space race is buttressed by significant developments in the global security environment over the past several decades. In the first place, the United States and Russia have revoked participation in a number of long-standing strategic stability treaties, including the 1972 Anti-Ballistic Missile Treaty, 1987 Intermediate-Range Nuclear Forces Treaty, and 1992 Open Skies Treaty. President Vladimir Putin recently stated that the United States and Russia have been in an arms race since the demise of the Anti-Ballistic Missile Treaty in 2002 (Johnson 2018). Second, the number of states testing ASATs is increasing and new weapons systems such as nuclear-capable hypersonic glide vehicles and cruise missiles, low-yield nuclear warheads, and autonomous intercontinental undersea torpedoes are also undergoing tests and deployment (US Congressional Research Service 2021; Weeden and Samson 2021). Third, the global number of deployed nuclear weapons, long-range delivery vehicles, and states with active nuclear weapons programmes is increasing (SIPRI 2021). Fourth, more states are acquiring launch capabilities and implementing launch-on-warning strategies (Goswami 2021; US Department of Defense 2021, 93). In this highly competitive security environment, the protection of leading advantages and extended deterrence commitments, nuclear proliferation safeguards, 'grey zone' conflict, and the ever-present threat of surprise attack restrict the possibilities for collaboration between spacefaring states.

The second major trend to consider is the growth of the private sector. In the past, only governments could afford to operate in space. Now commercial companies are driving innovation forward, private actors are entering space, and public–private partnerships are vital for national security (Butow et al. 2020; Lopez 2020). The flourishing private sector represents an opportunity for states to enhance defence capabilities through synergistic research and development. For example, China's 'Civil–Military fusion development strategy' aims to integrate the defence industrial base with the civilian technology and industrial base, leveraging innovation across both sectors (US Department of Defense 2021, 24). The close relationship between military and commercial capabilities complicates security in the sector as sensitive information is targeted by multiple competitors. For this reason, Public Law 112–10 Sec. 1340 (US Congress 2011) prohibits NASA from bilateral cooperation with the Chinese government and Chinese-owned companies without explicit authorization by a new law enacted by Congress. Under these conditions, the extensive cooperation envisaged in *Dark Skies* (Deudney 2020, 241) is unlikely to materialize. On the basis of current trends, competition over strategic regions of outer space is expected in the forthcoming decades (US Air Force Space Command 2019). Non-state actors will partake more in space activities, commercial operations will become military targets, and disruptive technologies such as 3D-printed rockets will bring pressure to bear upon non-proliferation regimes. The accelerated pace of technological development underlines the need for international law, governance, and cooperation to ensure the long-term sustainability of space activities.

The space security environment of the twenty-first century appears to be substantially different to the Cold War era which produced the OST. However, historical investigation reveals that the architects of the treaty grappled with fundamental problems of arms control which still remain to be solved.

4. Space Expansion and Disarmament

The first space lawyers faced the task of erecting a new legal framework to cope with the alarming threats created by the global reach of new space technology. It is perhaps surprising, therefore, to find in their work a sense of optimism that the governance of outer space through international law can lead to world peace (e.g. Chester 1961; Galloway 1961). Indeed, the phrase 'use of outer space for peaceful purposes' occurs twice in the preamble of the OST; 'exclusively for peaceful purposes' in Article IV; and 'peaceful exploration and use of outer space' in Article IX. Nonetheless, the meaning of 'peaceful' was disputed in the years immediately preceding registration of the treaty and these phrases continue to generate controversy (Cheng 1997; Maogoto and Freeland 2007; Blount 2021b). Debate in the Legal Sub-Committee of UNCOP-UOS focused upon a distinction between 'military' and 'aggressive' uses of space (Aoki 2017, 201–3). In the UN General Assembly First Committee it was argued that 'peaceful' denotes 'non-aggressive' rather than 'non-military' activity because 'military' activity is permitted by Articles 2(4) and 51 of the Charter of the United Nations (UN Doc. A/C.1/PV.1289, p. 13; UN Doc. A/C.1/SR.1422, p. 429). The contrary view holds that 'peaceful' denotes exclusively 'non-military' use of space because 'military' use is incompatible with the principle of common interest in Article 1 of the OST (Markoff 1976, 9–11). A third perspective regards these dichotomies as inappropriate because 'civil' space technology is indistinguishable from technology utilized as part of a 'military' mission (Sourbès and Boyer 1991, 74). In the absence of clearly defined criteria, we ought to wonder whether the language of 'peaceful uses' and 'peaceful purposes' in the OST possesses any legal content whatsoever.

The following analysis places the OST in the context of earlier UN negotiations, treaties, and resolutions. It shows that the articles of the OST incorporated specific terms and phrases with distinct connotations related to general and comprehensive disarmament. Consequently, the distinction between the 'peaceful use of space' and 'use of space for peaceful purposes' is not trivial with respect to the disarmament objectives of the Charter of the United Nations. This historical analysis points toward several long-standing, interrelated problems to be addressed before large-scale space expansion: the establishment of comprehensive and reliable means of treaty verification, and the prevention of surprise attack.

4.1 Origins of the Outer Space Treaty

The earliest debates on the peaceful use of space arose as part of a set of complex negotiations on general and comprehensive disarmament. Our starting point is the UN Disarmament Commission Sub-Committee meeting which took place in London from March to September 1957. Items on the agenda included (i) nuclear testing; (ii) conventional disarmament; (iii) nuclear disarmament; (iv) means of control;

(v) missiles and outer-space objects; and (vi) zones of limitation and inspection (NATO 1957a, 3). The initial Soviet proposal for a disarmament treaty stated that 'international control shall be instituted over guided rockets in order to ensure that all types of rockets suitable for use as atomic and hydrogen weapons will be exclusively for peaceful purposes' (UN Doc. DC/SC.1/49). In a working paper titled 'Proposals for Partial Measures of Disarmament', Canada, France, the United Kingdom, and United States proposed that 'parties agree that within three months after the entry into effect of the Convention they will cooperate in the establishment of a technical committee to study the design of an inspection system which would make it possible to assure that the sending of objects through outer space should be exclusively for peaceful and scientific purposes' (UN Doc. DC/SC.1/66). The proposal to cooperate on an inspection system was eventually incorporated in the first resolution on space adopted by the UN General Assembly on 14 November 1957 (UN Doc. A/RES/1148(XII)).

On 2 August 1957, the four NATO-allied powers presented a working paper on 'systems of inspection to safeguard against the possibility of surprise attack' (UN Doc. DC/SC.1/62/Rev.1). It proposed regions for aerial inspection; ground observation at major ports, railway junctions, main road networks and airfields; and mobile ground inspection teams. The NATO powers stated their aim of 'a world in which all nations and peoples may live free of the danger and fear of surprise attack' (UN Doc. DC/SC.1/PV.143, p. 9). The Soviet response emphasized that aerial photography would not prevent surprise attack by an aggressor possessing weapons of mass destruction and the latest vehicles of delivery (UN Doc. DC/SC.1/65/Rev.1). It is important to note that a Soviet news agency reported the first successful test of an intercontinental ballistic missile on 26 August 1957 (US Department of State 1957, 1311).

In September 1957, the Disarmament Commission Sub-Committee adjourned with the Soviet delegate, Mr Zorin, declaring it a futile endeavour since the position of the Western powers was determined by NATO, an organization which was not genuinely interested in disarmament (UN Doc. DC/SC-1/PV-151, pp. 27–30). Cross-referencing with declassified documents reveals that the Western powers asked NATO Supreme Commander Allied Forces Europe for recommendations before presenting specific proposals to the Soviet delegation (NATO 1957b–e). However, seeking the best available military advice in light of a disarmament proposal is a mark of common sense and not proof that the negotiations were simply marking time. Despite many technical and political obstacles to disarmament, constructive agreements transpired as a result of these initial discussions.

On 14 November 1957, UN General Assembly Resolution 1148 called for the suspension of nuclear weapons tests; terrestrial inspection posts under international control; cessation of the production of fissile material for weapons and future production under international control; the reduction of weapons stockpiles, armed forces, and armaments; ground and aerial inspections; and the joint study of an inspection system designed to ensure that the sending of objects through outer space shall be 'exclusively for peaceful and scientific purposes' (UN Doc. A/RES/1148(XII)). It requested the Disarmament Commission to investigate possible inspection systems for a staged disarmament programme. It referred to 'weapons uses' and 'weapons purposes' as well as 'non-weapons uses' and 'non-weapons purposes' of fissionable material. In a letter to Soviet Premier Nikolai Bulganin in January 1958, US President Dwight D. Eisenhower proposed a ban on the testing of military rockets in space and the manufacture of space weapons, stating that 'outer space ought to be used only for peaceful purposes' (Eisenhower 1958, 126). Eisenhower referred to the 'military purposes' of missile tests and rejecting 'purposes of war'. The proposal was rebutted on the basis that Soviet ICBMs functioned as a necessary deterrent against a surprise attack from military bases in countries surrounding the USSR. In March 1958, the USSR replied with a different proposal: prohibiting the use of space for military purposes, the elimination of foreign military bases, and international cooperation in the study of space (UN Doc. A/3818).

In the summer of 1958, a group of technical experts met in Geneva to discuss the verification requirements of an agreement to suspend nuclear weapons tests (NATO 1958). On the basis of their report, Eisenhower proposed a suspension of testing and the international control of fissile material on 22 August 1958 (Eisenhower 1959, 635–6). In a letter to the UN Secretary General on 2 September 1958, US representative Mr Lodge stated that international cooperation in the peaceful uses of outer space ought to be pursued in parallel with agreements on disarmament. Lodge proposed a UN committee on the peaceful uses of outer space, contrasting 'peaceful purposes' with 'destructive purposes' (UN Doc. A/3902). On 15 September 1958, the USSR stated that the question of the cessation of atomic and hydrogen weapon tests ought to be separated from the general disarmament programme and resolved independently (UN Doc. A/3915). On 18 September, US Secretary of State John Foster Dulles proposed an *ad hoc* committee to report specific projects of international space cooperation under UN auspices. Dulles explicitly referred to the US Memorandum to the General Assembly of 12 January 1957, proposing that the testing of space objects should be conducted under a system of international inspection, and international participation would ensure future developments devoted 'exclusively to peaceful and scientific purposes' (US Department of State 1960, 733). Dulles encouraged 'every effort to dedicate outer space exclusively to constructive pursuits ...] as we reach beyond this planet, we should move as truly "united nations"' (Dulles 1958, 529). It appeared as though the creation of a World Space Agency was a matter of serious consideration.

On 7 November 1958, the USSR proposed a UN agency for international cooperation in the study of space. It would formalize an agreement to launch intercontinental and space rockets under an international programme, coordinate national research programmes, and foster the exchange of research (UN Doc. A/C.1/L.219). The USSR proposed that this agency should be established in conjunction with other disarmament measures: the elimination of foreign military bases and prohibitions on the launch of rockets into space for military purposes. On 21 November, an alternative 20-power draft resolution incorporated elements of the USSR's revised proposal such as the mutual exchange of information and coordination of national research programmes, but not the elimination of foreign military bases, or a UN agency for international cooperative launches and collaborative research (UN Doc. A/4009, 6–7). The USSR withdrew its proposal and on 13 December 1958 the General Assembly adopted resolution 1348 (XIII), establishing the Ad Hoc Committee on the Peaceful

Uses of Space. It affirmed the principle of sovereign equality in Article 2 of the Charter of the United Nations and stated the desire to 'avoid the extension of present national rivalries into this new field', acknowledging the 'common aim that space should be used for peaceful purposes only' (UN Doc. ARES/13/1348(XIII)). UNCOPUOS was established one year later (UN Doc. A/RES/1472(XIV)). The mandate of the original committee to explore the ways in which international space cooperation could benefit states irrespective of their economic or scientific development was eventually incorporated in Article 1 of the OST.

In his address to the UN General Assembly on 22 September 1960, Eisenhower proposed banning weapons of mass destruction in space; no national appropriation of celestial bodies; no warlike activities on celestial bodies; no orbital weapons of mass destruction; advance verification of launches by the UN; and a programme of international cooperation in 'constructive peaceful uses of space' (UN Doc. A/PV.868, p. 48). Eisenhower suggested that outer space should be demilitarized, following the example set by the 1959 Antarctic Treaty, which prohibited 'any measures of a military nature, such as the establishment of military bases and fortifications, the carrying out of military manoeuvres, as well as the testing of any type of weapons' (UNTS 1959, 72). Eisenhower also suggested the international control of outer space on the model of the Acheson–Lilienthal Report and Baruch Plan, which proposed the transfer of fissionable material and nuclear technology to an International Atomic Development Authority under control of the UN (US Department of State 1946). Eisenhower connected the peaceful use of outer space to the control of armaments, affirming: 'Our aim is to reach agreement on all the various measures that will bring general and complete disarmament' (UN Doc. A/PV.868, p. 48). Soviet law professor Yevgeniy Korovin affirmed the analogous position of the USSR: 'it is not the space rocket as such that endangers the security of mankind, but the nuclear warhead which may be delivered by a space rocket, a rocket of any possible range, a military aircraft etc. Clearly the disarmament of outer space cannot be divorced from disarmament on Earth' (Korovin 1959, 82–3). In the ensuing years, the political and technical complexities of general and comprehensive disarmament proved too difficult but progress would be achieved by partial measures.

On 25 September 1961, President John F. Kennedy delivered a speech to the UN General Assembly in which he announced a new agreement between the United States and the Soviet Union on a set of negotiating principles for general and complete disarmament (Kennedy 1961, 620; UN Doc. A/4891, A/4789). The US draft Treaty on General and Complete Disarmament in a Peaceful World advocated the creation of an International Disarmament Organization in the first of three stages of comprehensive disarmament. It would verify progressive reductions in armaments and stockpiles of weapons of mass destruction, the cessation of weapons production, reductions in the number of military bases and troops, and caps on military budgets (US Arms Control and Disarmament Agency 1962). In the second stage, the parties to the treaty were obliged to provide staff and equipment for a UN Peace Force. The third stage entailed the total dismantlement of national military forces. The draft treaty encouraged all nations to negotiate a nuclear weapons test-ban treaty and international cooperation in the peaceful uses of outer space. The USSR presented a similar proposal to the Geneva Disarmament Conference in March and September of 1962 (US Arms

Control and Disarmament Agency 1963a, 103–27, 1963b, 913–38). Whereas the US proposed a reduction in the means of delivering nuclear weapons by one-third in the initial stage, the USSR proposed total elimination except for an agreed number to be retained until the end of the second stage. The USSR proposed that a limited number of rockets would be permitted for the 'peaceful exploration of space' (US Arms Control and Disarmament Agency 1963a, 107) but launch pads and rocket production would be subject to inspection by the International Disarmament Organisation to ensure launches are 'exclusively for peaceful purposes' (US Arms Control and Disarmament Agency 1963a, 113). Both treaty drafts stipulated non-placement of nuclear weapons into orbit and pre-launch inspections of all spacecraft.

The 1963 Treaty Banning Nuclear Weapon Tests in the Atmosphere, in Outer Space, and Under Water was the first international treaty to refer to outer space. The original parties to the treaty—the governments of the United Kingdom, United States, and Soviet Union—proclaimed as their principal aim 'the speediest possible achievement of an agreement on general and complete disarmament under strict international control in accordance with the objectives of the United Nations which would put an end to the armaments race and eliminate the incentive to the production and testing of all kinds of weapons, including nuclear weapons' (UNTS 1963, 45). Subsequently, General Assembly Resolution 1884 under the heading 'Question of General and Complete Disarmament' called upon states to refrain from placing into orbit objects carrying nuclear weapons or other weapons of mass destruction and installing such weapons on celestial bodies (UN Doc. A/RES/1884 (XVIII)). It was adopted on 17 October 1963 and incorporated in Article IV of the OST.

On 13 December 1963, the General Assembly adopted Resolution 1962 (XVIII), Declaration of Legal Principles Governing the Activities of States in the Exploration and Use of Outer Space. It affirmed that no sovereignty may be claimed in space and space should be used 'for the benefit and in the interests of all people' (UN Doc. ARES/18/1962). The UK, US, and USSR drafts (UN Doc. A/C.1/879; A/C.1/881; A/5181) did not include an explicit declaration that space would be used exclusively for 'peaceful purposes', prompting remonstration by a number of delegations in the Legal Sub-Committee of UNCOPUOS (UN Doc. A/C.1/SR.1343, p. 168; A/AC.105/C.2/SR.21, p. 9; A/AC.105/C.2/SR.22, p. 7). The United Arab Republic draft explicitly stated that UNCOPUOS should be guided by the principle that the activities of member states in outer space should be confined solely to peaceful uses (UN Doc. A/AC.105/12, p. 7). However, opinions were divided over the appropriate forum for discussion of the content of the expression 'peaceful purposes' (Jasani and Lunderius 1980, 62–3). The UK delegate, Miss Gutteridge, suggested that the Eighteen Nation Committee on Disarmament was the appropriate location for this discussion (UN Doc. A/AC.105/C.2/SR.24, p. 13). The United States and the Soviet Union maintained that prohibition of the use of space for military purposes was not within the competence of the Legal Sub-Committee (UN Doc. A/AC.105/C.2/SR.7, p. 4; A/AC.105/PV.12, p. 15). The Soviet representative, Mr Fedorenko, stated that the declaration of legal principles 'did not and could not' touch upon the issues of the prohibition of the military uses of space because that question could only be considered in the context of a general disarmament agreement (UN Doc. A/C.1/PV.1342, p. 161). The repeated assertion of this position by the leading nuclear powers effectively split the mandates of UNCOPUOS and the Geneva-based disarmament committee (UN Doc. A/AC.105/C.2/SR.29, p. 108). Whether it was due to the ambiguous scope of the functions assigned to UNCOPOUS (Schick 1964, 975); the problem of distinguishing clearly between 'peaceful' and 'non-peaceful' uses of space, or non-observance of the work which eclipsed this dichotomy (Froehlich and Seffinga 2020, 104–5), this cleavage between the different fora arguably prevented a satisfactory resolution in either.

The OST drafts presented by the United States and the Soviet Union in 1966 did not refer to 'peaceful uses' but stated that the Moon and other celestial bodies should be used exclusively for peaceful purposes, qualifying this by reference to the prohibition of military bases, weapons testing, and military manoeuvres (UN Doc. A/AC.105/35, pp. 4, 13). The US draft permitted military personnel, facilities, and equipment for scientific research or 'any other peaceful purpose' (UN Doc. A/AC.105/32, p. 6). The USSR representative, Mr Morozov, stated that military equipment on celestial bodies was impermissible, even for scientific research, as it could result in activities contravening the use of celestial bodies exclusively for peaceful purposes (UN Doc. A/AC.105/C.2/SR.70, p. 3). The US delegate, Mr Goldberg, insisted that 'a treaty on the peaceful uses of outer space should contain a provision giving any other State free access to all areas, stations, installations, equipment and space vehicles on celestial bodies' (UN Doc. A/AC.105/C.2/SR.70, p. 5). Goldberg argued that the principle of free access would ensure military equipment was not used for 'non-peaceful purposes'. The US revised the article to state: 'All stations, installations, equipment, and space vehicles on the moon and other celestial bodies shall be open at all times to representatives of other States Parties to this Treaty conducting activities on celestial bodies' (UN Doc. A/AC.105/35, p. 2). The USSR added: 'on a basis of reciprocity and subject to agreement between the Parties with regard to the time of visits to such objects' (UN Doc. A/AC.105/35, p. 12). Goldberg clarified that access to all areas of celestial bodies would be granted 'on the basis of reciprocity' if it was understood that denial of the right of access by a first state would be met with denial by a second state (UN Doc. A/AC.105/C.2/SR.70, p. 6). Japan and Italy proposed revisions to include conditions that access will be granted 'on the understanding that the time of the visit should not imperil the life of the personnel' or 'such representatives shall take maximum precaution not to interfere with the normal operation of activities therein' (UN Doc. A/AC.105/35, pp. 14, 16). These debates exposed the many complications involved in conducting inspections on the moon. Ultimately, the use of military personnel, equipment, and facility for the exploration of celestial bodies was permitted by the OST (see Article IV). However, it is significant that Morozov suggested the use of outer space for exclusively peaceful purposes was only partially addressed—in the prohibition on orbiting weapons of mass destruction—and a number of questions remained to be solved after the elaboration of the treaty (UN Doc. A/AC.105/C.2/SR.66, p. 6). The practical problems of comprehensive disarmament and access to vehicles on celestial bodies were explicitly mentioned (UN Doc. A/AC.105/C.2/SR.70, p. 8).

In summary, the OST stands at the centre of a web of atomic arms control treaties as one of the most enduring and successful partial disarmament measures of the Cold War era. In common with other treaties discussed in this section, it was consistent with the primary objective of the Charter of the United Nations to 'save succeeding generations from the scourge of war' (United Nations 1945). However, the problems of disarmament verification and the prevention of surprise attack raised during the negotiations continue to be unresolved. For instance, the 'Rumsfeld Commission' in 2001 encouraged the US government to establish military space forces in order to prevent a 'space Pearl Harbour' (US Commission to Assess United States National Security Space Management 2001, 8–9). The long-term sustainability of space expansion depends upon finding solutions to these problems.

Underlying the catastrophic predictions in *Dark Skies* is the belief that humans will be unable to govern the development and use of space technology successfully (Deudney 2020, 7, 369). Let us now consider two potential solutions brought to light by our historical investigation: an inspection system to verify compliance with treaty obligations and an international agency to coordinate space activities.

4.2 The problem of inspection

The institution of an effective system of inspection would significantly ameliorate the concerns raised by *Dark Skies* and promote the long-term sustainability of space settlement. However, difficult choices arise regarding the appropriate method and degree of inspection. First, there are advantages and disadvantages to international, neutral, or reciprocal inspection methods (Bull 1962). For instance, Article VII of the Antarctic Treaty 1959 provides for open inspections by designated national observers rather than an international organization. Second, inspection systems must be credible but excessive inspection can be as dangerous as none at all. Sensitive information could be passed to terrorist organizations (Schelling and Halperin 1961, 103) or inspection could be hijacked for espionage (Bailey 2019, 19), bringing its suitability as a confidence-building measure into disrepute. Third, is it necessary to inspect space objects prior to launch, in orbit, or only on celestial bodies? Inspection might be essential only on rare occasions, for example, for the purpose of disambiguation during intense or unusual crisis situations (Kahn 1965). On-site inspection might be impractical due to the large number of inspectors required. In extremely remote locations such as the surface of the Moon it may be difficult or impossible to conduct adequate inspection procedures without endangering the safety of the mission or crew.

In 1962 the United Kingdom conducted a preliminary study of problems related to the elimination of rockets as nuclear delivery vehicles (US Arms Control and Disarmament Agency 1963b, 701–5). It suggested that the formation of an agency for international collaboration on space projects would prevent the development of space technology which could be used to threaten aggression. It asserted that the only alternative assurance against aggressive developments in space would be the degree of supervision and inspection exercised by an International Disarmament Organisation (US Arms Control and Disarmament Agency 1963b, 705). The report recommended that spacecraft should be subject to inspection at all stages of design and production, with control exercised at assembly points and launching sites to ensure no illegal payloads are launched. In the absence of international outposts on celestial bodies, how can it be verified that permanent lunar bases will serve only peaceful purposes? In the forthcoming years, establishing a trustworthy system of pre-launch inspection may prove to be in the best interests of international peace and security.

Proposals for an inspection system to verify the non-deployment of space weapons were revived in the Ad Hoc Committee on Prevention of an Arms Race in Outer Space in 1987–1988. The Soviet Union proposed an International Space Inspectorate with pre-launch notifications and the permanent presence of inspection teams at launch sites, storage facilities, laboratories, and test centres (UN Doc. CD/817). Objections to continuous on-site inspection were grounded in the technical, political, and organizational difficulties, and the 'dual-use' problem (UN Doc. CD/786, p. 14). As more states acquire launch capabilities, on-orbit inspection methods may have greater appeal. The early proposal by Myres S. McDougal is perhaps the most pragmatic: states launching satellites could provide advance notification to an international agency on the basis that the launch would, in principle, be available for inspection to ensure the payload corresponds to the object registered (McDougal 1957, 77). This could be negotiated as part of a cislunar resources or space traffic management treaty. Unfortunately, inspection systems alone cannot guarantee treaty compliance and must be bolstered by complementary measures. For instance, formal reprisals for treaty violations uncovered by inspections.

The next subsection will consider an alternative approach to the challenges of verification in outer space. Would the establishment of a World Space Agency diminish the requirement for inspection and help to prevent the catastrophic scenarios envisaged by *Dark Skies*?

4.3 World Space Agency

Since the earliest years of the space age thinkers have proposed that a World Space Agency can mitigate the risks of conflict in space and promote the collective interests of humankind (e.g. Jenks 1956; Horsford 1958; McDougal and Lipson 1958, Clark and Sohn 1962; Brown and Fabian 1975; Diederiks-Verschoor 1978; Crawford 1981, 1995; Wolter 2006; Pinault 2015; Kealotswe-Matlou 2021). This section will review several early proposals.[3]

In 1952, Oscar Schachter raised the question: 'Who owns the Universe?' Schachter argued that a legal order protecting the free and equal use of space would strengthen the international community and lead to a peaceful world order. Schachter suggested that the Grotian principle of the 'free seas' could apply to outer space, including celestial bodies, which would belong to mankind as a whole. No state would have the right to acquire any part as national territory (Schachter 1952, 122). Schachter pursued the analogy further by suggesting that mineral deposits on the Moon could be analogous to sedentary fisheries on the open seas, belonging to the state developing them but subject to internationally recognized rules regarding ownership and control. Freedom of navigation and scientific investigation would be protected rights. Oversight would be required to ensure that exploration accords with a principle of common interest and future generations would not be deprived of potential benefits (Schachter 1952, 124). For example, the intentional sabotage of lunar mining operations would be criminalized under international law. Schachter also extended Werner von Braun's idea for a 'space sentinel' (von Braun 1952, 23) by suggesting that it could belong to an international space agency to detect violations of arms control treaties (Schachter 1952, 126).

In 1956, C. Wilfred Jenks proposed that sovereignty over unoccupied territory on the moon or other planets could be vested exclusively in the UN, which would issue licences for exploitation of natural resources. Alternatively, some territories could be nationally appropriated by applying rules of discovery and occupation in customary international law, while others would be partitioned for international appropriation (Jenks 1956, 111). Legislative authority over outer space would be invested in the General Assembly of the UN acting on the advice of a special agency which would give 'appropriate measure of influence to States in a position to make a positive contribution to the exploration and exploitation of space' (Jenks 1956, 108).

In 1958, Cyril E. S. Horsford anticipated that national sovereignty in space would lead to military rivalry and dangerous forms of exploitation. Horsford considered outer (void) space to be analogous to the high seas and free for all to use for lawful passage, but suggested this proposition would not be accepted with respect to celestial bodies (Horsford 1958, 75). Horsford did not reject the Roman legal principle of discovery and occupation, but suggested that a UN space agency could act as a trusteeship, administering territories through the states which discovered them and resolving disputes through arbitration or judicial settlement in the International Court of Justice. A UN Space Council would include the active space-faring nations and a representative number of smaller states not active in space programmes. Horsford emphasized that space activities can affect the whole of humankind and should be conducted with the greatest degree of consent by all (Horsford 1958, 76).

In 1959, Phillip C. Jessup and Howard J. Taubenfeld proposed a model of direct international administration but suggested that the responsible agency could be established either within or outwith the UN framework. The authors suggested that jurisdiction of outer space and trusteeship of resources could be vested in a 'Cosmic Development Corporation' representing all nations (Jessup and Taubenfeld 1959, 282). It would own, operate, and act as a licensing organization for all spacecraft, undertake responsibility for the governance of celestial bodies, and 'advance the welfare of all men through activities in outer space' (Jessup and Taubenfeld 1959, 280). The authors proposed the renunciation of national sovereignty in space and the maintenance of law and order in the neutral zone by this international organization (Jessup and Taubenfeld 1959, 276).

In 1962, Grenville Clark and Louis Sohn proposed a UN Outer Space Agency (OSA) as part of a plan for comprehensive disarmament and world federalist government. The agency would have two purposes: (i) to ensure that outer space is used for peaceful purposes only, and (ii) to promote to the fullest possible extent exploration and exploitation of outer space for the benefit of all (Clark and Sohn 1962, 296). The OSA would possess its own rockets, satellites, and spacecraft, and act as a licensing organization for states, organizations, and individuals possessing and operating space vehicles. The OSA would conduct research and development, and supervise the use of rockets, satellites, and other spacecraft. The OSA would prevent the use of space for military purposes. It would maintain a small fleet of spacecraft to police outer space or activate in emergency situations under the authorization of the UN General Assembly or Executive Council (Clark and Sohn 1962, 301–2). The OSA would exercise control over the Moon and other planets to ensure no individual nation or other alliance gained control of these territories (Clark and Sohn 1962, 30).

The establishment of a World Space Agency under the auspices of the UN or a world federal government would significantly ameliorate many of the concerns raised in *Dark Skies*. International space cooperation accompanied by authoritative global governance could ensure the pooling of the resources required to accomplish large-scale space expansion and reinforce international disarmament (Crawford 2015, 209; 2017). In future, effective planetary defence may necessitate the development of transnational institutions which promise to 'unify the human species' (Sagan and Ostro 1994, 72). However, if the twenty-first-century space race intensifies then addressing the problems identified in this chapter will become a matter of urgency and a world federalist solution appears distant. Space-faring civilization is at a critical juncture. In the coming decades, irreversible actions in space may lead to an increasingly divided and conflict-ridden world. As Eisenhower warned in 1960: 'National vested interests have not yet been developed in space or in celestial bodies... . Let us remind ourselves that we had a chance in 1946 to ensure that atomic energy be devoted exclusively to peaceful purposes... We must not lose the chance we still have to control the future of outer space' (UN Doc. A/PV.868, p. 48).

The final section proposes that space expansionism can be safely pursued and catastrophe averted with the assistance of co-ordinating international and transnational institutions. It responds to the problematic scenarios in *Dark Skies* with proposals which exceed traditional confidence-building measures but fall short of world federalism or hierarchical statism.

5. Long-term Sustainability of Space Expansion

Dark Skies suggests that the OST should be extended to accommodate a space traffic management regime (Deudney 2020, 246). Space traffic management is likely to be a prerequisite for the long-term sustainability of space activities. However, it seems to be an implication of an effective system that traffic and trade routes will need to be 'policed' in order to maintain safe and orderly transit. In 1957, Donald Cox posed the question: 'When space becomes a highway, who will control the traffic?' Cox argued that a UN 'space police force' patrolling interplanetary regions would ensure that 'any power-mad nation or nations does not usurp their free space prerogatives' (Cox 1957, 71). Cox suggested that the space vehicles could be supplied by Russia and the United States, but the crew would be recruited from member states of neutral nations. Ground-control stations would be constructed and owned by sovereign nations but monitored by UN teams. How might this 'Space Guard' scenario play out in future?

Space traffic management with tighter controls on the registration of spacecraft would serve to mitigate the problem of orbital debris. As active debris removal methods become available, an international treaty could be negotiated to ensure all objects launched are registered with tracking transponders (e.g. Abraham 2018). States party to the treaty could share space situational awareness data. If states, private companies, or individuals fail to officially register spacecraft, it could be treated as debris and removed from orbit by the Space Guard.

Space-mining is predicted to become part of a multi-trillion-dollar space industry (Bank of America Merrill Lynch 2017). If commercial operations exacerbate terrestrial wealth inequalities (Deudney 2020, 324) this will create a breeding ground for space pirates, terrorism, and guerrilla warfare. Robust security measures will be essential due to the inherent vulnerabilities of space operations. The OST affirms that 'the exploration and use of space ... shall be the province of all mankind' (Article I), thus it seems appropriate that the safeguarding role is performed by a peacekeeping UN Space Guard rather than national military forces.

Let us now turn to the problem of inadvertent escalation. Nuclear states collectively need to lengthen the time for decision-making in a crisis. Early warning satellites can detect the infrared signature of ballistic missile launch, providing a longer duration of time for decision-making than radar alone. If Russian military commanders only have 15 minutes decision-making time and US commanders have 30 minutes, then the chances for error and miscalculation appear to be greater on the Russian side. Russia's early warning satellites were non-operational between 2014 and 2015 while a new constellation was under deployment (Hendricx 2021). The significance of early warning satellites for global security is a sign that they ought to be continuously operational, highly accurate, and require international protection. An international space agency could provide the UN Security Council or General Assembly with independent monitoring information (UNIDIR 1983, 69–70; Dorn 1990; Deudney 2020, 236). Without replacing national space programmes, states could be trustees for the launch and maintenance of an early warning system designed, financed, and operated by a World Space Agency (McDougal and Lipson 1958, 431). The agency could act as a verification portal to defuse ambiguous situations and facilitate pathways to de-escalation below the nuclear threshold. In the most speculative scenario, peacekeeping early warning systems could be combined with space-based interception to enforce a World Peace Treaty, at the final stages of general and comprehensive disarmament.

Finally, how can Orbita be constructed and maintained without destabilizing international peace and security? A World Space Agency could be created for the purpose of delivering space solar power for the benefit of all. It could catalyse cooperative research programmes to ensure that orbital infrastructure meets international safety standards. It would be responsible for conducting inspections to ascertain these standards are upheld by national space programmes. Space solar power will not be completely risk-free, but sharing the burden of risk via a ground-level cooperative venture promotes the wide distribution of benefits. It would also place new limits upon the degradation of international relations. Is this not sufficient reason to pursue the project? International collaboration in this respect approaches a 'litmus test' for the future of civilization. Is humankind innovative and responsible enough to harness all the available free resources in the environment or are we doomed to live in the shadow of scarcity while gradually destroying our own natural habitat?

The proposals outlined in this section would require an unprecedented level of collaboration. Their chief merit derives from the potential to dial down tensions between nuclear-armed states, protect space assets against the threat of terrorism, and mitigate the harmful effects of space debris without limiting the commercial potential of space. For these reasons, they may be regarded as more attractive than the proposals for 'Whole Earth security' in *Dark Skies*.

6. Conclusion

In the first decades of the twenty-first century, technologies with global reach are providing an increasing number of the inhabitants of Earth with a nascent 'planetary consciousness' and awareness of problems which transcend national borders. Space systems not only support critical national infrastructures and the infrastructure of globalized society but also play a crucial role in finding solutions to the 'global problems' which affect us all. For instance, Earth observation satellites play a vital role in addressing the climate emergency by contributing data to the Global Climate Observing System (OECD 2016, 95–6). Protection of this critical orbital infrastructure which is to the enormous benefit of present and future generations is in the interests of everyone. However, there is a growing consensus that existing institutions are not prepared for the challenges of emerging technologies and the rising number of actors in space over the coming decades (US National Intelligence Council 2021, 100). Perhaps a new kind of research institute is required for the purpose of systematically addressing long-term, global problems. The individuals, organizations, and institutions contributing to this problem-solving effort will invariably promote the long-term, peaceful use of space.

In concluding, this chapter found the pessimism of *Dark Skies* to be unwarranted. Space technologies will improve our lives in ways we cannot foresee, and we should not presume that the worst aspects of history shall prevail over the admirable qualities in the human character. The chapter reconceptualized the catastrophic scenarios of environmental degradation and inadvertent escalation as challenges to overcome before large-scale space settlement. It proposed the integration of space expansion and disarmament initiatives through international and transnational institutions. It found that fundamental problems of space arms control remain to be solved. In approaching these problems, we have many lessons to learn from the history of space law.

Acknowledgements

The author wishes to express his sincere gratitude to the librarians of the United Nations Dag Hammarskjöld Library for their assistance with archival materials used in this research.

Notes

1. Space expansionists include Peter Glaser, Gerard K. O'Neil, Dandridge Cole, Hermann Oberth, Buckminster Fuller, Krafft Ehricke, Konstantin Tsiolkovsky, Freeman Dyson, Arthur C. Clarke, Wernher von Braun, G. Harry Stine, A. P. J Abdul Kalam, Elon Musk, and Jeff Bezos, amongst others.

2. Henceforth 'Outer Space Treaty' or 'OST' (UNTS, 1967).
3. The following sections build upon this author's research in McKenna (2021).

References

Abraham, A. 'GPS Transponders for Space Traffic Management' The Center for Space Policy and Strategy. The Aerospace Corporation, April 2018 <https://csps.aerospace.org/papers/gps-transponders-space-traffic-management> accessed 4 July 2021.

Acton, J. M. 'Why is Nuclear Entanglement So Dangerous? Carnegie Endowment for International Peace' (23 January 2019) <https://carnegieendowment.org/2019/01/23/why-is-nuclear-entanglement-so-dangerous-pub-78136> accessed 19 August 2020.

Aoki, S. 'Law and Military Uses of Outer Space' in R. S. Jakhu and P. S. Dempsey (eds), *Routledge Handbook of Space Law* (Routledge, 2017) 197–224.

Bailey, K. C. *The UN Inspections in Iraq: Lessons for On-Site Inspection* (Routledge, 2019).

Bank of America Merrill Lynch. 'To Infinity and Beyond—Global Space Primer' (October 2019) *Thematic Investing* <https://newspaceglobal.com/wp-content/uploads/imce/u3479/MerrillLynchSpace-Oct2017.pdf> accessed 17 August 2021.

Barrie, D. 'Emerging Challenges for European Security and Defence' International Institute for Strategic Studies <https://www.iiss.org/blogs/research-paper/2021/09/emerging-challenges-for—european-security-and-defence> accessed 21 September 2021.

Beckett, C. 'Getting to Grips with Grey Zone Conflict' Strategic Command blog (26 April 2021) <https://stratcommand.blog.gov.uk/2021/04/26/getting-to-grips-with-grey-zone-conflict/> accessed 4 July 2021.

Blount, P. J. 'Space Traffic Coordination: Developing a Framework for Safety and Security in Satellite Operations' (2021a) 9830379 *Space: Science & Technology* 1–10.

Blount, P. J. 'Peaceful Purposes for the Benefit of All Mankind' in C. Steer and M. Hersch (eds), *War and Peace in Outer Space* (Oxford University Press, 2021b) 109–22.

Butow, S. J., Cooley, T., Felt, E., Mozer, J. B. 'State of the Space Industrial Base 2020. Summary Report of United States Space Force, Defense Innovation Unit, and Air Force Research Laboratory workshop: A Time for Action to Sustain US Economic & Military Leadership in Space'. July 2020. https://apps.dtic.mil/sti/pdfs/AD1106608.pdf accessed 22 June 2021.

Brown, M. K., Lewis, H. G., Kavanagh, A. J., et al. 'Future Decreases in Thermospheric Neutral Density in Low Earth Orbit Due to Carbon Dioxide Emissions' (2021) 126(8) *Journal of Geophysical Research—Atmospheres* e2021JD034589.

Brown, S. and Fabian, L. L. 'Toward Mutual Accountability in the Nonterrestrial Realms' (1975) 29(3) *International Organization* 877–92.

Bull, H. 'Is International Inspection Necessary?' *The Spectator* (30 November 1962) 853–4.

Cheng, B. 'Definitional Issues in Space Law: the "Peaceful Use" of Outer Space, including the Moon and other Celestial Bodies' in B Cheng and Oxford University Press (eds) *Studies in International Space Law* (Oxford: Clarendon Press, 1997) 513–22.

Chester, W. 'Space Law as a Way to World Peace' in *Legal Problems of Space Exploration; a symposium prepared for the use of the Committee on Aeronautical and Space Sciences,*

United States Senate by the Legislative Reference Service, Library of Congress. United States Senate, 87th Congress, 1st Session (Washington, DC: Government Printing Office, 1961) 476–93.

Clark, G. and Sohn, L. *World Peace through World Law* (2nd edn (revised), New Haven, CT: Harvard University Press, 1962).

CNSA 'Russian-Chinese Joint Seminar on Cooperation in International' Lunar Research Stations. China National Space Administration (28 September 2021) <http://www.cnsa.gov.cn/english/n6465652/n6465653/c6812568/content.html> accessed 1 October 2021.

Cockell, C. S. *Extra-Terrestrial Liberty: An Enquiry into the Nature and Causes of Tyrannical Government beyond the Earth* (Edinburgh: Shoving Leopard, 2013).

Cole, D. *Beyond Tomorrow: The Next 50 Years in Space* (Amherst, WI: Amherst Press, 1965).

Cox, D. 'International Control of Outer Space' (1957) 2 *Missiles and Rockets* 68–71.

Crawford, I. A. 'On the Formation of a Global Space Agency' (1981) 23 *Spaceflight* 316–17.

Crawford, I. A. 'Space Development: Social and Political Implications' (1995) 11(4) *Space Policy* 219–25.

Crawford, I. A. 'Interplanetary Federalism: Maximising the Chances of Extraterrestrial Peace, Diversity and Liberty' in C. S. Cockell (ed.), *The Meaning of Liberty Beyond Earth* (Cham: Springer International, 2015) 199–218.

Crawford, I. A. 'Space, World Government, and a 'Vast Future' for Humanity' World Orders Forum. World Government Research Network <https://www.wgresearch.org/world-orders-forum> accessed 4 July 2022.

Crawford, I. A. and Baxter, S. 'The Lethality of Interplanetary Warfare: A Fundamental Constraint on Extraterrestrial Liberty' in C. S. Cockell (ed.), *The Meaning of Liberty Beyond Earth* (Cham: Springer, 2015) 187–98.

D'Onfro, J. 'Jeff Bezos Thinks That to Save the Planet We'll Need to Move All Heavy Industry to Space' *Business Insider* (1 June 2016) <http://www.businessinsider.com/jeff-bezos-on-blue-origin-and-space-2016-6> accessed 18 July 2021.

Deudney, D. 'Unlocking Space' (1983) 53 *Foreign Policy* 91–114.

Deudney, D. *Dark Skies. Space Expansionism, Planetary Geopolitics, & The Ends of Humanity* (New York, NY: Oxford University Press, 2020).

Diederiks-Verschoor, I. 'Observations on the International Civil Aviation Organization and an International Space Agency' in M. Schwartz (ed.), *Proceedings of the Twentieth Colloquium on the Law of Outer Space. International Institute of Space Law of the International Astronautical Federation. September 25–October 1, 1977, Prague, Czechoslovakia* (Littleton, CO: Fred B. Rothman and Co., 1978).

Dorn, A. W. 'The Case for a United Nations Verification Agency: Disarmament Under Effective International Control' Working Paper 26 (Canadian Institute for International Peace and Security, 1990).

Dulles, J. F. 'Problems of Peace and Progress. Address before the United Nations 13th General Assembly at New York, N.Y. on 18 September 1958' (1958) 39(1006) *US Department of State Bulletin* Publication 6709 (6 October 1958). Government Printing Office, Washington, DC.

Ehricke, K. A. 'Large-Scale Processing of Lunar Materials, North American Space Operations, Rockwell International, El Segundo, California' in D. Criswell (ed.), *Lunar Utilization: Abstracts* (NASA-CR-156167, Lunar Science Institute, 1976) 87–91.

Einstein, A. '1939 The Way Out' in D. Masters, and C. Way (eds), *One World or None* (New York, NY: New Press, 2007) 209–14.

Eisenhower, D. D. '1955 Statement on Disarmament Presented at the Geneva Conference' 21 July 1955. Gerhard Peters and John T. Woolley, *The American Presidency Project.* <https://www.presidency.ucsb.edu/documents/statement-disarmament-presented-the-geneva-conference> accessed 16 May 2020.

Eisenhower, D. D. 'Letter to Nikolai Bulganin, 12 January 1958' (1958) 38(970) *US Department of State Bulletin* 122–27.

Eisenhower, D. D. 'Statement by the President Following the Geneva Meeting of Experts Proposing Negotiations on Nuclear Controls, August 22, 1958' in United States Office of the Federal Register. *Public Papers of the Presidents of the United States. Dwight D. Eisenhower: 1958: containing the public messages, speeches, and statements of the president, January 1 to December 31, 1958.* National Archives and Records Service (Washington, DC: Government Printing Office, 1959) 635–6.

Eisenhower, D. D. *Mandate for Change. 1953–1956* (Doubleday, 1963).

Elvis, M., Milligan, T., and Krolikowski, A. 'The Peaks of Eternal Light: A Near-Term Property Issue on the Moon' (2016) 38 *Space Policy* 30–8.

Federation of American Scientists. 'Nuclear Test Ban, UN Control of Space Research, and UN Police Force—First Steps Toward Peace' (1958) 14(3) *Bulletin of the Atomic Scientists* 125.

Froehlich, A. and Seffinga, V. (eds), *The United Nations and Space Security: Conflicting Mandates Between UNCOPOUS and the CD* (Cham: Springer International, 2020).

Galloway, E. 'Introduction' in *Legal Problems of Space Exploration; a symposium prepared for the use of the Committee on Aeronautical and Space Sciences, United States Senate by the Legislative Reference Service, Library of Congress.* United States Senate, 87th Congress, 1st Session (Washington, DC: Government Printing Office 1961) xi–xxii.

Galloway, E. 'Nuclear Powered Satellites: the USSR Cosmos 954 and the Canadian Claim' (1979) 12(3) *Akron Law Review* 401–15.

Gaspari, F. and Oliva, A. 'The Consolidation of the Five UN Space Treaties into One Comprehensive and Modernized Law of Outer Space Convention: Toward a Global Space Organization' in G. D. Kyriakopoulos and M. Manoli (eds), *The Space Treaties at a Crossroads: Considerations de Lege Ferenda* (Cham: Springer International, 2019) 183–97.

Goswami, N. 'Statement of Dr. Namrata Goswami. US-China Economic and Security Review Commission Hearing on "China in Space: A Strategic Competition?"' 25 April 2019. https://www.uscc.gov/sites/default/files/Namrata%20Goswami%20USCC%2025%20April.pdf accessed 21 August 2021.

Goswami, N. 'Status of Existing and Emerging Asia-Pacific Space Powers Capabilities' (Asia Pacific Leadership Network, 2021) <https://cms.apln.network/wp-content/uploads/2021/08/APLN_Special-Report-1_Namrata-Goswami_V3.pdf> accessed 21 September 2021.

Goswami, N. and Garretson, P. A. *Scramble for the Skies: The Great Power Competition to Control the Resources of Outer Space* (London: Rowman and Littlefield, 2020).

Harrison, T., Johnson, K., and Roberts, T. G. 'Space Threat Assessment' (Center for Strategic and International Studies (CSIS), April 2018) <https://www.csis.org/analysis/space-threat-assessment-2018> accessed 1 December 2019]

Hendricx, B. 'EKS: Russia's Space-based Missile Early Warning System' (8 February 2021) <https://www.thespacereview.com/article/4121/1> accessed 8 September 2021.

Hill, J. H. 'Hypersonic/Highly-Manoeuvrable Weapons and Their Effect on the Deterrence Status Quo' in P. P. Cone (ed.), *Future Warfare Series No. 59 Assessing the Influence of Hypersonic Weapons on Deterrence* (United States Air Force. USAF Center for Strategic Deterrence Studies, 2019) 57–74.

Horsford, C. E. S. 'Principles of International Law in Spaceflight' (1958) 5 *St. Louis University Law Journal* 70–8.

IADC (2021). 'Space Debris Mitigation Guidelines' Inter-Agency Space Debris Coordinate Committee. IADC Steering Group and Working Group 4. IADC-01-02 Rev. 3, July 2021.

Jakhu, R. S. and Pelton, J. N. *Global Space Governance: An International Study* (Cham: Springer International, 2017).

Jasani, B. and Lunderius, M. A. 'Peaceful Uses of Outer Space—Legal Fiction and Military Reality' (1980) 11(1) *Bulletin of Peace Proposals* 57–70.

Jenks, C. W. 'International Law and Activities in Space' (1956) 5(1) *The International and Comparative Law Quarterly* 99–114.

Jenks, C. W. *Space Law* (London: Stevens & Sons, 1965).

Jessup, P. C. and Taubenfeld, H. J. *Controls for Outer Space and the Antarctic Analogy* (New York, NY: Columbia University Press, 1959).

Johnson, A. 'Putin Denies "New Cold War" but Says New Nukes Are on "Combat Duty"' *NBC News*, 2 March 2018.

Johnson, J. 'Catalytic Nuclear War in the Age of Artificial Intelligence & Autonomy: Emerging Military Technology and Escalation Risk between Nuclear-Armed States' [13 January 2021] *Journal of Strategic Studies* 1–41. DOI: 10.1080/01402390.2020.1867541

Kahn, H. *On Escalation: Metaphors and Scenarios* (New York, NY: Frederick A. Praeger, 1965).

Kardashev, N. 'On the Inevitability and the Possible Structures of Supercivilizations' (1985) 112 *Symposium—International Astronomical Union* 497–504.

Kealotswe-Matlou, I. 'The Rule of Law in Outer Space: A Call for an International Outer Space Authority' in C. Steer and M. Hersch (eds), *War and Peace in Outer Space* (New York: Oxford University Press, 2021) 91–108.

Kennedy, J. F. 'Let Us Call a Truce to Terror' Address to the UN General Assembly, 25 September 1961 (1961) 45 *US Department of State Bulletin* 619–25.

Kessler, D. J. and Cour-Palais, B. G. 'Collision Frequency of Artificial Satellites: The Creation of a Debris Belt' (1978) 83(A6) *Journal of Geophysical Research* 2637–46.

Kimball, D. and Reif, K. 'The Intermediate-Range Nuclear Forces (INF) Treaty at a Glance' (Arms Control Association, 2019) <https://www.armscontrol.org/factsheets/INFtreaty> accessed 10 September 2021.

Klinger, J. M. 'Environmental Geopolitics and Outer Space' (2021) 26 *Geopolitics* 666–703.

Kobe, D. H. 'A Theory of Catalytic War' (1962) 6(2) *Journal of Conflict Resolution* 125–42.

Kornuta, D., Abbud-Madrid, A., Atkinson, J., et al. 'Commercial Lunar Propellant Architecture: A Collaborative Study of Lunar Propellant Production' (2019) 13 *REACH* 100026.

Korovin, Y. 'On the Neutralization and Demilitarization of Outer Space' (1959) 5(12) *International Affairs* Moscow 82–92.

Kulacki, G. 'The Chinese Military Updates China's Nuclear Strategy' (Union of Concerned Scientists, 2015) <ucsusa.org/ChinaNuclearStrategy> accessed 7 March 2019.

Lopez, C. T. 'US Space Effort's Future Hinges on Private Industry' Department of Defense News (28 July 2020) <https://www.defense.gov/News/News-Stories/Article/Article/2291577/us-space-efforts-future-hinges-on-private-industry/> accessed 11 August 2020.

Mankins, J. C. and Kaya, N. (eds). *Space Solar Power: The First International Assessment of Space Solar Power: Opportunities, Issues and Potential Pathways Forward* (Paris: International Academy of Astronautics, 2011).

Maogoto, J. N. and Freeland, S. 'Space Weaponization and the United Nations Charter Regime on Force: A Thick Legal Fog or a Receding Mist?' (2007) 41(4) *The International Lawyer* 1091–19.

Markoff, M. G. 'Disarmament and "Peaceful Purposes" Provisions in the 1967 Outer Space Treaty' (1976) 4(1) *Journal of Space Law* 3–22.

McDougal, M. S. 'Artificial Satellites: A Modest Proposal' (1957) 51(1) *American Journal of International Law* 74–7.

McDougal, M. S and Lipson, L. 'Perspectives for a Law of Outer Space' (1958) 52 *American Journal of International Law* 407–31.

McDougal, M. S. and Associates. *Studies in World Public Order* (New Haven, CT: Yale University Press, 1960).

McDougal, M. S. 'The Emerging Customary Law of Space' (1963) 58 *Northwestern University Law Review* 618–42.

McKenna, A. 'In Search of Global Security: Everett C. Dolman's Astropolitik and Daniel Deudney's Dark Skies' (2021) 18(3) *Astropolitics* 199–222.

Mumford, L. *The Myth of the Machine. The Pentagon of Power* (New York, NY: Harcourt Brace Jovanovich, Inc, 1970).

NASA. 'Orbital Debris Quarterly News' (2012) 16(3). United States National Aeronautics and Space Administration <https://orbitaldebris.jsc.nasa.gov/quarterly-news/#> accessed 12 September 2021.

NASA. 'Artemis Plan: NASA's Lunar Exploration Program Overview' (September 2020). United States National Aeronautics and Space Administration <https://www.nasa.gov/sites/default/files/atoms/files/artemis_plan-20200921.pdf> accessed 12 September 2021.

NATO. Report by the delegations of Canada, France, United Kingdom, United States on the discussions of the disarmament sub-committee of the United Nations disarmament commission. 24 April 1957. C-M (57) 69. NATO Archives.

NATO. Disarmament—Possible questions to be put by the Council to SACEUR. 2 July 1957. PO/57/786. NATO Archives.

NATO. Disarmament. Military Committee Standing Group Liaison Office Messages. 22 July 1957. LOSTAN 2141. NATO Archives.

NATO. Disarmament. Military Committee Standing Group Liaison Office Messages. 21 August 1957. LOSTAN 2158. NATO Archives.

NATO. Disarmament. Military Committee Standing Group Liaison Office Messages. 23 August 1957. LOSTAN 2159. NATO Archives.

NATO. Report of the Conference of Experts to Study the Methods of Detecting Violations of a Possible Agreement on the Suspension of Nuclear Tests. 1 September 1958. RDC(58)300. NATO Archives.

NATO. 'London Declaration' Issued by the Heads of State and Government participating in the meeting of the North Atlantic Council in London, 3–4 December 2019. The North Atlantic Treaty Organization <https://www.nato.int/cps/en/natohq/official_texts_171584.htm> accessed 10 December 2019.

Newlon, C. and Howard W. E. (eds). 'John F. Kennedy Reply to Open Letter: "If the Soviets Control Space—They Can Control Earth—Kennedy"' (1960) 7(15) *Missiles and Rockets*(Harrisburgh, PA: Telegraph Press, 1960) 12–13.

Oberg, J. 'Planetary Climate Modification and the US Space Command–As-yet Unrecognized Missions in the post-2025 Time Frame.' presentation at Futures Focus Day Symposium, US Space Command, Colorado Springs, Colorado, 23 July 1998. https://www.phoenixat.com/~vnn2/PDdebate.html accessed 29 December 1999.

OECD. *Space and Innovation*. Organisation for Economic Co-operation and Development (Paris: OECD Publishing, 2016).

O'Neill, G. K. *The High Frontier* (3rd edn, Ontario: Space Studies Institute and Apogee Books, 2021).

Pardini, C. and Anselmo, L. 'Evaluating the Impact of Space Activities in Low Earth Orbit' (2021) 184 *Acta Astronautica* 11–22.

Pelton, J. N. *New Solutions for the Space Debris Problem* (Cham: Springer International, 2015).

Pence, M. 'Remarks by Vice President Pence at the Fifth Meeting of the National Space Council' Huntsville, AL, 26 March 2019 <https://www.whitehouse.gov/briefings-statements/remarks-vice-president-pence-fifth-meeting-national-space-council-huntsville-al/> accessed 22 July 2019.

Pinault, L. 'Towards a World Space Agency' in C. S. Cockell (ed.), *Human Governance Beyond Earth: Implications for Freedom* (Cham: Springer International, 2015).

Reif, K. and Bugos, S. 'UK to Increase Cap on Nuclear Warhead Stockpile' [2021] *Arms Control Today* <https://www.armscontrol.org/act/2021-04/news/uk-increase-cap-nuclear-warhead-stockpile> accessed 4 July 2022.

Ross, M. and Vedda, J. 'The Policy and Science of Rocket Emissions' (Center for Space Policy and Strategy. The Aerospace Corporation, 2018) <https://csps.aerospace.org/papers/policy-and-science-rocket-emissions> accessed 15 March 2020.

Russell, B. 'Only World Government Can Prevent the War Nobody Can Win' (1958) September *Bulletin of the Atomic Scientists* 259–61.

Sagan, C. *Pale Blue Dot: A Vision of the Human Future in Space* (New York, NY: Ballantine Books, 1997).

Sagan, C. and Ostro, S. J. 'Long-range Consequences Of Interplanetary Collisions' (1994) 10(4) *Issues in Science & Technology* 67–72.

Schachter, O. 'Who Owns the Universe?' in C. Ryan (ed.), *Across the Space Frontier* (New York, NY: Viking Press, 1952) 118–31.

Schelling, T. C. 'The Retarded Science of International Strategy' (1960) 4(2) *Midwest Journal of Political Science* 107–237.

Schelling, T. C. and Halperin, M. H. *Strategy and Arms Control* (New York, NY: The Twentieth Century Fund, 1961).

Schick, F. B. 'Problems of a Space Law in the United Nations' (1964) 13(3) *The International and Comparative Law Quarterly* 969–86.

Schwarz, J. S. J. *The Value of Science in Space Exploration* (New York, NY: Oxford University Press, 2020).

Shalikashvili, J. M. *National Military Strategy of the United States of America* (Washington, DC: Government Printing Office, 1995).

SIPRI (Stockholm Peace Research Institute). *SIPRI Yearbook 2021. Armaments, Disarmament, and International Security* (Oxford: Oxford University Press, 2021).

Sourbès, I. and Boyer, Y. 'Technical Aspects of Peaceful and Non-Peaceful Uses of Space' in B. Jasani (ed.), *Peaceful and Non-Peaceful Uses of Space* (United Nations Institute for Disarmament Research, Taylor & Francis, 1991) 57–75.

Steer, C. 'Global Commons, Cosmic Commons. Implications of Military and Security Uses of Outer Space' (2017) 18(1) *Georgetown Journal of International Affairs* 9–16.

Syal M. B., Dearborn, D. S. P., and Schultz, P. H. 'Limits on the Use of Nuclear Explosives for Asteroid Deflection' (2013) 90 *Acta Astronautica* 103–11.

Taylor, D. M. 'Plutonium: Environmental Pollution and Health Effects' in J. Nriagu (ed.), *Encyclopedia of Environmental Health* (London: Elsevier, 2011) 264–73.

UN Doc. A/3818 Soviet proposal for banning of the use of cosmic space for military purposes, the elimination of foreign bases on the territories of other countries, and international co-operation in the study of cosmic space. Note dated 15 March 1958 addressed to the Secretary-General by the Permanent Representative of the Union of Soviet Socialist Republics, 17 March 1958.

UN Doc. A/3902 Letter from the Permanent Representative of the United States of America to the General Assembly. Programme for International Cooperation in the Field of Outer Space, 2 September 1958.

UN Doc. A/3915 Letter dated 15 September 1958 from the Chairman of the Delegation of the Union of Soviet Socialist Republics to the United Nations, addressed to the Secretary-General. Request for the inclusion of an additional item in the agenda of the Thirteenth Regular Session: The Discontinuance of Atomic and Hydrogen Weapons Tests, 15 September 1958.

UN Doc. A/4009. Report of the First Committee to the United Nations General Assembly, 28 November 1958.

UN Doc. A/4078 Report of the Conference of Experts for the Study of Possible Measures which might be Helpful in Preventing Surprise Attack and for the Preparation of a Report thereon to Governments, 5 January 1959.

UN Doc. A/4789 Letter from the Permanent Representatives of the Union of Soviet Socialist Republics and the United States to the United Nations addressed to the President of the General Assembly 16th Session, 20 September 1961.

UN Doc. A/4891 Letter from the Representative of the United States of America to the United Nations addressed to the President of the General Assembly. 25 September 1961.

UN Doc. A/5181 Union of Soviet Socialist Republics: draft declaration of the basic principles governing the activities of States pertaining to the exploration and use of outer space, 16 April 1963.

UN Doc. A/AC.105/12 United Arab Republic: Draft Code for International Co-operation in the Peaceful Uses of Outer Space. United Nations Committee on the Peaceful Uses of Outer Space: Report of the Legal Sub-Committee on the Work of its Second Session (16 April–3 May 1963), 6 May 1963.

UN Doc. A/AC.105/32 Letter dated 16 June 1966 from the Permanent Representative of the United States of America addressed to the Chairman of the Committee on the Peaceful Uses of Outer Space. Draft Treaty Governing the Exploration of the Moon and Other Celestial Bodies, 17 June 1966.

UN Doc. A/AC.105/35 Report of the Legal Subcommittee on the work of its 5th session (12 July–4 August and 12—16 September 1966) to the United Nations Committee on the Peaceful Uses of Outer Space, 16 September 1966.

UN Doc. A/AC.105/C.2/SR.7 United Nations Committee on the Peaceful Uses of Outer Space. Summary Record of the Seventh Meeting, Held at the Palais des Nations, Geneva, 7 June 1962.

UN Doc. A/AC.105/C.2/SR.21 United Nations Committee on the Peaceful Uses of Outer Space Legal Subcommittee, Summary Record of the Twenty-First Meeting, 25 April 1963.

UN Doc. A/AC.105/C.2/SR.22 United Nations Committee on the Peaceful Uses of Outer Space Legal Subcommittee, Summary Record of the Twenty-Second Meeting, 26 April 1963.

UN Doc. A/AC.105/C.2/SR.24 United Nations Committee on the Peaceful Uses of Outer Space Legal Subcommittee. Verbatim Record of the Twenty-Fourth Meeting, held at headquarters, New York, 22 November 1963.

UN Doc. A/AC.105/C.2/SR.29 United Nations Committee on the Peaceful Uses of Outer Space Legal Subcommittee, Summary Records of the 29th to 37th Meetings Held at the Palais des Nations, Geneva, from 9 to 26 March 1964, 26 March 1964.

UN Doc. A/AC.105/C.2/SR.66 United Nations Committee on the Peaceful Uses of Outer Space. Summary Record of the Sixty-Sixth Meeting. Held at the Palais des Nations, Geneva on 25th July 1966, 21 October 1966.

UN Doc. A/AC.105/C.2/SR.70 United Nations Committee on the Peaceful Uses of Outer Space. Legal Sub-Committee. Fifth Session. Summary Record of the Seventieth Meeting. Held at the Palais des Nations, Geneva, on Wednesday, 3 August 1966, at 3.50 p.m., 21 October 1966.

UN Doc. A/C.1/879 United Kingdom of Great Britain and Northern Ireland: draft declaration of basic principles governing the activities of States pertaining to the exploration and use of outer space, 4 December 1962.

UN Doc. A/C.1/881 United States of America: draft declaration of principles relating to the exploration and use of outer space, 8 December 1962.

UN Doc. A/C.1/PV.1342 General Assembly. Verbatim Record of the 1342nd meeting, 2 December 1963.

UN Doc. A/AC.105/PV.12 United Nations Committee on the Peaceful Uses of Outer Space. Verbatim Record of the Twelfth Meeting, Held at Headquarters, New York, 12 September 1962, 21 February 1963.

UN Doc. A/C.1/SR.1343 Report of the Committee on the Peaceful Use of Space. United Nations General Assembly, First Committee, 1343rd meeting, 3 December 1963.

UN Doc. A/C.1/SR.1422 United Nations General Assembly. 1422nd meeting. First Committee. International co-operation in the peaceful uses of outer space: reports of the Committee on the Peaceful Uses of Outer Space, 20 December 1965.

UN Doc. A/C.1/L.219 USSR: Draft resolution. 7 November 1958.

UN Doc. A/C.1/L.220 Australia, Belgium, Bolivia, Canada, Denmark, France, Guatemala, Ireland, Italy, Japan, Nepal, Netherlands, New Zealand, Sweden, Turkey, Union of South Africa, UK, USA, Uruguay, and Venezuela: Draft resolution, 13 November 1958.

UN Doc. A/AC.105/C.1/111 National research on space debris, safety of space objects with nuclear power sources on board and problems relating to their collision with space debris. International Association for the Advancement of Space Safety. United Nations Committee on the Peaceful Uses of Outer Space Scientific and Technical Subcommittee 54th session, 23 October 2016. 7–10.

UN Doc. A/AC.105.C.1/2011/CRP.14 Towards Long-term Sustainability of Space Activities: Overcoming the Challenges of Space Debris. A Report of the International Interdisciplinary Congress on Space Debris. United Nations Committee on the Peaceful Uses of Outer Space Scientific and Technical Subcommittee. Forty-eighth session, Vienna, 7–18 February 2011.

UN Doc. A/AC.105/C.1/2012/CRP.16 Active Debris Removal—An Essential Mechanism for Ensuring the Safety and Sustainability of Outer Space. A Report of the International Interdisciplinary Congress on Space Debris Remediation and On-Orbit Satellite Servicing. United Nations Committee on the Peaceful Uses of Outer Space Scientific and Technical Subcommittee. Forty-ninth session, Vienna, 6–17 February 2012.

UN Doc. A/AC.105/C.1/2017/CRP.12 'National research on space debris, safety of space objects with nuclear power sources on board and problems relating to their collision with space debris', Report by Space Generation Advisory Council. United Nations Committee on the Peaceful Uses of Outer Space Scientific and Technical Subcommittee 54th session, 17 October 2016 pp. 9–14

UN Doc. A/AC.105/2018/CRP.20 Guidelines for the Long-term Sustainability of Outer Space Activities. United Nations Committee on the Peaceful Uses of Outer Space. 61st session, 27 June 2018.

UN Doc. A/RES/1148(XII) United Nations General Assembly Resolution 1148(XII). Regulation, Limitation and Balanced Reduction of All Armed Forces and All Armaments; Conclusion of an International Convention (Treaty) on the Reduction of Armaments and The Prohibition of Atomic, Hydrogen and other Weapons of Mass Destruction. 12th session, 14 November 1957.

UN Doc. A/RES/13/1348 United Nations General Assembly Resolution 1348 (XIII). Question of the Peaceful Use of Outer Space. 13th session. 13 December 1958.

UN Doc. A/RES/1472(XIV) United Nations General Assembly Resolution 1472(XIV). International Co-operation in the Peaceful Uses of Outer Space. 14th session, 12 December 1959.

UN Doc. A/RES/1884 (XVIII) United Nations General Assembly Resolution 1884. Question of General and Complete Disarmament. 1244th plenary meeting, 17 October 1963.

UN Doc. A/RES/18/1962 United Nations General Assembly Resolution 1962(XVIII). Declaration of Legal Principles Governing the Activities of States in the Exploration and Use of Outer Space. 1280th plenary meeting, 13 December 1963.

UN Doc. A/RES/2222 (XXI) United Nations General Assembly. Treaty on Principles Governing the Activities of States in the Exploration and Use of Outer Space, including the Moon and Other Celestial Bodies. 21st session, 19 December 1966.

UN Doc. A/PV.868 Address by President Dwight Eisenhower to the United Nations General Assembly. 22 September 1960. pp.45–50.

UN Doc. A/PV.1208 United Nations General Assembly, 18th session: 1208th plenary meeting, 19 September 1963, New York.

UN Doc. CD/786 Report of the Ad Hoc Committee on Prevention of an Arms Race in Outer Space. UN Conference on Disarmament, 24 August 1987.

UN Doc. CD/817 Letter Dated 17 March 1988 from the Representative of the Union of Soviet Socialist Republics addressed to the President of the Conference on Disarmament, Transmitting the Text of a Document entitled Establishment of an International System of Verification of the Non-Deployment of Weapons of any kind in Outer Space. UN Conference on Disarmament, 17 March 1988.

UN Doc. DC/SC.1/49 Union of Soviet Socialist Republics: Proposal on the Reduction of Armaments and Armed Forces and the Prohibition of Atomic and Hydrogen Weapons, 18 March 1957.

UN Doc. DC.SC.1/62/Rev.1 Canada, France, United Kingdom of Great Britain and Northern Ireland and United States of America: working paper on systems of inspection to safeguard against the possibility of surprise attack, 2 August 1957, 11 September 1957.

UN Doc. DC/SC.1/65/Rev.1 Statement of the Government of the Union of Soviet Socialist Republics on the disarmament talks communicated by the Chairman of the delegation of the USSR at the 151st meeting of the Sub-Committee, on 27 August 1957.

UN Doc. DC/SC.1/66. Western Working Paper Submitted to the Disarmament Subcommittee: Proposals for Partial Measures of Disarmament, 29 August 1957.

UN Doc. DC/SC.1/PV.143 United Nations Disarmament Commission. Sub-Committee of the Disarmament Commission. Verbatim Record of the 143rd meeting, held at Lancaster House, London, on 2 August 1957.

UN Doc. DC/SC-1/PV.151 United Nations Disarmament Commission. Sub-Committee of the Disarmament Commission. Verbatim record of the 151st meeting, held at Lancaster House, London, on Tuesday, 27 August 1957.

UNIDIR. 'The Implications of Establishing an International Satellite Monitoring Agency' United Nations Office for Disarmament Affairs. Disarmament Study Series, 9 (United Nations, 1983).

United Nations. *Charter of the United Nations and Statute of the International Court of Justice,* signed on 26 June 1945. Entry into force on 24 October 1945 (New York, NY: United Nations, Office of Public Information).

UNTS. Antarctic Treaty, opened for signature at Washington, 1 December 1959. United Nations Treaty Series, 402, 5788, 71.

UNTS. Treaty Banning Nuclear Weapon Tests in the Atmosphere, in Outer Space, and Under Water, opened for signature at Moscow on 5 August 1963. United Nations Treaty Series, 480, 6964, 45.

UNTS. Treaty on Principles Governing the Activities of States in the Exploration and Use of Outer Space, including the Moon and Other Celestial Bodies, opened for signature at Moscow, London, and Washington on 27 January 1967. United Nations Treaty Series, 610, 8843, 205.

UNOOSA. 'Space Debris Mitigation Guidelines of the Committee on the Peaceful Uses of Outer Space' (Vienna: United Nations Office for Outer Space Affairs, 2010).

US Air Force Space Command. 'The Future of Space 2060 and Implications for US Strategy: Report on the Space Futures Workshop. 5 September, 2019 (US Air Force Space Command, 2019).

US Arms Control and Disarmament Agency. 'Blueprint for the Peace Race: Outline of Basic Provisions of a Treaty on General and Complete Disarmament in a Peaceful World' (Washington, DC: Government Printing Office, 1962).

US Arms Control and Disarmament Agency. *Documents on Disarmament*, Vol. 1 (Washington, DC: Government Printing Office, 1963a).

US Arms Control and Disarmament Agency. *Documents on Disarmament*, Vol. 2 (Washington, DC: Government Printing Office, 1963b)

US Commission to Assess United States National Security Space Management. 'Executive Summary Report. Report of the Commission to Assess United States National Security Space Management and Organization, pursuant to Public Law 106-65, 11 January 2001' (US Congress. Washington, DC. 20033-0633, 2001).

US Congress. 'Space law: A Symposium, Prepared at the Request of L.B. Johnson, chairman' Senate Special Committee on Space and Astronautics. 31 December 1958 (Washington, DC: Government Printing Office, 1959).

US Congress. H.R. 1473 (112th). Public Law 112–10. 'Department of Defense and Full-Year Continuing Appropriations Act, 2011.' 125 Stat. 38. United States Statutes at Large (Washington, DC: Government Publishing Office, 15 April 2011) 38–212.

US Congressional Research Service. 'A Low-Yield, Submarine-Launched Nuclear Warhead: Overview of the Expert Debate, by A. F. Woolf.' Report IF11143 (US Congressional Research Service, 5 January 2021).

US Department of Defense. 'National Security Space Strategy—unclassified summary' (United States Government, January 2011) <https://www.dni.gov/files/documents/Newsroom/Reports%20and%20Pubs/2011_nationalsecurityspacestrategy.pdf> accessed 11 June 2020.

US Department of Defense. 'Nuclear Posture Review. February 2018' (Washington, DC: Office of the Secretary of Defense. United States Government, 2018).

US Department of Defense. 'Military and Security Developments Involving the People's Republic of China 2021' (Washington, DC: Office of the Secretary of Defense. United States Government, 2021).

US Department of State. 'A Report on the International Control of Atomic Energy', Publication 2498, 16 March 1946 (Washington, DC: Government Printing Office, 1946).

US Department of State. 'Successful testing by the Soviet Union of an intercontinental ballistic missile: Communiqué issued by the Soviet Government News Agency TASS, 26 August, 1957' in *American Foreign Policy, Current Documents* (Historical Office, Historical Division, Bureau of Public Affairs, Washington, DC: Government Printing Office, 1957).

US Department of State. *Documents on Disarmament 1945–1959. Vol. II. 1957–1959* (Washington, DC: Historical Office, Bureau of Public Affairs, Government Printing Office, 1960).

US National Intelligence Council. 'Global Trends 2040', National Intelligence Council Report No. 2021–02339. Office of the Director of National Intelligence, March 2021. https://www.dni.gov/files/ODNI/documents/assessments/GlobalTrends_2040.pdf accessed 10 June 2021

Vedda, J. 'Reviving Space Futurism: A New Focus on Long-Term Strategic Planning', presentation at AIAA Space 2008 Conference & Exposition. 9–11 September 2008. San Diego, California, USA. Published online 15 June 2012 (American Institute of Aeronautics and Astronautics, Inc.) 1–10. https://doi.org/10.2514/6.2008-7873 accessed 14 July 2021.

Viikari, L. *International Environmental Law and the Space Sector: Assessing the Present and Charting the Future* (Leiden: Martinus Nijhoff, 2008).

Viikari, L. 'Environmental Aspects of Space Activities' in F. G. von der Dunk and F. Tronchetti (eds), *Handbook of Space Law* (Cheltenham: Edward Elgar Publishing, 2015) 717–68.

von Braun, W. 'Editorial: What Are We Waiting for?' (1952) 22 (March) *Collier's* 23.

Walker, R. S., Sattenspiel, L., and Hill, K. R. 'Mortality from Contact-related Epidemics among Indigenous Populations in Greater Amazonia' (2015) 5 *Scientific Reports* 14032.

Weeden, B. and Samson, V. 'Global Counterspace Capabilities: An Open Source Assessment'. Secure World Foundation. <https://swfound.org/counterspace/> accessed 20 July 2021.

West, J., (ed). *Space Security Index 2019* (Ontario: Project Ploughshares, 2019). https://spacesecurityindex.org/ssi_editions/space-security-2019/ accessed 24 July 2021.

Wolfe, T. W. *The SALT Experience* (Cambridge, MA: Ballinger Publishing, 1979).

Wolter, D. *Common Security in Outer Space and International Law* (Geneva: United Nations Institute for Disarmament Research, 2006).

14

Indigenous inclusion within the democratization of space

Tony Milligan

When the Perseverance Rover landed on Mars in 2021, the first prominent feature that it examined was then named 'Máas', a Navajo/Diné borrow word from English (NPR 2021). One of an initial list of 50 names put together with the help of Jonathan Nez, then president of the Navajo Nation, with a view towards use by NASA. The latest development in a Navajo–NASA relationship that goes back to the very beginnings of the US space programme, with some predictable ups and downs in the early days. In more recent times it has helped to create pathways for the use of satellite data to monitor water supplies on Navajo lands, and for individual Navajo to make their way into NASA posts. At the time of the landing, Aaron Yazzie was called upon regularly by the media, as an engineer on the Perseverance team based out of the NASA Jet Propulsion Laboratory. Yazzie is Navajo/Diné. The Navajo/Diné relationship is also something of a model for a broader series of cooperations and moves to include Indigenous groups within the space programme. Although admirable, there remains a good deal of distance between cooperations of this sort and the idea of a distinctive Indigenous role in our current processes of space expansion, or within processes that we might refer to as 'the democratization of space'. What I will be trying to identify in this chapter is a role of this more distinctive sort, a role that (in some sense) only Indigenous agents can play.

The approach will be broadly philosophical and analytic, in the sense that it will be heavily weighted towards concepts and argument building rather than anything like ethnographic studies or anthropological overviews of Indigenous peoples. Section 1 will provide some overall framing by situating the question of a distinctively Indigenous role within the broader idea of 'democratizing space'. This is not just a preliminary, the chapter has an arc that returns, towards the end, to democratic commitment as an important constraint. Section 2 will outline my core argument, the 'argument from belonging', and will clarify the scope of what is claimed. My argument is not that all indigenous peoples can or should have a distinctive contribution but merely that space expansion would benefit in a distinctive way from distinctive Indigenous input. A way that goes beyond the gain in legitimacy for space programmes that generally results from greater inclusion. This distinctive input will turn upon Indigenous knowledge about belonging. Section 3 will introduce a further 'liberties argument' as a foil to help clarify the argument from belonging. Section 4 will appeal to an understanding of science that is constrained by commitment to democracy in order to form a clearer conception of the kind of knowledge about belonging

Tony Milligan, *Indigenous inclusion within the democratization of space.* In: *The Institutions of Extraterrestrial Liberty.* Edited by Charles S. Cockell, Oxford University Press. © Tony Milligan (2022). DOI: 10.1093/oso/9780192897985.003.0015 that is at stake, and an understanding of the science/Indigenous knowledge distinction that reduces (but may not entirely remove) worries about inclusion as appropriation.

1. Democratizing Space

The idea of 'democratizing space' is one of the recurring concepts appealed to in contemporary public discussions of human space expansion (Baiocchi and Welser (2015); Welser 2016; Kim 2019). To help broaden it out beyond a largely commercial conception, I want to situate two minimal commitments at the heart of the idea. The first is commitment to the permissibility of human expansion into some broader region of space. The second is a commitment to the democratization of space in a sense that relates to our regular understanding of democracy as inclusive and value based rather than merely procedural. These are minimal commitments in various respects. They concern activities only in 'some broader region of space' rather than grandiose plans. Indeed, this region of space could be cislunar and need not reach out any further than the orbit of the Moon. Alternatively, it could extend as far as Mars and beyond, out towards the main asteroid belt. The commitments are also minimal in the sense that they are constrained in terms of content as well as location. Saying that expansion into space is 'permissible' does not actually require anyone to go there. (For convenience, I refer to space as 'there' as if we were not ourselves already part of it.) An appeal to the permissibility of going is just a way of saying that there is no fundamental ethical barrier in the way of our doing so. We might do bad things in space, or good things in space. We will probably do both. But going is not itself a bad thing.

By embracing 'democratization', we are also committing to much less than might be imagined. It may involve as little as the following: as far as reasonable, actual expansion into space should be in line with inclusion and the political liberties that we associate with a modern democracy whatever form it may take. Note, I have connected democracy strongly to values rather than adopting a procedural conception. Values of this sort are more basic to democracy than any particular type of electoral system. In a sense, the electoral systems that we adopt are themselves expressions of values such as liberty and equality (community too), which is one of the reasons why we ordinarily regard its flaws and occasional paradoxes as something secondary. After all, government within a modern democracy rarely depends upon securing the votes of an actual majority of the population. US presidents sometimes come close, UK governments never do. Yet both enjoy democratic legitimacy. Democracy is about something more than procedure. And so, an inability to shift all our familiar procedures from here out into space does not mean that we cannot democratize space in a sense that focuses upon values as much as mechanisms.

Building a utopia in space is, however, unlikely. Nonetheless, if we are committed to democratization, and if we take our earthly democracies as a benchmark, then we should try to do at least as well there, and perhaps better in obvious respects. A useful analogy may be made with moving to a new town. Sometimes we need to move to a new place in order to become a better version of ourselves. Something similar may apply to humans collectively as we reach up into space. There are entrenched failings down here. And a kind of sociopolitical inertia that makes widely desired changes difficult to bring about in the absence of war, state collapse, and processes of reconstruction. We simply do not need to reproduce all such failings up there, especially when they can pose problems for the sustainability of what we are trying to bring about. Here, we may reflect upon the Apollo Program and its more or less abrupt end. In retrospect, basing space expansion on white men from the military was hardly a recipe for sustained and diverse public support in the face of demographic change, social unrest, and economic difficulties. There are some failings, such as gender bias and racism, both personal and institutional, that we would do well to leave behind. Partly, this is because it is the right thing to do. But it is also sometimes the most practical and sustainable thing to do; not always, but sometimes and perhaps often. Ethical commitment to robust forms of inclusion can enable space programmes which might otherwise be compromised by failings of these familiar sorts.

Perhaps, even if successes over such matters are counterbalanced by other failings, we might find that overall we are no better in our treatment of one another in space, but we should minimally try not to be worse. (Again, in some overall sense.) A qualification to this picture is that there are significant physical obstacles in the way of any attempt to reproduce political liberties in space (Cockell 2022). Freedom in space would be what Charles Cockell has called 'freedom in a box' (Cockell 2015) Yet, the sheer physical obstacles facing life in an extreme environment are very different from the more easily variable socially generated barriers that we might take with us as legacies of prejudice and exclusion. Recognition of such dangerous legacies will also help us to understand just what it is that we are trying to secure when we talk about 'democratization' if the latter is to be understood in a sufficiently pragmatic sense as something attainable. There are liberties and entitlements that many of us enjoy, from which many other humans are unfairly excluded for reasons of colour, ethnicity, gender, nationality, wealth, and religion. Reflection upon these liberties, entitlements, and unfair exclusions can help to shape our grasp of just what it is that we ought to be doing in space and anywhere else if we are genuinely committed to democratization.

All of this does, of course, involve an understanding of 'democratizing space' that is constrained but still significantly broader than the way that the concept is sometimes used in discussions which centre upon commercialization (Cobb 2021). Within the latter, democratization is thought of in terms that lean upon an idea of the democracy of the market, the avoidance of commercial monopoly, and of excessive or exclusive state control. These things may be integral to our best available systems of democracy, but democratization in the sense that concerns me here will be broader and extend to the valuing of liberties and entitlements of other sorts. Liberties such as certain kinds of freedom of movement and freedom to withdraw from projects in some reasonable staged way. Entitlements such as the entitlement to having one's religious and cultural group fairly represented at the level of personnel and planning, as well as some kind of 'equal entitlement' to access opportunities to go into space. The latter is a particularly complex kind of equality, one that will be difficult to bring about across large populations only a tiny fraction of which could ever be sent into space. But it is an entitlement that we should take seriously and try to realize in some form if we are serious about democratization in the enlarged sense that I have in mind.

Democratization of this sort is also a selling point for space tourism as a way of making access to space available to more and different groups of humans. It is regularly appealed to by Elon Musk, Jeff Bezos, and Richard Branson, and has shaped their participant choice. Particularly during the 2021 race into space tourism between Blue Origin and Virgin Galactic. For example, the Virgin Galactic cabin was split between four men and two women, one of whom was Sirisha Bandla, the second woman from India to go into space, and the first to come back alive. Bandla hoped that this would pave the way to access for 'people from different backgrounds, different geographies, different communities' (Pandey 2021). The Blue Origin crew included Wally Funk, one of the survivors of the 11-strong Mercury 13 group of women who underwent the same training as NASA's first (all male) astronauts, and the only one to have actually made it into space; a belated instalment of justice.

Mae Jemison, who became the first Black woman to go into space in 1992, as a member of the Space Shuttle Crew for the STS-47 mission, reinforced the importance of such steps when reflecting more broadly upon the engineering accomplishment of Blue Origin: 'Space exploration has been part of human history' (CNN 2021). A thought missing from a good deal of the critical commentary. The obvious implication being that any exclusion from exploration is itself an exclusion from an important part of our shared history. Something to be taken seriously. When Gil Scott-Heron delivered his satirical 'Whitey on the Moon' in 1970, actual travel into space really was exclusionary in this way. A case of 'Whites only'. By the time of the 2021 space race, this was no longer the case.

In brief, amongst other things, democratizing space on the conception at work here involves a multi-sided inclusion of individuals and collective groups who have historically been excluded, marginalized, or subject to one of the many forms of subordination to greater power and greater powers. This covers a considerable territory from persons of colour and Indigenous peoples whose interests are the focal point of ongoing political protest; small states such as Luxembourg attempting to carve out a role in space expansion; states such as Nepal trying to access satellite technologies to deal with earthquake vulnerabilities; and more powerful states such as India and Japan with wealth and robust space ambitions but without the advantages of the historic Euro-American space powers. Finally, democratization when understood in these inclusive terms is not only an attempt to be fair. It is also a way to give content to the idea that space is the common heritage of mankind, an idea at the heart of the Outer Space Treaty (1967). This need not involve the claim that space itself is 'a commons' in the formal sense, just so long as it commits us to inclusion on the grounds that all have claim.

2. Distinctiveness of Role

Making sense of the multiple potential roles of Indigenous peoples within such an inclusive approach to democratization poses something of a problem. Obviously, it will not be good enough to regard those who go into space as automatically representatives of all of us, in which case Indigenous peoples would always have been included by proxy, just like everyone else. Assuming that the issues of participation and inclusion are much the same for Indigenous peoples and all persons of colour also looks problematic when we reflect upon the distinctiveness of claims made within terrestrial politics. A classic exemplar here is the contrast between the struggle for equality and inclusion on equal terms by the Civil Rights Movement and Black Lives Matter and the focus upon autonomy and sovereignty, outside the uniform federally enforced rules, by political activists and representatives of First Nations in the United States (Deloria 1969; Walker 1999).

The point is, however, cautionary rather than a claim that this same contrast must mark all aspects of democratization. There are, after all, forms of inclusion such as crew membership, which is much the same in both cases. John Bennett Herrington of the Chickasaw Nation flew on the Space Shuttle to the International Space Station in 2002 as part of the STS-113 mission, becoming the first Indigenous American in space, but inclusion of this sort is still rather different from making sense of a distinctive Indigenous role. It is the kind of thing that we might underpin simply by an appeal to justice and as a recognition of historic injustice. But if we want to move from inclusion to democratization in the fuller sense, it is not clear that an appeal to considerations of justice alone will be enough. It may do some work but will require sturdy allies if we are to make headway. Elsewhere (Milligan 2022), I have suggested one such ally, 'the argument from belonging'. This is a pragmatic line of argument for a distinctive role for Indigenous peoples within space programmes based upon a certain kind of knowledge about belonging. Here, I will try to clarify the argument and will reflect upon its distinctiveness by contrast with another pragmatic argument for Indigenous inclusion, one that flows more directly from the focused discussions about 'extraterrestrial liberty' which have been going on for more than a decade (Cockell 2008, 2016).

As my concern here will be with clarification of the argument, the approach will be broadly analytic rather than interwoven with multiple quotes from Indigenous agents that readers can find elsewhere, and which help to support the view that there are already a multiplicity of attitudes towards space across and within different Indigenous communities. In other words, considering Indigeneity in the context of space expansion need not lead us to reaffirm some old assumptions about uniform Indigenous hostility to space programmes (Young 1987). And it need not lead us directly into some manner of hostility to current or projected human activities in space as a playground for wealthy White colonial settlers and their scientific and technological representatives. Critical engagements from Indigenous perspectives can, instead, follow the approach of Deondre Smiles of the Ojibwe, which fuses critique and openness (Smiles 2020).

As indicated at the outset, the argument that follows is not that all Indigenous peoples can or should have a distinctive contribution, but merely that space expansion would benefit in a distinctive way from distinctive Indigenous input, and benefit in a way that goes beyond the gain in legitimacy for space programmes that might result from greater inclusion. This distinctive input could, in principle, come from some small cluster of Indigenous peoples, and need not involve any particular cross-section of the Indigenous 5 per cent of the world's population (World Bank 2021). It does not, therefore, attempt to capture all aspects of democratization but only a particular advantage of a certain limited aspect of democratization. And it does not assume that all Indigenous agents are bearers of the knowledge about belonging that is in question.

If we think in terms of establishing a stable human presence on the Moon or Mars, or about further forms of space exploration (the kind that may only be possible in science fiction), there is something intuitively appealing about this idea of a distinctive Indigenous role. After all, if we were to identify some ideal crew, we would probably want to include group members with experience of living in extreme environments and not only agents who might have visited such environments, written about them as non-Indigenous anthropologists, or participated in simulations inside analogues. In a sense, it is this intuition about knowing how to cope that I am trying to tease out and present in a more formal and precise way with emphasis placed upon the knowledge and not just the coping. So, let us call it 'the background intuition'. I want to draw upon this background intuition without any problematic reliance upon claims concerning actual space settlements which may or may not happen, and without appeal to scenarios in which humans boldly go beyond the bounds of the solar system, to explore other places. Instead, I will stick to the opening commitment to the permissibility of expansion into some broader region of space. Put in other terms, the argument relies only upon the prospect of a level of space expansion that is extensive enough to make us think of ourselves (i.e. humanity at large) as now occupying more than the ground. We might then think of this as a matter of humanity's future involving a continuous movement between sky and ground, a future in which familiar concerns with belonging persist.

With these qualifications in place, the argument from belonging runs as follows. Space expansion is a cluster of multi-generational projects, some of which will have a sequel but many of which will be abandoned. Expansion will occur, but not necessarily in the ways that we want, or with the goals that we currently pursue. This places a question mark over the meaningfulness of our actions. Especially so, given that space programmes and expansion are not simply goods in their own right but rather are goal directed activities. Realization of goals, or of sequels to them that we ourselves might buy into, will depend upon the actions of others in the future, actions that we cannot guarantee or control. Much of what we do might dead end. One way to deal with the problem, as a problem about the meaningfulness of our actions, is to try and make sense of 'what runs deep within our human attitudes towards space'. That is to say, we can draw a distinction between matters of depth and a much more volatile set of concerns: idiosyncrasies, fashions, historically transitory views about sex, gender, and the family, economic and political doctrines rooted in particular phases of social development, and so on. If we place too much emphasis upon these more volatile and transitory concerns, then our plans and projects are more likely to be set aside. At least this is what our own attitude towards the projects of previous generations suggests. We have set aside core plans and projects of the nineteenth and twentieth centuries and few of us desire to live in ways that our predecessors would have had us live. Similarly, basing multi-generational space projects on shifting fashions in economic theory seems like a poor way to proceed.

Whilst many of our attitudes are volatile, and economic theories are transitory, a concern with belonging is not. Modern ideas of belonging in Western cultures tend to have a ground bias, a sense of disconnection between sky and ground. Even a tendency to rewrite insights which depend upon both as if they were sourced in an exclusively earthly way. Here, we may even think of theorizing about ecosystems whilst overlooking the fact that the very concept appeals to the Galilean concept of a system developed to make sense of planetary motion (Olson 2018). (A qualifier to this is the uneasy appeal to Western cultures, which is at best a shorthand for a far more complex story.) By contrast, Indigenous cosmologies have an inbuilt sense of belonging that reaches beyond the Earth. A sense of continuous movement between sky and ground, a sense of belonging to a larger order of things than the Earth alone. These cosmologies are more than ecologies and, whatever their limitations if we mistakenly attempt to read them as primitive precursors of modern science, they do not suffer from the same kind of ground bias that our prevailing forms of belonging have so often fallen into.

For clarity, I do not think that Kepler or Tycho Brahe or Dante or Galileo fell into this same ground bias, but that it is something more recent and not a kind of carelessness but rather connected to our ways of being. Many of us have had no great need to see life on Earth in a larger context, although we now face the need for something more than ecologies. And whilst the concept of the 'cosmological' can sometimes sound a little too grand, the idea of a more cosmological perspective as a more than ecological perspective may help us to understand the challenges of coming to terms with belonging to a larger region of space. These are, again, matters of depth, matters that concern 'what runs deep within our human attitudes towards space'. And so the argument is that Indigenous cosmologies and the knowledge about belonging associated with them can help us to position ourselves more effectively to develop projects with sequels, rather than projects which end up being displaced. So ends the argument.

This is an argument with a significant number of moving parts, but none are especially complex or reach beyond familiar ways of understanding the world. Appeal is made to an idea of the meaningfulness of actions, and this might be disputed as a sort of fiction. However, the option of dismissing this concept is lost for those who have a commitment to space expansion on grounds that appeal to its importance for humanity. That, after all, would be a kind of meaningfulness. Besides which, disputing the concept of meaningfulness generally trades upon a conflation of (i) the idea that the universe is indifferent to what we want, and (ii) our ineradicable practices of treating some things as more important than others. The latter is almost certainly something that goes with being an agent of any sort at all. It certainly goes with any familiar way of being human. In other words, it is not optional. It is part of the canvas on which we paint our pictures about ethics and about human life. No picture would hold together without it.

One of the argument's distinctive features is that it is broadly pragmatic. This has been emphasized from the outset. The argument also does not depend directly upon considerations of justice. This does not mean to say that the importance of justice can be set aside but rather that we may need to bring more than one argument to bear when we try to understand the complexities of the democratization of space, and of Indigeneity in relation to space. Besides which, appeals to justice may motivate arguments but often they are not enough to motivate action. Effective arguments may have to appeal to practical advantages as a selling point if they are to be taken seriously within processes of policy formation for space programmes. The argument also leaves a good deal about the concept of 'belonging' unsaid. Further detail needs to be added. Nonetheless, the above sets out the core of a line of argument, and the line of argument does point towards the possibility of a distinctive Indigenous contribution. This marks it out from more familiar ways of arguing for inclusion.

3. Comparison with an Appeal to Liberties

The argument has distinctive features which we can see if we compare it with another plausible argument for inclusion that also addresses the idea of belonging, but in a slightly different way. Let us call this second argument the 'liberties argument' and regard it as a foil which can be used to highlight some good and bad features of the argument from belonging. The liberties argument draws upon an appeal to the special vulnerabilities of space and to the difficulties of sustaining anything approaching conditions of liberal democracy within space settlements. Again, for the sake of clarity, I suspect that both arguments are sound, or can be reformulated in terms which are sound.

The liberties argument runs as follows. Freedom in space will be just what Cockell (2015) has suggested: freedom in a box, subject to persistent pressures towards authoritarian control. Whoever controls the air supply will also have leverage over settlements. Reducing such dependencies, for example through modularity of settlement design, and multiplying the number of suppliers of the immediate requirements for human life, are key considerations. Avoiding monopoly in space really is an aspect of the democratization of space. So too is attitude towards place. In the light of multiple vulnerabilities and the adversities of living in extreme environments, the temptation will be to think of settlers existing in a state of war with their surroundings. Pioneers trying to conquer the places that they occupy. There is a familiar literature drawing upon such metaphors, from Cosmism, which situated space exploration in the context of a larger conflict between man and nature (Tsoilkovsky 1963; Yefremov 2004), through to Hannah Arendt's paper, 'The Conquest of Space and the Stature of Man' (Arendt 2006). Such ideas are likely to play into political vulnerabilities. They are likely to reinforce the pressures towards authoritarianism in space. After all, wartime conditions justify curtailment of liberties in near unique ways. Rather than drawing upon such a pool of dangerous metaphors, it would be better to draw from ways of being at home in a place, ways of belonging which acknowledge the extreme nature of space environments in order to avoid any lapse into more belligerent, warlike ways of being and seeing. No doubt, both have a role, but the balance between them is important. And one of our best storehouses of ideas about how to be at home in extreme environments is Indigenous knowledge. Indeed, it is remarkable that we have yet to tap into such ideas more fully in order to understand the challenges of living in the difficult conditions of space. As indicated previously, if we were planning for an actual settlement in an extreme space environment, we would do well to draw upon knowledge and perhaps also some personnel with direct experience of life in another extreme environment such as the Arctic, or desert conditions, as well as personnel with more brief experience in terrestrial analogues.

Considerations of justice may well motivate both arguments. In my own case, they certainly do. I am looking for ways to show the practicality of justice. However, neither make a direct appeal to justice. They appeal instead to pragmatic considerations and to the sustainability of space projects. Both arguments are also consistent with the continuation of certain forms of terrestrial injustice. They perform some of the work which is required in order to support a fuller democratization of space, but they do not presuppose any leap into a set of ideal circumstances. They are also contributory arguments. That is to say, they give strong reasons in support of inclusion and the factoring in of Indigenous traditions, however these reasons could in theory be outweighed by other countervailing reasons. Someone might also reject either or both arguments without being guilty of a failure of rationality. Neither are conversation stoppers. Of the two, the liberties argument has fewer moving parts and makes no appeal to ideas such as the meaningfulness of our actions. That may be an advantage of sorts. Someone might also argue that more weight should be placed upon the liberties argument than upon the argument from belonging. However, it may turn out that the placing of emphasis upon one or the other is a matter of philosophical sensibilities rather than anything else. My own approach places more emphasis upon the argument from belonging, but this may draw upon nothing more than a fondness for asking questions about meaningfulness and about what goes deep within a human life.

Be that as it may, the argument from belonging has at least some advantages over the liberties argument. First, it does not presuppose actual settlement but only the emergence of a significant larger than Earth human presence. Second, both involve an appeal to our human vulnerabilities, but in the case of the argument from belonging, the vulnerabilities in question are ineradicable, even in principle. With settlements, we can always imagine wonderful technologies and a point in time at which the box within which freedom is enjoyed happens to be so big and reliable that it will present few limitations beyond those of human life on Earth. What we imagine may not be fully realized, but at least in principle many of the vulnerabilities might eventually be removed. By contrast, our reliance upon future generations to continue multigenerational projects involves a vulnerability that can never be removed. Or, at least, it cannot be removed in the absence of some manner of 'conquest of time' of a sort that the Cosmists also believed in but which looks like a strictly fictional scenario.

Both arguments are also liable to face a set of standard objections, for example that Indigenous inclusion of almost any sort can all too easily lapse into cultural appropriation. The plundering of Indigenous traditions and appropriation of cosmologies in a search for something that we can use. Arguments with a pragmatic edge may be particularly vulnerable to a charge of consuming Indigeneity as a resource. The boundaries between some manner of cultural imperialism and democratizing may at points become blurred. There is danger as well as opportunity. However, the greatest danger of all may be that Indigenous peoples become 'superfluous', a concept that I borrow from Hannah Arendt and her discussion of the banal evil of Nazism (Arendt and Jaspers 1992). 'The Jew' simply had no place within the Nazi economy and within its conception of a revival of Germany. Whilst its application to Indigenous peoples needs considerable further development (and I am certainly not making a claim about actual genocide), what matters in the present context is the idea of lacking a place within the human community, and within our shared projects. And so, when Mae Jemison speaks about space exploration as part of our history, I take this idea very seriously and see it as having implications of an important sort. Those outside the process or rendered superfluous to the process are rendered superfluous to a part of our history, a part already of growing importance and one which is likely to shape all other civilization-level changes in the not too distant future.

Finally, there are practical questions in both cases about exactly how inclusion can work. But it seems likely that reflection upon Indigenous cosmologies and the sense of belonging that both arguments appeal to will only be viable against the backdrop of a multiplicity of other kinds of inclusion within space programmes. From inclusion as project personnel, to cooperation when drawing down satellite information about water supplies in Indigenous territories, and so on. In other words, the kinds of inclusion that we see NASA and the Navajo working towards. Without this, it is difficult to imagine how Indigenous knowledge about belonging and space exploration could ever be brought together in meaningful ways. Again, the concept of the meaningful may play a role.

4. Democratic Constraining

If we are satisfied that the argument from belonging has its problems but is no more problematic than many arguments which we have good reason to accept, this will still not end our difficulties. We will still have the task of trying to make sense of what it shows. For example, Indigenous knowledge about belonging has been appealed to, above, in a rather loose way. There is no indication of whether the knowledge in question involves information or knowhow, or some special combination of the two. This ambiguity about the kind of knowledge that I am suggesting we might benefit from also applies in the case of both arguments. However, by using an understanding of science which is constrained by the opening/framing commitment to democracy, we can form a clearer conception of the kind of knowledge about belonging which is at stake. One that reduces worries about appropriation. This is the point at which the arc of this story returns to the theme of democracy and how it impinges upon our ways of thinking about other matters, such as science as well as justice.

This problem about the kind of knowledge at stake may be framed in terms of a traditional philosophical distinction between knowing how and knowing that, but the terminology need not matter, just so long as we allow that there are different kinds of knowledge, and that one kind of knowledge need not reduce to any other. If the relevant and useful kind of Indigenous knowledge is a matter of knowing how, then it will not reduce down to propositional knowledge. It is tempting to think about knowledge of this sort, instead, as a kind of practical wisdom. And if our motivations are those of justice and inclusion, we will have strong reasons to favour more of a practical wisdom reading. After all, if it is only information that is at stake, the knowledge in question might be acquired from some or other groups of Indigenous people, whose role could then be ended. There would be no further need for them. That really would look like appropriation, or the extraction of knowledge as a resource.

If, instead, practical wisdom is involved, then it will be much harder to set aside the involvement of actual Indigenous agents as knowledge bearers. Wisdom of this sort arises within a set of practices from which it might not easily be disentangled. The relevant agents cannot be used and then pushed aside.

However, we can hardly settle this matter simply by appeal to considerations of justice if we are to insist upon the pragmatic standing of the argument. And we may want the knowledge in question to be (at least in part) a matter of practical wisdom. But that hardly makes it so. Instead, I will suggest that there is a non-arbitrary way to choose between these two competing readings of the relevant Indigenous knowledge about belonging, a way which aligns with our concern for justice but does not depend upon it, a way which is shaped by our understanding of science in the context of democracy.

To be more precise, at least for policy purposes, we should maintain a distinction between regular science and Indigenous knowledge and, more specifically, we should not appeal to Indigenous science as simply one more regular scientific discipline. A more problematic way to make the point would be to uphold the distinction between 'Western' science and Indigenous science. More problematic because the idea of Western science can itself be a little misleading. There are many things that Indigenous peoples may know which are of interest to science, and there are many things known which can figure as data for science (in the sense of regular science), and which may shape its practices. But accepting the possibility of weaving together knowledge systems is rather different from saying that Indigenous storytelling is itself nothing other than science as we ordinarily think of it, in spite of the multiple deep metaphysical commitments and quasi-religious commitments built into such storytelling. In so far as we are thinking about science at a policy level, we have good reasons to separate out deep metaphysical and quasi-religious commitments from science in the regular sense, or at least to do so within a broadly liberal democratic political setting in which policy formation may be directly informed by science, but not by any special religious or metaphysical viewpoint (Duhem 1991). On matters of the latter sort, we expect our democratic political systems to be, up to a point, neutral. When it comes to policy formation, as opposed to other contexts where we might address matters in a less constrained way, we embrace something close to 'methodological naturalism' as a hallmark of regular science.

For the sake of clarity, methodological naturalism is not the same as a generalized method of science of the sort once sought after by Karl Popper and others. It is an overall adequacy condition that lots of different theories (scientific and other) might satisfy. And it involves official neutrality about more specialized religious and metaphysical viewpoints which are unlikely to command assent across large political communities. It does not involve hostility towards such viewpoints. It is not, for example, a coded assumption of atheism, but a genuine neutrality over a variety of matters taken to sit outside of regular science. The latter is then treated as a limited discourse. One that does not try to speak about everything or claim to account for every kind of knowledge. Finally, methodological naturalism is embraced for reasons, and not because science has to be thought of in this way under all circumstances. A concept of science will track other social practices. Science may talk about natural kinds but it is not itself a natural kind, fixed by the structure of nature itself. It is entirely conceivable that some society may do things differently, and that we might still recognize their way of doing things as science, yet we ourselves live in broadly liberal democratic societies, with a plurality of religious and metaphysical commitments, as well as shallower ideas of dangerous sorts which sometimes aspire to be classified as science but which serve only some special set of interests which, if embraced, would violate a democratic commitment to citizen equality. In brief, people living as we do have reasons to use the concept of science in more restricted ways. If our goal is the democratization of space then this will also be the approach towards science that we should adopt.

This will automatically exclude the possibility that a good deal of Indigenous knowledge, and Indigenous cosmologies in particular, count as regular science. Yet, we may still remain committed to the view that they contain a good deal of knowledge and could constitute science of a special sort which ought not to be reduced to regular science. Indigenous cosmologies in particular happen to be especially relevant to the argument from belonging as they involve narratives about belonging that might help us to make sense of what it is that runs deep within our human attitudes towards space. But a desire to make sense of them simply as regular science would lead us to regard the knowledge in question as information rather than practical wisdom, or some combination of the two within which practical wisdom has a diminished role to play. After all, ease of transmission, and the possibility of results being checked by others, are integral to practices of regular science. This is not to deny that there is also a good deal of knowhow involved in scientific practices, but theories, models, and claims must be readily understandable to anyone with the relevant technical expertise, irrespective of their personal history or background. In the absence of any idea that Indigenous knowledge should be classified as science of this more strictly limited sort, there is less pressure to make a similar set of assumptions. There is no reason to read the knowledge in question as information rather than practical wisdom. Concerns about appropriation and what happens once the knowledge is transferred may also be reduced.

5. Conclusion

The arc of this chapter has moved from and back to a concern with democratization, and the possibility that some manner of democratization may be possible as we expand into space. There are, as we would expect at this stage of the discussion, a good many ambiguities that need to be resolved, and a need to examine Indigenous knowledge and ways of belonging in ways which are particularly geared to the needs of expansion into space. The analytic approach, with its focus upon concept development and argument building, also has its limits. All of these things are true. Nonetheless, the approach has its strengths. One is that the distinctiveness of the role assigned to Indigenous knowledge, as well as its pragmatic underpinnings, may take us further than an appeal to justice alone. Another is the clear direction of travel for further research if the central argument is to be upheld. A 'good-making' feature of the approach is that we can draw a larger programme of research out of it.

Acknowledgements

This article is part of a project that has received funding from the European Research Council (ERC) under the European Union's Horizon 2020 research and innovation programme (Grant agreement No. 856543).

References

Arendt, H. 'The Conquest of Space and the Stature of Man' in H. Arendt, *Between Past and Future* (London: Penguin, 2006) 260–74.

Arendt, H. and Jaspers, K. *Hannah Arendt/Karl Jaspers: Correspondence 1926–1969*, eds L. Kohler and H. Saner, trans. Robert and Rita Kimber (New York, NY: Harcourt Brace, 1992).

Baiocchi, D. and Welser, W. IV. 'The Democratization of Space. New Actors Need New Rules' May/June 2015 *Foreign Affairs* <https://www.foreignaffairs.com/articles/space/2015-04-20/democratization-space> accessed 5 July 2022.

CNN. 'Amazon's Bezos Completes Historic Flight to Edge of Space' (2021) <https://twitter.com/i/web/status/1417623739072663559> accessed 5 July 2022.

Cobb, W. N. 'Commercialization and Space: Democracies Can Fly in Space' (2021) 19 *Astropolitics* 145–64.

Cockell, C. S. 'An Essay on Extraterrestrial Liberty' (2008) 61 *Journal of the British Interplanetary Society* 255–75.

Cockell, C. S. 'Freedom in a Box: Paradoxes in the Structure of Extraterrestrial Liberty' in C. S. Cockell (ed.), *The Meaning of Liberty Beyond Earth* (Cham: Springer, 2015) 47–68.

Cockell, C. S. *Dissent, Revolution and Liberty Beyond Earth* (Cham: Springer, 2016).

Cockell, C. S. *Liberty Beyond: Building Free Worlds in the Cosmos* (Oxford: Oxford University Press, 2022).

Deloria, V. *Custer Died for Your Sins: An Indian Manifesto* (New York, NY: Macmillan, 1969).

Duhem, P. *The Aim and Structure of Physical Theory* (Princeton, NJ: Princeton University Press. 1991).

Kim, D. 'The "Democratization of Space" and the Increasing Effects of Commercial Satellite Imagery on Foreign Policy' (2019) Center for Strategic and International Studies Ava<https://www.csis.org/democratization-space-and-increasing-effects-commercial-satellite-imagery-foreign-policy> accessed 5 July 2022.

Milligan, T. 'Indigeneity and the Three Body Problem' in J. S. J. Schwartz, L. Billings, and E. Nesvold (eds), *Reclaiming Space* (New York, NY: Oxford University Press, 2022).

NPR. 'Perseverance Rover Will be Naming Mars Landmarks in the Navajo Language' (2021) https://www.npr.org/2021/03/17/978288277/perseverance-rover-will-be-naming-mars-landmarks-in-the-navajo-language?t=1618051955783> accessed 5 July 2022.

Olson, V. *Into the Extreme: US Environmental Systems and Politics beyond Earth* (Minneapolis, MN: University of Minnesota Press, 2018).

Pandey, A. 'Sirisha Bandla's Family in Andhra Watches Virgin Galactic's VSS Unity Take Off for Space' *India Today*, 11 July 2021 <https://www.indiatoday.in/india/story/sirisha-bandla-family-andhra-virgin-galactic-vss-unity-take-off-space-1826736-2021-07–11> accessed 5 July 2022.

Smiles, D. 'The Settler Logics of (Outer) Space' (2020) *Society + Space.* <https://www.societyandspace.org/articles/the-settler-logics-of-outer-space> accessed 5 July 2022.

Tsoilkovsky, K. *Call of the Cosmos* (Foreign Languages Publishing House, 1963).

Walker, T. 'The Black and the Red: Responding to Sioux and other Native American Instructions on Red-Black Solidarity' (1999) 55(1) *Journal of Religious Thought* 73–86.

Welser, W, IV. 'The Democratization of Space' *The Rand Blog* (2016) <https://www.rand.org/blog/2016/03/the-democratization-of-space.html> accessed 5 July 2022.

World Bank. 'Indigenous Peoples' (2021) <https://www.worldbank.org/en/topic/indigenouspeoples#1> accessed 5 July 2022.

Yefremov, I. *Andromeda: A Space-Age Tale* (Amsterdam: Fredonia, 2004).

Young, J. M. 'Pity the Indians of Outer Space: Native American Views of the Space Program' (1987) 46(4) *Western Folklore* 272–3.

15

Decoupling physical and spiritual ascent narratives in astronomy and biology

Lucas John Mix

1. Introduction

From a planetary perspective, rockets fly outward, not upward. The traveller departs but does not ascend. The shift in perspective from Earth-centred to Sun-centred cosmos had a profound impact on human imagination by decoupling two concepts. The heaven of gods, moral perfection, and human salvation could no longer be identified simply with the sky of Sun, Moon, and stars (Midgley 1992, 55). This radical reorientation, called the Copernican revolution, permeated popular culture as well as science.

The universe did not change, but our relationship to it did. To be more precise, the paradigm shift within science became a symbol for a larger cultural shift associated with science. Galileo summed it up admirably, asserting the independence of science from scriptural interpretation, and thus ecclesiastical oversight: 'That the intention of the Holy Ghost is to teach us how one goes to heaven—not how heaven goes' (Sobel 1999, 65).

The idea of separating heaven and sky would not have occurred to many of Galileo's contemporaries, for whom the moral and physical cosmos aligned perfectly, radiating outward from base earth to exalted stars. Learned discussion consistently represented a spherical world in which 'up' was interpreted as outward from the centre. Ascent described a journey in spiritual space from gross earthly impermanence to eternal perfection amongst the stars.

Ancient and medieval versions described this as divine intervention, heavenly powers reaching down to lift humans up. Renaissance and Enlightenment thought turned towards human agency, particularly through science and technology. And yet, the language of salvation and apotheosis remained. Journeys into space represented escape from physical and moral limitations, elevation to divinity. In the popular imagination, spaceflight remains associated with narratives of ascent, development, and salvation (Midgley 1992, 183–94; Noble 1999, 115–42; Kilgore 2003; Traphagan 2021).

The physical decentring of a human world came to be associated with a metaphysical decentring of humanity, a shift in human significance (Midgley 1992; Scharf 2015; Vainio 2018; Mix 2021). At the same time, the shift is celebrated as the pre-eminent

Lucas John Mix, *Decoupling physical and spiritual ascent narratives in astronomy and biology*. In: *The Institutions of Extraterrestrial Liberty*. Edited by Charles S. Cockell, Oxford University Press. © Lucas John Mix (2022).
DOI: 10.1093/oso/9780192897985.003.0016

sign of human advancement through science and technology, a rational ascent. Spaceflight is interpreted as an adolescence for humanity. Drawing on the language of biology, proponents speak of a predictable sequence of developmental stages as the species matures: radio astronomy, spaceflight, planetary colonization, interstellar travel, and galactic expansion. This technological progression is associated with intellectual and ethical advances as well as regular, usually exponential, population growth. Situated between spaceflight and planetary colonization, humanity can be viewed as entering cosmic adulthood.

Familiar from science fiction, the developmental narrative has, nonetheless, a profound impact on astronautics and astrobiology, where it is presented as science or common sense. Indeed, it does reflect views of evolution common a century ago. Both biology and sociology, however, have moved away from this approach. Biologists have become deeply sceptical of developmental narratives applied at the species level and actively antagonistic to theories of progressive evolution. Sociologists, meanwhile, have recognized the dangers of developmental narratives, which were frequently used to justify exclusion and even harm towards individuals and groups deemed less advanced.

Developmental narratives do normative work, distinguishing desirable and undesirable futures, in a way that takes them beyond empirical science. What constitutes maturity? And who decides? Carrying through with the Copernican revolution will require refuting the developmental narrative of predictable ascent and replacing it with a dispersal narrative of movement outward into space. It is a frontier, but not the 'final' frontier. Strange new worlds, new life, and new civilizations will appear through unpredictable innovation and diversification, losses as well as gains, adaptation to local environments, and regular reorientation. Every star, every planet, and every moon has its own 'up' and 'down'.

This chapter explores the long association between space science and space fiction and the role of ascent narratives woven through both. It starts with the classical model of the cosmos and the place of humanity within it (section 2). The physical location of the world placed humans within a moral hierarchy defined by up/heavenward and down/Earthward. Heavenly journeys involved moral as well as physical movement on this scale. The new models of early modern astronomy embraced heliocentrism but continued to associate planetary journeys with spiritual ascent (section 3). The lines between science, fiction, and prophecy were blurry as scientists used stories to share their findings whilst storytellers worked with the best of contemporary logic and observation. Early modern biology also developed ascent narratives associated with development and progressive evolution, but these models were rejected in the mid-twentieth century (section 4). Ascent narratives remain popular in astronautics, astronomy, and astrobiology, drawing on a long history of planetary journeys (section 5). Speculation on the future of humanity, narratives of salvation, and blatant evangelism have consistently been associated with science. Contemporary thinking, both public and scientific, still links the two questions: how to go to heaven and how the heavens go. This calls for care and caution by those who wish to complete the Copernican revolution and uncouple the two (section 5).

2. The History of Up and Down

The relationship between heaven and sky has always been complicated. The simple narrative that science decentred humanity—through Copernican and Darwinian revolutions—does not hold up (Lewis 1994, 116; Lovejoy 2001, 102; Mix 2016, 2017). Earlier cosmologies made humanity medial—poised the between extremes of heaven and hell, rarefied atmosphere and gross earth, ideal form and changeable matter—or peripheral—cast out of blessedness onto the trash heap. The disorientations and reorientations of modern science reflect, instead, a break in the classical association between elevation and salvation.

The earliest myths from Mesopotamian and eastern Mediterranean regions presented a three-tiered cosmos, with heaven over Earth over the land of the dead, but even these often represented the world as a bubble with up and down only existing within a larger chaos.

By late Antiquity, scholars situated humanity on a globe. The world—at this time, the range of humanity, not yet a planet—covered part, though probably not all, of a sphere. Plato's *Phaedo* (fourth century BCE) describes a spherical cosmos (Pender 2012). Tartarus, a chaotic pulsing heart, circulated fire, air, and water from the centre of the underworld. Plato's use of up and down makes it clear he was referring to out and in. Layered on Tartarus were Hades, earth, water, and air respectively, with the human world at the bottom of air. Most souls, upon death of the body, circulated in the underworld until returned to the earth. A few, truly evil souls, were cast forever into the turbulent flux of Tartarus, whilst a few, truly philosophical souls, journeyed beyond the air to the pure clarity of gods and planets. For Plato, down referred to imprisonment within an eternally changing material body, whilst up referred to escape into ideal space. Plato's *Timaeus*, influential for medieval Christian and Golden Age Islamic thinkers, also spoke of a spherical world. The Platonic association of up/out with freedom, mind, and the good—and down/in with imprisonment, body, and evil—shaped much subsequent thinking.

Aristotle disagreed with Plato on several key issues, but affirmed the spherical cosmos, paying close attention to both logic and observation (*On the Heavens*; Johansen 2009). In usual Aristotelian fashion, he provided a disambiguation of three senses of heaven (*ouranos*): first, the utmost height or circumference of cosmos; second, that which stretches inward to the circle of the Moon; and third, the totality including Earth. For all three, he spoke of both location and composition. Even in Classical Greece, scholars recognized the ambiguity of up and out and situated the human world within the heavens.

In place of Plato's organismic metaphors (pumping heart and cosmic soul), Aristotle described three kinds of simple motion. Within the circle of the Moon, particles incline upward/outward (air and fire) or downward/inward (earth and water)—the sublunary cosmos. These inclinations provided a coherent explanation of both gravity and the curvature of the world, as earth and water settle inward. Beyond the Moon, particles of ether followed ideal circles. Those circles explained the movement of stars and planets. Thus, there were two types of physics, terrestrial and heavenly. This defined the great boundary broken conceptually in the Copernican revolution and physically by rocketry.

2.1 Cosmology as science

In the third century BCE, Eratosthenes of Cyrene estimated the circumference of earth by observing the angle of sunlight in distant wells. Subsequent geometers correctly surmised that the known world covered nearly half the surface of the terrestrial globe. In the second century BCE, Claudius Ptolemy wrote the *Algamest*, which became the authority for cosmology around the Mediterranean for 17 centuries. The Aristotelian, Ptolemaic, or geocentric model became the standard for astronomy. Though tradition was highly valued in this period, observations were still made and compared against the standard.

Modern astronomers tend to deride Aristotle and valorize alternative Hellenistic cosmologies (e.g. Sagan 2013[1980], 187–90; Dick 1982, 11–12; Scharf 2015, 13–14). Democritus, Epicurus, and Lucretius denied geocentrism, appealed to universal physical laws, and remained open to other worlds. They affirmed flat earths falling through infinite space carrying their own sun and moon with them—attached by mysterious means. David Grinspoon (2003, 8) provides the most sympathetic account. 'Both theories—Epicurean and Aristotelian—are elegant and logical, but they lead to opposite conclusion about other worlds and extraterrestrial life. This, in a nutshell, is the problem with philosophizing in the absence of evidence.' Aristotle was not, however, theorizing in the absence of evidence, nor were his followers. The movement of the Sun, the shadow of the earth in lunar eclipses, and the regular motion of falling bodies (at all locations on earth) provided concrete empirical evidence in support of the Aristotelian view. Copernicus and Galileo would eventually provide a better alternative, one that unified terrestrial and heavenly physics. It overcame the Aristotelian picture only gradually, however, for scientific as well as religious and aesthetic reasons; the chief scientific complaint was that it undermined traditional understandings of gravity (Pellegrin 2009; Graney 2017).

Arguments about a plurality of worlds are somewhat murkier as they depend on ontological questions about how the universe is ordered. The question can be answered now only because it has been redefined. After Copernicus, the planets (transcendent loci of meaning and movement) became rocks orbiting stars, earth (an element and location) became the Earth (a planet), and *the* world (the realm of humans or the larger entirety of the ordered universe, the cosmos) became *a* world (a synonym for planet). Comparisons can be made across the centuries, but only with careful attention to language and shifting concepts. Astronomy and astrobiology can, for the first time, answer the question of whether there are rocks orbiting other stars. It remains in the dark about a plurality of inhabited rocks or a plurality of universes. New questions can be answered by new means, but the ancient questions remain unresolved.

2.2 Cosmology as salvation

Heavenly up has long been associated with physical movement outward. The Copernican revolution, however, brought into question another association, between heavenly up and salvation. Following Plato and Aristotle, Christian cosmologies respected observations of a spherical earth. In the fourth century CE, Augustine noted the roundness of the earth. Discussing the word 'day' in Genesis, he noted that the sun cannot shine on all earth at once (Augustine *De Genesi* 1.10; *Confessions* 11). The term, then, cannot be taken literally and must reveal something important about time and space and God's relation to them. Augustine introduced a distinction between salvific and scientific interpretations of scripture, insisting that the first was more important, even though it may be figurative. When the science speaks clearly against a literal interpretation, then a figurative interpretation is called for. Science and scripture cannot disagree; any insistence that they do must only dissuade people from being Christian. Galileo would later latch onto Augustine's distinction (Sobel 1999; Dowe 2005).

A thousand years after Augustine, the spherical earth and outwardness of heaven appeared in Dante's *Divine Comedy* (1320). For Dante, the centre of the known cosmos was Lucifer, who had been cast out of heaven and, like a speck of grit at the centre of a pearl, accumulated fallen matter around himself. The world of humans occupies one hemisphere of this globe. The story ends with Dante (author as protagonist) ascending towards God in the *Paradiso*. Travelling outward from the atmosphere, he passes through the seven 'planets'—the Moon, Mercury, Venus, Sun, Mars, Jupiter, and Saturn—to the sphere of fixed stars, the prime mover, and the empyrean height (the highest heaven made of pure light).

The key to understanding Dante's cosmos is a fundamental inversion in the human perspective. As Dante passes beyond the fixed stars to the prime mover, he sees the universe as it really is, with God at the centre (canto 28). In the material universe, the prime mover is the largest of the nested spheres, most worthy and most powerful, and thus, necessarily, the biggest. But in the true, spiritual universe creation revolves around God. The 'low' things of Earth are really at the periphery.

Islam provides earlier examples of the planetary journey to God. The Qur'an (surah 17, *al-Isra*) points to the Mi'raj, a spatial and spiritual ascent by the Prophet Muhammed (570–632). From the site of Al-Aqsa Mosque in Jerusalem he went up the ladder of heaven to the close presence of Allah. The word *mi'raj* in Arabic means ladder or ascent and ninth century hadith specify the seven heavens of the classical model (see *Sahih al-Bukhari* and *Sahih Muslim*).

Classical authors described a cosmos centred on chaos, confusion, and hell. Human habitation in the suburbs reflected both suffering and low status. Thus, ascent into heaven—almost universally conceived as outward, not linearly upward— reflected salvation. Humans escaped into the larger universe. The classical cosmos could be described as a spherical coordinate system, with an origin at the lowest point in hell. Polar and azimuth angles were determined by the ecliptic and equinoxes, which held their own spiritual and astronomical significance (Kleiner 1994). Questions of salvation could be reduced to radial distance alone. Spiritual up and down had definite, concretely physical analogues, not because of a flat earth but because of a spherical earth concentric with a spherical cosmos. Humans were never central, but they were reliably located on a continuum of value.

The Copernican revolution challenged the salvific narrative of spiritual ascent as well as the scientific model of physical centrality. Copernican heliocentrism set the Earth—now a planet—adrift. Humans were never physically central, so science could not de-centre them. Instead, the Copernican revolution breached the divide between terrestrial and heavenly physics, enabling free movement between them, both physical and spiritual. It also disoriented humanity, making concepts of up and down relative.

3. Planetary Voyages

Natural science and the novel arose together in the early seventeenth century. The interplay of observation and imagination had been strong for at least a century as both scholars and publics contemplated astronomical discoveries and the possibility of plural worlds (Dick 1982; Crowe 2008). Scientists wrote and cited fantastic stories, whilst fantasists and poets drew inspiration from science.

Francis Bacon set the foundations for empiricism, chiefly in his *New Organon* (1626). He believed that God had given humanity two instruments to restore their place in the cosmos after falling from grace (Noble 1999; Dowe 2005). Though made in the image of God and set to rule the earth, Adam and Eve's disobedience resulted in an alienation from God and nature for all humanity. Bacon believed that Christians could be re-elevated, with scripture restoring the relationship with God and science restoring dominion over nature.

The *New Organon* was organized around a worship metaphor, calling for humans to reject four 'idols of the mind', erroneous ways of seeing the world. Bacon unequivocally called science true worship and compared it with the journey to heaven.

> So much concerning the several classes of Idols, and their equipage: all of which must be renounced and put away with a fixed and solemn determination, and the understanding thoroughly freed and cleansed; the entrance into the kingdom of man, founded on the sciences, being not much other than the entrance into the kingdom of heaven, whereinto none may enter except as a little child.
>
> (Book 1, aphorism 68; Bacon 1858, 69)

Many early scientists were motivated by a belief that salvation occurred through the active participation of humans. They saw science and technology as both sign and substance of this process. Bacon did not map it onto a physical ascent but he did set forth a speculative account of the future in *The New Atlantis* (1626).

The first science fiction novel about space may be *Somnium* (1634) by astronomer Johannes Kepler. The story incorporates contemporary scientists, data, and theories in a fabulous tale about a daemon capable of carrying humans to the Moon. Thus, it worked to promote scientific insight in an entertaining and speculative context.

3.1 Planetary travel as epic poetry

Somnium is only one example of a Moon craze in early seventeenth-century Europe. Writings in this time drew readily from epic poems in Antiquity, the Middle Ages, and the Renaissance. Amongst the most influential ancient sources were Plato and the accounts of the underworld in the *Odyssey* and *Aeneid*. More heavenly—and more expressly fictional—journeys were described by Lucian of Samosata (second century CE) in *Menippus, Icaromenippus*, and *True Story*. The last is a satire about kingdoms on the Sun and Moon fighting over who would colonize Venus.

In addition to Dante's *Divine Comedy*, two of the most famous Renaissance epics mention inhabited planets. *Orlando Furioso* (1506–1532) by Lodovico Ariosto describes a journey to the Moon whilst the *Faerie Queen* (1590) by Edmund Spenser briefly touches on the possibility of other inhabited worlds, arguing from the recent discoveries in astronomy (and in America) to the possibility of worlds like Faerie.

Inspired by new telescopes and the discoveries of Copernicus, Galileo, and Kepler, many authors wrote of the Moon as an inhabited country, visitable by humans. They mixed scientific, fantastical, and religious themes. Prominent English works include *The Man in the Moon* (1638) by Francis Godwin, *The Discovery of a World in the Moon* (1638) by John Wilkins, and *Anatomy of Melancholy* (1621) by Robert Burton.

3.2 Planetary travel as science

Together, these works demonstrate a tight connection between how one goes to heaven and how the heavens go. Humans remained relatively passive, however, limited by chance or heavenly intervention. Modern science changed that. Drawing on Baconian ideas of human salvation through knowledge and technology, space fiction slowly acquired themes of human elevation by human artifice.

Alex MacDonald (2017) suggests that *The Man in the Moon* was the first example of travel to the moon by human invention. The protagonist, Domingo Gonsales devises a machine whereby 25 swans can lift him. Though he only intends air travel, he soon finds himself lofted above the clouds to a realm of perpetual light, where he is freed from earthly needs and discomfort.

> [T]he Air in that Place I found without any Wind, and exceeding temperate, neither Hot nor Cold, where neither the Sun Beams had any Object to reflect upon, nor the Earth and Water appear to affect the Air with their natural Quality of Coldness; as for the Philosophers attributing Heat and Moisture to the Air, I always esteemed it a Fancy: Lastly, I remember that after my Departure from the Earth, I never felt either Hunger or Thirst, whether the Purity of the Air, freed from the Vapours of the Earth and Water, might yield Nature sufficient Nourishment, or what else might be the Cause I cannot determine, but so I found it, though I was perfectly in Health both of Body and Mind, even above my usual Vigour.
>
> (Godwin 1768[1638], 23)

Gonsales notes the incorrectness of traditional cosmology and suggests that the Moon has its own attractive power, its own up and down. Descending to the surface of the Moon, he discovers both people and plants, though quite different from their earthly equivalents. Notably, his hunger returns in a novel way that attracts him to Moon food. Thus, this may also be the first account of a new value system associated with travel to a new planet.

Technological advances slowly provided engines of ascent, quickly adopted by fantasists for planetary travel. The first ascent by hot-air balloon in 1783 inspired Edgar Allan Poe and imitators to write about balloon trips to the Moon (Poe 1835; Sessarego 2020). Jules Verne sent men to the Moon with a cannon (*From Earth to the Moon*, 1865). Astronomer Garrett P. Serviss wrote a sequel to H. G. Wells' *The War of the Worlds* in which Thomas Edison and other contemporary technologists used electricity to develop anti-gravity drives (*Edison's Conquest of Mars*, 1898). Wells himself wrote about newly discovered nuclear energy empowering human flight to the Moon and beyond (Wells 1914). Wells tellingly speaks of salvation from earthly limits through his character Marcus Karenin: 'We can still keep our feet upon the earth that made us. But the air no longer imprisons us, this round planet is no longer chained to us like the ball of a galley slave' (Wells 1914, 303) Karenin also speaks of escaping the limitations of earthly bodies. (Wells dedicated his book to radiochemist Frederick Soddy and attributed the science to *Interpretation of Radium* (1909). For more on Soddy, their relationship, and the origins of synthetic biology, see Campos 2015.)

These stories were particularly influential in the history of spaceflight. Rocketry pioneers Konstantin Tsiolkovsky and Hermann Oberth cite Verne as a key influence. Robert Goddard found Verne interesting but was more taken with Wells and Serviss. Nuclear pioneer Leo Szilard was likewise moved by Wells' vision of space travel.

It is hard to overstate the importance of speculation—including popular fiction—in the history of astronomy and spaceflight. Galileo lectured on Dante's *Inferno* (Kleiner 1994, 25–34). Kepler wrote the *Somnium*. Christiaan Huygens speculated on the inhabitants of planets and exoplanets in *Cosmotheoros* (1698). The trend continued through the twentieth century with leading scientists writing space fiction and speculation. Influential examples include *The World, the Flesh, and the Devil* (1929) by J. D. Bernal, *The Black Cloud* (1957) by Fred Hoyle, and *Contact* (1985) by Carl Sagan.

3.3 Planetary travel as salvation

Whilst scientific ascents waited on new technologies, spiritual ascents proliferated, often appealing to spiritual travel or astral projection, leaving the body behind. Classical stories involved bodily ascent into heaven, seemingly ruled out by modern science. Mental travel, however, remained a respectable scientific enquiry well into the nineteenth century. Thus, many authors, still operating at the boundaries of science and fiction, discussed planetary travel as a movement of minds.

New religions included planets and exoplanets in their reflections on salvation. Influenced by 20 years of work in the sciences, Emmanuel Swedenborg, founder of the New Church, described conversations with spirits from other planets in *Planets in Our Solar System and in Deep Space, their Inhabitants, and the Spirits and Angels There* (1758, republished as *Other Worlds* 2018). Joseph Smith included a hierarchy of planets in the scriptures of the Latter Day Saints (*Book of Abraham*, 1842). Their followers have debated how literally to take these accounts, in recent years erring on the side of figurative interpretation, but they reflect a close link between salvation and planetary voyages.

Brian Stableford provides a compelling survey of planetary voyages in early modern fiction in his introduction to *Lumen* (2002[1887]) by Camille Flammarion. Working briefly with Urbain Le Verrier, Flammarion was an astronomer chiefly remembered as a science writer and popularizer of astronomy; his astronomical speculations, however, came entwined with spiritualist beliefs. The spiritualists spoke of communication with the spirits of the dead and of a transmigration of souls from one life to the next. *Lumen* and other works describe successive incarnations on different planets, with the inhabitants of each world shaped by its place in the planetary hierarchy. The sentient species of a given planet would be more or less free, intellectual, and dignified based on that place. *Lumen* purports to trace Flammarion's own successive and ascending incarnations. Flammarion's past incarnations include time as a plant on Cygnus and as a woman on Venus. Meanwhile, in *Urania* he describes Mars as above us:

> Terrestrial humanity is young ... and still in primitive ignorance ... you cannot live without wresting your daily bread from the Earth ... The lowest animal on Mars is better, finer, gentler, more intelligent, and greater than the god of the armies of David, Constantine, Charlemagne, and all your crowned assassins. There is therefore nothing surprising in the coarseness and stupidity of terrestrial humanity. But the law of progress governs the world. You are more advanced than at the period of your ancestors of the stone age, whose wretched existence was spent in fighting night and day with ferocious beasts. In a few thousands of years you will be more advanced than you are now.
>
> (Flammarion 1890, 252–4)

Other spiritual journeys to the planets include *The Voyage of Lord Ceton to the Seven Planets* (1765) by Marie-Anne de Roumier and *Consolations in Travel* (1830) by chemist Humphrey Davy, both of which include successive, evolutionary incarnations on multiple worlds. Edgar Rice Burroughs' Barsoom stories (1912–1943) likewise features spiritual travel to a more advanced Mars. Along with Flammarion's work, they set the stage for the planetary romances of the twentieth century. Notable examples include *A Voyage to Arcturus* (1920) by David Lindsay, the *Space Trilogy* (1938–1945) by C. S. Lewis, and the *Hainish Cycle* (1969–1974) by Ursula K. LeGuin.

Throughout the modern period, the heavens provided a blank screen on which to project theories of human salvation. Appearing in almost every literary genre, the heavens remained consistently associated with ascent, advance, and escape from limitation. Whilst earth became the Earth and planets devolved from celestial spheres to localized rocks, narratives of salvation were consistently mapped onto them. Humans began to design their own engines, but they still spoke of journeying upward to freedom, harmony, and god-like power.

4. The Path to Perfection

Upward narratives of astronomy came to be associated with upward narratives in biology. The two have important differences, however. Whilst planetary ascents remained popular throughout the history of modern astronomy, evolutionary ascents were only popular in biology from the late eighteenth to early twentieth centuries. Biologists rejected evolutionary progress in the Modern Synthesis. More than simply outdated, it is now considered unscientific, disproven, and even dangerous (Ruse 1996; Mix 2022).

4.1 Up and down in biology

Ancient and medieval biologists described a continuum of increasing freedom and dignity running from simple or 'lower' life forms, literally pond scum, upward to humans and potentially even 'higher' creatures such as angels. This hierarchy came to be known as the *Scala Naturae* or Ladder of Nature (Lovejoy 2001; Ruse 2010; Mix 2018). Considered from the bottom up, as nutrition empowering the whole system, or from the top down, as God drawing all things heavenward, the system reinforced the continuity of human and non-human life.

The *Scala Naturae* served many philosophical purposes including justifying hierarchical forms of government, displaying God's creativity (by filling all possible niches), and undergirding ethics. It explained a preference for higher organisms and higher activities. It explained why intellectual pursuits were more important than aesthetic pursuits and why both were more important than domestic activities. It validated human use and consumption of animals (and use of plants by both), but not the converse. Comparison between higher and lower forms of life provided a key principle for religious and secular ethics.

Biological and cosmic hierarchies aligned in literal ways. Upright posture and head elevation were seen as signs of human nobility, contrasted with the lowness of worms and snakes. Plants, rooted in the earth, were liminal between mineral and living worlds, whilst fish (in the water) and creeping things (in the earth) were barely animal. Birds became signs of freedom and grace, not closer to salvation than humans but nonetheless elevated spiritually as well as physically.

Enlightenment thinkers tried to break the *Scala* in two. What Aristotle did to space—separating terrestrial and heavenly physics—Rene Descartes did to biology—separating nutrition from reason. He distinguished minds (thinking things) from bodies (extended things) allowing only humans to bridge the divide. A gradual separation of body and mind can be traced back to the first century CE and Philo's two-creation interpretation of Genesis. His work contributed to strong ontological and epistemological dichotomies that developed in Western thought. Most theologians and philosophers, however, continued to think of spirit acting in matter. Descartes would be the first major thinker to definitively exclude mind and spirit from nature (Mix 2018, 143–57). This dualism proved crucial to the mechanical philosophy, empiricism, and modern science (Foucault 1994; Osler 1994, 2010). Physics and astronomy benefitted greatly from bracketing questions of the mind. Biologists, however, never fully embraced the scheme.

Early modern biologists attempted to maintain a continuity of human and non-human life whilst respecting the epistemic bounds of science (Mix 2022). Evolution and genetics changed thinking about formal and final causes (identity and ends). They naturalized teleology, reimagining form, fitness, and function as products of natural selection. But they also finalized nature, granting identity and ends to molecules (e.g. genes and enzymes) as well as organisms.

4.2 'Evolution' as development

Prior to Darwin, evolution and development had similar meanings; etymologically they both meant 'to unroll' suggesting a scroll that is already written but only gradually revealed. Considering fossils and the mutability of species, they turned the ladder into an escalator. Species moved upward over time, resulting in the growth or maturation of species: developmental, or progressive, evolution. The hierarchy remained eternal, but species moved upward in it.

Nineteenth-century biologists began to consider genuine change, but most still believed that evolution occurred along a fixed trajectory from lower to higher organisms. Jean-Baptiste Lamarck (1914, 56–61) described a path to perfection from simple, passive, and undignified micro-organisms to complex, active, and noble humans. Lamarck and geologist Robert Chambers imagined parallel ascents for plants (lichen to orchids) and animals (animalcules to humans). At the same time, philosopher John Locke and naturalist George-Louis Leclerc, Comte de Buffon, championed a single trajectory.

Tree of life diagrams retained the lower to higher orientation through the mid-twentieth century, with phylogenies rooted in bacteria and growing up to the 'crown' eukaryotes (fungi, plants, and animals). Comparing individual change-through-time and population change-through-time, they retained the developmental analogy and its associations with physical and spiritual ascent.

4.3 Evolution by natural selection

Darwin showed that evolution works differently, producing organisms adapted to local conditions. Nor did this reliably produce change for the better, whether measured in complexity, freedom, nobility, or any other universal criterion. Natural selection was the biological equivalent of the Copernican revolution. It bridged the Cartesian divide, reunifying explanations of nutrition and reason. It also relativized life; every niche has its own value system, its own up and down. Modern phylogenies reflect this by branching outward in all directions, emphasizing the radial quality of evolution.

Darwin was fascinated by organisms that defied the traditional pattern of lower and higher life. He wrote as much on the animality of plants (sensation and motion) as on the animality of humans. He was intrigued by parasitic wasps that lay their eggs in insect hosts, keeping them alive to better nourish the young that hatch and eat their way out. This led Darwin to reject moral ascent narratives in biology.

Later biologists reformed evolutionary theory with the Modern Synthesis in the early twentieth century, integrating natural selection with Mendelian genetics and statistical methods (Ruse 1996; Mix 2022). Labelling earlier views teleological, vitalist, or orthogenetic, they restricted evolutionary theory to stochastic variation and local adaptation. Whilst probabilistic prediction may be possible in rare cases of convergence—where a biological innovation is within reach and the environment well understood—development is usually a misleading analogy.

Biologists avoid the generic use of 'evolution' as any change through time, and repudiate the progressive use of 'evolution' as development. Ascent narratives have become anathema, partly because they were so popular—yet unproductive—in the nineteenth century and partly because they were associated with anti-evolution movements in the twentieth. Twenty-first-century biology still struggles to avoid both Cartesian dualism and Lamarckian progressivism—both of which might be considered anti-Copernican errors. Cartesian dualism divides the world into two realms, each with its own rules. Progressive evolution conflates random physical changes with moral improvement. It is as anthropocentric as geocentrism, if not more so. Geocentrism places humans near (though not at) the nadir of physical space, but progressive evolution places humans at the apex of biological development.

4.4 Developmental narratives and eugenics

Darwin's contemporaries readily interpreted his theories in a competitive way that justified the success of one culture or race at the expense of others. 'Survival of the fittest' became a moral axiom as well as a biological observation. In this normative sense, it was used to justify the suppression of 'inferior' races morally through re-education and physically through eugenics.

Darwin's cousin, Francis Galton, applied his ideas about variation and selection to humans, arguing for the artificial selection of favoured traits. He coined the term 'eugenics' for intervention in human breeding to 'improve the stock' and speed up the natural process of competition. A key figure in the origins of statistics, his probabilistic methods and his convictions about engineering the species were passed on to many prominent biologists, including R. A. Fisher and J. B. S. Haldane, key figures in the Modern Synthesis. Oddly, the statistical methods of these theorists made possible the strongly anti-progressive biases of later biologists. They showed that natural selection could occur probabilistically, eliminating the need for progress and intent.

Whilst the rejection of progressive evolution is usually presented on the more abstract grounds mentioned above, the link with social engineering, eugenics, and racism cannot be avoided. The idea that some populations were more advanced and more fit than others pervaded discussions of evolution throughout the early twentieth century. Colonial expansion and suppression of reproductive liberty were presented as moral imperatives revealed by biology.

4.5 Developmental narratives and sociology

Parallel thinking in sociology applied developmental narratives to human 'evolution'. Auguste Comte and Lewis Henry Morgan inherited the scientific ascent narratives of

Bacon, applying them to the comparative study of societies. Each proposed a three-stage reliable trajectory of civilizations analogous to the maturation of individuals (Bordeau 2021; Traphagan 2021). Their benchmarks were scientific insight, technological sophistication, and power over the environment, leading to a concept of 'higher' or 'more advanced' civilizations.

Comte described his theory of positivism as 'the final religion', which 'will afford the only possible and utmost possible, satisfaction to our natural aspiration after eternity' (quoted by Noble 1999, 84). He centred the role of the engineer as applied scientist, vanguard of both physical and spiritual advance for the species. Arguably the first philosopher of science, his thought had a huge impact on generations of scientists and how they viewed their interaction with the world. Comte and Morgan remained committed to a view of societal advance as inherently Christian, but they set the stage for a generation of progressive beliefs that included atheist thinkers, including Karl Marx and Friedrich Engels, devoted to leading humanity along a path to perfection.

Ideas of social, technological, and moral engineering of the human species appeared famously in the work of Wells (1893, 2005[1895]), who provides a fascinating example of interaction between biology and science fiction. As a student, he took a class with famed biologist T. H. Huxley, whom he cites as an inspiration. He formed a picture of long-term evolution towards large brains, atrophied muscles, and moral sense, dependence on technology, and imperial aspirations—a trope still manifest in common depictions of 'advanced' civilizations. Wells went on to author a popular science series, *The Science of Life*, with zoologists G. P. Wells (H. G.'s son) and Julian Huxley (T. H.'s grandson). Stiles (2009) traces links between early biological doubts about progress and Wells' dystopias. Aldous Huxley (brother to Julian) had a deep interest in theology of science and tackled eugenics in one of the most famous biological fiction novels, *Brave New World* (1932).

The works of Huxley and Wells (and G. P. Wells and T. H. Huxley and Julian Huxley) helped to inspire the transhumanism of chemist J. D. Bernal, physicist Freeman Dyson, and cosmologists John Barrow and Frank Tipler—each influential in astrobiology (Midgley 1992). The same process, expanded to a galactic scale provided the key ingredient for Isaac Asimov's 'psychohistory' in the *Foundation* series (1942–1993) and Frank Herbert's 'Golden Path' in the *Dune* series (1965–1985), which in turn inspired a generation of astronomers and economists to attempt long-term prediction.

5. Modern Spaceflight as Human Ascent

Ascent narratives have been popular in astronomy, partly due to the continued—and fruitful—interplay between science, science fiction, and value narratives. The vast scope of cosmology and astrobiology readily lends itself to discussion about the Copernican revolution, human significance, and speculations about the future. It invites commentary on humanity's physical and moral grounding on the planet Earth, its potential to rise above that context, and its future in the cosmos. The closely related search for extra-terrestrial intelligence calls for speculation about commonalities and differences with other sentient species. Thus, the plurality of worlds plays out in contemplated futures as well as observed systems.

5.1 Growing up in space

Developmental narratives describe a predictable path to maturity as normative, but not necessary. Deviation is possible but regarded as tragic and often fatal. Individual human development provides numerous examples, easily expressed in a single word: pre-term (pregnancy), pre-natal (care), pre-K (education), pre-adolescent (development). Nominally neutral—referring only to a period of time—they carry normative weight because of the prospect for continuing to the named stage and beyond.

Sociology has produced similar descriptors for human societies: pre-Columbian, pre-scientific, pre-industrial, and pre-modern. Once again, they are nominally neutral but rapidly acquire problematic ethical consequences when paired with developmental assumptions that the end stage is more advanced. They fit too easily with an idea that lower societies can and should be sacrificed for the sake of higher—indeed this must be the case for growth to occur. Even the term pre-Cambrian was dangerous in this sense as it discouraged research into evolution, diversity, and metabolic innovation prior to the rise of multicellular organisms.

The same idea appears in science fictional tales of pre-spaceflight and pre-interstellar travel ('pre-warp' in *Star Trek*) societies. It carries with it an assumption that sentient life will naturally develop along a predictable path labelled 'science' that begins with a Copernican Revolution, a realization that multiple worlds and multiple reference frames exist. It passes through radio-telescopes, spaceflight, interstellar travel, and stellar colonization, leading (surprisingly often) to galactic dominion. The trope of human development to galactic empire—or at least a human dominated confederation spanning a large portion of the galaxy—is a plot point in some of the most popular science fiction universes. Asimov's *Robot, Empire*, and *Foundation* series (1939+), Frank Herbert's *Dune* (1965+), Gene Roddenberry's *Star Trek* (1966+), Ursula K. Le Guin's *Hainish Cycle* (1969+), and Daniel Abraham and Ty Franck's *Expanse* (2011+), provide only a few examples.

Michael Kilgore (2003, 115) describes the developmental narrative in Arthur C. Clarke. 'Clarke believes that the continued evolution of civilization depends on the conquest of space. The use of the term "evolution" in this context is deliberate for Clarke believes human history to be the expression of a biological imperative toward perfection.' Kilgore also explores the importance of imperial and frontier aspirations for Heinlein, and other prominent figures in 'hard' science fiction.

The idea appears often in scientific speculation on the future. Noble (1999, 121–6) describes the developmental, Christian, and baldly soteriological (salvation-oriented) beliefs of rocket pioneers Konstantin Tsiolkovsky and Wernher von Braun, both of whom saw spaceflight as the first step towards a cosmic destiny. Midgley (1992) follows a similar trend in Bernal, Dyson, and Barrow and Tipler, each of whom starts with colonization of space and ends with dominion over the cosmos.

Carl Sagan returned again and again to developmental metaphors linking space-flight with astronomical, biological, and spiritual ascent.

The human species is now undertaking a great venture that if successful will be as important as colonization of the land or the descent from the trees. We are haltingly, tentatively breaking the shackles of Earth—metaphorically, in confronting and taming the admonitions of those more primitive brains within us; physically, in voyaging to the planets and listening for the messages from the stars.

(Sagan 2013, 338)

Sagan (1979, 368) quotes Tsiolkovsky: 'The Earth is the cradle of mankind. But one does not live in the cradle forever.' The theme persists in later writings as well. 'And it is clear that if we do not destroy ourselves, we will continue this progressive outward movement until there will be human settlements on neighbouring worlds' (Sagan 2006, 214).

More recent examples include David Grinspoon's language of the 'mature Anthropocene' and 'galactic destiny' (Grinspoon 2016). His earlier and grander discussion in *Lonely Planets* (Grinspoon 2003, 402) echoes the cosmic and soteriological language of Bernal and company: 'You must only admit the possibility that immortality is occasionally the outcome of an evolutionary process to conclude that the universe must be headed in this direction.'

Examples could easily be drawn from other prominent astrobiologists. In *God and the New Physics*, Paul Davies argues that 'science offers a surer path to God than religion ... science has actually advanced to the point where what were formerly religious questions can be seriously tackled' (Midgley, 1992, 178). Midgley (1992, 159) goes into further detail and explores his concept that humanity will develop until we become 'lords of the universe'. Astronomer and NASA historian Steven J. Dick (2003, 72) describes a similar apotheosis:

AI may not be the ultimate emergence of cultural evolution, and Morowitz (2002) has suggested that 'spirit' could be an emergent phenomenon beyond AI. Where cultural evolution would ultimately lead one cannot say, except that ultimate entities might have characteristics approaching those we ascribe to deities ... .

The developmental narrative may be most clear in discussions of the Fermi Paradox that take for granted planetary colonization as a normal feature in the development of intelligent species. In one of the most influential papers on the topic, Michael Hart (Hart 1975, 127) writes:

If ... there were intelligent beings elsewhere in our Galaxy, then they would have eventually achieved space travel, and would have explored and colonized the Galaxy, as we have explored and colonized the Earth. However, ... they are not there; therefore they do not exist.

More recent authors have doubled down, arguing that we now have more data to support the trend citing Moore's law and advances in computing.

This is closely related to the issue of astroengineering: the energy limitations will soon cease to constrain human activities, just as memory limitations constrain our computations less than they once did. We have no reason to expect the development of technological civilization elsewhere to avoid this basic trend.

(Ćirković 2009)

The label 'more advanced' begs the question of 'towards what?' It remains an ascent narrative and, in this context, both a planetary journey and a form of salvation. The steps are clear: science, radio-telescopes, Moon colony, Mars colony, interstellar travel, and galactic expansion. The inevitability, normativity, and desirability of the path remain in question. It would be too easy to fall into the Marxist trap of moving from inevitable future to a call to action—and conflating both with scientific insight. Humans have only achieved two of these steps, providing an inadequate sample for meaningful induction. Contrary to popular belief, biology does not support exponential growth of populations (regardless of resources or barriers) and adamantly rejects predictable evolution (regardless of local environments).

5.2 Spaceflight as transcending physicality

The physical ascent of leaving Earth to colonize the galaxy remains coupled with the spiritual ascent of escaping Earth and earthliness for a Platonic existence in space. Here the biological aspects of the process cannot be ignored, but speculation continues to be driven by scientists (though not biologists) interested in space. Flammarion (2002, 97) described progressive evolution as integral to the cosmic development of humanity: 'The earthly animal kingdom has followed, from its origin, this continuous and progressive march toward the perfection of its typical forms of mammalia, freeing itself more and more from the grossness of its material.' The unearthed, debiologized brains of Wells' Moon men, Martians, and future humans set the stage for a century of scientific speculation (Wells 1893, 1914; Stiles 2009).

Chemist J. D. Bernal was one of the earliest proponents of space stations ('Bernal spheres') and wrote the mid-twentieth century entry on 'Life' for the *Encyclopaedia Britannica*, commenting on the relative chemical composition of space, Earth, and the biosphere. A pioneer in origin and definition of life studies, he also seeded developmental and soteriological narratives in astrobiology. His 1929 book proposed three obstacles to human flourishing: the physical 'world' of Earth, the physiological 'flesh' of embodiment, and the psychological 'devil' of our 'lower' selves. The first would be overcome by spaceflight, the second by creating cyborgs (as in Wells), and the third through social engineering (as in Marx).

His ambivalence about evolution reflected a materialist desire to remain scientific alongside a belief that human development transcended biology. 'The progress of the future depends no longer on physiological evolution but on the reaction of intelligence on a material universe' (Bernal 2017, 49). His equivocation between biological and social evolution allowed him to use science as both epistemic foundation for his developmental beliefs and as a developmental step itself, a tool for future 'evolution' of the species.

This ambivalence would be picked up by later writers, even though the rejection of progressive evolution by biologists made the first leg false and the second problematic. Sagan (1979, 47) writes, 'We are the first species to have taken our evolution into our own hands'. Dick (2003, abstract) suggests that evolution has pushed us beyond natural selection into a 'post-biological' universe: 'Here I propose another option—that we may in fact live in a post-biological universe, one that has evolved beyond flesh and blood intelligence to artificial intelligence that is a product of cultural rather than biological evolution.' It has become even more important to distinguish cultural evolution from biological evolution to maintain the developmental expectations and salvific power of science whilst still acknowledging the scientific findings of biologists.

Carl Sagan's *Contact* resolves, unsurprisingly in light of this discussion, by a non-physical journey to another planet. The protagonist, Ellie Arroway, is conveyed in spirit to a distant star, where she converses with spirits whilst her body remains, for all empirical purposes, on Earth. The journey is made through ambiguous means, a joint product of heavenly intervention and human engineering. And yet the message is clear: advanced aliens show her the way for humans to advance along the same path, through scientific exploration and spaceflight to transcendent life.

6. L'Envoi

The highest aspiration for all these writers is a spiritual ascent for humanity, one that escapes Earth, simultaneously the gravity well of a specific rock, the heaviness of embodiment, and the shackles of irrationality. Their work represents a continuous interplay of ideas around how one goes to heaven and how the heavens go. Humans cannot help but map earthly, base, lowness in body or spirit onto the coordinate system of the Earth, conflating upward with goodness and downward with harm. And yet, the Copernican revolution represents a significant insight, echoed in Darwin. Up and down are local, and we are moving outward in a way that must reorient humanity.

To complete the Copernican revolution, we must decouple spaceflight and physical elevation from narratives of spiritual ascent and developmental progress. Models of higher and lower life, more and less advanced societies have been rightly critiqued as non-biological and prone to dangerous dualisms. They reinforce sexual, environmental, and racial injustices (Midgley 1992; Miller 2002; Nealon 2016; Keel 2018). Kilgore (2003) in particular looks at the symbiosis of science, policy, and fiction in twentieth-century 'hard' science fiction, creating a utopia more suited to some than to others. Our next step will be a recognition of how deeply embedded we still are in geocentrism and salvific ascent narratives. The time has come for other narratives, including diversification, conservation, and kenosis.

The impersonal end of technological development and astral ascent is no less dangerous than the more personal end of God and heaven. And it is equally amenable to abuse when we start viewing some populations as farther along than others. With a fixed path to improvement, there is a moral obligation to help groups advance—often at the expense of individuals. Within a society, this becomes social engineering and, when tied to biological development of the species, eugenics. From without, it becomes paternalist colonial oversight.

The ascent narrative continues to shape astrobiology, chiefly through the influence of astronomers who present a developmental picture of species ascent. Couched vaguely in biological terms, it nonetheless falls into the area of introductions and conclusions where the speaker can emphasize it as the societal import of their work whilst simultaneously defending it as common sense or popularization and, therefore, below the need for scholarly defence or review. Midgley (1992, 15) sums this up brilliantly:

> Scientific reviewers, when discussing writings of this kind, often treat the myths as a side-issue. Concentrating on what is acceptable as science, they expect the rest to fade away harmlessly into the general culture. But it does not necessarily do this. It can hang around like a fog, changing the atmosphere of thought and influencing ideas quite strongly. It tends to be the part of the book that people remember. In particular, it can be expected to have a strong effect on students.

Astrobiology, as a transdisciplinary endeavour producing narratives of deep time and cosmic breadth, has inherited the mantle of capstone science from Bacon's physics and Comte's sociology and, thus, is open to some of the same temptations to overreach. We have an obligation, therefore, to be on the lookout for ascent narratives and ask how they shape both scientific and public thinking. Science should continue free from ecclesiastic oversight, but it still requires critical thinking about the relationship between how one goes to heaven and how the heavens go.

Acknowledgements

This chapter was developed in connection with the Institutions of Extraterrestrial Liberty Conference and as part of my research on human futures in space as the Baruch S. Blumberg NASA/Library of Congress Chair in Astrobiology, Exploration, and Scientific Innovation at the Kluge Center 2021–2022. Special thanks are due to Julia Dougherty for research assistance and to Charles Cockell, Kathryn Denning, and Lynn Rothschild for discussions on progress and astrobiology.

References

Augustine. *Confessions*, trans. H. Chadwick. *Oxford World's Classics*. (Oxford University Press, 2008).

Augustine. *On the Literal Meaning of Genesis*, trans. J. H. Taylor, S. J. *Ancient Christian Writers Series 41–42*. Mahwah, NJ: (Paulist Press, 1982).

Bacon, F. *The Works of Francis Bacon*, vol. IV, eds. J. Spedding, R. L. Ellis, and D. D. Heath (London: Longman and Co., 1858).

Bernal, J. D. *The World, the Flesh and the Devil: An Enquiry into the Future of the Three Enemies of the Rational Soul* (Brooklyn, NY: Verso, 2017).

Bourdeau, M. 'Auguste Comte' in *The Stanford Encyclopedia of Philosophy* (Stanford, CA: Stanford University, Spring 2021) <https://plato.stanford.edu/archives/spr2021/entries/comte/> accessed 5 July 2022

Campos, L. A. *Radium and the Secret of Life* (Chicago, IL: University of Chicago Press, 2015).

Ćirković, M. M. 'Fermi's paradox—The last challenge for Copernicanism?' (2009) arXiv preprint arXiv:0907.3432.

Crowe, M. J. *The Extraterrestrial Life Debate: Antiquity to 1915* (Notre Dame, IN: University of Notre Dame Press, 2008).

Dick, S. J. *Plurality of Worlds: The Origins of the Extraterrestrial Life Debate from Democritus to Kant* (New York, NY: Cambridge University Press, 1982).

Dick, S. J. 'Cultural Evolution, the Postbiological Universe and SETI' (2003) 2(1) *International Journal of Astrobiology* 65–74.

Dowe, P. *Galileo, Darwin, and Hawking: The Interplay of Science, Reason, and Religion* (Grand Rapids, MI: Eerdmans, 2005).

Flammarion, C. *Urania*, trans. A. R. Stetson (Boston, MA: Estes and Lauriat, 1890).

Flammarion, C. *Lumen*, trans. B. Stableford (Middleton, CT: Wesleyan University Press, 2002).

Foucault, M. *The Order of Things: An Archaeology of the Human Sciences* (New York, NY: Vintage Books, 1994).

Godwin, F. *The Strange Voyage and Adventures of Domingo Gonsales, to the World in the Moon* (London: John Lever, 1768).

Graney, C. M. *Mathematical Disquisitions: The Booklet of Theses Immortalized by Galileo* (Notre Dame, IN: University of Notre Dame Press, 2017).

Grinspoon, D. *Lonely Planets: The Natural Philosophy of Alien Life* (New York, NY: Harper Collins, 2003).

Grinspoon, D. *Earth in Human Hands: Shaping Our Planet's Future* (Boston, MA: Grand Central, 2016).

Hart, M. H. 'An Explanation for the Absence of Extraterrestrials on Earth.' *Quarterly Journal of the Royal Astronomical Society* 128–35.

Johansen, T. K. 'From Plato's *Timaeus* to Aristotle's *De Caelo*: The Case of the Missing World Soul' in A. Bowen and C. Wildberg (eds), *New Perspectives on Aristotle's* De Caelo 9–28 (Boston, MA: Brill, 2009).

Keel, T. *Divine Variations: How Christian Thought Became Racial Science* (Stanford, CA: Stanford University Press, 2018).

Kilgore, D. D. *Astrofuturism: Science, Race, and Visions of Utopia in Space* (Philadelphia, PA: University of Pennsylvania Press, 2003).

Kleiner, J. *Mismapping the Underworld: Daring and Error in Dante's 'Comedy'* (Stanford, CA: Stanford University Press, 1994).

Lamarck, J.-B. *Zoological Philosophy: An Exposition with the Natural History of Animals*, trans. Hugh Elliot (London: Macmillan, 1914).

Lewis, C. S. *The Discarded Image: An Introduction to Medieval and Renaissance Literature* (New York, NY: Cambridge University Press, 1994).

Lovejoy, A. O. *The Great Chain of Being: A Study of the History of an Idea* (Cambridge, MA: Harvard University Press, 2001).

Lucian. *Certain Select Dialogues of Lucian: Together with His True Historie*, trans. Francis Hickes (Oxford: William Turner, 1634).

MacDonald, A. 'The Narrative Origins of Space Flight' *TEDxAukland* (2017) <https://www.ted.com/talks/alex_macdonald_the_narrative_origins_of_spaceflight> accessed 5 July 2022.

Midgley, M. *Science as Salvation: A Modern Myth and its Meaning* (New York, NY: Routledge, 1992).

Miller, E. P. *The Vegetative Soul: From Philosophy of Nature to Subjectivity in the Feminine* (Albany, NY: State University of New York Press, 2002).

Mix, L. J. 'Life-value Narratives and the Impact of Astrobiology on Christian Ethics' (2016) 51(2) *Zygon* 520–35.

Mix, L. J. 'Philosophy and Data in Astrobiology' (2017) 17(2) *International Journal of Astrobiology* 189–200.

Mix, L. J. *Life Concepts from Aristotle to Darwin: On Vegetable Souls* (New York, NY: Palgrave, 2018).

Mix, L. J. 'Play Among the Stars: Astrobiology and Intra-action in Pryor's *Living with Tiny Aliens*' (2021) 60(1) *Dialog* 86–93.

Mix, L. J. *The End of Teleology in Biology* (New York: Palgrave, 2022).

Morowitz, H. *The Emergence of Everything: How the World Became Complex* (Oxford University Press).

Nealon, J. T. *Plant Theory: Biopower and Vegetable Life* (Stanford, CA: Stanford University Press, 2016).

Noble, D. F. *The Religion of Technology: The Divinity of Man and the Spirit of Invention* (New York, NY: Penguin, 1999).

Osler, M. J. *Divine Will and the Mechanical Philosophy* (Cambridge University Press, 1994).

Osler, M. J. *Reconfiguring the World: Nature, God, and Human Understanding from the Middle Ages to Early Modern Europe* (Baltimore, MD: Johns Hopkins University Press, 2010).

Pellegrin, P. 'The Argument for the Sphericity of the Universe in Aristotle's *De Caelo*: Astronomy and Physics' in A. Bowen and C. Wildberg (eds), *New Perspectives on Aristotle's* De Caelo (Boston, MD: Brill, 2009) 163–85.

Pender, E. 'The Rivers of Tartarus: Plato's Geography of Dying and Coming-back-to-life' in C. Collobert, P. Destrée, and F. J. Gonzalez (eds), *Plato and Myth: Studies on the Use and Status of Platonic Myths* (Boston, MD: Brill, 2012) 199–233.

Poe, E. A. 'Hans Pfaall—A Tale' (1835) 1(10) *Southern Literary Messenger* 565–80.

Ruse, M. *Monad to Man: The Concept of Progress in Evolutionary Biology* (Cambridge, MA: Harvard University Press, 1996).

Ruse, M. *Science and Spirituality: Making Room for Faith in the Age of Science* (New York, NY: Cambridge University Press, 2010).

Sagan, C. *Broca's Brain* (New York, NY: Ballantine, 1979).

Sagan, C. *Contact* (New York, NY: Pocket Books, 1985).

Sagan. C. *The Varieties of Scientific Experience: A Personal View of the Search for God* (New York, NY: Penguin, 2006).

Sagan, C. *Cosmos* (New York, NY: Ballantine, 2013[1980]).

Scharf, C. *The Copernicus Complex: Our Cosmic Significance in a Universe of Planets and Probabilities* (New York, NY: Farrar, Straus and Giroux, 2015).

Sessarego, C. '"The Moon's a Balloon": Hot Air Balloons and Airships in Speculative Fiction' (2020) 168 *Clarkesworld* <https://clarkesworldmagazine.com/sessarego_09_20/> accessed 5 July 2022.

Sobel, D. *Galileo's Daughter: A Historical Memoir of Science, Faith, and Love* (London: Bloomsbury, 1999).

Stiles, A. 'Literature in "Mind": H. G. Wells and the Evolution of the Mad Scientist' (2009) 70(2) *Journal of the History of Ideas* 317–99.

Swedenborg, E. *Other Planets, the Portable New Century Edition*, trans. G. F. Dole and J. S. Rose (West Chester, PA: Swedenborg Foundation, 2018).

Tipler, F. J. 'The Anthropic Principal: A Primer tor Philosophers' (1988) 2 *PSA: Proceedings of the Biennial Meeting of the Philosophy of Science Association* 27–48.

Traphagan, J. W. 'SETI, Evolutionary Eschatology, and the Star Trek Imaginary' (2021) 19(2) *Theology and Science* 120–31.

Vainio, O.-P. *Cosmology in Theological Perspective: Understanding Our Place in the Universe* (Grand Rapids, MI: Baker, 2018).

Wells, H .G. 'The Man of the Year Million: A Scientific Forecast' *Pall Mall Budget* 18 November 1893.

Wells, H. G. *The Time Machine* (New York, NY: Penguin, 2005).

Wells, H. G. *The World Set Free: A Story of Mankind* (New York, NY: E. P. Dutton and Co., 1914).

16

The law of Mars: The problem of violence mitigation in the development of extraterrestrial political institutions

Ethan Morales

1. Introduction

What is different about space?

Part of the methodological approach of the social sciences is generalization—the search for insights into human interaction (whether it takes the form of governance, economics, or culture) that transcend the individual. The core question is: how do humans interact with each other in aggregate?

Circumstances and context matter significantly in this inquiry, but not all context is equal. The institutional background of an interaction is incredibly important, but other elements less so. If space settlement is merely a shift in aesthetics, where politics is now occurring on a red planet and not a green one, then this volume should not exist. Politics would be as usual, and all the previous insights of political science would still apply, as people are still people; we would probably not need a whole new political science if Antarctica was ever settled. The allure of considering novel constitutional designs or institutional frameworks would then derive from the novelty—the ability to work on a blank canvas to pass policy solutions to problems we would love to solve at home as well. Space is not contributing anything in such a consideration, other than being a fantasy location where we are free to design a government from scratch exactly as we want it—a continuation of the venerable tradition of Utopia.

Space is often posited as a solution to future problems like nuclear war or over-population, and the technological challenge of arriving is presented as the key issue (society will seemingly just work itself out as it does on Earth). Yet, in the future history of a space settlement the actual arrival will be a small note at the beginning of a textbook—millions if not billions of lives will have to live in the new society that follows (eventually far outstripping Earth if multiple planets are settled), and as such there is an imbalance between where attention is most concentrated and where attention needs to be concentrated.

Space is a real place, and it may be settled within a relatively short time period. What if there is something legitimately different about how politics will operate in space? What if standard political science cannot apply? What if a lack of consideration of this question, and the resulting lack of solutions to important problems, dooms those that settle there to some terrible outcome? The number of lives potentially impacted, and the low cost of consideration now, creates a modern Pascal's wager

Ethan Morales, *The law of Mars: The problem of violence mitigation in the development of extraterrestrial political institutions.* In: *The Institutions of Extraterrestrial Liberty.* Edited by Charles S. Cockell, Oxford University Press. © Ethan Morales (2022). DOI: 10.1093/oso/9780192897985.003.0017 that justifies a deep consideration of space politics (despite how ridiculous it may sound to those who only engage with space through fiction).

This isn't to say that all political questions will be decided for the future settlers in advance. That would be both impossible and immoral. Rather, those settlers need to be equipped not only with the technical tools needed for their survival but also with a conceptual toolbox that can allow them to fashion proper policy responses to problems that arise. Ideas and institutions are just as important as any other invention of humankind. To prepare them, we need to consider what those problems might be.

So, what, if anything, *is* different about space?

The social sciences are premised on a generalizable human experience. We are able to find insights into how humans interact because humans, in aggregate, act in similar ways. Perhaps the most important insight is that they are bound by the incentives around them. These incentives may arise from the technology they have access to, the ways in which they are able interact with other humans, the extent to which they are able to use force against them, the environmental constraints around them, and the cultural, economic, and political institutions that bind them. The alternation of any of these is a change that requires a new analysis.

Fundamentally, a civilization that develops on Mars will be just that—a civilization. It will be just as concerned with the pursuit and distribution of power as any other society. It will be composed of people—people that have the same cares, wants, and incentives as humans everywhere have. As such, a Martian society is able to be studied in the same way political science studies modern day civilizations. On Mars, or anywhere that people congregate, all of the old problems of human political organization will still be present. This chapter and this volume will offer no new solutions to the questions of old, questions such as those concerning how a democracy should be properly constructed and defended.

However, Mars is not just another landmass. It carries the traditional issues as well, and those will need to be approached within the existing political science tradition in the same way, resulting from the same academic journals, as on Earth. However, it also shifts the calculus of political science in dramatic ways. The situation and environment of a Martian settlement is sufficiently different that it brings up whole new questions about the nature of human society that deserve serious analysis and consideration. It is, quite literally, a whole new world for the social sciences.

Human behaviour provides the outer edges of human political interactions. Whether we have a human nature or not (or whether there are multiple human natures), there is still a range of behaviours that humans typically fall into. We are affected by identities, incentives, interests, and institutions—the last of which often defines the first two and, to an extent, the third. How we identify ourselves affects how we relate to others and to our place in society; the incentive structures around us influence the choices we make and what our preferences are; what we identify as our interests shapes the list of choices we face; institutions guide our entire lives and shape our identities and the incentives we are faced with. As humans, there are baseline behaviours that we follow on average—we prefer gains to losses (though who is gaining and losing, and by how much, matters). We tend not to be suicidal, barring extreme circumstances. We tend to enjoy interacting with other humans, and often do so via the easiest means available.

What makes the technologies that we consider 'emerging' (such as artificial intelligence, bioengineering, and of course, space travel) so important is that they alter those baselines of human behaviour. By shifting our identities, incentives, and institutions, they shift all of political and social reality in really deep, fundamental ways. We can certainly see how this might apply to artificial intelligence or bioengineering, technologies that will change the society we know, but we can see proof of such impacts when we look back at 'emerged' technologies. For example, the Internet so dramatically changed the way humans interact, and thus how political ideas are created and spread, to the point where we legitimately can no longer undertake political science studies regarding a modern society without including how the Internet shapes everything. The same holds for the Industrial Revolution.

The successful settlement beyond Earth is another epochal event alongside the Industrial Revolution and the Internet—one of those moments in time still visible on the timeline on the greatest possible macro scale. Like with other emerged or emerging technologies, the way politics functions in its aftermath will be different. Old rules will not apply, and new thinking will be necessary to make sense of the changes. Following both the Industrial Revolution and the arrival of the Internet, we have done that work in order to respond to novel challenges. For artificial intelligence (AI) and bioengineering, we are attempting to do that work far in advance in order to pre-empt those challenges before they arrive. Space exploration falls into that latter group, yet as a field it is not directly comparable. It will present novel political challenges by altering the way politics is conducted and the incentives that act upon political actors, but it is potentially much more imminent, yet it also has the least amount of theoretical work already done.

Space exploration is different, however, from those other technologies. Primarily, this is because the technology itself doesn't shift incentives that might lead to unusual challenges when placed on the meta-structure of baseline human behaviours. Rather, space exploration places humans in a condition that is dramatically different from Earth. In particular, the lived environment becomes significantly more constrained and limited. All vital goods must be produced (including oxygen) and failure of those systems means death; likewise, the outside environment kills you upon exposure and guards against that possibility must be constantly maintained.

The environment changes, and with it changes the incentives that people face. The nexus of incentives that one faces in an extreme environment is unlike anything on Earth. At the most basic level, consider how usable force is in such a constrained and fragile environment. These are completely new and unique incentive structures, ones that probably couldn't be replicated on Earth without severe ethical violations (such as mandating that destruction of public infrastructure will result in the instant death of all members of the society). As such, these new incentives will cause humans to act differently. The baseline of human behaviour, acting in order to maximize interests and influence by institutions and identities, will lead to different decisions in an extreme environment due to the dramatically different incentives. These will have both positive and negative consequences for politics. Those negative consequences in particular should be explored and understood long before we arrive on Mars—not to write any laws or constitutions for the Martians, but to inform them and make sure that they don't get trapped in a terrible scenario that cannot be effectively changed.

Imagine if the first creators of the Internet had access to the wealth of social science research about the Internet that we have now. What would be different?

2. The Changes of Space

Politics is not a simple function of environment, but the problems that politics attempts to solve through collective action are often intertwined with the fixed conditions under which a community lives. Likewise, the range of possible systems that can form in response to those problems, and the shape, nature, and limits of such systems, are influenced by external factors, including environmental constraints. On Earth, this is not usually a necessary observation, as most human environments are similar enough in the fundamentals (such access to breathable air) that differences in environmental conditions only result in tangible changes to specific policies (such as how water is managed), and not how the system itself forms such policies (the question of how the power to determine "who gets what, where, and why" is distributed, which is the barebones structure of a political system). Beyond Earth, that assumption does not hold. The environment's alterations of incentives shifts from the scenario to the systemic, and creates new boundaries on the types of communities we can form, not merely the actions those communities can take.

2.1 Space is death

Consider what it would be like to live in a functioning Martian settlement. Past the science fiction and the majesty of exploration, what would day-to-day life be like for an average individual who was either born on Mars or arrived there once the settlement advanced enough to be more than a scientific expedition?

The reality is that if we successfully travel to Mars, it will be to the most dangerous place that we have ever lived. There is not a single environment on Earth that compares. In order to even match the lethality of Mars, you have to pick and choose environmental factors from the most extreme places on Earth. However, there is nowhere on Earth that can be as lethal simply due to the fact that in all areas on Earth where a society can be built there is always air. The default state for life on Earth is life—if an external stimulus or eventual lack of supplies doesn't kill you, you can simply continue being. The default state in extremis is death. An environment must be meticulously created and maintained, forever, in order to accomplish the simple act of living, and who has the power and responsibility of such creation is an important question.

Ignoring the lack of oxygen for a moment, the lack of air means a lack of air pressure. This means that your blood would boil even at ambient temperature, as the boiling point would be different. Speaking of boiling, if you somehow survived the lack of oxygen and the low air pressure, you wouldn't survive the −80 degrees Fahrenheit nights.

All of this requires protection from the outside. This is one axis of the dangers that extremis poses. Consider its implications on day-to-day life, outside the constant fear it might generate. What if a mistake caused a minor electrical fire, which in turn caused damage to the thermal regulation of the habitat? The population freezes to death that night. What if someone accidentally puts a weak bolt in an airlock door and it can't close properly? The whole settlement, or at least a significant part, asphyxiates. Thus far we have been considering benign freak accidents. But what if you wanted to kill everyone? What if you were a deranged individual that sought to harm others? How easy would it be to simply pop a hole in the wall of the habitat at a vulnerable point? You could use simple tools to do it. What if you simply opened all the airlock doors? Additionally, similar results can come from unintentional human action. What if a fight somehow leads to observation glass being shattered?

Whilst safeguards can be made, accidents resulting from human conflicts, assuming for a moment no weapons or firearms, can kill significant populations. Intelligent and premeditated human action, aware of the presence of safeguards? That could kill everyone. No safeguard can protect against all possible threats because the environment is so dangerous and the settlement is so vulnerable. The diversity and simplicity of ways that an individual could destroy a settlement's external protection far surpasses ways to defend those protections.

Consider an additional axis of the extreme environment, not the external threats but the internal ones. Mars has no naturally occurring water (at least, not in large enough quantities as far as we know), air, or food sources. All of that must be produced. However, remember that you are doing such production thousands of kilometres from Earth. Relying on Earth for supplies in an emergency will not work. Reliance on Earth would additionally be likely to mitigate one of the key purposes of space settlement—to create a backup in case something goes wrong on Earth.

Construction on Mars is supremely difficult, and the space is limited, but you must find a way to generate those resources there, primarily relying on recycling those resources once you have them. Those systems would be incredibly fragile. A quirk in the system or the coding could lead to a harvest being ruined—one harvest being ruined means everyone starves. A burst in a pipe leads to significant amounts of water being lost that cannot be recovered. One of the external events before, like a broken airlock, might not kill everyone, but it could do enough damage to a vital system to kill everyone slowly—a frozen pipe means death by dehydration.

What about human agency applied to these systems? These complicated systems of production will be vulnerable to attack. A series of vents closed up, a water tank punctured, a couple of pipes broken, a few gears smashed, a crop of fertile land trampled or covered in chemicals, or an oxygen tank exploding would be the end of the settlement, and a gruesome one at that. The centralization of these systems, due to the incredible difficulty of producing things in space, means that the very systems that keep humans alive also put them at risk. A system that recycles all the water in the settlement is vulnerable to poisoning with chemicals. An air regulation system, required as all oxygen must be produced, could be misused to remove oxygen or raise the level high enough that a single spark could set the whole settlement alight. As the fictional astronaut Mark Watney remarks in the book *The Martian* regarding NASA safety systems, 'they're designed to work against technical faults, not deliberate sabotage (bwa ha ha!)'.

Space is life, in the sense that travelling to Mars may be an effective way to guarantee that we will survive significantly longer into the future. But we must always remember, despite all the techno-optimism of science fiction, that space is also death.

3. How Space Constrains

There are several constraining factors that any extreme environment creates, but for the purposes of this chapter I want to focus on just two: containment and fragility.

3.1 Containment

Containment of an entire society, without the 'containers and contained' dynamic resulting from human malice but rather containment from environmental indifference, is not easily paralleled in the historical record.

An analogy to prison is fair, if not in terms of punishment but in terms of the practical experience of humans. An extraterrestrial settlement is a prison of sorts. The land space available for humans to move around and interact with others is dramatically smaller than any space we have tried to build a full society before. On Mars, outside is death, and as such, the idea of the outside itself shall die. This is a society with no open space, where flight is rarely a reasonable response to threat or challenge. That last part is especially important—conflicts that arise cannot be avoided and distance cannot be built; they must be dealt with, one way or another. That applies to interactions between people and between the people and the state. What is conflict like in a contained world?

Note also here the idea of construction. This world is not self-creating. In the absence of intention, Mars is a barren and desolate wasteland. That contained nature of the human component of extraterrestrial environments will hold so long as that wasteland cannot be permanently converted into a safe environment, and such safety that does exist is human made. The walls need to be built and maintained. Extraterrestrial settlements are constrained because their physical space is contained, flight is less of an option for most conflicts, and that physical space itself is not a naturally occurring phenomenon—it must be created and maintained by somebody.

3.2 Fragility

Fragility is the state of nature for any extraterrestrial settlement. Indeed, this may be the most fundamental dynamic of any extreme system. Every day is a constant balancing act of the highest stakes. A settlement is a tiny bubble of life within an ocean of annihilation, and the slightest error could cause the whole system to collapse. A window breaks and the settlement depressurizes. A wall tears or breaks, and oxygen is evacuated from the area. Temperature regulations fail and the population boils or freezes. Food production fails, or even underperforms, and the population starves. Water production, storage, or reclamation bursts, and the population dies of thirst. Atmospheric stabilizers fail and the population chokes to death. Atmospheric stabilizers overcompensate and the population dies of oxygen poisoning, or a single spark causes the entire settlement to explode. An airlock door is all that stands between a determined person and major destruction.

Fragility in this case is not the mere fact that life is dependent on a series of complicated systems where a break anywhere in the system means death (though that is true). Rather, it refers to how easy those breaks are to make. A city on Earth might be considered to be in a 'fragile' system if you consider how a bombing campaign would cause it to collapse. However the difficulty and cost of running a bombing campaign is on a different scale than poking a hole in a wall, especially how even in that extreme example it takes a campaign, not one bomb (depending on the bomb), and it is still not likely to wipe out the entire population. On Mars, there are several systems on which life depends, most importantly the physical architecture, and irreparable damage to any of them is enough to do devastating damage. There should be redundancies and ways to separate the systems from each other, but even if the damage isn't existential for the whole system, it is absurdly easy to do a lot of damage to parts of it.

It simply isn't possible to devastate a town on Earth as completely as the smallest piece of sabotage on Mars, other than a nuclear bomb—and even then, the cost and difficulty of doing that damage is way higher. On Earth, you need to spend billions on developing nuclear weapons, mostly likely with a state apparatus, alongside all the other complications that might reasonably come with nuking a city. On Mars, you need a determined person (or group) and basic tools. The same level of destruction can be made for much less cost or difficulty—that is the nature of a fragile system.

That of course discounts the natural fragility—those systems could independently fail, and life itself depends on those systems not failing in order to continue. Ensuring that those systems don't fail is a technical problem, and one that a lot of time and effort needs to be dedicated to solving.

The challenge I am addressing in this chapter is how to design political institutions and governance with that ease of destruction in mind. How does a society function in a fragile system? How can a government run when individual actors are capable of such destruction? The effects of isolation and containment can be explored by historical examples, but not so for a fragile system (with some minor exceptions). Here I hold constant the idea that actors may want to sabotage such systems deliberately for their own gain, and how that reality shapes the political institutions around them.

An extraterrestrial settlement is fragile because there is a constant, non-negligible and non-removable chance of death, massive destruction, or societal collapse, which can occur both by random acts of nature and by intentional human triggering. The size of this risk can grow and decrease, but it can never be removed entirely whilst the environment remains extreme.

4. The Problem of Violence Mitigation

Space settlements will be filled with thousands of completely average people. Within the boundaries of normality is the possibility for there to be actors who wish to use violence to achieve certain ends. On Earth, the role of violence is fundamental to states. States are identified within the classic formulation as entities with a monopoly on legitimate violence in a given territorial space. States use the threat of physical violence to make state commands enforceable (violence gives states the ability to punish, and thus coerce, conduct), and they use their control over violence to constrain bad actors who wish to do the same for their own ends or seek to challenge the power of the state.

Yet what happens if violence is not as easy to use? What if the risks of any violence are dramatically higher? How can states compel behaviour in such a scenario, where the risks of any application of violence (including a violent response from the population) outweigh the benefits of any specific coercion of behaviour? How does a state deal with a population where any individual has the equivalent destructive power of the state? How does society function with a state that is more able to inflict damage than ever before? Is the resulting system dictatorial, with the vulnerability of space exploited to give the state unimaginable coercive power? Or is it anarchical and self-destructive, where any state pressure is defanged due to the inherent instability of ruling a population where revolt could kill the entire project?

This is the problem of violence mitigation. In settlements located in extreme environments, the barrier, in terms of ease and accessibility, of large destruction, devastating acts of violence, and mutually assured destruction is dramatically lowered, down to the level where individual actors possess the capacity to inflict damage relative to the size of the settlement in a way that is proportional to the capacity of individual states to inflict damage in a non-extreme environment. This state-level violence is brought down to the level of the individual, and this alters the dynamic between individual people, between people and the state, and between states in terms of conflict resolution. This problem is the fundamental dynamic, and the fundamental change, of space societies.

The analysis in this chapter will not be a complete reckoning with this idea. Instead, this chapter will narrowly focus on how this issue impacts how states function, namely in the area of how political dissent is channelled without collapse, and how states are able to govern effectively in such an environment. This approach will be macro in scale, leaving the specific policy details aside and instead focusing on regime type— whilst leaving other issues, such as how to constrain irrationally violent individuals (like serial killers) for another time. Politics is a constant give and take, a negotiation amongst all of society as to how power is distributed. The sole question of this chapter is how that negotiation is altered when violence's utility is altered in a fundamental way.

As mentioned, a useful way of defining a state is an actor with a monopoly on the use of legitimate violence across a geographic area. In other words, a state holds power over a region when it is the only actor with the rightful ability to deploy force. All state actions require some level of compulsion somewhere down the line, if only in terms of collecting funding for an otherwise benign action. In modern societies, where the government should act is hotly debated, but it does so under the backdrop understanding wherein it is generally understood as a principle of policy-making that the state does not need to be concerned when designing its penalty structures that an individual or even groups of dedicated individuals are capable of destroying the state if they were so dedicated (when designing a crackdown on speeding limits, one does not typically consider as an element of that policy design the possibility that a coordinated group of speeders could bring down the government). The focus is solely on properly aligning incentives to encourage proper behaviour, mitigate the harms of bad behaviour, and do so in a way that comports with moral considerations of justice as well as the political constraints of the system. On Earth, force typically runs one way—the state can act on individuals, but other than targeting individual officers the average subject of state power cannot act on the state as a whole. That does not hold in extremis—the directionality of force develops from an imbalance in power and scale. The amount of damage a state can do is dramatically more than that which a citizen could do, on Earth, and each citizen is a small and isolated target, whilst the state is an amorphous leviathan. In extremis, scale matters much less due to the dramatically increased power of each individual.

5. State Limitations on Standard Police Force

On Mars any use of violence slightly increases the risk of a systemic collapse, both as an accidental result of the deployment and escalation of violence, or because such an act inspires such an escalation from the accused or those that identify with the accused and fear policing, or because the act inspires a general political pushback of widespread instability. Such is a principle of nature in extremis, and it is not concerned with the morality of the act. All of a sudden, state action now risks more than just the accused, in almost every case. Even if a collapse is not caused, the usage of police force inherently runs such a risk with the lives of others. A missed shot means a breached airlock. Police action is no longer isolated to one individual, and as such the morality of the action needs to be judged not merely in terms of whether the criminal deserves force applied to them, but whether the criminal's actions and the need to stop them or bring them to justice justifies running a risk with the rest of the nearby population, married with whether letting the criminal off would lead to more criminal activity that is either more violent or provokes a more dangerous violent response.

To simplify, the increased dangers of violence as used in normal policing raises the costs of enforcement for every crime across the board, making the deployment of state power more difficult. Since the risk of collapse is totalizing, there is not much of a distinction between violence employed to resolve a greater or lesser crime—when the costs of enforcement actions can no longer be effectively differentiated with the scale of the crime, and the cost runs to the extreme point, the ability of policy to effectively coerce individuals through conditional responses (such as a greater punishment for robbery + murder than just robbery) begins to go haywire, and the average criminal may be incentivized to act as if the costs of all criminal acts are roughly equivalent. Additionally, the average civilian has the power to transition debates about law and enforcement into the realm of violence in a much easier way, and this can effectively coerce the state in a manner not seen on Earth, raising the risks of enforcement as well as the power of civil society.

Political arrangements on Earth assume a power asymmetry between the people and the state, where the goal of political reform projects is to constrain the massive power of the state and bring it under the control of the people. In extremis, although the state is massively more powerful, so too are the people, and their increase in power exceeds the increase to the state as they end up roughly equivalent. Given the people are so powerful, and the state so capable of destruction, how do you manage the decision-making process in such a world? How does conflict work?

6. Conflict in Extremis

What dynamics are in play when conflict arises? Whether it is between two people that cannot resolve a serious dispute, between the people and the state over a major public issue, or between states locked in the standard gauntlet of international competitions, the dynamic should be similar. That is because if everyone has the ability to do the same amount of proportional damage, which typically means damage in excess of the entire population of the settlement so comparative power (usually the arbiter of all conflict) is not really relevant, then the dynamic should be the same between all parties, regardless of whether it is two individuals or groups of individuals calling themselves a state.

Specifically, all major conflicts in extremis match the game theoretic concept of brinkmanship. The idea behind brinkmanship is that there are two or more parties that are attempting to coerce each other to do something. Whether it is between people or states, both parties want the other to concede in whatever their dispute is. The way they attempt to prevail in the competition is by threatening to impose a cost so severe that it outweighs any and all possible gain by refusing. On Earth, this is seen in nuclear arms races. The threat of a nuclear war is a higher cost than anything one could lose by giving in. On Mars, any and all violence contains a proportional amount of risk equivalent to a nuclear bomb. This perhaps does not hold in person-to-person conflicts involving only two people (and other small-scale conflicts, depending on the physical infrastructure of the settlement), such as a domestic dispute, though it is certainly present in extremely emotional conflicts and there is always a risk of accidental damage. However, it is incredibly high in person-to-person conflicts amongst groups, such as rival political parties or two groups of protestors. It is especially prevalent between the people and the state, or between two states, as there are few acts of violence that can be done effectively on a group of individuals that does not risk collapse.

Why doesn't every conflict end with the other party giving up when the threat is made? The problem is that the threat is inherently incredible. In extremis, and in the case of nuclear war between superpowers, such a threat does not act upon one side alone. It imposes the same cost on both parties, or at least a similar cost. The problem is thus that you have no incentive to believe the threats because just as the threat poses a higher cost than any gain you might get by refusing, so too does the threat pose a higher cost on the one who threatens the harm than anything they might gain from the dispute.

The issue we thus face is the fact that mutually assured destruction makes no logical sense, and yet the concept of brinkmanship is an empirical reality. How can such conflicts happen?

The key is to think of brinkmanship as two people on a cliff bound together with rope. If either of them falls off, they both fall to their death. Such a scenario reflects mutually assured destruction, and gives the metaphor existential stakes.

Now, if the cliff was a sharp drop, with the only possible movement towards the edge a push off, then the threat would be legitimately incredible, as any movement towards the cliff would be instant death for both parties. However, that is not the way to think about brinkmanship in nuclear politics, and it is certainly not the way to think of brinkmanship in extremis conflicts. In extremis, violent acts are not guarantees of destruction. They are simply increases to the baseline risk, a concept explored further shortly. By initiating violence, you do not instantly kill everyone but rather you create an atmosphere where accidental or intentional damage to critical infrastructure is more likely. Certainly, individuals and states can directly threaten an existential attack—the important note is that it doesn't necessarily have to be such a threat as small conflicts can increase the risk of death and move society closer to that point.

Thus, don't picture brinkmanship as a sharp cliff but rather a rolling one. A movement a step down the cliff doesn't necessarily cause you to fall, but it certainly increases your risk of falling, and every step could in theory be your last. Both players don't know which step would be the last, but they assume that risk when they take a step. The analogy for extremis conflicts is that by one group attacking another group, they may not intend to collapse the settlement, but they are increasing the risk something will go wrong, even if they are careful specifically to avoid critical areas. That group might retaliate strongly, and the risk compounds.

If the cliff is rolling, smooth not sharp, then it suddenly becomes possible to use your opponent's fear of going over the cliff to get them to back down, even if you would not want to go over yourself. You simply need to be willing to take on one more step then they are. As you are bound together, each step increases the risk equally for you both. Jumping off is something neither of you are willing to do, which means at some point there is a step you are not willing to take. If your step is one further than your opponents, then you have a credible threat. You are willing to take the step because you are not jumping off the cliff. You are leaving something to chance and are willing to increase the risk higher than they are willing to go. By being more risk tolerant, you can get them to back down, not by threatening to push you both over the cliff but by threatening to take steps that raise the risk to unacceptable levels.

The proper model for extremis conflict is dynamic brinkmanship, contrasted to static brinkmanship games like chicken where both players move at the same time. In dynamic brinkmanship, players alternate in moves that increase the risk of disaster somewhat. The first actor might initiate violence, and the second might respond in kind but to a greater degree. The first player may raise the level of danger by threatening to aim at a vital target, which raises the risk inherently since the other party may call their bluff and force them to act.

The players don't actually need to know how many moves they are away from disaster to calculate the risks they are running. All they need to know is an abstract idea of how much risk they are adding to the system with each escalation, as the collapse point can be some far point that the risk asymptotically approaches.

What the players need in order to solve these problems is an understanding of both players' resolve (a game theory concept). Resolve is (roughly) the maximum risk they would be willing to run.

There are two ways such a conflict can be resolved.

First, in theory if both players are aware of each other's resolve (which might be possible in this setting if, for example, the state was much more resilient to such threats than an average person, or vice versa if the person had 'nothing to lose'), the conflict would be over at the start. If you know for a fact that the other party has more resolve than you, there is no reason to run the risks at all if you will eventually give in anyway. This scenario may properly be termed 'the tyranny of the resolute', for the entity with the most resolve has all the power, including the power to compel and constrain behaviour, and the other party is required to follow. If a state has greater resolve, you have an absolute dictatorship, likely to be worse than anything on Earth given the absolute power a state might possess over air and water. If citizens possess this resolve, then you are likely to have a form of anarchy, or at minimum a state without the ability to compel behaviour. The types of conflicts that have this dynamic that are especially relevant here are political conflicts between the people and the state, and this makes conflicts of that nature tyrannical—as they are dominated by the party most immune to such threats, with little ability to shift the balance. The power symmetry of extremis may actually be quite helpful here in mitigating the risk of tyranny, despite the seemingly obvious power of the state.

The alternative is not much better. An authoritarian case is much more stable. The alternative is that the parties cannot guess the other's resolve, or only have a rough (false) approximation. Without knowing resolve, the conflict runs through cycles of escalation before the resolve point of one of the players is hit. That point might be at the very beginning of the game, which is sad for that actor but good for the settlement. If it is higher, or if the players believe they know the other's resolve and that theirs is higher, it could continue for many rounds of escalation. Each round risks extreme damage to the settlement; it puts the end of the future on the table. Any of those risks could actually be the one that ends the settlement (something neither party wants but is impossible to fully predict). Any one conflict might be lucky enough to go many rounds until the resolve point is hit and the game ends without issue, but every major conflict will have tyranny of the resolute dynamics. A settlement has to survive, through luck or whatever physical barriers it can impose (with physical durability improvements being one of the only non-political tools available to increase the likelihood of the settlement weathering the escalation), through an infinite number of such conflicts for the indefinite future (pre-terraformation). Can any society survive that? What incredible luck would be required for that?

If resolve is known, then every conflict can be dominated by the resolute. Tyranny like this, probably held by the state, is extremely stable. Luckily, resolve is hard to determine in complex environments, but that absence creates new problems. If people think their resolve is greater, then a series of risks will be run in an escalating conflict until a resolution is reached. However, society may not survive until that point and it must do so through an unknown number of such conflicts. A society such as this is inherently unstable. Stability depends on information.

We can see an incredibly important inverse relationship here. Efforts to make a society less vulnerable to tyranny, such as through creating redundant necessary systems (like oxygen generators) that reduce the power of the state, make it increasingly unclear who has the higher resolve and this increases the likelihood of escalation and collapse.

What this consideration has not yet reached is how this maps on to institutions, including how institutions shape this problem, and how the knowledge of this problem impacts the institutions people choose.

This dynamic is the curse of extremis. It, more than the inhospitable environment or the technical difficulties, is the greatest challenge that humankind must overcome if it wants to settle beyond Earth, at least before terraformation and in anywhere where terraformation is impossible. In one case we have absolute tyranny, but luckily that is unlikely and should be relegated to only a few clear conflicts. In the other case, we have a constantly escalating existential risk. If tyranny of the resolute conflicts cannot be effectively resolved, then life on Mars will be hell, and unlikely to last long at all. A way to effectively mediate such conflicts is required.

It is *the* problem of space settlement.

A third party method of arbitrating disputes is absolutely required in space. No settlement will last without it. But that creates two problems: first, how can a third-party make its decisions binding? And what institutional design for such a party, which in case it is not evident is equivalent to a state, is the most stable?

And to reach those questions, we need to have an idea of risk and stability.

7. Risks and Collapse

In this chapter I frequently use the term 'collapse' to refer to the end of a settlement, as depicted in the various ways death can arrive in extremis that I explored earlier. This is not necessarily a specific event. Certainly a collapse is a moment in time one can point to, but collapses are inherently unpredictable. A large violent clash, such as a riot or a civil war, could, by chance or careful action, avoid a collapse entirely, despite the wide-ranging use of violence. Likewise, a small altercation, such as a drunken brawl too close to a vital agricultural resource line, could cause the settlement to collapse despite the relatively limited extent of the violence.

Rather, state design should be thought of in terms of stability. Every settlement, no matter how well designed, has some level of inherent risk of collapse. This baseline collapse risk can be dramatically altered by stronger walls, redundant systems, or predictive methods, but it can never be eliminated. There is always some chance of a freak event—an unexpected meteorite strike or a technical glitch in computer software that ruins a vital crop yield.

The baseline risk is environmental in nature—it grows and shrinks in direct proportion to the threat of the environment and the defences of the settlement. It exists without human presence or action (though presumably a human is needed to build the settlement and man some of the redundancies). Humans can lower the baseline risk, but it is not their presence that does so, but rather the technical solutions they put in place to mitigate risk. The presence of humans inherently raises the risk of collapse to a level above baseline—with humans comes the possibility of human error, which can cause a collapse outside of the ability of the non-sentient structure to guard against. This includes the errors that can occur due to miscalculated violence—it is rare that a conflict intends to kill everyone, but a violent clash in space can do just that. Yet the risk does not rise merely because of human error but also because of human malevolence, which must be factored in, especially as generations pass. A settlement can be so well designed that the possibility of collapse due to some freak event becomes almost negligibly small—yet nothing has yet been built that dedicated enough humans cannot destroy if they wish. Intentional destruction must be kept in mind.

Humans are constrained by the incentive structures and institutions around them. These funnel human action through channels of possibility that are wider or narrower depend on the institutional structure. Thus, different state structures affect the extent to which a settlement is prone to collapse and can dramatically limit or enhance that additional human risk. How states manage the human element is equivalent to their stability, given that the presence of humans by default increases risks above the baseline. Some state forms may keep the additional risk to merely a few percentage points. These would be the most stable state forms in extremis. This stability could come from its ability to effectively limit certain actions, or by the way that it doesn't aggravate certain events and cause certain actions. An unstable state is the reverse of this, where due to its institutional structure the human factor aggravates the risk to much higher levels. One can view this as directly analogous to violence on Earth—there we can see that different institutional designs (be they state, city, local, regional, private, or charitable; economic or political) can lead to differing levels of violence.

This is nominally independent from the question of liberty. Here, the most stable state is the one most likely to survive the longest period due to it constantly running a lower risk than other state forms. It could, by chance, collapse first—such is the nature of probability. But that is not the statistically likely result, something that would be seen the more settlements (and thus more experiments) are made. If one were to create thousands of settlements, and distribute state institutional design randomly amongst them, the 'stable' institutions would last the longest. I address the topic of stability because it is an empirical one. Though it relates to human action, it requires no moral statement other than 'I want the settlement to exist'. A stable state may not be a free state, and tyranny may be more likely to endure in space than democracy. Now, in some cases death may be preferable, but I think it is fair to assume that for most settlers they unanimously agree that the settlement should continue to exist; for the political questions of liberty to be answered and fulfilled, there must be people. It does not mean that the most stable form is the most desirable—one may be perfectly willing to accept a slightly higher risk of collapse in return for better living standards and more freedom. Yet as we shall see, the questions of liberty (once operationalized into something empirically comparable, regardless of the moral hangings placed upon it) are not that separate from ones of stability.

The law of Mars is the law of stability—over a sufficiently long period of time, states that are stable in the sense that they mitigate collapse risks the most are the ones that will survive in an extreme environment.

Although a settlement can collapse from a wide range of things, including criminal action or an inability for vital goods to be produced, for this chapter I am going to focus on the risk from political conflict.

Governments change and adapt over time to new problems and new outlooks—in modern democratic states, it is the voting process and the representative structure through which this change flows. Nominally power derives from the people, and the government is meant to be representative of that. Typically the desire for change is mediated through an institution, but to cause such a change to be reflected in the institutions, or when those institutions fail or the issue at hand is so egregious, the people dissent. Dissent is the embodiment of disagreement with the government or a particular policy direction, which takes physical form in non-violent protests and boycotts on the peaceful end, and riots or rebellion on the violent end.

Dissent is also important for authoritarian states. Such states attempt, often violently or through inhuman pressure, to stifle dissent, and inversely dissent is often the only recourse of the people who seek change in a system that contains no formal methods for enacting that change.

On Earth, the discourse around dissent is varied and complex. On Mars, it takes on a new aspect, some of which has already been discussed. Dissent is now high risk. Any mass public mobilization, whether peaceful or not, contains a latent element of disorder and the potential for escalation. There are two ways to approach this: one might want to preserve the ability to dissent akin to Earth in an extreme environment due to the fundamental human rights involved, or one might approach dissent concerned about making sure the settlement survives. These can lead to the same policy recommendations, but right now we are focusing on the latter. How do brinkmanship dynamics intersect with dissent?

Dissent is a high-stakes game, which can discourage people from playing. The people can use the threat of dissent in a similar way as a nuclear weapon, threatening its deployment with the understanding that there is a risk of accidental mutual annihilation, in order to get policy concessions that a state might otherwise not give. The state might likewise use the fear of a dissent collapse to tamper protests from the population in order to maintain the status quo. In any case, a dissent collapse is a collapse that results from a mismatch between the people's desires and the government's direction, which results in violence. Now, on a surface level this is similar to any other act of violence, but the reason it is listed distinctly is because it has a distinct matrix of solutions. These are solutions related to the political process—the institutional framework to allow for the peaceful change in leadership and the peaceful act of collective decision-making. In other words, whether this is a tyranny of the resolute scenario, or an anarchic collapse scenario, or a mediated and managed system depends on the relevant institutions.

Dissent-based collapse can result from the actions of protestors—a growing civil conflict—or from the state cracking down on the protestors. The key linkage is the matrix of solutions. If the flow of political power is badly designed, the risk of a dissent collapse rises. Dissent collapse also calls for some level of formal proceduralism in social construction—whilst an informal anarchic private system may or may not work on Earth, on Mars we need to be certain that the decisions being made that affect the group (especially those related to the maintenance or usage of protection and production) are done in a procedurally acceptable way. This again calls for third-party arbitration: a state that has an institutional framework for managing conflict in a way that makes it a lower cost for people to join the system and accept the result than escalate by increasing the systemic risk.

In simple terms, anarchic brinkmanship conflicts are no way to live life. Without uncertainty, gained by one party having an absolute upper-hand, you have a dictatorship but you also have relative stability (assuming the certainty of both parties remains). Assuming there is a situation of uncertainty (which is likely to be common), there is room for the existence of a different kind of state: all actors should prefer a safer and more reasonable alternative to running countless brinkmanship contests. A fair, open, popularly controlled, and comprehensible system of third-party arbitration, bought into by all parties, can present that more reasonable alternative. Conceding to such a system for resolving disputes (both amongst individuals and more importantly between individuals and the state regarding the deep questions of how to lead society) risks losing any one individual conflict, but consent is maintained because if the system is legitimately fair (a requirement for everyone to have bought in at all) those loses will be balanced by victories. Such a scenario, with more loss but also more stability, should be preferable to anarchic collapse.

The overall question we will need to resolve is what institutional design is the most stable: is it better to achieve complete stability through some form of tyranny, or is it possible to manage these conflicts in a non-tyrannical fashion? Both require some ability of the state to make binding decisions.

8. The Source of State Power

The tyranny of the resolute and the problem of violence mitigation establishes the need for a centralized authority that can temper violence and coerce particular actions. Such a need exists on Earth, and it is enhanced and made more pressing as the threat from uncontrolled violence becomes existential in a way that it is not on Earth. Yet the primary mechanism of the state—the usage of force—is limited, morally and effectively. In addition, even if a need was not present, authority tends to come into existence anyway—and if we want to avoid the risks that uncontrolled authority presents, we need to understand how that authoritative power develops if violence is less of an option. So where does state power in extremis come from, if force is less able to be used?

There are two other sources of power that can only exist in extremis, due to the contained and fragile nature of the civilization. The new environment that politics will play out within leads to two loci of power in extremis: production and protection. These are the lever points on which an individual or state can act in order to compel behaviour and control society.

I should note at the introduction here that small-scale physical violence will absolutely still be employed, especially in extremely large constructed settlements, contained to smaller issues like theft or other issues of complying with state directives. The important question is how large conflicts, those bearing higher risk, are fought, focusing here on political conflicts over the future of state policy. In other words, how does a state, when large-scale physical violence is not as usable, maintain its authority in the presence of challenges to its very ability to make decisions?

8.1 Production

The production locus refers to the production of the goods necessary to survive, these primarily being food, water, mechanical supplies (to keep the systems running), and oxygen. These cannot be found in an extreme environment, almost by definition, and are difficult to produce. Each good varies in two key areas: the ease in which an alternative system can be established and their ability to be excluded from certain individuals.

Let's start with establishing alternatives. Unlike on Earth, it is not easy to gain these goods outside the settlement, and it is likewise difficult to set up an alternative method of production (especially given the available land area and the difficulties in building food and water production in the first place). A key characteristic of these essential goods is that any one of them gives you power—even if food and water can be produced independently, control over the oxygen flow alone would give you power because the nature of these goods are that they are necessary for survival. Power revolves around the least available necessary good. Oxygen and water are also powerful in the sense that they are goods designed to be recycled, to varying degrees. Oxygen is pumped around the whole facility, and recycled from everywhere. Water is drawn from the atmosphere to be recycled whenever any water passes out of the direct reclamation system, and that water could go anywhere to be recycled—it is intangible and fungible. Since operating an oxygen and water reclamation system requires access to the entire settlement to be efficient (a requirement in some cases due to the extremity of the circumstances), as the goods to be recycled travel freely following their consumption, it is very difficult for an alternative system to be established, especially if those with power seek to stop it—such a system would be difficult and time-intensive to construct, and hard to hide.

The strength of each good varies with the amount that they can be excluded from others. Food comes in the most discrete packages. Each food item is consumed by one individual or divided into smaller food items, each of which is then consumed by one individual. The targeted nature of food, alongside its tangibility, allows it to be controlled more directly. It is easier to refuse food to specific people, compared to oxygen where you could turn off the flow to particular areas, a brutal and effective way of deploying power without physical force, but it cannot be isolated to an individual.

Thus each good in the production locus comes with advantages and disadvantages. All of them give you power since all of them are required, but which is more useful depends on what is needed. There is an inverse relationship between how excludable the good is and how easy it is to establish an alternative system.

It starts with food, which can be denied to specific individuals but does not require the entire settlement to generate, allowing alternative systems to be established easier than other goods (although any production of goods in extremis is likely to require intense effort).

Next is mechanical supplies. Their production is likewise intensive, and the end result is tangible goods that can be excluded. However, their excludability is less than with food, as the benefits of mechanical repair cannot be isolated to help one individual and not help another. Alternative production is possible, as their creation comes in the form of discrete goods that do not require access to the whole settlement, however their utility can only really be applied to the settlement as a whole, with the exception of local machine failures where there is a level of power in denying their repair, but also less strength behind that power since if the problem is only local it is unlikely to be life or death.

Water is somewhat similar but inverted. Its creation is partially non-discrete (as it, for reasons of efficiency, will tend to involve a production unit that involves large parts of the settlement), whilst its utility is partially excludable (it does come in concrete units that can be controlled, but recycled water rapidly diffuses into the environment and can be pulled from multiple locations, including from recycling lines, in a way food cannot)—the opposite of mechanicals supplies. To recycle water effectively, you need access to the whole settlement as the water is recycled from the atmosphere with no ability to control where that water, once in the form of water vapour, goes. You also would want to be able to recycle any water specifically put into a water reclamation system. Thus an alternative production system is difficult to establish (though not impossible if you are very careful—one non-atmospheric water reclaimer could reclaim water placed intentionally within it, but that is a fragile system to depend on). The end result is a good that can be excluded, both to places and individuals.

Finally, oxygen is the most difficult to build an alternative system to produce, but is the most difficult to exclude. Air flows where it wills, and it is nearly impossible to have an oxygen production system without access to the whole settlement. An alternative system would be very difficult to create, especially if someone intentionally acted to stop the creation of such a system. An alternative system may also overproduce oxygen. Yet the good produced cannot be excluded from individuals, only from areas. The power one gets from controlling oxygen is thus a blunt instrument, but one you can be assured will be only yours.

Thus for a would-be power wielder, the benefits of each good need to be weighed, but the key understanding is that control over the production of goods gives you power—this comes not only from specifically refusing access to those goods to people (the totalitarian approach) but also by your simple importance to the survival of the settlement. Any vital good gives you power simply by the fact that they are vital. This is the production locus of power in extremis.

Now, despite these limitations it is certainly possible to create alternative systems for oxygen and water. The question is whether we would want to. A state would be likely to attempt to prevent such a thing, either for its own tyrannical desires or because without *some* way of controlling the population the state has no real power. Violence is less of an option, and control of production may be necessary if we are to have a state at all. Additionally, a large amount of alternative systems makes the brinkmanship resolve values increasingly unclear, thus potentially empowering individual members of the society to challenge the state in a way that risks total collapse.

As an important note here, it may seem inhumane to suggest weaponizing control of production to wield against citizens of the settlement. However, I am assuming as a baseline a universal acceptance of the basic principle that states have the right to use physical coercion to prevent law breaking, which is inherent to the existence of states and associated with a range of state actions most people accept as a requirement of society (such as the usage of state violence against criminals).

Humans in general have been socialized to accept the usage of large amounts of physical force by the state (coming in many forms, from beatings, to shootings, to chemical deterrents) to enforce compliance, and such activities are generally accepted without raising a comparable moral revulsion (though individual usages can certainly be unjust, and one can reasonably dislike the usage of violence whilst recognizing its utility). States have to have *some* way of coercing behaviour, and we have thus far accepted physical violence as the way, which can be used for both good and bad ends. In extremis, that option is much more dangerous and much less available. Weaponizing production is not different in kind, certainly in terms of outcome and possibly also in terms of physical discomfort; rather, any moral revulsion present results from the unusual and uncommon nature of the act, and because human history does not typically require such acts (indeed, denying food is usually solely used by dictators) resulting in the development of human rights norms regarding the cruelty and off-limits nature of the acts. However a Martian state with similar socialization in the opposite direction may regard our easy acceptance of state violence as a violation of a similar norm—*of course* the state can deny food to criminals, but a state could never *shoot* someone. These are differences in human social constructs, and this chapter does not endorse any of them in any direction. Instead, it is purely descriptive; assuming that a state will try to exist, and a state, in order to exist, will have to coerce, how will it do so in extremis?

Returning to the production question, whether we democratize production seems to be a choice between the presence of a state or anarchy. However, that is not the same as a choice between tyranny and freedom. We can avoid tyranny of the resolute style conflicts by having a state with the clear ability to enforce, but there are also ways to avoid that becoming a tyranny. Once we have a state at all, we can consider ways of controlling it—in the same fashion, although violence can be used to enforce tyranny on Earth, we do not call for the state having no ability to use violence as we recognize that a state needs some ability to compel, and living in a world with a state is preferable to the alternative. In this case, we may want to concede the existence of the state, with a much more infirm ability to control (although limiting access to vital goods seems totalitarian in power, it is actually much less direct than normal violence and thus weaker). A monopoly on production of vital goods (everything else, by the way, can be democratized and produced in a market setting) does not arise from economic needs, but political ones, and as a result the key challenge is finding a way to manage that monopoly through the hands of the people.

Here we see a similar idea to the third-party arbitration point with tyranny of the resolute conflicts: for stability, the option presented has to be preferable to a brinkmanship conflict and the resulting risk of destruction. Production in the hands of an arbitrary dictator may not deter brinkmanship conflicts if they cannot effectively deploy it—indeed, they may trigger conflicts. However, if production is controlled by a third party recognized by all actors as relatively fair and bound to popular control through regular and comprehensible processes, it would provide more utility to acquiesce to production being controlled by this entity (a state), accepting the possibility of loss in a given conflict supported by the recognition that a fair process will lead to victories in later conflicts and preferring that production be controlled by such an entity as opposed to any one single actor (as well as preferring that to an attempt to overthrow or challenge an actor with control of production).

8.2 Protection

The protection locus of power deals with the physical integrity of the settlement. It is protection against the latent death of an extreme environment that is ever-present. The dangers of an extreme environment can never be removed, only mitigated, and the extent in which you control that mitigation gives you power.

Before there can be Martians there must first be Mars. Mars exists independent of human action, but Mars as it is does not welcome a human presence. The Mars that we are talking about in this book is an artificial construct—it is the human Mars, which does not exist in nature and must be created. The playing field must be built for the game to be played out. Consider the contrast to Earth. States use force to control a geographic region, a region that existed by default. In extremis, the effective latitude of force is constrained (force is no longer generally usable) and the environment does not exist by default. The state no longer occupies an existing space and holds it—the space must be created before it can be occupied, and when it is occupied, force is less able to be used to defend it. Given that the living environment is created, the human Mars is less constrained by space and the power of geography is somewhat lessened beyond access to natural resources—everywhere equally kills you and the environment must be equally created to hold outside that ever-present death.

There is power here. The entire world will be a man-made creation. That protection against extremis endows great power upon those that design and control it. This comes first and foremost from the existential nature of protection. Without the walls, death is certain. Those in charge of maintaining those walls are given power over those that depend on them, assuming that they are sufficiently removed from the walls that they themselves do not risk annihilation (that assumes a rational leader, but some dictatorial regimes may be willing to risk themselves to maintain power or rather die than lose it, which must be considered). This control lies not necessarily in the purposeful destruction of walls but control of their maintenance—and control of the airlock doors. The targeted destruction of protection will always be a threat, giving additional power to those in control of protection as they can selectively refuse to protect their own protective apparatus, allowing brinkmanship conflicts to play out in contained scenarios.

Control alone does not give power to protection, but also its design. The concept of 'place' itself is state controlled in its entirety. The boundaries of all human possibility is the place in which they can act, which is entirely manmade. There is nowhere you can go that was not designed by someone else. The places you go, the places you work, the places you gather with others—all constructed. The flow of human society is directed, and that design has power. There are legitimate worries about how furthering AI dependency in human society gives the AI a lot of discretionary power to influence human action. Consider the extent of directional power given to those who design every element of human society. One may consider that there will be some aspects of an extreme society designed independently from the rest, which is true. But it won't be independent in the way we view the term on Earth. It has to fit as part of an integrated whole or risk the whole system collapsing, which gives huge amounts of control to those in control of the overall design. The life-or-death nature of design encourages centralization.

The same critiques about democratization apply here. Additionally, it should be noted that a sufficiently advanced settlement may rapidly move beyond the point where either of these can compel—production may be effectively democratized due to market forces and protection may become irrelevant due to technological advances. But at that point, normal state power over violence will return to prominence, and political science as usual can progress.

Additionally, it should be noted that mere settlement design cannot fully eliminate state power. For example, if one built a settlement sufficiently far underground, or protected by layers of shielding impervious to human attempts to dismantle it, protection would no longer be an effective lever of state power. But due to that loss, the role of production escalates dramatically in importance, both as the thing a state might withhold to coerce, and because it is damage to production lines that would be the key risk of brinkmanship conflicts. There is a direct relationship between all elements of the loci of power, and the more elements taken off the table the more the remaining ones grow in value (as they can no longer be effectively counter-balanced).

It also isn't impossible that these two elements of power could be held by separate entities. The dynamics that would resolve from such a 'split state' are outside the purview of this chapter, however we can consider that such a scenario might be simply another brinkmanship conflict—two entities that may escalate conflict if they are unaware of each other's resolve, or capitulate to the other if they are. One may want to divide such powers intentionally, but doing so would in practice mean a single state entity with (literal) powers divided amongst branches or across some other division. Unity is assumed to be wanted to avoid brinkmanship conflicts.

In any case, we are assuming that these fall under a single state here for simplicity—I simply wanted to draw attention to the fact that states can take weird shapes when they rely on things other than violence for authority. The same goes for religious or social methods of control.

So a state in extremis has a monopoly on the legitimate control over production of vital goods and protection, with the loci of power balanced or centralized. This is the entity that can determine the political process. What that entity offers as an alternative to conflict determines what happens. If the entity is tyrannical and offers destruction in response to dissent, and the people do not believe they can effectively coerce it, tyranny results. If the state cannot effectively threaten with these, there exists no state or there is an anarchic risk of collapse. If the state offers a mediated political process, it may be possible for the costs of joining such a system to be lower than the costs of risking escalation.

9. Regime Types in Extremis

We can finally reach a conclusion here on the political question. I want to note two things first.

First, there exist other problems created by the extraterrestrial environment, and needed aspects of state design to respond to those, other than the political issues listed here. For this chapter, I am focusing only on violence and politics. Those other issues, such as constraining violent outliers and criminals, directly impact state design and may also impact the political question (cracking down on violent outliers, such as the criminally insane, impacts the normal political process if you cannot effectively differentiate them from the population). However it simply cannot all be touched on here.

Second, I am dramatically oversimplifying regime type for philosophical reasons. Political scientists are aware that regime type is a spectrum, and there are many variations of democratic and authoritarian regimes with different definitions, but for now I am simply trying to resolve a basic question: given what we know about state power (states have full control over the air we breathe, giving incredible tyrannical power in theory) and civilian power (due to the fragility of the state in extremis, average civilians can cause incredible amounts of damage), and how the risk of escalation and collapse in a system where it is unclear who is stronger necessitates some form of state that can mediate or crush these conflicts, what institutional design is the most stable?

I am going to divide democracy and authoritarianism into two versions each: strong and weak democracy, and authoritarianism/totalitarianism, reflecting how formalized and developed the systems are, as well as how closely they approximate the ideals of the system.

There are two obvious steady states: strong democracies and totalitarian states. Each poses a unique solution to the risk of dissent-based collapse. The totalitarian state avoids it by brute force and the ability to weather threats. It has such a control over society, most importantly the air they breathe and the very walls around them (covered in digital eyes), that dissent and criminal activity is impossible. In the terminology we have been using thus far, it is able to avoid brinkmanship conflicts by having an undisputed higher resolve whilst being able to weather any threats—power is disproportionately divided between the society and the state.

In contrast, a strong democracy answers each of the collapse risks by creating solutions to the inherent game theory problems in extremis. It has power through its control over the loci of power but avoids tyranny through public control. It remains stable because it presents a relatively lower-cost alternative to conflict—a mediated political process that is fair to all. The fairness element is key and is the reason I divided this into strong and weak democracies. A strong democracy is stable because the people buy into the system—they view the results as fair, and thus believe agreeing to the system is lower cost than starting a brinkmanship conflict. This also constrains tyrannical impulses—the cost of starting a conflict to take absolute power (during the period where you are striving for dictatorial power, you don't have the ability to weather all threats and you run the risk of collapse from the brinkmanship conflict you started) from such a constrained position is lower than attempting to work within the political setting. A strong democracy is thus a stable one as its procedures are always preferable to conflict.

When the strength of these systems degrades, then stability decreases.

Consider a weak democracy. If the system is flawed or unfair, and if democratic ideals are not maximized, there may be some political actors out there that believe that it is a better option to attempt a brinkmanship conflict, which risks collapse. An authoritarian (not a complete tyrannical system) state is similar. It only survives to the extent to which it either convinces the population that it is totalitarian, or it doesn't impose costs on the population high enough to encourage dissent, but it cannot stifle dissent once it occurs without starting a brinkmanship conflict, and risking collapse.

The primary survival strategy of a weak democracy and an authoritarian state in extremis should be to move as quickly as possible to whichever steady state is closest. An authoritarian state should attempt to become totalitarian and a weakly democratic one should attempt to become a strong democracy. Unfortunately, the exact line where we can say that one state is closer to one steady state or the other is difficult.

The line should not be based simply on conceptual links but also ease. Shifting to totalitarianism is easier as it simply requires a would-be dictator to control the air, food, water, and airlocks to assume power quickly. That is easier to do for a state that is already authoritarian. Of course, this jump is extremely dangerous due to the risk of inspiring escalating dissent before the transition can be fully made (and assuming you cannot convince the population the tipping point is already reached).

An authoritarian state can attempt to survive as long as it can with its current level of oppression, attempting to control a population with imperfect tools. Risks of dissent and rebellion will always be high, but an extended game theoretic stability can hold so long as the costs of the oppression don't outweigh the risks of dissent. Unfortunately, the reason that this is unstable is it is not a binary game. It is not the state and the populace in a contest; rather, it is the thousands if not millions of individuals and the thousands of state workers, all of which can set off a cascade in the absence of powerful and constraining methods of peaceful resolution of disputes. In other words, the continued survival of an authoritarian state can either be based on full stagnation (a refusal to increase oppression, whilst maintaining authoritarian control and thus having little accountability mechanisms with the populace and continually risking dissent), fabricating the totalitarian nature of the state, or either of the first two whilst moving towards actual totalitarian control.

It may be lower cost for an authoritarian state to attempt to shift to a weakly democratic state, which is likely to be marginally more stable as it provides an alternative to violent dissent that an authoritarian state does not. A democratic state probably won't move to totalitarianism either, as it probably has greater controls against such a jump, and moving to authoritarianism risks a violent backlash without achieving a more stable system. It is a much further journey to make whilst risking all the dissent brinkmanship collapses along the way from those fearful of dictatorial control. Thus, improving the system is a better strategy—expanding equality in the political sphere to cover all actors with grievances is important, as is making the process fairer and more transparent. Such a move to a strong democracy is required as weak points in the system (unequal access, lack of transparency, or other issues) risk escalation.

Now, not all political leaders act with survival in mind; some seek power and glory. However, in extremis survival will be a much more pressing concern. It will be an ever-present issue, and reminders of the stakes should also be ever-present. Assuming the risks are never fully mitigated, political action should take place with survival in mind. And that survival necessitates a movement towards totalitarianism or strong democracy. A weaker system creates confusion, which risks a collapse due to the equally situated population and state.

However, I do believe that whilst these two systems, totalitarianism and strong democracy, are uniquely steady, a strong democracy is likely to be marginally more stable.

A totalitarian state, despite its absolute control, is threatened by the potential overuse of its powers, or the accidental employment thereof. Additionally, in a system of centralized control and little need for public legitimacy, contests for power and the transition of leadership between officials at the time of the death of the current leader are acute weak points. One of the benefits of a democracy is that power can transition without bloodshed, and groups within political leadership that desire power have a transparent, fair, and publicly legitimate method of gaining that power.

Shifting to a form of weak authoritarianism and then collapsing should be marked against the stability of totalitarianism if the slip occurs easily. Maintaining totalitarian control requires incredible brutality maintained forever, against a powerful society. A totalitarian state is also riddled with threats that may not be credible. Now, in most cases the loci of power can be deployed without risk to the state, but the continued survival of the stake depends on deploying those threats. Yet, it is impossible to claim that there will never be a conflict that a totalitarian state will respond to that risks the state itself. A strong democracy shouldn't need to deploy the loci of power except for in extremely limited enforcement circumstances, such as with violent outliers or for a small number who believe they can never achieve their goals through the political process, or in cases where small-scale physical violence does not work.

In fact, totalitarian control may be more stable at any given moment compared to a weak democracy, but it faces greater instability over the long term. It can absorb individual challenges better but it has a weaker response to systemic shifts and repeated incidents.

Thus we can rank the systems as follows in terms of stability: strong democracy > totalitarianism > weak democracy > authoritarianism. We can also predict institutional design flows. Authoritarianism systems are likely to attempt to shift to totalitarianism, but if that is too costly it will attempt to shift to a weak democracy. A weak democracy will attempt to shift to become a strong democracy.

And notice how a strong democracy behaves. On Earth, the only pressure to become a stronger and more equal democracy is political in nature. On Mars, it is a question of survival. The more unequal, the less transparent, the more unfair, the less stable the system becomes. If the democracy does not try its absolute best to be as inclusive and fair as possible, it risks some people deciding that the system is not worth the costs (although that is a low bar given the risks of escalation, but the risk of someone doing so also inspires further democratization). Because of how powerful the average person is in a space settlement, democracy there will be more complete than we are likely ever to reach on Earth. There is a constant survival pressure towards a complete democracy. That is an absolutely beautiful conclusion of game theory and environmental logic: the stars demand equality.

Yet it isn't the only outcome. Which path is set upon, towards totalitarianism or a strong democracy, is entirely dependent on the starting point. That determines which direction the state will evolve. As such, it really, really, really matters what institutions are present at the very beginning of the settlement. Given that the default might be some form of military rule, as that is how space missions typically operate, and such a rule combined with the need for centralization and the risks of ineffective governance in an extreme environment may jump easier to more authoritarian systems, which could cascade down the line to a totalitarianism greater and more complete than anything on Earth, where the very air is held captive. As always, Mars is a place of extremes, in governance as well as in nature, as such extremity is required to be stable in such an extreme environment. This conclusion justifies the whole endeavour of analysing space politics in advance—by identifying these principles before we arrive, we can properly design our institutions to maximize the greatness of extremity.

17

Anarchy and authority

Summary justice on long-term space missions

Simon J. Morden

1. Introduction

Every culture without exception has a collection of norms which guide daily behaviour (Sumner 1906). Some cultures possess highly codified norms that they express as a formal legal system with police and judges, whilst others rely solely on societal pressure to enforce norms. Formal legal structures imply a hierarchical structure of those who judge, whilst more informal systems tend to be enforced by the whole community acting in concert (Durkheim 1893). Many norms arise from religious beliefs and carry further societal weight as taboos (e.g. Cook and King 1784). There are also mores and folkways to consider in this context as well: expected behaviours which are more informal, but also more pervasive.

Along an orthogonal axis is the concept of whether a society is guilt-based, shame-based, or fear-based (Hesselgrave 2002). Norms are enforced by internalized guilt, externalized shame, or by fear of retribution. But however a society is organized, or how norms are enforced, the requirement for a way of correcting unwanted behaviour is universal.

As new societies form in novel situations, a unique collection of norms coalesce, along with enforcement mechanisms to correct transgressive behaviour (Creanza, Kolodny, and Feldman 2017). Space programmes are simply one example of this phenomena.

2. Existing Command Structures on Spaceships

Because state space programmes originated and evolved from within state military programmes (Gimbel 1990; Andrews and Siddiqi 2011), their command structure was essentially that of the military organization from which they grew. Initially, space crew were overwhelmingly recruited from either former or serving military personnel, or inducted into the military after recruitment: subsequently, increasing numbers of civilian technicians, scientists and educators with no military experience have been rostered as crew (Day 1996).

However, a de facto quasi-military hierarchical structure of authority still exists. On board is the Commander, who is responsible for the overall mission success and the safety of the crew. On the ground is the Flight Director, whose role is the same. Flight Directors would assert that only they have the authority to 'overrule the rules'—this

Simon J. Morden, *Anarchy and authority*. In: *The Institutions of Extraterrestrial Liberty*. Edited by Charles S. Cockell, Oxford University Press. © Simon J. Morden (2022). DOI: 10.1093/oso/9780192897985.003.0018 position is problematic since it should be obvious that the crew can act independently of Mission Control or any other oversight, whether they have explicit authority to do so or not. This seeming duplication of duties may give the appearance that confusion might occur as to who has seniority. It does not appear to have done so yet.

Much of this is due to the current model for crewed missions. They are universally high cost, high prestige, infrequent events, initiated by state actors for national pride. Missions are of short- and medium-term duration in cis-lunar space with carefully selected crews and carefully selected ground support. Training of both crew and support is intensive and long-term—the personal cost to an individual is high, and that of the commissioning body likewise in both monetary and technical investment. Crew are likely to be in training for far longer than the duration of a single mission, and also longer than their total time in space (Seedhouse 2010).

On a spaceship, the commander gives orders—it is moot to the discussion as to whether the flight director outranks the commander—and expects total and immediate compliance from the crew. The crew will have been trained, either through previous military service or through inculcation during astronaut training, to obey without question. It is unlikely, if not unknown, that a commander has given an order at odds with their responsibilities, and equally unlikely that a crew member would have had sufficient reason to question either the validity or the efficacy of the ordered action.

It is also unlikely that a situation arises where orders need to be given, as only novel situations require decision-making. The nominal condition is one where everything that happens has been exhaustively trained for: everyone knows the correct action in any given situation to affect the desired outcome (Spohn 1907).

This paradigm represents the norm. The evidence suggests that it is effective in maintaining shipboard safety (Only three astronauts have ever died whilst in space— the crew of Soyuz-11, (Snelgrave 1971).) Astronauts are drawn from a small pool of highly motivated, intelligent people who have also undergone extensive physical and mental tests before launch, and have assented to this model. Space flight is not safe but it is subject to rigorous safety checks: it is as safe as it can be, given the constraints of physics, technology, and resources.

Whilst crews are united by a common purpose, and the commissioning organizations motivated by national pride rather than profit, this model of command is unlikely to change, nor does it have any need to. Crews that have trained together grow to trust each other, accept their roles, and perform their required tasks without reminders or warnings—either that or any inadequacies in an individual or poor team dynamic becomes evident during selection and training, and changes to the crew roster are made pre-launch.

3. A Paradigm Change Due to Commercialization

But there is a future envisioned where astronauts become employees, and the profit motive of space flight moves from the periphery to the centre—becoming the reason for the mission rather than an incidental, if welcome, aspect. An expansion in the number of missions, and the types of missions, will lead to an expansion and privatization of astronaut training, and what a state-run astronaut training programme considers 'the right stuff' (Conrad 1898) is likely to differ in a commercial training programme, run for profit.

As space flight becomes more common, and more routine, then the more likely it becomes that crews will become autonomous, without a mission control, in the same way merchant shipping currently operates. A company will hire pre-trained individuals, drawing from a large and changing pool of applicants. Individuals will be rostered on to a crew on an ad hoc basis: they will be chosen for their technical competencies rather than their interpersonal dynamics. Hiring decisions will be commercial. Crew members are unlikely to have trained together. If they know each other, past experiences may be bad as well as good. Work-for-hire will become a parallel, and then dominant, model to that which is exhibited by state-run missions, with a primary emphasis on working for the organization who commissions the crew, and towards the goal of completing the mission.

4. The Purpose of Command Structures

Command structures exist to streamline and simplify decision-making, to define responsibility and accountability. In hierarchical structures, they also exist to provide a clear and unambiguous line of communication from the top of the structure to the bottom (Piellusch 2018).

There are acknowledged points of weakness in a hierarchical command structure. Because power rests in positions (or ranks), abuse can occur unchecked if the abuse is carried out by, or with the consent of, the head of that command structure. There are few, if any, external avenues of complaint open to those below the abuser. Likewise, because commands are promulgated from the top, a break in communications at any point in the chain can cause a potential mistake to be transmitted to those whose duty it is to follow orders.

As has been noted, such failure modes have not appeared so far within active space missions. But as also has been noted, access to space is carefully controlled by state actors, the crews hand-picked and extensively trained. If we posit that access to space becomes widely available to commercial operations and commercial crews, then a brief overview of the current operations of merchant ships indicates that the exemplary disciplinary record enjoyed by crewed space missions is unlikely to continue.

Merchant ship captains, like their military counterparts, are responsible for discipline aboard ship (e.g. UK Chamber of Shipping 2016). In the Age of Sail, punishments were routinely corporal—ranging from beating with tarred ropes, through whipping, running the gauntlet, and keelhauling—all the way through to capital punishment. Monetary fines in the form of withholding pay were also imposed, but physical punishment of crew was generally how the captain, through his officers, kept order (Byrn 1987).

But ship-board punishments have always followed land-based laws (Leeson 2007)—physical punishments have been superseded by, depending on jurisdiction, fines and escalating warnings, and ultimately to dismissal from the ship. A crew member may also be subject to a police investigation, depending on the nature of the incident. In an emergency situation, the ship's captain is permitted to 'take such action to preserve the safety of the ship, persons or cargo on board, or the marine environment, as he[sic] considers appropriate' (UK Chamber of Shipping 2013). The severest penalty under this situation is the 'immediate suspension' of a crew member: the crew member is suspended from duty, may be confined to their quarters or other suitable accommodation, and is likely to be put ashore at the next appropriate port for shore-based authorities to deal with.

The duties of a ship's captain are essentially the same as for a spaceship's commander—to ensure the safety of the vessel and the crew, and the timely execution of the mission. Disciplinary measures are part of the captain's suite of behaviours and actions to allow them to discharge their duty. But it can be argued that a spacecraft's construction and situation are such that the existing scale and nature of punishments for disciplinary infractions are both inappropriate and inadequate.

5. The Problem of Enforcing Discipline Aboard Spaceships

A spaceship (for the purposes of this discussion, the term will also cover all non-terrestrial habitats) has some similarities with a ship at sea: the voyage is between two destinations, and between times is surrounded by a hostile environment; the vessel itself provides the only habitable conditions within reach; on board is a working crew, usually sufficient and no more to see the ship to its destination, each with their own specific duties which may or may not overlap with other crew competencies; the crew, including the captain, are usually employees with no specific financial interest in either the ship or the cargo, with the owners of both absent from the vessel.

A crew member's duties will be both specific to their role and general to being crew. Merchant ships are permitted to have enforceable rules about drink and drugs on board, possession of offensive weapons, flammable items, unauthorized personnel, stealing from the manifest, strictures regarding interpersonal conduct, health and safety regulations, and doing anything that might endanger the ship (UK Chamber of Shipping 2013).

Assenting to the rules is, individually and corporately, an important pre-voyage action and part of the contract of employment. In assenting, the crew member is agreeing to be bound by the command structure, the discipline aboard ship, and the punishment regime as set out. It is entirely likely that any particular voyage will pass without incident. However, the existence of the rules acknowledges that they may be broken, even as they set out the possible punishments: the penalties for transgressing shipboard rules act—as they do portside—both as a deterrent and as a practical jurisprudential framework, setting out infractions and their tariffs. The nature of long voyages in confined quarters, surrounded by an implacably hostile environment, are acknowledged to be a stressor for interpersonal conflict and a cause of degradation in functional capacity (Bhattacharjee 2020).

Ocean-going ships, be they commercial or military, will sometimes have a purpose-built brig in which to confine transgressors until they can be put off at the next port or otherwise wait out their period of incarceration (Ackerman 2011). Even temporary brigs are secure not because they are specially designed but because they form part of the existing infrastructure of the ship—solid walls and deck—and anything that a captive might do within that space is not going to be able to harm any of the crew, affect the seaworthiness of the vessel, nor imperil its passage to its destination.

Incidents involving unruly passengers aboard commercial airliners are usually of limited duration and can be effectively dealt with by an ad hoc coalition of cabin crew and other passengers acting in concert until the pilot can effect an emergency landing. But as armed hijackers have shown—repeatedly and tragically—the vulnerability of the airframe to damage risks the lives of all those conveyed inside it (Marchiafava 2016).

The problems faced by aircraft are multiplied in spaceships. Spaceship design is currently predicated on two mutually dependent factors. One is mass and the other is thrust. Until spacecraft propulsion technology enters the realm of 'magical', the efficiency of a drive will always be greater with lower gross mass. Thus, spaceships are constructed to be of lowest possible mass whilst still being able to carry out its core functions safely. The presence of a brig implies a built superstructure robust enough to withstand intentional physical disruption without catastrophic failure. As a consequence, unless and until spaceship designs radically alter to include secure detention facilities, there will be no space prisons (*pace* Kutner 1968). Moreover, this is where the problems start, not where they end—a shipboard space prison suggests that there is somewhere to transfer a prisoner. Any port of call may not have the detention facilities, nor the legal and paralegal infrastructure, to deal with a prisoner brought there.

To understand the environment in which spaceship crew discipline must be exercised, we must consider three factors.

The first is fragility: a spaceship is both easy to damage, be it accidentally or on purpose, and for that damage to be injurious or fatal to the crew and the mission. The integrity of a life support system and a pressure vessel keeping the atmosphere in—and vacuum out—must be maintained at all costs. Rocket fuel is volatile, oxygen is an oxidant, cooling systems contain asphyxiants, vital computers and displays are easily broken or rendered inoperative, and so on. Because every kilogram of mass needs to be boosted from Earth (currently) and, in any event, weight affects the efficiency of propulsion systems, spacecraft and prefabricated habitats are likely to remain as lightweight as possible whilst still maintaining their functions. A failure in any single system puts the crew at severe risk of death, be it quickly or slowly. Moreover, because the crew and their life support make up a significant portion of the payload, their number will be kept to a bare operating minimum. Each crew member will be a specialist: in an ideal mission, there will be redundancy between specialisms, with more than one person being able to function in any given role, but there may be critical roles which are not duplicated.

The second factor to consider is isolation: whatever lofty ideals mission control might impute onto the crew, or the ethical and command paradigms the crew are used to operating under, the key concept is that there is only the crew. The crew might be able to seek guidance from others—potentially with a considerable time delay—but they are the only possible actors, and no-one else has agency except the crew. This is acknowledged to apply to critical technical decisions—Alexei Leonov's decision to partially deflate his suit on the Voshkod 2 spacewalk, Valentina Tereshkova's in-flight reprogramming of the Vostok 6 flight controller, the first Moon landing by Neil Armstrong on almost-empty fuel tanks, and the return of the Apollo 13 crew are all early examples of crew agency. That it applies to every decision taken is less well regarded.

The third factor to consider is immediacy. Decisions that might have been able to be delayed or otherwise deferred until a more convenient time elsewhere (i.e. not on a spaceship) have to be made in a very short time frame. Technical issues—a failure of critical systems, for example, and remembering that almost all systems are critical—need to be dealt with before the crew suffer harm. Human factors are no less urgent, exacerbated as they are by close proximity forced on the crew by a restriction of habitable space and the knowledge that the crew are the only people for potentially millions of kilometres. If there are serious failures in carrying out duties, or if interpersonal conflict arises which might endanger both the success of the mission and the safety of the crew, these will rapidly become the presenting issue, and a resolution necessary before any escalation or further offence is possible.

It follows that individually, the crew are under a high degree of psychological stress for the entire duration of the mission as they come under the scrutiny of their peers. The necessary corollary to this is that any crew member will continually assess—judge, even—the conduct of others.

6. Ship Culture

The crew will carry with them their own cultural norms, but they will also recognize that ship norms take total precedence. In this context then, a crew or colony can be viewed as its own culture, with its own law. Enforcement of ship-board norms will be by the whole crew, not just the person nominally 'in charge'. Ship-board norms will encompass all behaviours and actions that impact on ship systems, but will necessarily evolve to ignore behaviours that are odd, eccentric, even antisocial in non-ship environments. Any crew member who appears to be either neglecting their ship-critical duties or be incapable of carrying them out correctly will face—and sooner, rather than later—judgement.

Certain cultures may be more predisposed towards self-enforcement of punishments. In guilt-based cultures, where breaking absolute, externally imposed norms brings external punishment, members may see their punishment as necessary in order to restore community. Likewise, in shame cultures, it is possible that those who have brought shame on themselves through transgressing the norms will see punishment and loss of honour as the first step towards regaining their previous status. Fear-based cultures fare less well here: hiding transgressions—something that is likely to be fatal on a spaceship—becomes as, if not more, important than the actual transgression itself (Hesselgrave 2002).

If a transgressor agrees to their punishment, and acts in such a way that they do not require management, then the punishment essentially becomes self-imposed. The transgressor accepts the community penalty for their transgression, and moreover accepts that the penalty is just (Garvey 1999). This self-imposition is in part performative, an act of contrition and penitence which placates the community, and in part an acknowledgement that if they do not comply with the community rules on relatively minor matters, escalation can be swift and severe.

Just how severe is something that needs to be explicitly stated. Keeping in mind that a commander is authorized to 'take such action to preserve the safety of the ship, persons or cargo on board, or the [marine] environment, as he considers appropriate', almost all situations where a crew member commits an action which, or by inaction, imperils ship—and crew safety, whether by negligence or malice, the event can be punishable by death. There is nowhere to maroon someone, nor a boat on which to cast anyone away, nor a port to strand them at, nor prison to contain them safely. Given that there is no space prison, and with current technology no way of incapacitating or otherwise restraining a person for the length of a voyage, then and only if every transgressor accepts, and is trusted to comply fully with, a 'confined to quarters' order, is a non-lethal, top-tier punishment a valid option. The three factors of fragility, isolation, and immediacy automatically create a crisis situation. All that is otherwise left is the modern equivalent of walking the plank—casting the transgressor out of the airlock.

A system of rules which has the death penalty for mistakes and errors of judgement, let alone deliberate acts, needs to be carefully considered by all those it affects. This is simply not just for those who might operate under it but for wider society. There has been a gradual and worldwide removal of corporal punishments from statute, and whilst capital crimes are sometimes the last to go, a large majority of nation states no longer execute criminals. Those states which do retain capital punishment still usually have due process. Trials can be lengthy, with lawyers, juries, judges, and multiple appeals to higher courts.

This situation is not what is envisaged for shipboard executions. Due to the circumstances, the process will be more akin to a drumhead court martial. The crew will necessarily serve as judge, jury, and executioner, and must be willing to do so.

There are psychosocial penalties to such an *ad hoc* system of enforcement. As a further consideration, risk aversion to the point of paralysis could be just as destructive as the vagaries of any onboard legal system (Roberts 2010). It is at this point that the hierarchical command structure comes under greatest strain.

7. The Nature of Leadership, and the Command Structure

In the context of a minimally crewed space ship, where there is close proximity of the ruler to the ruled, the authority structure is necessarily flattened—orders are given directly without intermediaries (Teubner 2001). Due to the value of each crew member to the success of the mission, and the crew's awareness of their irreplaceability, a leader's ability to be the sole enforcer of shipboard rules is degraded.

If we posit a scenario where a commander imposes a punishment that a crew member does not immediately and fully accept, then without the other crew actively reinforcing the commander's decision—and the punishment—the commander's authority is effectively broken. The reverse is also true, if a commander refuses to impose a punishment that the rest of the crew believe necessary. However they are appointed, a leader can only command by custom and consent (Suchman 1995). If sufficient of the led reject a leader's authority, they are no longer a leader.

Given everything that has preceded—the specific constraints of spaceship construction, the isolation of the crew, the necessity for summary punishment that is agreed to and enforced by the whole crew—a hierarchical quasi-military command structure appears particularly unsuited as a model for future commercial space flight.

An alternative, and potentially applicable, model of leadership presents itself from the Age of Sail: that of the pirate ship.

Piracy in the seventeenth and eighteenth centuries was a phenomena predicated on multiple factors. Some of it was state-sponsored: privateers received 'letters of marque' from one country which permitted them to attack, loot, and seize or sink vessels from other countries. Some of it was not, and independent piracy grew in size and importance—some privateers moved seamlessly into piracy, other pirate ships appeared as the result of mutinies aboard armed merchant shipping, and others still were deliberately conceived as piratical companies (Leeson 2007).

Wealth was an obvious incentive: an ordinary seaman could receive many multiples of an average yearly salary if their ship was able successfully to seize a single rich prize. But so was the autonomy afforded by what became known as 'the Pirate Code' (Fox 2013).

Discipline and order onboard military and commercial shipping of the time was usually maintained by a strict hierarchical structure imposed by external authority. The captain of the vessel received their authority from the owners of the ship—be it the government in the case of naval vessels or the share-holding merchants in the case of a cargo ship. The result for the implementation of rules was more or less the same: the codified legal system onshore was exported offshore, with military craft governed by act of parliament, and merchantmen by very similar regulations. The captain was the sole authority whilst at sea. Not following an order would be harshly punished— corporal punishments were severe and common, often resulting in permanent injury and sometimes death. Capital punishment was applicable to many shipboard crimes.

There was little preventing a domineering captain from doing what he wanted. Through his officers, he was able to punish any and every infraction. Such strict discipline was greatly resented, and ordinary seamen, at least some of whom had been pressed into service, were at the very bottom of the social strata—poorly paid but unable to gain useful employment shoreside, in a highly dangerous environment and suffering under a regime that was harsher and more inescapable than on dry land.

The harsh treatment of seafarers was deemed necessary to the safe passage of the ship: a sailor working for their wages did not share in the success of a voyage, and there was no carrot for them, only a stick (Benjamin and Thornberg 2007). For the captain, however, victory against an enemy would lead to promotion, and having a share in the cargo would lead to wealth. Since the captain was responsible for punishing transgressors, the practice of stealing sailors' pay by imposing fines and under-victualling the ship on purpose would enrich them further.

The hierarchy was class-based, and discriminatory, but the ordinary crew had little or no remedy to alter their treatment save that of the high-risk strategy of attempting a mutiny. Mutiny was always an option, but it happened very rarely. Not only was the action perilous—successful or not, the punishment at the hands of state authorities was death—but the outcome was uncertain (Rediker 1987). Convincing an entire ship's crew to mutiny, or enough of them to overpower the others, was no easy task, and if the plot failed, the mutineers' end would be swift and final.

It is unsurprising, therefore, when seamen could order their own affairs they did so in a radically different way. Pirate ships were both republics and democracies, before either republics or democracies were usual on shore. Each member of a pirate crew had a vote and a voice. A pirate company drew up its own code of conduct, covering behaviour on board ship, how to share provisions and treasure, and the payments that were to be paid to crew members injured or maimed during their voyage. Moreover, the pirates then freely assented to be bound by the code.

Pirates operated under an entirely different economic model to their military and merchant contemporaries. Because of the 'no prey, no pay' paradigm of piracy, every crew member had an incentive to seize as much prize as possible, weighing their actions against the risk of death at the hands of governmental justice, either at sea or on land. But rather than resulting in a free-for-all, it was quickly and universally realized that piracy could only work if the crew didn't prey on each other—this principle extending to the senior crew and the captain.

Consequently, pirate crews chose not just their captains by popular vote, but also chose to dilute a captain's power by also voting for a quartermaster (Snelgrave 1734). Pirate captains would lead during engagements with other ships, and during those times, their authority was unquestioned. But it would be the quartermasters who would take control of the prize, strip it of goods, and ensure that a fair share went to every pirate—one share each, with the captain usually only taking two shares, and senior officers and craftsmen somewhere between (Johnson 1726–1728). Bonuses for exemplary action, and being first to spot a prize ship, were awarded. Consequently, the watch on a pirate ship was first-rate. Quartermasters were also responsible for supplying the crew with equal shares of food and drink—captains ate what the crew ate.

Discipline, whilst authorized by the captain, was maintained by the whole crew who had positively assented to be bound by it (Rankin 1969). Historically, there was no more ill-discipline on pirate ships than on military or merchant ships, and anecdotally less: crew had a monetary stake in the success of any given action, and in the voyage overall. Stealing from the (unlocked) treasure chests was punishable because it was stealing from all. Having a naked flame around flammable or explosive material was punishable because it endangered the ship. Gambling onboard was forbidden (contrary to expectations) because of its potential to cause arguments and fights. The exact nature of the rules of each pirate ship differed, whilst coalescing around a general practice of fair shares, democratic control, personal accountability and self-discipline, and insurance payments.

Far from being lawless vessels, pirate ships were self-disciplined to a large, and highly effective, degree. Pirates acted for no-one else but themselves. There was no higher authority to appeal to. They were, essentially, independent countries afloat on the sea. Their pay was likely to exceed what they could earn legitimately. They suffered less arbitrary punishment, and overbearing officers—the captain and the quartermaster—could be removed by a majority vote, rather than a bloody and possibly ineffective mutiny. This flat, democratic structure would not be mirrored on land for some time after pirates discovered its efficacy.

8. A New Model of Spaceship Command

Whilst not suggesting that piracy would be either welcome or feasible in the solar system of the twenty-first century and beyond, nor eliding over the wholesale murder, theft, and outlaw status that piracy necessarily entails, there is much that is laudable and effective in their model of governance that could be applied to commercial spaceships.

Becoming crew using this model means assenting to a code of conduct and a way of self-rule that will dissuade many otherwise suitably qualified candidates. However, it is critical that, for their own safety, and that of the ship, that any potential crew be fully cognizant and accepting that the code includes carrying out disciplinary actions—including capital punishment—before signing on to a ship. This acknowledges that only the crew will have agency on any voyage, and that a hierarchical command structure is inappropriate, ineffective, and potentially dangerous in these circumstances.

Crews should be able to organize themselves: all spaceship personnel are highly trained specialists who understand both the risks and the rewards of their labour, their duties whilst on board, and their responsibilities to both the hiring company and each other. Whilst it would be a commercial decision on which crew to hire, the *ad hoc* crew ought to be able to choose their own captain from amongst their number—they might also wish to elect a quartermaster to oversee supplies and act as the crew's conduit to the captain, if they feel the crew roster is sufficiently large to warrant that.

Crew pay ought to be based on shares—not of prize money per se but still based on the principle of equality. Every crew member is just as vital as the next. If there is variation in pay—to a captain or an engineer, for example—then a piratically inspired limit of twice the pay should be agreed beforehand. Bonuses for exemplary conduct could also be employed as incentives. Commercial operations may also wish to further induce the crew with a share in the profits of a mission.

By laying out the crew's rights and responsibilities beforehand, rather than applying them in a *post hoc*, arbitrary way, crews of commercial spaceships will be better protected from poor working practices and exploitation. This model will also enable commercial operators to operate in an unambiguous legal framework that is fit for purpose, far beyond any Earthly jurisdiction.

References

Ackerman, S. 'Drift: How This Ship Became a Floating Gitmo' (2011) *Wired* <https://www.wired.com/2011/07/floating-gitmo/> accessed 7 June 2021.

Andrews J. T. and Siddiqi A. A. (2011). 'Introduction: Space Exploration and the Soviet Context' in J. T. Andrews and A. A. Siddiqi, *Into the Cosmos: Space Exploration and Soviet Culture* (Pittsburgh, PA: University of Pittsburgh Press, 2011) 1–12.

Benjamin, D. K. and Thornberg C. 'Organization and Incentives in the Age of Sail' (2007) 44 *Explorations in Economic History* 317–41.

Bhattacharjee, S. '18 Serious Offences under Merchant Navy Code of Conduct Resulting in to Dismissal from Job at Sea' (2020) *Marine Insight* <https://www.marineinsight.com/life-at-sea/18-serious-offences-under-merchant-navy-code-of-conduct-resulting-in-to-dismissal-from-job-at-sea/> accessed 7 June 2021.

Byrn, J. D. Jr, *Crime and Punishment in the Royal Navy: Discipline on the Leeward Islands Station, 1784–1812 (England)*. Lousianna State University, LSU Historical Dissertations and Theses, 1987, 4345.

Conrad, J. *Youth* (Blackwood's Magazine, September 1898).

Cook, J. and King, J. *A Voyage to the Pacific Ocean* (London: W. and A. Strahan, 1784).

Creanza, N., Kolodny O., and Feldman M. W. 'How Culture Evolves and Why It Matters' (2017) 114(30) *Proceedings of the National Academy of Sciences* 7782–9.

Day, D. A. 'Invitation to Struggle: The History of Civilian-Military Relations in Space' in J. M. Logsdon (ed.), *Exploring the Unknown: Selected documents in the History of the US Civilian Space Program* (The NASA history series (NASA SP: 4407), Washington, DC: US Government Printing Office, 1996) 233–82.

Durkheim, É. *The Division of Labour in Society*, trans. G. Simpson (New York, NY: The Free Press, 1993).

Ehrenfried, M. von. *The Birth of NASA: The Work of the Space Task Group, America's First True Space Pioneers* (New York, NY: Springer Praxis, 2016).

Fox, E. T. (2013). *'Piratical Schemes and Contracts': Pirate Articles and their Society, 1660-1730*, Ph.D. thesis, University of Exeter.

Garvey, S. P. 'Punishment as Atonement' (1999) 46 *UCLA Law Review* 1801–58.

Gimbel, J. (1990). 'Project Paperclip: German Scientists, American Policy, and the Cold War'. *Diplomatic History*, 14, 343–65.

Hesselgrave, D. J. *Counseling Cross-Culturally: An Introduction to Theory and Practice for Christians* (Eugene, OR: Wipf and Stock Publishers, 2002).

Johnson, C. *A General History of the Pyrates, from Their First Rise and Settlement in the Island of Providence, to the Present Time*, ed. M. Schonhorn) (London: Dover, 1999).

Kutner, L. (1968) 'A World Outer Space Prison: A Proposal' (1968) 45 *Denver Law Journal* 702–708.

Leeson, P. T. 'An-arrgh-chy: The Law and Economics of Pirate Organization' (2007) 115 *Journal of Political Economy* 1049–94.

Marchiafava, G. 'The Montreal Protocol 2014 and Current International Regulation Issues on Aviation Security' (2016) 83(2)330 *Rivista Di Studi Politici Internazionali* 235–62.

Piellusch, M. 'Is the "Chain Of Command" Still Meaningful?' (2018) *War Room—* U.S. Army War College <https://warroom.armywarcollege.edu/articles/chain-of-command/> accessed 7 June 2021.

Rankin, H. F. *The Golden Age of Piracy* (Williamsburg, VA: Colonial Williamsburg, 1969).

Rediker, M. *Between the Devil and the Deep Blue Sea: Merchant Seamen, Pirates and the Anglo-American Maritime World, 1700–1750* (Cambridge: Cambridge University Press, 1987).

Roberts, L. 'Analysis Paralysis' (2010) Jan–Feb *Defence AT&L Magazine* 18–22.

Seedhouse, E. *Prepare for Launch: The Astronaut Training Process* (New York, NY: Springer Praxis, 2010).

Snelgrave, W. *A New Account of Some Parts of Guinea and the Slave-Trade* (London: Cass., 1971).

Spohn, K. von. *The Art of Command*, trans. The General Staff, War Office (London: HMSO, 1907).

Suchman, M. Managing Legitimacy: Strategic and Institutional Approaches' (1995) 20 *The Academy of Management Review* 571–610.

Sumner, W. G. *Folkways: A Study of the Sociological Importance of Usages, Manners, Customs, Mores, and Morals* (New York, NY: Ginn and Co., 1906).

Teubner, G. (2001) 'Flattening the Organizational Structure: Encouraging Empowerment or Reinforcing Control?' (2001) 13 *Research in Organizational Change and Development* 147–68.

UK Chamber of Shipping. *Code of Conduct for the Merchant Navy* (UK Chamber of Shipping, 2013) <https://www.ukchamberofshipping.com/latest/code-conduct-merchant-navy/> accessed 7 June 2021.

UK Chamber of Shipping. *Master's Guide to Shipboard Disciplinary Procedures—Code of Conduct for the Merchant Navy* (Eugene, OR: Witherby, 2016).

18

Making history cosmic, making cosmic history

Waking up to the richness of life's potentials beyond Earth, or, how consequence and contingency became astronomical in scope

Thomas Moynihan

'... the great shadows that futurity casts upon the present ... '.
—Percy Bysshe Shelley (1840 57)

'...but it is not the number of years we have behind us, but the number we have before us, that makes us careful and responsible and determined to find out the truth about everything ... '.
—George Bernard Shaw (1921 183)

'The highest conception of altruism which I am able to form is built on the thought of doing things that are sound enough and true enough to last and be doing good ages after they have lost the name and superscription of those who set them going.'
—Thomas Chrowder Chamberlin (1910)

1. Introduction

1.1 Ethical motivations for settling space

Long-termism is the view that the future ahead of humanity could potentially be vast, that this future could be full of intense and expansive good, and that we should thus act now to safeguard this immense potential. In other words, we ought to ensure our very early place in the total history of humanity. We want to have been the very most distant ancestors future generations can look back on because if things go well, our descendants are likely to have available to them a fuller range of what intelligent, ethical actors can do and achieve within this world than the range at our current disposal.

Compared to how many generations have preceded us, the number of generations there could be ahead is staggering. Science has found no reasons as to why it would

Thomas Moynihan, *Making history cosmic, making cosmic history.* In: *The Institutions of Extraterrestrial Liberty.* Edited by Charles S. Cockell, Oxford University Press. © Thomas Moynihan (2022).
DOI: 10.1093/oso/9780192897985.003.0019

be impossible for humanity to persist into its Methuselah years. Accidents aside, we might currently be a very young species (Ord 2020).

Let 'humanity' be defined, in a suitably capacious manner, as the cumulative project initiated by our ancestors of bettering ourselves and our world. Defined this way, humanity is shot through with colossal potential: the potential to explore a truly vast space of possible achievements and meaningfulness; the potential to persist for eons within which to explore this terrain prudently, patiently, fairly, and fully.

It would be short-sighted to suggest that in the eons ahead, there will be nothing possibly new to discover, in the arts as much as the sciences, and no new possible intensities of meaning and value to apprehend. The peaks and preponderances of achievement are likely to be lying, awaiting their potential uncovering, in the time ahead.

But possibility does not at all entail realization. The wider ranges of this zone of possibility—indeed, overwhelmingly the lion's share of it—all rest upon humanity one day leaving Earth behind.

In the long run, our planet will become uninhabitable for complex, multicellular life due to an expanding sun, in something of the order of 1.0 billion years (e.g. O'Malley-James et al. 2013). However, the wider universe will remain capable of supporting complex life, and meaningful experience, for many orders of magnitude more than this (Adams and Laughlin 1999). Thus, the amount of future left ahead for civilization is, in the longest-term, dependent, critically and crucially, upon *migration to and settling of space beyond the solar system*. In the longest view, it is like a bottleneck sorting the set of truly remarkable, grand futures from all the paltrier, more truncated ones. The predominance of life's longest-term potential lies in space.

Leaving Earth, and doing it in a way that is just and egalitarian, is a task for the future. There are pressing risks facing the present, but these two insights don't exclude each other. We ought now to protect this potential proactively, for future generations to decide what to do with, rather than allowing it to be squandered and irreversibly lost, as is otherwise the case with most potentials left to the capriciousness of our independent nature.

(Here it is important to note that most of our potential remains in outer space, but there is not any rush to reach it. Sometimes it is argued that there is some rush to become multiplanetary inhabitants as a failsafe against the unsustainable levels of extinction risk in play on Earth, such as the risk of thermonuclear war or worst-case climate change. But achieving this in the near term would not necessarily guard against *all* such threats: as long as there is communication between planets, certain threats—for example, pandemics or information hazards—could spread between them. So, rushing to become inhabitants of more planets than Earth might not provide a foolproof failsafe against every worst-case risk (Ord 2020, 194). In the immediate term, focusing on achieving 'existential security'—a state of resilience and security, for everyone, sustainable to survive on Earth—is probably much wiser than some select few sprinting to space. Nonetheless, it remains undeniably true that for humanity to fulfil its longest-term potential, people must *eventually* reach beyond Earth. Therefore, our immediate priority should be safeguarding this potential by charting a course, here and now, that ensures that humanity does not—whether now or later—permanently lose the option.)

Ultimately, if humanity wants to find its place in the universe, then we must also steadfastly remember that humans have recently discovered that 'most' of the universe is not to be found here on planet Earth. This 'most' applies not just to space, nor time, but also to *what is possible*. Put differently, only the most lilliputian fragment of the full affordances that this universe proffers for self-conscious animals, such as ourselves, will be discovered—and enjoyed—by fettering our entire future to that of our birthplace.

1.2 Extinction, the ultimate heteronomy

Another way of thinking about this is that the lion's share of liberty, yet to be attained, lies off-world. The species is not yet even close to being as free as it could be. There is a liberty in leaving the Earth behind.

What does liberty mean here? Liberation is the increase in autonomy, or the ability to maintain and expand options and fulfil ends; whereas conversely, heteronomy is the decrease in liberty, alongside loss of options and capacity to fulfil ends. Remaining Earth-locked brings severe heteronomies for the human species, compared to what could be achieved *beyond* Earth. Liberating ourselves from these impositions can be thought of in the following terms:

- As liberation from scarcity:
  - Whether of resource—as Earth only intercepts a fraction of the Sun's total available energy output;
  - Or of affordance—as we ought not assume Earth's environment has afforded the most optimal possible forms and modes of life, sentience, or mind;
  - Or of time—as Earth has a finite future window of habitability;
- And, ultimately, therefore as liberation from the premature extinction that would truncate exploration of intelligent life's full potential.

Extinction, after all, is the *ultimate heteronomy*: the loss of *all* options and opportunities for humanity's collective future, forever. Once a species enters this state, it cannot leave. It is the destruction of any potential that the species once had.

The following is a history of the idea that space and survival are conjoined but, crucially, that the attainment of survival is pursued for precisely *ethical reasons*. It is about making the best, impartially, of our unlikely endowment of intelligence: giving our descendants the fullest time available within which to explore this improbable endowment and truly become worthy of the gift.

Such an attitude is radically new: out of all the thousands of generations that came before us, only the past handful have recognized such notions. Such an attitude was not merely unthought. It was—until comparatively very recently—unthink*able*. Making it thinkable required the unlocking of three other nested insights, upon which the above view is dependent:

- Through our actions, we can make the whole of humanity's future meaningfully longer or shorter.
- The differences—between 'shorter' and 'longer'—may be overwhelming large in magnitude, rather than trivial or inconsequential.

- That this difference—regarding a 'truncated' versus 'plentiful' futurity—may even, at least potentially, be sizable enough to make some ethical difference *from the point of view of the universe.*

The last is the key. It rested on growing *scientific* understanding that wider and wider ranges of nature exhibit genuine history: from the biological, to the planetary, to the solar, and beyond. Because only if nature itself exhibits genuine historicity, at some suitably cosmic scale, can we then realize that *there is at least the potential* for our descendants to have made a positive impact on the amount of life and value that will have been realized within this cosmos. Combine this with the similarly historicizing insight that it appears no-one else has previously achieved this, and that wider and wider ranges of the universe appear to remain unpopulated, this puts into relief the idea that eventual settlement of space might be impacting the cosmos in the impartial sense of making it a better, richer place.

2. The Expanding Horizon of History

2.1 Making history cosmic enables making cosmic history

Pursuing liberation, of course, first requires recognizing that one's situation is one of bondage, penury, privation, or frustrated potential (because, after all, from the perspective of exploring options and autonomy, 'opportunity squandered' is just as antithetical as 'opportunity snatched').

However, for the majority of Western history, people could not see that leaving the planet—and thus expanding the foothold of intelligence and life—would represent a *gain* for intelligence or life considered impartially. This was because they did not yet apprehend the seeming *scarcity* of intelligence and life within the wider cosmos. (As we shall see, for many, scarcity was not yet even a conceptual option.) Neither did they recognize that from the perspective of being used for the sustenance and promotion of life and mind, the resources of the wider cosmos are—at present—titanically underutilized. Nor was it acknowledged that leaving the planet would represent, specifically, a gain in *options* or *opportunities*, given that it was also widely assumed that the conditions—and thus also affordances for life— upon the Earth were more or less representative of those everywhere else in the universe.

This applied as much to conditions in space as in time: meaning that current terrestrial, biophysical conditions were assumed to be representative of all other cosmic moments to a more or less exhaustive degree. Moving into the endless past, tracking forward into the limitless future, large-scale directional change—at the level of the physical universe or some meaningfully consequential portion of it—was not commonly apprehended as possible.

Of course, Abrahamic religions and other cosmogonies inject obvious linearity, narrating a cosmogonic beginning, middle, and end. But in these schemes, nature does not have its *own* history and epochal structure, stretching prodigiously beyond human experience and intuition: a history that is open-ended and can conceivably take branching paths; that can—and has—unfolded various long episodes beyond the presence of conscious observers; and one within which the conditions of possibility for life and moral agency itself emerge and can disappear.

Not yet recognizing, therefore, that life and ethical agency is not coeval with the cosmos and is itself *through and through* a historical phenomenon—with a genuine origin, emerging first at a certain place and at a certain time—it was not yet possible to recognize that such agencies *may* not have already achieved absolutely everything they can possibly achieve within the wider cosmos *nor* that it is also possible that they might never achieve all that they can within the wider cosmos.

Importantly, it could, thus, also not be recognized that it is even plausible that we could be living during that cosmic epoch *before which* life has fulfilled its full potential to spread and flourish. Neither could it be recognized that should our descendants not achieve this, then there is no guarantee that such fulfilment might otherwise come to pass. Only with genuine history can there be a finitude of opportunities as well as windows of opportunity and opportunities lost. Because the cosmos had not yet gained its own history, within which life emerges and can disappear, it was not yet possible to recognize that the future of the cosmos could either be full of life or relatively devoid of it and the difference may just rest, no matter how miniscule the chance, on what happens here and now.

It is important to note that, of course, we don't now know any of these aspects of our cosmic situation with certitude. Each one is still highly uncertain and revisable—we don't know much with confidence about the biggest picture—but each of them is now a variously probable option feeding into that picture.

2.2 The growing arena of consequence

Each of these plausible aspects of our wider cosmical situation was facilitated by a progressive expansion of the scope of that spatiotemporal region within which *genuine history* is acknowledged to operate meaningfully. What is denoted by 'genuine history' here is that arena wherein the trajectory of the past persistently molds and constrains the possibilities of the future: closing off some areas of possibility and opening others such that not every possible state can or will be passed through or returned to, and there is thereby room for both 'irreversible lasts' and 'genuine firsts' as seen from the perspective of the whole of the arena. It is that conception of history as having a space of possibility—more expansive than current actuality and past precedent—wherein unexplored regions can be locked off by prior events, whilst explored regions can latterly become incapable of being retraced, creating what are known as path dependencies. A system with these features is known as 'non-ergodic'.

But what happens *beyond* this horizon of genuine history? Here, change and mutability can very comfortably be acknowledged as taking place, but crucially they are conceived of as taking place within a system that reliably traces through all its possible states, and then keeps returning through all of them, with zero likelihood of *never* returning to any particular region of the state space. This means that, over a long enough period of time, the overall system has no discernible developmental direction.

A system with these features is known as 'ergodic'. Much of the Ancient world, for example, seemingly assumed this kind of ahistoricity applied for nature and nations (Morris 1986): whereby each provided an ever-changing, but fundamentally ahistorical or cyclical, backdrop to the evident directionality within our own, individual biographies. If I lose a limb, it is permanent, but nations rise and fall and rise again, and nature moves, everlastingly, in compensatory cycles.

Speaking in terms of how human ideas have progressed throughout the centuries, the story of recognizing how liberating it would be to leave Earth is tied to the story of how science and empirical study have expanded the circumference of the 'genuinely historical' from the autobiographical, outward to the national, to the anthropological, evolutionary, planetary, solar, galactic, and beyond.

Something like the Copernican revolution in space has also been taking place to the arena of the truly historical: it's not just that the cosmos has become increasingly bigger since Copernicus, it's also become increasingly historical in the sense that what is observably actual has been sequentially decentred and parochialized within wider spaces of historical possibility—both in terms of radically different past epochs and vastly divergent potential futures—a decentring of present actualities which has, in turn, been revealed to apply at larger and larger spatial scales. Waking up to the truth of how consequential going to space could be, in the long term, is tied with the story of how we have learnt that 'history matters' within wider and wider domains of reference.

This is because only within the horizon of the truly historical can there be genuinely *unfulfilled potentials* and *squandered opportunities*. (Because, in a truly directionless system, all things lost will be eventually returned such that potentials can never be truly wasted.) Thus, history's horizon had to expand to cosmic catchments before people could recognize that there may be signs that our descendants may be able to have meaningful impacts on the cosmic whole and, therefore, it becomes at least believable that they may be able to make a consequential difference to the total amount of consciousness, and thus meaning, *that will have been realized* in our reachable cosmos throughout its entire history. If this is dependent upon intelligences going to space, then that is what intelligences ought eventually to pursue.

In its grandest, most speculative, form, the suggestion is this: if the 'universe' (or, at least, the practically relevant part of it) has a genuine history, then our descendants can potentially make a genuinely astronomical and impartial impact upon it by being the catalyst that tips it from its currently barren and abiotic state into a richer, biotic state. But if the cosmos doesn't exhibit genuine history, and is ahistorical in some deep manner, then it seems more likely that there can only be more 'partial' arguments for long-term human survival via human space settlement, given that it won't actually meaningfully alter the total amount of value in the cosmos (given that other predecessor civilizations will have already achieved all possible outcomes—positive and negative—and posterior ones will achieve them again, uncountably many times, such that the aggregate amount of value within the whole remains invariant and invariable across time).

So, before it was even thinkable that humanity, or its distant descendants, could make a genuinely consequential and impartial impact 'from the perspective of the universe', and that this evidently is gated upon the spilling of life from its terrestrial cradle, the universe had to itself become subsumed within what was acknowledged as the horizon of history: that arena within which concepts such as 'impact', 'consequence', and 'window of opportunity' all gain their traction and bite, and beyond which they are defanged; defanged, that is, by eternal returns of the same wherein everything gained is later lost and everything lost, later returned.

Put differently, through eventually subsuming the wider cosmos within the horizon of history, it became possible to so much as propose, and hence to assess and explore, that our descendants could, at least potentially, alter its overall course in terms of how 'good' its overall arc may end up having been (no matter how unlikely or difficult this may, in fact, turn out to be). This, then, is the requisite outlook for the argument that attempting to move to outer space may not just be a good thing to do but an overwhelmingly good thing to do.

In short, the cosmos itself had to gain a history before it became even feasible that our own history could ever have a genuine and impartial positive impact upon it, or even make a dent upon how positively it goes, by our far-off descendants leaving Earth behind.

3. A Directionless Cosmos

Prior to apprehending these aspects of our wider cosmical situation, there could no ethical motivation to spread life beyond its place of birth so that there may be more of it, and that it may persist for much longer, and explore more of its potential for development or diversity than it otherwise might have done. In other words, life's current limitations had to be illuminated and identified as arbitrary before people could become motivated to overstep them and propose methods for liberating life's forward-running lineages of them.

Because, from the Ancient Greeks to the medieval Christians, there was little awareness of the range and reach of genuine history throughout nature in its independence from observers and moral agents. This meant there was no awareness that moral agency—dependent upon intelligence and life—is, in its entirety, a historically emergent phenomena contingent upon an origin or common ancestor.

3.1 Self-suffusing good

From Aristotelian to Epicurean to Keplerian cosmology, regardless of the particulars, there was simply not a recognition that vast regions—in both time and space—could be, or had been, totally abiotic or unpopulated (Moynihan 2020).

This was because of an underlying assumption that everywhere that was unpopulated had been populated or would be populated again. For example, the Roman philosopher Lucretius (Lucretius 2001a, 62) was convinced that 'in other parts of the universe there are other worlds inhabited by many different peoples'. Though such speculations on 'other worlds' became controversial for patristic Christianity, their geocentric cosmos was nonetheless maximally populated with moral patients, agents, beneficiaries, or sufferers: from the tortured basement of hell through the angels of the celestial spheres to God's outer loft of heaven. There was nowhere simply unpopulated. Moreover, with monotheism's stress on omnibenevolence, this morphed into the intuition that everywhere that *could* be usefully populated *was*, in fact, usefully populated. There was nowhere that *could be* populated but simply wasn't, for no good reason or justifiable warrant.

Scholastic thinkers captured this with maxims such as '*bonnum diffusivum sui*' and '*ens est bonum convertuntur*'. The former expresses the idea that cosmos is fully suffused with goodness; the latter expressed the assuredness that whatever exists does so because it is of value. As the non-existence of any one thing would thus be bad, so too does every region of the cosmos exist for some purpose or some good. And if good is naturally self-suffusing, there is simply no demand to use artifice to spread it beyond any particular sphere.

Thus, when Lucian (in the second century) and Nicholas of Cusa (in the fifteenth century) speculated upon the populations of other planets, and Lucian even imagined trips to visit them, they simply could not and did not arrive at the idea of spreading life to *otherwise unpopulated* regions of the cosmos and, more so, were not even close to acknowledging that there might be an ethical argument—connected to survival—that might give this utmost importance in the longer run.

How could they? If moral activity is everywhere already, it cannot gain anything— let alone survival—from finding new environments. Cusa, indeed, initiated a tradition, shockingly persisting into the nineteenth century, of assuming that *even the surfaces of the suns themselves* house 'intellectual' inhabitants (Cusanus 2007, 115–17).

3.2 Copernican confidence

By the time of the Copernican revolution, this habit of thinking stubbornly persisted. The expansion in the acknowledged size of the universe, triggered by the revelations of Copernicus and others, did not lead to stipulations that there were more 'empty' or 'unpopulated' spaces than was previously thought. Expanding the acknowledged size of the universe simply expanded the region of physical space within which people invariably assumed goodness must be self-suffusing.

Giordano Bruno, the Italian philosopher, was among the first whose system very prominently and explicitly spread life to all the planets of all the stars beyond our Sun. He argued that each of the 'fixed stars' is itself its own sun, girdled with populated planets, like ours.[1,2] Bruno argued this because, he thought, more planets—with more moral beneficiaries on them—would represent a work more adept at reflecting the largesse and limitless magnanimity of its maker.

According to Bruno, 'as it would be ill were this our space not filled, that is, were our world not to exist, then [it] would be no less ill if the whole of space were not filled [with meaningful activity]'. Behind this was the assumption that within a rationally ordered universe, the 'abolition' (i.e. destruction) of our Earth must be *naturally impossible* because it is *morally impermissible*; thus, we must extend the same 'obligatory existence' to all the other stars and their planets, in so far as they *could be* populated by moral beneficiaries. 'Even as the abolition and nonexistence of this world would be an evil, so would it be [for] innumerable others', he concluded (Bruno 1968, 256–9).

The other Copernicans were not so quick to make such grandiose claims about extra-solar populations. But the disagreement was not how *full* of purpose the cosmos was, rather it was whether 'the cosmos' extended meaningfully beyond our own heliocentric solar system. So, the contention was not about the fullness but about the physical extensity of the receptacle—or canvas of existence—to be assumed as filled.

Galileo, however, was cagier on the populousness of other celestial bodies. Reaching insights ahead of his time, he professed he thought it unlikely that terrestrial 'plants' and 'animals' could subsist on the Moon—given its different conditions. But nevertheless, he still asserted that he thought 'its regions' cannot be 'idle and dead'. Regardless of whether 'life and motion exist there' (as we know it), Galileo was still certain it existed *for something*. There is no part of the universe that is not valuable, and doesn't pay rational dividends, in some way (Galileo 1967, 99–100).

Because of this background belief, many *did* simply assume that other planets were populated with creatures essentially identical to us. Even when exotic physical differences were admitted, there was nevertheless a stress that the extraterrestrials remained equals or peers (and possibly even superiors) to us in all the features that morally matter: rationality, ethics, aesthetics, even religion (Moynihan 2021a).

As St Bonaventure (1864, 245) wrote, in the thirteenth century, the world cannot 'exist without men, for in some sense all things exist for the sake of man'. When the Danish astronomer Christiaan Huygens (1698, 21) was writing, by the close of the seventeenth century, he was saying essentially the same thing when he insisted that we should not 'allow the Planets'—of other suns—to be 'nothing but vast Deserts, lifeless and inanimate Stocks and Stones, and deprive them of [rational] Creatures'.

Due to this assumption, no-one acknowledged that the end of the Earth could also simultaneously spell the end of human life, civilization, or history within the cosmos at large. As the 1700s dawned, such confidence only grew. This was because of growing acceptance that there *are* extra-solar planets and, thus, each must be the receptacle of life. The canvas kept growing. This even made possible belief that the destruction of a populated system, rather than representing a severe loss, could somehow be *purposed* towards the greater good. This was achieved by assuming the essentially ahistorical nature of everything in the cosmos.

3.3 Ergodic universe; ineliminable life; invariable value

In a directionless, or ahistorical, system there can never be a true loss or true gain: all gains are eventually lost, all losses are eventually returned, the whole remains invariant across long enough timespans. Combined with a dose of teleology, this even allowed people to believe that whenever some quantum of life, no matter how big or small, is destroyed here and now, this destruction merely 'clears room' or 'frees up resource' for an equal and compensatory amount to be generated at some other place or time. As with absences, even annihilations must have a just reason. Because of this, any local destruction of life is *mere circulation* rather than *genuine loss*, such that the aggregate amount is conserved across time. This belief, seemingly alien now, was widespread in early modern natural philosophy.

To take an illustrative example, look to the French natural historian Benoît de Maillet. In the 1720s, he wrote that planets die and their life annihilated but that this is in fact 'good' because such destruction fuels the recreation of new worlds and their repopulation with life. If life cannot die on any one particular planet, then why bother even proposing the decoupling of life from that planet?

But, we might now ask, the destruction of some planet would *surely* extinguish the species of the perished planet? The answer is no, because there was not yet an understanding that species are the unique product of an unrepeatable and unique planetary history (such that they will not appear again beyond it). Indeed, de Maillet (1968, 225–6) hypothesized that each species exists fundamentally as indestructible 'seeds', and that the 'entire space' between stars is 'full of seeds of everything which can live in this universe'. Because of this, even the 'immense spaces'—those 'oceans of void by which the globes are separated'—are pregnant with life that is composed, unwaveringly, of all the species that can possibly exist. The potential for any species to exist is, thus, not something that can be *lost*.

Imagining that these 'seeds' are attracted to planets as 'filings of iron' to a magnet, de Maillet stated that this entails that there is never an 'instant' during which seeds are not 'capable of reaching life'. Indeed, the destruction of a populated planet is a *good thing* because it facilitates the means by which seeds can be 'transferred from one' system 'to another'. Good is self-suffusing, after all. (This Panglossian speculation was entertained as late as the 1800s, when none other than Lord Kelvin resurrected it (Thompson 1871, 200–2).)

Indeed, de Maillet was perversely ahead of his time in trying to deny what little anecdotal empirical evidence had, at that time, begun accumulating that life—and its species—might be included within the horizon of the genuinely historical. He used his panspermic theory as a way of conceptually denying the reality of species extinction and evolution (long before almost everyone else was even aware that this might be a theoretical possibility that they might want to deny or forbid). Claiming that '[t]here are few countries in which characteristic species of animals, trees, or plants have not disappeared', de Maillet (1968, 228–30) conjectured that the species nonetheless persist as 'seeds' after their disappearance. They merely relapse, temporarily, to their invisible, germinal form, dispersed universally throughout the ether.

> These species which we know have been lost for our globe [and] vanished from earth, certainly survive [as] their seeds still occur in the air surrounding it and therefore could reappear again any day.

If something is lost, it will return; if something is extant, it has been elsewhere before. Though idiosyncratic in its superficial expression, the basic assumptions behind de Maillet's system were widespread during the seventeenth and eighteenth centuries (and precursors can be found stretching back to the patristic and Augustinian notion of '*rationes seminales*'). The most important basic assumption was that everything beyond the trajectories of our own lives, and possibly the fate of nations, is beyond the horizon of history (i.e. beyond which where there can be no truly missed opportunities or lost potentials). This early theory of panspermia was conceptually attractive because it enabled belief that, even in a universe where planets and suns die, life—and the biodiverse species that make it up—are not *historical* entities capable of being *irreversibly* lost.

In this outlook, there is no future of life to safeguard because life is not a historical phenomenon. It has neither a specifiable beginning nor end. There is no genuine origin that it could beneficially spread outward from, and no end in sight that such migration could save it. Nor could such spread award life with an expansion in options and autonomy, given that all potentials and pathways have been fulfilled uncountably many times before, in another time and place.

At the time, there simply was not yet conceptual demand for explanations of a historical 'origin of life' in the sense we now recognize it. There were theories such as spontaneous generation and de Maillet's 'airborne seeds', but these explain how species emerge independently of any past history, divorced of any dependency on succession or descent. Ideas like spontaneous generation preclude awareness that life as a whole can become extinct because species are continually renewing independently and concurrently. Life didn't begin long ago; it is continuously beginning anew; fresh starts are not a remarkable event. As Augustine (1954, 143) declared, seeds of species are 'primordially woven into the texture of the world': species and entire orders can emerge and disappear but their *potential for being* cannot be permanently lost.

Without awareness that accumulated biodiversity depends on an uninterrupted backward chain of succession, there can be no sense that if the forward chain of reproductions is interrupted, this spells the end of species forever more. In a wider sense, this leads to the idea that the destruction of some planet-locked species at some planetary locale matters little, because there could be no apprehension that its existence and persistence strongly depends upon that planetary locale and its potentially unique history. Or, as de Maillet (1968, 232–3) put it, the 'vicissitudes' that destroy planets are, in fact, 'largescale circulation' just like 'that of blood in our veins'. Destruction is mere diastole, followed by systolic suffusion, which is an operation that we can leave up to nature. He stressed that this is applies to humanity, its history, and its achievements too. A 'hundred thousand other globes may contain better' examples, and when 'these things disappear in one globe, they reappear in another, perhaps with more perfection'.

> If a sun is extinguished, it is replaced by another. If a globe like ours is set on fire, and all its living creatures destroyed, new generations will replace them on another globe.

If nature is going to disseminate life for you, and painstakingly preserve all its myriad morphologies, then why bother even proposing that the plausibility of achieving this yourself? If life and its species are ahistorical and invariant, then why bother trying to prolong its history and protect its diversity?

3.4 Interplanetary jaunts; cosmic claustrophilia; celestial censuses

Because of this, up until the very late 1800s, vanishingly few explicitly acknowledged that the end of the Earth could simultaneously spell the end of humanity (and whatever potentials it might retain). There was accordingly no motivation to imagine humanity migrating to bring *life* to an otherwise *non-living* universe. In this situation, where the cosmos is not just maximally but also invariantly full of life—independently of the actions of moral agents—there was little reason to make ethical arguments for life spreading beyond Earth. It would produce little consequential benefit, since the cosmos was already *full*.

Of course, as mentioned, there were, ever since Lucian, speculative visions of interplanetary visits, but no concern for permanent migration nor prolonging humanity's future. As far back as April 1610, none other than Johannes Kepler (1610, 23) wrote about travelling to the Moon or Jupiter.

> Certainly, as soon as somebody demonstrates the art of flying, colonists from our nation of humans will not be in short supply.

With revealing choice of language, Kepler shows that he, like many at the time, is imagining inhabitants of other planets simply as other pedigrees of the rational animal that is humanity. He specifies it as a meeting of different *human nations* rather than of different *species* (the original Latin is '*ex nostra hominum gente*' where '*hominum*' means human and '*gente*' means tribe, clan, nation, or people). The German astronomer thus frames it, in clearly colonialist terms, as a meeting of geographically distinct human peoples. This, then, is a trip to another settlement, and thus another episode in the 'circulation' of nations, rather than the settling of an entirely unpeopled space, which would be a historic first for an alien ecosystem. Though the astronomer's usage of the word '*colono*' most likely conceals much darker ambitions—in line with the atrocious prejudices of the time—this would still entail the annexing of an *already peopled* region. Writers afterward, from Wilkins to Dafoe, imagined trips to other celestial bodies and their curious (but ultimately all-too-human) occupants, but never the permanent settlement of otherwise uninhabited spaces. Trips to space remained trips to other peoples—or, worse, the subjugation of other peoples—but had not yet taken the possible form of the peopling of unpeopled spaces.

Moreover, as mentioned, the conviction that the cosmos has been invariantly populated across past time instilled a sense that—in a given time and place—all that could intelligently be accomplished had already been accomplished. This was even used by Bernard de Fontenelle (1688, 68) as an argument for the implausibility of space flight. He wrote that, if it were possible, then 'the Inhabitants of the Moon would have been with us before now'. Here the attribution of a directionless history to the cosmos (wherein everything achievable has already been achieved and will be again) combines with the obvious lack of evidence or remnants from lunar invasions to create the inference that travel to and from the Moon is impossible.

Even more importantly, one finds rare examples—prior to the 1900s—of imaginative depictions of *artificial interstellar travel*. There were trips to other planets in our solar system but scant few to regions beyond. Teleological and wishful thinking allowed early panspermists, like de Maillet, to imagine nature passively accomplishing interstellar feats of organic fertilization, but this hardly counts as 'artificial interstellar travel'. An exception is Voltaire's *Micromégas*, published in 1752, which imagines the protagonist, a giant from Sirius, embarking on an interstellar 'Grand Tour'—using the 'laws of gravity', comets, and sometimes 'the aid of a sunbeam'—to propel himself to our solar system (Voltaire 2002). Nonetheless, this remained a quixotic trip, accomplished by a humanoid, and not a real migration. Crucially, it presents just another meeting of peoples, not a peopling of unpeopled space.

As the eighteenth century wore on, confidence in a fully populated cosmos became even more intense. The economist Adam Smith (2009, 276–7) in 1759 wrote that the cosmos maintains 'at all times the greatest possible quantity of happiness' for all 'inhabitants of the universe'. Smith recoiled from the idea that 'all the unknown regions of infinite and incomprehensible space may be filled with nothing'. Robert Chambers, in the 1840s, wrote of nature's 'aim' that it seeks 'to fill up every vacant piece of space with some sentient being to be a vehicle of enjoyment' (Chambers 1844, 367). To celebrate this joyous saturation of space, the theologian Thomas Dick attempted, in 1838, to provide estimates of the total population of the solar system—generously including the surface areas of Saturn's rings alongside the satellites of other planets. Betraying his conviction that his own nation is omphalos of the known universe, Dick extrapolated from England's population density of '230 inhabitants per square mile' to output a total circumstellar population of '21,891,974,404,480' (Crowe 1988, 199). Later, he guesstimated the total population of the *entire visible universe* and put it at an ample '60,573,000,000,000,000,000,000' or 'sixty quartilions' (see Figure 18.1; Dick 1840, 290).

Demonstrating a strange persistence of this kind of wishful thinking the astronomer Kenneth Heuer (Heuer 1953, 156) repeated this celestial directory, this

	Square Miles.	Population.	Solid Contents.
Mercury	32,000,000	8,960,000,000	17,157,324,800
Venus	191,134,944	53,500,000,000	248,475,427,200
Mars	55,417,824	15,500,000,000	38,792,000,000
Vesta	229,000	64,000,000	10,035,000
Juno	6,380,000	1,786,000,000	1,515,250,000
Ceres	8,285,580	2,319,902,400	2,242,630,320
Pallas	14,000,000	4,000,000,000	4,900,000,000
Jupiter	24,884,000,000	6,967,520,000,000	368,283,200,000,000
Saturn	19,600,000,000	5,488,000,000,000	261,326,800,000,000
Saturn's outer ring	9,058,803,600		
Inner ring	19,791,561,636	8,141,963,826,080	1,442,518,261,800
Edges of the rings	228,077,000		
Uranus	3,848,460,000	1,077,568,800,000	22,437,804,620,000
The Moon	15,000,000	4,200,000,000	5,455,000,000
Jupiter's satellites	95,000,000	26,673,000,000	45,693,970,126
Saturn's satellites	197,920,800	55,417,824,000	98,960,400,000
Uranus's satellites	169,646,400	47,500,992,000	84,823,200,000
Amount	78,195,916,784	21,894,974,404,480	654,038,348,119,246

Figure 18.1 Dick's circumstellar census, 1840.

World	Square Miles	Population
MERCURY	27,573,000	312,440,000
VENUS	179,225,000	2,030,862,000
EARTH	196,950,000	2,231,716,000
MARS	55,146,000	624,880,000
JUPITER	23,634,000,000	267,805,925,000
SATURN	16,543,800,000	187,464,144,000
URANUS	2,954,250,000	33,475,740,000
NEPTUNE	3,348,150,000	37,939,172,000
PLUTO	40,694,000	447,638,000
CERES	723,000	7,858,000
PALLAS	290,000	3,192,000
JUNO	45,000	497,000
VESTA	181,000	1,990,000
MOON	14,574,000	165,147,000
IO	13,980,000	153,776,000
EUROPA	10,980,000	120,783,000
GANYMEDE	33,576,000	369,333,000
CALLISTO	30,959,000	340,551,000
TITAN	39,572,000	435,290,000
TRITON	105,630,000	1,161,926,000
(TOTAL)	47,230,298,000	535,092,960,000

Figure 18.2 Heuer's circumstellar census, 1953.

time taking into account the 'discovery of new worlds' since Dick's time. With the addition of Io, Europa, Callisto, Titan, Triton, and other minor bodies, Heuer updated the total surface area of the solar system to '47,230,298,000', outputting the total population as '535,092,960,000' (his overall smaller estimate based on his—marginally—more sensible idea of extrapolating Earth's total population density rather than merely that of England) (see Figure 18.2). By the time of the 1950s, Heuer was probably not being entirely serious: the book he supplied the numbers in is self-consciously a 'scientific fantasia'. However, back in the 1840s, Dick was, it seems, entirely serious about his celestial census.

In this situation, there was simply no penury—no planetary specificity, and thus existential fragility—that could serve as bondage to be liberated *from*.

4. Penuries of Being Planet-Locked, #1: Scarce Habitability

One can only want to extend the future of life if one recognizes that this is meaningfully possible. For many generations, people broadly assumed that human history couldn't have an end, that it would go on, in directionless variations on the present, forever, no matter what. (And even if humanity itself disappears, eternity entails that it will one day return to repeat the process all over again.) Or, otherwise, it was assumed that we live nearer to history's end than its beginning and that this has been divinely ordained and is unchangeable. For devout Christians, Judgement Day was anticipated sooner, rather than later. Aside from the millenarians, it was believed that there was little that could be done to change this.

The eloquent words of Sir Thomas Browne, writing in 1658, illustrate this well. He said, on religious grounds, that we live in the 'latter scene of time'. For him, this meant that we shouldn't concern ourselves with posterity, given there just isn't the time left to make such concern worthwhile (it is a 'vanity almost out of date', a 'superannuated piece of folly'). Even the pagan 'vainglories' (i.e. pyramids) were less vainglorious than any present attempts to build for the future, because the ancients 'acting early, and before the probable Meridian of time, have by this time found great accomplishment of their designes'. But we, living latterly, must 'decline the consideration of that [future] duration, which maketh pyramids pillars of snow, and all that's past a moment'. We 'cannot expect such mummies unto our memories'.

> 'Tis too late to be ambitious. The great mutations of the world are acted, or time may be too short for our designes.
>
> (Browne 2014, 542–7)

Again, in this scenario, why bother attempting to extend the future of life, particularly if its duration is delimited by divine fiat?

4.1 First forecasts on habitability

However, almost a century later, the question of the length of the future began being offloaded from scriptural prophecy to physical evidence. It also became a question of finite, yet large, timescales rather than uninformative eternity. In the 1700s, scientists started making estimates regarding how long our Earth would remain habitable. Working on assumptions about Earth's dissipation of internal heat, the French naturalist Georges Buffon first provided a quantitative estimate. He forecast the Earth was 74,832 years old and would remain habitable for another 93,291 years before freezing (Moynihan 2020, 177).

Not long after, Jean-Baptiste-Claude Delisle de Sales—declaring that the Earth 'contains within itself multiple principles of dissolution'—predicted that, if we perfected our telescopes sufficiently, we could see enough other planets such that we could observe suitably many of their beginnings and endings to derive a sample size representative enough to put a 'probable' date on our own planet's death. But Earth, he added, is likely to perish before we perfect our instruments. Nonetheless, he surmised that 'probably more than ten thousand centuries will elapse before it touches decrepitude' (de Sales 1777, 198).

With these developments, a sense of a forward time limit for terrestrial humanity had been set. Moreover, it was now a question of *blind and amoral physical process* rather than divine orchestration or diktat, such that it could be scientifically measured and, as some even began suggesting as early as the late eighteenth-century, even altered or prolonged. By bequeathing a beginning and an end to the planet as a habitable entity, the nascent sciences of geophysics expanded the horizon of history to include the planetary entirety and, by implication, therefore also the whole of human civilization as something that emerges, develops, and potentially ends within it.

4.2 Limits to growth

Moreover, these early models of planetary evolution put an end-stop on past habitable time as much as on future time (e.g. Buffon's planetary history arced from molten accretion, through a window of lukewarm habitation, unto frozen decrepitude). By gaining a finite past, this helped dispel the previous sense that, *at least on Earth*, everything achievable had already been achieved throughout the endless past, and that therefore 'achievement' is just revisiting peaks of the past. In other words, such ideas provided a suitable physical backdrop for emerging theories of social progress and improving material conditions. Combining these insights, the Marquis de Condorcet, writing in 1794, proposed that humanity could keep inquiring, inventing, improving, and that this cumulative process has 'no other limit than the duration of the globe upon which nature has placed us' (Caritat 1795, 4).

But so too was early sensitivity to scarcity of natural 'resources': voiced most vociferously to Thomas Malthus in his 1798 response to Condorcet's optimism. But, in response, some utopians refused to believe that nature is already optimally populated whilst also refusing to acquiesce to hard Malthusian limits barricading further optimization. In 1833, one wrote bombastically on bold techno-utopian schemes to ensure that '100 millions of square miles might nourish 1,000,000,000,000 of human individuals, that is, about 1,000 times as many as there are actually living on Earth; and all might enjoy all the happiness that [this] world may afford' (Etzler 1833, 98).

Malthus (1992, 4) himself declared that that 'germs' of life, 'with ample food and ample room to expand', would 'fill millions of worlds in the course of a few thousand years'. He intended this *per impossibile*, as an outcome, so intuitively unattainable, that it firms up the strength of his claim that consumption *on this planet* will always overshoot carrying capacity. Nonetheless, such quips probably inspired another visionary thinker, the Russian prince Vladimir Odoevsky who in the late 1830s conjectured that future humanity would extract resources from the Moon (which, he stressed, is 'uninhabited') in order to offset depletion back home, and remedy Malthus's problem of 'overpopulation' (Moynihan 2021b). But Odoevsky's was a suggestion of lunar prospecting, not habitation. Moreover, he does not appear to have looked beyond Earth, to other planets, as a means to prolong and promote civilization's flourishing. Amusingly, the first suggestion of interplanetary migration and settlement, for this purpose, appears to have come from a satire of such anti-Malthusian visions.

In an 1843 satire of such grand designs, the American essayist Henry David Thoreau (1848, 451–63) decided to beat the utopians at their own game. He wrote, sarcastically, that 'perchance, coming generations will not abide the dissolution of the globe' but use 'future inventions in aerial locomotion' and 'the navigation of space' to 'migrate from earth, to settle some vacant and more [clement] planet [and] people the shining isles of space'. Though intended in jest, and scorn, as a *reductio ad absurdum*, this is perhaps the first place someone suggested *interplanetary* settlement as a means for human survival.

5. Penuries of Being Planet-Locked, #2: Solar Senility

In mid-1800s, the new science of thermodynamics heightened the sense of scarcity of future time. Physicists theorized, falsely, that the Sun produces light by gravitationally collapsing under its own weight. This meant that it would eventually contract entirely and become terminally dim. In a sense, this expanded the horizon of history to include the entire solar system: a luxuriantly, languorously imploding Sun can only expend so much life-giving radiance.

5.1 Early thermodynamic forecasts

In 1854, Lord Kelvin calculated that Earth has been 'efficiently illuminated' by the Sun for around '32000 years' but that this 'cannot last' for another '300,000 years' (Thomson 1854, 79–80). By 1861, he revised his numbers upwards by not all that much, claiming that Earth's 'inhabitants' cannot have 'many million years longer' (Thompson 1861, 28). Nine years after this Helmholtz (1995, 270) said we cannot expect our Sun 'to maintain for an additional 17,000,000 of years the same intensity of sunshine as that which is now the source of all terrestrial life'; the Canadian-American astronomer Simon Newcomb (1879, 521) prophesied that in about '12,000,000 years' it will 'rapidly cease to radiate the heat necessary to support life on the globe'; by 1881, US spectroscopist Charles Augustus Young (1881, 276) wrote that it is unlikely that the Sun can offer 'sufficient heat to support life on the earth [for] ten million years from the present time'. Forecasts varied but by the closing decades of the 1800s, the community converged on the very low tens of millions. The solar system's future was not all too long.

Some ignored this prospect entirely, calling it 'repugnant to the moral sense and against all law', claiming 'old sol' will shine 'forever' (Swagman 1892, 3). Others retreated to the comfort of some afterlife. Newspapers quipped that '[i]t may be news to know that the world will wind up in ten million years, but the news will become stale before that time' ('Alarming Situation' 1892 2). Others accepted the prospect but offered the craven comfort of temporal discounting: that extinction would not befall 'our children nor our children's children' (Winchell 1870, 413).

Many produced lurid visions of the last humans, in the frigid future, retreating from 'the dread empire of frost' as '[t]he face of the sun is veiled' (Winchell 1870, 412). Expressing the mood in a widely circulated passage from 1892, Irish astronomer Robert Ball pronounced that there 'is [a] distinct limit to man's existence on the earth': our 'sun has already dissipated about four-fifths of the energy with which it may have originally been endowed; it may hold out for 4,000,000 years, or for 5,000,000 years, but not for 20,000,000 years'. He wrote:

> The race is as mortal as the individual, and, so far as we know, its span cannot under any circumstances be run out beyond a number of millions of years which can certainly be told on the fingers of both hands, and probably on the fingers of one.

(Ball 1892, 490)

Of course, this prospect was larger than that of the century prior, much more commodious than Buffon's ~90,000 years., but expectations of human potential had also grown in the meantime. More importantly, it was the ratio, of 'time spent' to 'time ahead', that proved distressing—not the brute number of zeroes—as highlighted by Ball above. Evolution had come thus far; it didn't have the time to go much further. There wasn't room for grand optimism regarding humanity's far future on Earth.

The prospect troubled evolutionists like Charles Darwin and Thomas Huxley. In an 1893 lecture, Huxley (1893, 190) looked forward miserably to a slow, inglorious squander of the accumulated complexity of eons of past evolution, as our Sun slowly shuts down after its lifespan's meridian. One attendee at the lecture was explicit about the depressing implications of the perceived past-to-future ratio:

> [all the] millions of years [spent] bringing us up to the present point [and] no sufficient time [remains] to evolve all the possibilities of our actual state! This is sad indeed. Waste, sheer waste on the most gigantic scale ... .
>
> (Klein 1893, 598)

He saw no way out or round or through: '[t]he most daring imagination will hardly venture [that] the power and the intelligence of man can ever arrest the procession of [this end]' (Huxley 1893, 190). (Happily, the next generation proved this professed ceiling on 'daring imagination' wrong, with Thomas Huxley's own son, Julian, announcing in 1942 that 'adventures to other planets and stars are possibilities', albeit 'for the remote future' (Huxley 1942, 573).) Darwin, in 1865 letter of 1865, professed that he hated the suggestion that after 'the progress of millions of years' the planet will nonetheless freeze (Darwin 1865). Rejecting this, he couldn't help but appease himself by putting faith in the inevitability of some 'fresh start' after 'our planetary system has been again converted into red-hot gas' (and, presumably, the whole evolutionary drama repeats over, from the nebular accretion of planets all the way up to the re-evolution of civilized life). Like many before him, when faced with the termini made possible by genuine history, Darwin retreated to speculations about repetitions allowable within the scales beyond history's acknowledged horizon, in order to find solace.

Nonetheless, any confidence in the rerun of humanoid civilization—here or elsewhere in space—was being troubled by growing awareness that the intervals between such recurrences must reasonably be vast enough to seriously dampen any solace they might have once provided. In 1884, Simon Newcomb looked back on the earlier works of Thomas Dick, and his Panglossian population censuses of the universe, and took aim at the naivety of such reasoning. Dick's naivety, Newcomb wrote, may be seen 'by reflecting on the brevity of civilization on our globe, when compared with the existence of the globe itself as a planet'.

> [Earth] has probably been revolving in its orbit ten millions of years; man has probably existed on it less than ten thousand years; civilization less than four thousand; telescopes little more than two hundred. Had an angel visited it at intervals of ten thousand years to seek for thinking beings, he would have been disappointed a thousand times or more. Reasoning from analogy, we are led to believe that the same

disappointments might await him who should now travel from planet to planet, and from system to system, on a similar search, until he examined many thousand planets.

(Newcomb 1879, 527)

5.2 Solar squander or panspermic purpose?

Given this heightening sense of the potential barrenness of space (given the plausible temporal intervals between civilizations), the notion that the destruction of Earth might well spell the end of civilization in universe (or at least some significantly large portion of it to be consequential) began to coalesce. So too, therefore, would motivation to migrate also come increasingly into focus. Relying on cycles and repetitions was becoming less comforting, so proactive solutions started to come into focus.

During the latter half of the 1800s, scientists began conclusively realizing that almost all energy 'derived from the solar rays, directly or indirectly' (Gregory 1857, 371), that the 'sun is the spring that drives all [and] maintains all human life' (Tesla 1900). But, additionally, it had started to become apparent that the Sun is incredibly wasteful: that the Sun's finite fund of energy is being pumped into space without compensation. Prior generations would have responded with assurance that for every Sun that dies, another is necessarily born, such that global equilibrium is conserved. But this panacea was becoming less convincing due to building awareness of the sheer ratio of energy being squandered every second. Given such manifest profligacy, could one really have faith in some unobservable compensation elsewhere?

From this, it becomes possible to infer that remaining Earth-locked means acquiescing to an arbitrary finitude of not just future time but also of available energy (which could otherwise be used to routing towards intelligent, constructive ends).

Against this background, in 1900 Nikola Tesla penned an essay, 'The Problem of Increasing Human Energy', in which he asserts that 'the great problem of science' is to 'increase the force accelerating human movement [and] work', which 'means to turn to the uses of man more of the sun's energy'. Interestingly, what frames Tesla's awareness of our currently suboptimal interception of the Sun's available funds—and thus also enables his innovative identification of it as a *problem that can be addressed*—is acute awareness of the finitude of the solar lifespan and his citing of the 'short span of life' granted by Lord Kelvin's estimate of 'six million years' (Tesla 1900, 175). You can't squander an infinite resource, after all. And even if it isn't infinite, assuming that the outpourings are somehow part of a wider compensatory system—where loss here becomes gain elsewhere—obstructs any urgency or sense of lost opportunity.

Nonetheless, assumption that *everything* in nature must be teleological—and that good is thus self-suffusing throughout sidereal space—persisted. Some invented ingenious ways to refuse the reality of solar wastage rather than attempt to find a fix. Svente Arrhenius, a Swedish physicist and one of the discoverers of global warming, suggested in 1906 that this solar squander is not in fact squander at all but is in fact the mechanism behind passive panspermic interstellar germination. Returning to ideas first championed by de Maillet, Arrhenius suggested that the 'radiation pressure of our sun' provides propulsion that circulates 'germs' out 'into space' to other stars. Arrhenius calculated the journey to neighbouring star systems could take 'nine thousand years'—but was bullish that organisms could survive the trip.

> In this manner, life may have been transplanted for eternal ages from solar system to solar system ... .

The Swedish scientist positioned this as disproof that life ever had a beginning at any one particular time or place. As with de Maillet, long before, this underwrote the further view that, since it didn't have a spatiotemporal beginning, neither can life have an end in time or space:

> ... life may exist and continue forever undiminished.
>
> (Arrhenius 1908, 211)

Even Tesla ultimately resorted to this type of thinking. Despite adroitly identifying the problem of solar scarcity, and going so far as to prioritize it explicitly as a 'great problem', Tesla evidently didn't think it feasible that we could do anything about it in the longer run. So, he retreated, rhetorically, back to relying on deliverance from blind processes of nature. Demonstrating wishful thinking rivalling Arhennius, he claimed that even after the Earth is sterilized by cold, the 'glimmering spark of life' will remain on it, such that if Earth ever enters into the orbit of a new star, there is 'a chance to kindle a new fire'. Supporting this, he cited the confidence of certain scientists that organic 'germs' can be 'transmitted through the [cold of] interstellar space' (Tesla 1900, 175).

Others, during the turn of the century, resorted to even stranger schemes. The German parapsychologist Karl du Prel, writing in 1880, noted that 'the work of one generation is always only for the profit of succeeding ones' such that the 'collective product' of humanity rests in its 'latent generations', but if it is so that humanity is doomed to perish by a dying sun, then 'the futility of the whole game' is 'exposed'.

> The life of the earth will have an end ... Of what use is our culture-history [if] Earth is always to remain an isolated star, and the history of man is never to debouch into the general stream of cosmic history?

Unwilling to accept this, du Prel sought supernaturalism as an escape. Via a grab-bag of interstellar clairvoyance, telepathy, and the transmigration of souls, he said that human history might 'receive a cosmical extension' and transmit its cultural heritage to other stars. Believing such 'cosmic intercourse' possible, du Prel (1889, 257–91) felt hopeful that 'every after the dying out of man upon the earth when it has become wholly cold and uninhabitable, the attainments of human culture might be preserved'. Humanity 'would find its historical heir in the kosmos', thus gaining immortality 'even though the earth should plunge into the sun'. Similar attempts to find a kind of secular 'afterlife' in the stars—by way of restitutions, repetitions, or transmigrations of our identities across other populated planets—remained popular during the decades buttressing 1900 (e.g. Blanqui, 1872).

6. Huddling Around the Solar Ember

But against this persistent wishful thinking, others were arriving at more proactive conclusions. Rather than relying on blind nature to keep life's flame alive—or resorting to spiritualist fancy—some had already suggested that as the Sun contracts and cools, humanity should take technological action to migrate, inward, to the inner planets of the solar system. In other words, humanity might be able to prolong its future by huddling around the solar ember.

One of the first to propose such an idea was, unsurprisingly, H. G. Wells. During the epilogue of his 1898 novel, *The War of the Worlds*, he suggested that

> there is no reason to suppose that [settling Venus] is impossible for men, and when the slow cooling of the sun makes this earth uninhabitable, as at last it must do, it may be that the thread of life that has begun here will have streamed out and caught our sister planet within its coils.
>
> (Wells 1898, 286)

Even more boldly, only a few years earlier, in 1895, the Russian rocket-pioneer and cosmist Konstantin Tsiolkovsky had offered a speculative answer to Tesla's 'great problem' of solar wastage. He had bombastically mused that humanity's descendants might settle the asteroid belt, remould and combine the matter of the celestial bodies so as to make some kind of vast Sun-girdling ring or sphere that would also be able to intercept magnitudes more of the Sun's rays (alongside offering more ample *surface area* for supporting a greater population). He argued that this would vastly increase the number of souls that can be supported within our solar system. The Irish crystallographer J. D. Bernal independently arrived at identical conclusions around 30 years later in the 1920s (Moynihan 2020, 371–5).

But huddling around the solar ember—whether by migrating inward or by creating Sun-girdling spheres to sap its scarce light—would only delay the inevitable. Wells clearly realized this, imagining life's continued fight against the dying of the light.

> Dim and wonderful is the vision I have conjured up in my mind of life spreading slowly from this little seed bed of the solar system throughout the inanimate vastness of sidereal space. But that is a remote dream.
>
> (Wells 1898, 286)

The vagueness of this vision speaks volumes. It isn't even clear what Wells is envisioning here: crewed spaceflight? Directed panspermia? Something else entirely?

All in all, as the 1800s closed, it seemed that the prospects for humanity's far future in the physical cosmos were not bright. There was seemingly not any available power source to plausibly propel humanity itself beyond the solar system, nor was there any known power source to prolong the Sun much beyond 10 million years. Aside from Wellsian 'remote dreams', the outlook was somewhat bleak.

7. Radium to the Rescue

Then, just as the 1900s dawned, radioactivity was discovered. Uranium rays were first noticed, by fortuitous mistake, by Henri Bequerel in 1896. Two years later (not many months after Wells published *The War of the Worlds*), Marie Skłodowska-Curie and Pierre Curie discovered radium. This changed everything.

7.1 Atomic stoves

Measures of the amount of energy corked up inside these atoms proved truly astonishing. In 1902, it had been established—by Frederic Soddy and Ernest Rutherford—that this energetic wealth was being emitted by radioactive decay. Previously assumed indestructible and impenetrable—the most elementary elements of the universe—radioactive atoms were shown to be expelling energy as they *slowly disintegrated*. And slow it was: some elements seemed to be decaying over a period stretching into the billions of years. The British physicist Robert Strutt, in 1903, was inspired to make a clock, powered by a miniscule amount of radium, projected to keep ticking for multiple millennia ('A Radium Clock' 1905).

Time scales couldn't help but expand. The revelations of the Curies, Rutherford, and Soddy almost immediately triggered a flurry of excited responses from scientists across fields ranging from geology to astronomy. Within months, it was suggested by Irish astronomer William Edward Wilson (1903) in an article in *Nature* that radioactivity may 'afford a clue to the source of energy in the sun'. By September of 1903, the geophysicist George Darwin (son of the biologist) celebrated Curie's 'unexpected' discovery of this 'new source of energy'. If the Sun powers itself by 'liberating atomic energy', he conjectured, rather than mere gravitational collapse, then we may have to expand the 'cosmical time-scale' by 'some such factor as ten or twenty' (Darwin 1903).

John Joly (1908), the Irish geologist, similarly wrote that radioactivity 'in the materials of the sun would suffice to account for many millions of years of solar heat'. Rutherford (1907) concurred, writing that we can expect solar 'brilliancy' for orders of magnitudes more than Kelvin's considerations of no 'more than 12 million years and probably much less'. Curie's 'new source of heat', he wrote, would suffice to power the 'sun for a much longer period', thus vouchsafing a much greater 'time for habitation of our globe'. Robert Stout, former prime minister of New Zealand, summed up the situation in 1914:

> Before radium was discovered, the greatest age the physicists would give our earth was a hundred million years; now it may be many thousands of millions. Nature is on a vaster scale than our imagination can grasp.
>
> (Stout 1914, 10)

By 1920, Arthur Eddington—an authority of physics—was venturing that our Sun, powered by atomic reservoirs, would be able to 'maintain its output of heat for 15 billion years' more (1920, 298). (To hammer home his point, he referred to the stellar theory of Kelvin and Helmholtz as a 'corpse'.) From meagre millions, the expected future had swollen to multiple billions. In the following years, another prominent physicist, Sir James Jeans, produced an even more eye-watering forecast. Declaring atoms are 'pure bottled energy', Jeans (1928) produced his own theory for how they disintegrate in the Sun and boldly estimated that the star retains enough 'unbroken bottles' for *a trillion years more sunlight*. (This was cut to 10 billion years in the 1940s, and then again to 5 billion years in the 1960s.)

7.2 Enormously finite futurity; upping the upward bound

Artist Wassily Kandinsky proclaimed, in 1913, that 'the disintegration of the atom was, in my soul, like the disintegration of the whole world'.

> Suddenly the thickest walls fell. Everything became precarious, unsteady, and pliant.
> (Kandinsky 1913, x)

But this was simply the feeling of history creeping into domains where previously it had not been allowed. One scientist, writing in 1909, captured this mood when writing that, even from the 'early days', humans have readily acknowledged that everyday objects at the mesoscale are mutable—having originations, durations, and terminations—but that this mutability has traditionally been forbidden of the supposedly 'ultimate entities', occupying the smallest and largest scales (such as atoms, the human species, 'and the Earth on which it dwells'). But the expanding circle of history, it was mused, will continue to include more and more of these previously 'ultimate entities' as humanity learns more about wider swathes of the universe and 'the horizon of its vision is enlarged' (Mather 1910, 50).

Indeed, the arena of history had not just expanded outward, to encompass our planet and our Sun, but it now had also seeped downward, to inject irreversibility into the very basement of matter. What were previously considered the most immutable constituents of the universe—assumed to be sealed off from vicissitude—were gaining a veritably directional history.

More importantly, the ticking away of atomic decay now allowed scientists to read this 'cosmical time-scale', to get a glimpse, for the first time, of some sense of its scale and proportions beyond our own Earth and even our own Sun. Nuclear physics, that is, enabled truly 'cosmical' questions about history and chronology to begin to become scientifically tractable in the early 1900s.

And the proportions were huge. Some atoms decayed at rates extending into the billions of years, thus giving great support to the nascent sense that the whole cosmos was billions of years old and had billions of years left to live.

As early as 1904, Frederick Soddy (Soddy 1904, 188) was conjecturing that if all matter radioactively decays, then, regardless of how glacial the rate, we have to entertain that a definite 'beginning of the universe' is therefore 'also fixed'. This would be when, for whatever reason and no matter how distant in the past, the unspooling of all atoms got underway. By the 1930s, Eddington (1931) was also speculating the 'end' of the universe would be something like the dissolution of all matter—even the most stable—into a ball of radiation that expands into ever greater wavelengths. All of this was conjectural, of course, but new vistas and means of investigating them had opened. As Henry Norris Russell (1925) put it: 'within the last few years, vistas have been opened by new discoveries that extend the range about which we can make predictions by a thousand-fold'.

Of course, many scientists still stubbornly tried steadfastly to cling to the idea that there must be some compensatory process that would resolve this vast but directional history into some kind of directionless cycle (e.g. Milikan 1931). (For example, one author—in an article responding to radium's revelations, with the self-explanatory title 'Is the Human Race Mortal?'—decided to assert confidently that the lost energies of the atom must somewhere in the cosmos be 'gathered up again and thus again become available', allowing the universe to still go round in cycles, such that humanity will never *truly* disappear (Saleeby 1906).) But, despite such persistence, the atom instilled a keen sense that the future could be *enormously finite*; that is, no matter how generously long it could be, it will still have had a definite upper bound. It was adroitly remarked that radioactivity had extended the expected future 'indefinitely ... but not infinitely'; a Sun powered by subatomic process will shine for eons but still undeniably 'exhaust itself at last'.

> No matter how large your capital, you cannot live upon it forever.
>
> (Saleeby 1906)

Similarly, as H. N. Russell wrote in 1925:

> Amazement at this prodigious sweep of time is naturally every one's first reaction, and a life of tens of thousands of billions of years at first thought seems eternal. But, as the first surprise passes, and our minds become adjusted to the new idea, it is plain indeed that the life of a star, long as it is, is very decidedly finite.
>
> (Russell 1925)

And this is crucial: here, for the first time, was concrete evidence of an upward limit on the future that was enormous yet still fundamentally finite. Not infinite, such that it cannot be meaningfully made shorter or longer, but neither was it meagre, such that prospects for progress are defanged and desires for prolongation demotivated.

Indeed, just as a paltry expected future discourages responsible action and concern for posterity, so too does an infinite one. Because again, in eternity, everything achievable has already been achieved, and every achievement lost will one day be eventually returned. Within eternity, one cannot make any positive thing persist for longer or shorter through action, inaction, or misdeed. As Jeans (1930) put it: 'it is hard to see what advantage could accrue from an eternal reiteration of the same theme, or even from endless variations of it'.

Previously, these two attitudes to futurity—of either an imminent end or an infinite deferral—were more or less the exclusive options on the table. For generations, the assumption in Europe had been that we lived nearer to history's end than its beginning. But now there was an upper limit on forward time that was finitely enormous, yet *not* infinite, such that it appeared that, accidents aside, we could living at history's very beginning. And, more importantly, it could eventually come into view just *how much comparatively better* a limitless future (wherein humanity persists for as long as possible) is when compared to a truncated one (wherein somehow this gift of time is squandered or prematurely lost).

Giving an enormously finite limit created a sense that the human future could be meaningfully shorter or longer and, moreover, that the differences between the minimal and maximal allotments could be seriously 'cosmical' in extent. In other words, it brought into crisp relief just how bad an *early extinction* for humanity could be, but also just how good an elongated and complete futurity might be. This is relevant to the ethical argument for going to outer space because it made it clear, for the first time, that the longest possible futures could be *much* longer—and thus, potentially, also *much* better—than the shorter futures.

It revealed the sheer scale of potential differentials involved, and moreover facilitated the first serious thought on how the near-term actions of humanity might foreclose or help facilitate this entire elongated futurity. By extending the upper bounds on humanity's potential life should it remain on Earth, it made people clearly think about how much better a longer future might be, and thus made much clearer the motivation behind attempts to explore the possibility of decoupling humanity from Earth—and the Sun—such that it could persist for *even longer*.

However, this realization came in multiple parts. By giving proportions and boundaries to 'cosmical time', both past and future, atomic physics allowed people—for the first time—to orient themselves as to humanity's position within it. It became obvious that, compared to the cosmological and geological past, humanity had existed for only a vanishing sliver of time. Our species is a 'very recent apparition indeed', Jeans (1928) commented. Moreover, it simultaneously became obvious that—compared to the huge terrestrial habitability ahead—the entire past of our species could be dwarfed by its future.

Jeans spoke of the 'fresh glory of the dawn': writing of 'unimaginable opportunities for accomplishment' (Jeans 1930, 353) and 'unexplored potentialities' ahead (Jeans 1928). It became common for writers to state that human history is a split second of Earth history, and civilization a fraction of that fraction, with the scientific endeavour (and, with it, the project of materially improving humanity's lot in the world) making up only a further splinter of that splinter. From here it wasn't far to reason that if we have achieved 'so much in the first few moments [of] existence', then we can only wonder at what might be achieved 'in the long future ages' within which humanity is 'destined to labor on [Earth's] surface' (Jeans 1928).

The geologist Kirtley F. Mather (1924, 142) claimed that if we suppose humanity unique in its capability to respond to moral reasoning, then (fallible and error-prone though this faculty of ours may remain) the epoch of ethical agency on Earth may *only just be dawning*. Joseph McCabe made similar comments in 1921, claiming that the era of mind is only just beginning, and that the geological prospect ahead 'becomes almost stupefying ... if we pay attention to the modern rate of progress'. We can but 'dimly conceive' what might be accomplished in the 'millions of years' ahead if the 'rate we are [currently] going continues to even be minimally maintained', he marvelled (McCabe 1921, 114–17).

Others, such as American glaciologist Thomas Chrowder Chamberlin, began arguing that the protracted future means that, the 'ulterior' impact of irreversibly damaging actions is correlatively increased. Chamberlin (1910, 5) wrote:

> If [the extended future] be true, it is eminently fitting that our race should give a due measure of thought to the ulterior effects of its actions.

Having thus realized that the majority of potential value exists latently in the elongated future, Chamberlin was also nonetheless well aware that possibility or probability does not 'ensure actual realization'. Protracted habitability 'does not necessarily carry the actual realization of the future opportunities thus open to our [species]', he wrote (Chamberlin 1911, 388). Catastrophes and cataclysms happen, species go extinct. Jeans (Jeans 1930, 353), similarly, clearly realized that '[a]ccident may replace' a vast future for humanity with only a 'truncated ... fraction'.

So, here were three key ideas that would combine to provide ethical motivation for the project to liberate humanity from its parent star: first, there is a new sense that it is at least possible that we are living at the very opening of the entire human story; second, the inference that therefore the peaks and preponderances of achievement lie ahead in the longest of all possible futures; and third, that the premature foreclosure of the longest future could be an enormous tragedy, given that the comparison between longer and shorter is now potentially 'cosmical' in scope.

'No previous [species] has shown clear evidence that it was guided by moral purpose seeking distant ends', Chamberlin (Chamberlin 1911, 388) mused, yet in humanity 'such moral purpose has risen to distinctness'. He continued:

> And as it grows, beyond question it will count in the perpetuity of the race.

This is significant. It is amongst the first clearly explicit claims that humanity *ought* to safeguard its survival. It represents the *granting of a reason* for what was previously blindly assumed: survival here rises from instinct into ethic, becoming a self-conscious goal with justifying warrants. His reasoning is simple: the future of ethical agency could be commodious, or truncated, and the difference in magnitude is potentially overwhelming in scope; thus, we ought to act now, to do what we can to safeguard our potential to achieve the longest, most fully fulfilled future possible.

Now that the ethical precept of human survival had been made explicit, and so too the idea that a longer future is much better than a truncated one, it fell to others to articulate the next logical step: the future of the Sun may be long but it remains finite, and there are other Suns. Moreover, humanity has recently gained unexpected headroom in the departments of expected future time and untapped available energy; surely, with all that at its disposal, there must be some way of decoupling the future of humanity from that of the Sun, liberating it from solar finitude.

In the light of the extended future, the desire for extending it further, and the revelations of energetic wealth initiated by nuclear physics, the first serious proposals for interstellar exodus could emerge.

As Jeans (1928) wrote:

Looked at on the astronomical time-scale, humanity is at the very beginning of its existence—a new-born babe, with all the unexplored potentialities of babyhood; and until the last few moments its interest has been centred, absolutely and exclusively, on its cradle and feeding-bottle. It has just become conscious of the vast world existing outside itself and its cradle; it is learning to focus its eyes on distant objects, and its awakening brain is beginning to wonder, in a vague, dreamy way, what they are ... Its interest in this external world is not much developed yet, so that the main part of its faculties is still engrossed with the cradle and feeding-bottle, but a little corner of its brain is beginning to wonder.

It was Tsiolkovsky (1967, 83) who famously wrote in 1911 that 'the planet is the cradle of mind, but one cannot remain in the cradle forever'.

8. Interstellar Exodus

Chamberlin and many others of his generation had warned that Earth's resources (such as fossil fuels) are finite, yet the discovery of radium, aside from provoking expectations on Earth's habitable future to boom, also seemingly revealed new methods to catapult civilization off-world.

Victorian thermodynamics came with a ceiling on future time that, judged against subsequent subatomic revelations, instils claustrophobia. That ceiling now seems very low. The same applied also to assumptions of available energy. For someone living at the close of the Victorian era, it might have seemed that human civilization—burning coal, damming rivers—had pretty much uncovered all available sources of work. But here, in mundane matter—of which we are all awash in abundance—were revealed energetic coffers 'of a magnitude of which we have no experience'. So wrote Frederick Soddy (1920, 33), co-discoverer of radioactive decay. 'The energy is there. The knowledge that can utilize is not—not yet.'

With the discovery of radium, he often claimed, it became clear that humanity had not yet come *anywhere near* to the ceilings of energy use imposed by physics. We were only just setting out. This keyed directly into expectations on humanity's hopes to decouple its future from that of the Sun.

As previously mentioned, trips to other celestial bodies—like Mars or the Moon—had been envisaged before, but it is hard to find people anticipating crewed travel to *other* stars before 1900. This is because of there simply wasn't the recognized headroom, with regards to as yet untapped energy sources, to think seriously about propelling crewed vessels beyond the Sun's gravity. But, by revealing extremely spacious headroom, in expected future and untapped energy, nuclear physics made interstellar travel suddenly seem like it may be within the realms of possibility (at least, eventually).

Already in 1904, the French writer Maurice Maeterlinck was commenting on the unexpected discovery of the 'radium', and pointing to the fact that it gives good reason for 'having confidence in the destinies of our kind'. For most of past human history, people had toiled under ignorance of the natural world whereby 'a mere nothing could have destroyed our human future at the groping hours when our [intellect] was [just] forming'. But today, humanity seemingly stands on the cusp of unlocking the base powers of nature, such that some further discovery 'may lead us straight to the very sources of energy and the life of the stars'. He imagined oncoming break-throughs that would emancipate humanity from Earth, such that it could 'flee from worn-out suns' and seek 'virgin and inexhaustible worlds' (Maeterlinck 1920, 340–2).

None other than Soddy was likely one of the first to make explicit connection between radium and space travel. In his classic of 1909, *The Interpretation of Radium*, he imagined that a civilization that has unlocked the atom would be able to 'explore the outer realms of space, emigrating for more favourable worlds as the superfluous emigrate to more favourable continents' (Soddy 1909, 244).

The first to put all the pieces together, however, was Konstantin Tsiolkovsky. In 1911, he wrote the following:

> [I]f it were possible to sufficiently accelerate the decomposition of radium or other radioactive bodies, which are probably all bodies, then its use could give, under the same other conditions, such a speed of a jet device at which the reaching of the near-est star would be reduced to 10–40 years. A pinch of radium would be enough for a ton rocket to sever all ties with the solar system.
>
> (Tsiolkovsky 1967, 96)

He clearly saw that achieving interstellar exodus would decouple the future of human-ity from the future of the Sun. Even though our Sun was now expected to afford us sunlight in the billions of years, the ability to migrate to other suns would mul-tiply humanity's longevity again. Most likely taking aim at Kelvin, he wrote that his calculations on a 'pinch of radium' invalidate 'the gloomy views of scientists about the inevitable end of all life on Earth'. 'Cooling due to the loss of solar heat' should no longer be acquiesced to as an 'immutable truth', he continued: in the future, interstellar humanity will migrate from 'Sun to Sun', persisting for cosmo-logical timescales (Tsiolkovsky 1967, 100). (In 1912, French astronautics pioneer Robert Esnault-Pelterie (1913) independently arrived at the conclusion that atomic power makes flight much more feasible. He even calculated that '400 [kilograms] kg of radium' in a '1000 kg vehicle' would be enough 'for the journey to Venus and back', thus allowing humanity to 'visit' its 'immediate neighbours'.)

Not many years after this, Tsiolkovsky's American counterpart, Robert Goddard, independently arrived at similar conclusions. By 1913, he was using his notebooks to record the reasons why he had decided to dedicate his life to rocketry. He wrote that 'the navigation of interplanetary space must be effected to insure the continuance of the [species]', that 'the continuance of life and progress must be the highest end and aim of humanity, and its cessation the greater possible calamity' (Goddard 1970, 117). Five years later, he penned a bold plan to achieve this aim, entitled 'The Final Migration'. Goddard asked:

> [w]ill it be possible to travel to the planets which are around the fixed stars, when the sun and the earth have cooled to such an extent that life is no longer possible on the earth?

Goddard answers yes, if it is 'possible to unlock, and control, intra-atomic energy'. He even imagined the use of a 'radium alarm clock' to wake up pilots intermittently throughout the multigenerational flight ('at intervals perhaps of 10,000 years for a passage to the nearer stars, and 1,000,000 years for the greater distances'). In terms of destination, Goddard had recommendations: an Earth-like planet around a comparatively young sun, where 'further development' of humanity may 'take place for many ages' before the sun becomes 'too cool'.

> The destination should [also] be in a part of the sky where the stars are thickly clustered, so that further migration would be easy.

But should the unlocking of atom-propelled flight not prove possible, and crewed interstellar flight by other fuels and means remain infeasible, Goddard proposes a contingency plan: he suggested creating organic 'protoplasm' that 'can withstand the intense cold of inter-stellar space' and can be shot, slipshod, into sidereal space. The hope, as ever, being that it finds purchase elsewhere and produces 'human beings eventually by evolution' (Goddard 1978).

The issue was that, from the perspective of the 1910s and 1920s, it was unclear as to whether atomic energy could ever be uncorked and, moreover, how long this could take to master. But remember that the atom's energy had, of course, radically extended the time left ahead within which humanity could perfect such an art. Several years afterward, Bertrand Russell (Russell 1949, 217) made exactly this claim. Citing Jeans' estimates of a trillion years left of terrestrial habitability, the philosopher said that this 'gives us some time to prepare for the end' and 'we may hope that in the meantime' aeronautics improves enough to be able to send people to settle other stars. 'It is perhaps a slender hope', the philosopher concluded, 'but let us make the best of it.'

Precisely this vision was depicted in a narration on the future history of humanity by British-Indian geneticist J. B. S. Haldane in 1927. In his *Last Judgement*, Haldane set up the familiar story: a 'generation ago it seemed very plausible' that a cooling sun would make our Earth uninhabitable in 'a few million years', but now '[m]odern physics suggests ... that it will last for at least a million million years', so he attempts to speculate upon what might happen over this time scale. In the story, an inspiralling Moon threatens Earth in the far future, such that some people deign to undertake diaspora. These people make it to other planets in the solar system, and eventually develop into branching forms of post-humans. (Haldane explicitly celebrates the idea that, in other non-terrestrial settings, 'the human brain may alter in such a way as to open up possibilities inconceivable to our own minds'.) At the end of the tale, Haldane looks outward and forward, claiming that in the long swathes of time available, another star system may pass close enough our own, such that the leap will be attempted. He states this might take 'thousands of years' and 'only a very few projectiles per million would arrive safely' (Haldane 1927, 35).

However, what is truly remarkable about Haldane's piece is that he gives what is almost certainly the first *quantitative estimate* of just how long the protracted future for humanity could be, should it achieve interstellar exodus. Using what little was known to the physics of the time, he surmised that '[o]ur galaxy has a probable life of at least eighty million million years'. He remarked that within that time, the 'heirs of the species' might have achieved almost as much as they could possibly physically achieve, reaching a state of maturity and advancement unrecognisable to us now. He closed the piece, piquantly adding: 'And there are other galaxies' (Haldane 1927, 36).

Other scientists soon echoed such visions, often making the connection with the discovery of atomic energy. Haldane's friend, J. D. Bernal, similarly remarked in 1929 that our descendants may eventually leave behind the Sun and fill 'the sidereal universe' (Bernal 2017, 27–8). Tsiolkovsky, likewise, imagined 'artificial biospheres' being used to travel between stars. Thinking of these as 'creatures' in their own right, he imagined them spreading out throughout the universe:

> A million billion childless … suns, young and old, will tirelessly emit [solar energy] for many trillions of years… This abundant radiant energy must be used by such creatures. They [will] surround all suns, even those without planets, and use this energy in order to live and to cogitate. The energy of the stars should exist for something!

(He also curiously argued that gravity puts limits on the size of our brains, which would be left behind in zero gravity, thus eventually allowing the evolution of greater encephalization.) Soon, such visions made their way into fiction. One of the first depictions of a generation ship in science fiction arrived in the form of Laurence Manning's 1934 'The Living Galaxy', which depicts a far future humanity engaging in 'long trips of exploration [which] in a few million years spread [humanity] over the planets [and] throughout all [the] galaxies in space'. Manning even imagines humanity using planets themselves as spaceships, annihilating their radioactive internal matter to turn them into gargantuan spherical rockets, powered by 'atomic motors', that can propel entire populations across the cosmic expanses (Manning 1934).

Such bold visions were instilled by the sense of confidence in the future that radioactivity had fuelled. But the period of 1900–1930 also saw a huge expansion of the known cosmos itself: impelled by improved telescopes and methods, observational astronomy had increased 10-fold the acknowledged size of the Milky Way (Smith 1982). So too had it led to the eventual acceptance of the definite existence of *other* galaxies. (This was no doubt the context for Haldane's comments.) Thus, humanity's cosmos had expanded in three monumental ways: not only had the expected future grown massively but so too had the headroom grown commodiously with regards to untapped energy and available space to explore.

But, again, mere possibility does not ensure actual realization at all. Even some of the most boldly far-reaching authors of this boosterish period portrayed interstellar exodus as a project thwarted. Indeed, Olaf Stapledon's *Last and First Men* (a 1930 novel narrating the future history of humanity, billions of years into the future, directly motivated by the elongated predictions of Eddington and Jeans) imagines a troubled, stop-and-start course for 'subatomic energy'. Stapledon chronicles how, in around 120,000 AD, an industrial accident sets off a nuclear reaction, igniting radioactive ores throughout the planet, causing the collapse of civilization. Only much later, in around 400,000,0000 AD, after a slow recovery, do future generations eventually use radioactive decay ('annihilation of matter') to propel 'ether ships' and '[i]mmense rockets' to Venus, which they terraform and settle (Stapledon 1968, 184–5). Even further into the future, in 800,000,000 AD, the Sun has continued to shrink and plans are afoot to jump, again, to Mercury so as to chase our star's dwindling heat. However, it is discovered that the solar system is on track to pass through a gas cloud that will cause the Sun to 'flare up and expand prodigiously', thus making the inner system uninhabitable (Stapledon 1968, 203–4). Another diaspora is hastily planned, but this time in the reverse direction: outward, to Neptune. Here, humanity persists, whilst planning to use further 'annihilation of matter' to accomplish the remarkably longer journey to other stars. But, before it can undertake this generations-long project, the sun unexpectantly begins to go nova. As a last-ditch, the last humans set about 'the forlorn task of disseminating among the stars the seeds of a new humanity': using 'pressure of radiation', and the potent blast given off by the oncoming nova, to propel seeds in all directions. The hope is that 'by extremely good fortune man may still influence the future of this galaxy', Stapledon (1968, 238) concludes; he is at pains, however, to clarify that this will only be the *indirect* influence of starting some new evolutionary saga, from protoplasmic scratch, given the sheer unlikelihood of evolutionary repeats of humanity on other stars.

9. Subsuming Astrobiology within History's Horizon

Indeed, important developments in life sciences during the early decades of the 1900s—alongside growing interchange between life sciences and other, more 'cosmical' sciences like physics and chemistry—had brought to further clarity the legitimacy, even importance, of the question of how rare human-like minds, and life as a whole, may be across the wider cosmos. Space was getting bigger, but not less empty.

First, more biologists and evolutionists than ever before were explicit that Darwinism stresses the contingency of the human organism. Harlow Shapley, in 1922, wrote that humanity will be 'unduplicated elsewhere in the sidereal universe':

> Astronomically speaking, he represents such a transient, fortuitous, and uncertainly poised combination of circumstances, that it would be surprising if the laws of probability exactly reproduced him.
>
> (Shapley 1922, 722)

Two years afterward, K. F. Mather clearly wrote that paleontological evidence shows that humanity is by-product of 'the unique succession of felicitous events' that is Earth's history. Moreover, increasing complexity is 'by no means a necessary consequence of life'. There is only 'change':

> Variation may be in any direction, for better or for worse. Environment and environmental changes seem to be the directive influences. It is hardly to be expected that another planet, even among the myriad planets which may pertain to the stellar galaxy, should have had a geologic history sufficiently similar to that of our earth to have developed life along similar channels.

Hence, humanity will be 'unique both in time and space'. The 'law of probabilities almost excludes the possibility that anywhere else or at any other time' evolution has or will take the 'long chain' from 'protoplasm' to anything 'which would be even an approximate duplicate' of *Homo sapiens* (Mather 1924, 140–2).

Second, there was movement towards a fuller, and deeper, subsumption of life within history's horizon. This was heralded by the novel theories of 'abiogenesis' in the 1920s. Prior to this, things had not changed too much from the days of de Maillet or Buffon when it came to life's 'ultimate' origins. Panspermia remained prevalent. For example, in the 1880s, Helmholtz (1995, 275) asked whether, after the Sun and Earth have become 'solidified', the 'germs of life' could be scattered from here to 'a new world', which would mean that Earth's lineage would continue long after Earth's death. Others applied the same backwards: as late as the 1920s, Thomas Hardy was pondering on whether the 'germ of Consciousness' had arrived on Earth, 'aions ago', on a meteorite from 'some far globe' (Hardy 1925, 145).

Particularly since Pasteur's 1859 experiments had seemingly proved that life only comes from life, it seemed that the only viable options were some kind of supernatural special creation or panspermic eternalism. Both are equally ahistorical: the latter because it transposes the question of temporal origin to spatial circulation; the former because it claims that life is miraculous, lacking any *natural historical* cause, and thus not amenable to analysis as an emergent part of nature's history.

Of course, Darwin (1871) had flirted gingerly with the proposal of a common origin for life—from inorganic matter—and had even offered a properly historical explanation as to why this event hasn't seemingly happened again since (i.e. because subsequent life would thereafter have eaten any new glimmerings). However, he remained cautious and quiet on the matter.

Only following developments in biochemistry did the 1920s see the first emergence of a viable theory as to terrestrial life's origins as a singular event, in the deep past, that has not since repeated. These came, separately, from Haldane and Oparin (the former of which explicitly identified this as a 'historical problem' (Haldane 1929, 10).) What provided new impetus was their innovative stress on the drastically different atmospheric conditions of a young Earth (i.e. its lack of oxygen), allowing them to claim life could have appeared once, when the atmosphere was very different, and is impossible (or extremely unlikely) to have happened again (in part due to life itself and its oxygenation of the atmosphere).

With this, it became more theoretically plausible to believe that life, *as a whole*, is a historic event within Earth's biography requiring conditions that haven't been extant since. From this, it became clearer to ask the question of what these conditions are and, given their transience within Earth's own history, also their transience within the wider galaxy. In other words, by clarifying that the origin of life may have been a one-off that requires potentially transient conditions, the question of the stringency of such conditions—in both space and time—could start more clearly to be assessed. Prior debates (i.e. the pluralism debate from Whewell to Wallace) had focused on questions of habitability, but now another, additional issue had become clear: the potential historical unlikelihood of life's origin. Habitability and inhabitation had come apart.

Third, and finally, theories from physics had intensified this sense of stringency. At the time, popular theories for planet origin implied that the 'distribution' of terrestrial planets in the galaxy 'is astonishingly sparse' (Barnes 1931, 721). (This has, of course, since been proven wrong.) Moreover, increased understanding of the noxiousness of various radiations to life—and the discovery of 'cosmic rays'—facilitated increased understanding that outer space was not only deadly cold but also full of life-destroying emissions. This can be seen in the discovery of the ozone layer in the early 1900s, and the following announcements that this 'one-eighth inch' layer is 'all that stands between the human race and destruction by "death rays"' ('1/8 Inch Ozone Alone Saves Life' 1933). Such insights led James Jeans (Jeans 1928, 470) to remark that 'life must be limited to an exceedingly small corner of the universe'; that it is the cosmical 'accident, and torrential deluges of life-destroying radiation the essential'.

Surrounded by an environing cosmic ocean of death rays, and the likely loneliness of other suns, the pageant and procession of terrestrial life seemed all the more precious. It was the Russian physicist George Gamow who, in the 1940s, made this clear. One of the first to understand that the Sun will expand—not contract—as it ages, Gamow suggested future life will 'be forced to emigrate' (Gamow 1948, 189), but he clearly envisioned this as an exodus outwards to outer planets where 'the heat will not be so intense' (Gamow 1940, 118). And he gave a good reason why only 'highly intelligent beings' could steward this future feat of resilience, and extremely unlikely that nature would, left passively to its own devices (Gamow 1948, 189). That is, he nodded to Arrhenius's Panglossian optimism regarding solar radiation pumping seeds wide across space, and explained how new knowledge destroys this theory:

> It is now known that the ultraviolet rays of our Sun, which are almost entirely absorbed by the terrestrial atmosphere, will rapidly kill any micro-organism that ventures beyond this protective shield. Thus, life must be inevitably extinguished in such travelling spores long before they are able to reach even the nearest planet.

But there is another, worse objection 'quite apart' from ultraviolet rays: panspermia, he reasoned, 'becomes rather senseless in the light of modern knowledge concerning the age and origin of the stellar universe':

> [I]t seems at present rather certain that the stars themselves are not eternal and were born ... This 'physical creation of the universe' must have taken place during the epoch immediately preceding the formation of our Earth and other planetary systems, and since it is obvious that *at that time no life could exist anywhere in the universe*, the problem of the origin of life has to be faced anew.
>
> (Gamow 1948, 155–7)

No longer is there any 'sense' in 'efforts to push the important question' of life's origins 'into the distant corners of space that are in no respect different from our own good Earth', Gamow declared. Why? Because, lately, the universe itself had seemingly gained a history such that no how many times it may have independently originated, we must now accept that biology, *as an astrobiological whole*, also had a beginning. There was once a lifeless cosmos; now life inhabits a corner (or, some corners). From this, it quickly becomes feasible to further speculate upon whether it may eventually spill out of any one of these corners to become a much wider factor in the cosmos. Previously this option had been occluded by the prominence of eternalism, which implies that everything possible has already been previously actualized: such that the universe, at the widest level, doesn't exhibit epochal shifts past or future. But before people could start speculating about how life's history may influence the universe's history, they first had to gather evidence for the universe's own history.

10. Disturbing the Universe

Due to advances in early observational cosmology, throughout the 1920s, the redshift of other galaxies—or, at the time, 'spiral nebulae'—had been noticed by astronomers such as Hubble. Combined with issues in theoretical cosmology, this led to a situation where the acknowledgement of an 'expanding universe' was 'in the air' (Kragh 1996, 21).

10.1 A universe hinting towards its exordium

Astronomers began speaking tentatively of a 'universal evolutionary process' (Tolman 1929, 304), and thus started circumspectly acknowledging that if expansion is true, the 'observable' universe cannot be 'unlimited' (Robertson 1933, 847) and, moreover, 'all the galaxies' may have 'originally' been 'close together' (Russell 1929)—because, tracking expansion backwards, this leads to the idea of an initial beginning. These tentative suggestions were given proper theoretical embodiment in the theories of Georges Lemaître who, in 1931, made 'the first proposal of what can reasonably be called a big-bang theory' and a universe 'with a finite past' (Kragh 1996, 22). Echoing the recent revelations in nuclear physics—which, as explored, revealed an explosive amount of energy in the heart of matter—Lemaître soon proposed that the universe itself had begun as a 'primeval atom' that had thereafter explosively disintegrated, and decayed, into the receding galaxies we observe today. (He even compared the process of creation of the cosmos to the decay of a 'highly unstable atom' by a 'kind of super-radioactive process' (Kragh 1996, 47).)

During the 1930s, the idea of a definite beginning remained unacceptable to most, but the idea of forward expansion began its road to acceptance. Almost immediately, the link to practical matters was made. In the early 1930s, Eddington postulated upon the predicament that this expanding universe, with its receding island galaxies, puts humanity in. Eddington (Eddington 1933, 123) exclaimed:

> I suppose that the distance of one galaxy from the next will ultimately become so great, and the mutual recession so rapid, that neither light nor any other causal influence can pass from one to another.

This galactic 'bursting of the bubble' will come one day and though it might seem a 'nightmare', it 'does not threaten any particular disaster to human destiny', Eddington surmised. He explained that 'only the intergalactic distances expand' and 'galaxies themselves' (along with 'all lesser systems'—i.e. 'star clusters, stars, human observers and their apparatuses, atoms') remain 'unaffected'. The 'world will [not] come to an end sooner than we have been expecting', he concluded: we still anticipate a 'complete running-down of the universe by a slow degradation of [available] energy', but that 'distant day is not brought noticeably nearer [by] cosmical repulsion'. He concluded that we still have 'vast domains of space and vast periods of time to dispose of': the process of 'cosmical repulsion' will merely have 'lopped off' a 'few 0s ... here and there' (Eddington 1933, 174).

But here was the first suggestion that physics could finally tackle a previously vague and speculative question: the domain of ultimate forward causal influence, and thus the complete arena of historical impact and consequence, for intelligent lifeforms and their descendants, had *finally come within the range of scientific assessment*.

Moreover, in Eddington's conclusions there is the seed of the further insight that there might be *windows of opportunity* for certain accomplishments and achievements, from the perspective of the ultimately reachable universe, such as reaching or otherwise influencing those 'other galaxies' which Haldane had so jovially hinted towards.[3] In a historical cosmos, potentials—for achievement or for the wider ranges of flourishing—can be lost. Most importantly, this was the first explication of the foundations of the idea that the reachable or affectable universe, in the insuperable sense bequeathed by relativity, is (like many other aspects in natural history) *enormous but finite*, in both space and time. From this it is possible to infer that from the perspective of the ultimately reachable universe, it is at least plausible in principle to assert that intelligent actors may be able positively to influence—no matter how small—the amount of good *that will have been* achieved across its entire backward and forward history.

Previously such a thing couldn't even be suggested, but now it can be put upon the table, and thus could form a background belief motivating more ambitious visions of settling space.

10.2   The affectable universe is enormously finite

Demonstrating that no matter how rudimentarily, people were beginning to think in these terms, Haldane responded in the 1940s to the growing support for non-static and evolutionary models of the cosmos. Assessing various competing options, Haldane wrote that in neither of the static and non-static models can we have *any strong hope* that the universe 'will be a better place a million [years hence]'. But at least in non-static ones it becomes *possible*. If the 'universe has a real history, not a series of cycles of evolution', then 'life could not have started much before it did, or have got much further than it has at the present date'. He arrived at a far-reaching conclusion:

> If this is so, human effort is worthwhile and human life has a meaning.
>
> (Haldane 1945, 132)

In other words, intellectual and ethical achievements *may* actually make an impact from the perspective of the whole, now that the entire reachable cosmos had been subsumed within the horizon of history. It was this *plausibility*—no matter how astronomically unlikely or impracticable it may be to verify or actually achieve—that simply was not clearly on the table before evidence began accumulating that the cosmos itself had a finite past and was still evolving. Though Bondi, Hoyle, and Gold soon proposed steady-state explanations for accumulating evidence of expansion (and it needs to be remarked that it is no coincidence Hoyle was architect and stubborn advocate of *both* steady-state cosmologies and theories of panspermia), ahistorical alternatives have largely receded since the 1960s and the discovery of the cosmic microwave background. The 'Big Bang' theory won out: now it is widely accepted that the universe itself has a history.

In a universe with a history, and thus a finite past, you can postulate that life *can* become a large-scale, observable factor whilst simultaneously holding that this simply hasn't happened yet. Contrarily, in an eternalist universe, the belief that it is possible is precluded by the seemingly obvious observation that life hasn't left its mark on the universe at a blatant scale: if it could happen, it would have already happened at some point in the eternal past, and yet we appear to find ourselves in a universe which seemingly isn't teeming with life. (Since the 1960s, and the beginning of serious SETI endeavours, this has importantly become a matter of active assessment.) Moreover, if large-scale spread of biology throughout the wider cosmos has already happened, at least once, then there would presumably remain *some* evidence of 'past dynasties'.[4] (The wider question has only intensified, largely since the 1970s and 1980s, after scientists calculated *just how rapidly* we could sensibly expect an spacefaring species to explore the entire galaxy (Tipler 1980).)

Hence, against a background in which it had become feasible to hold that life can become a large-scale factor but also it simply hasn't yet, then this raises the further question as to why *this hasn't already happened.* Responding to precisely this, Loren Eiseley wrote in the early 1950s that:

> [s]ince we now talk ... endlessly of space rockets, it is no surprise that this thinking yields the obverse of the coin: that the rocket or its equivalent may have come *first to us* from somewhere 'outside'. Surely, in the infinite wastes of time, in the lapse of suns and wane of systems, the passage, if it were possible, would have been achieved. But the bright projectile has not been found ... .

To answer this, he rallied the Big Bang theory (which was currently gaining the upper hand over Hoyle's alternative) to propose that it is at least plausible that we may simply be the *first* on the scene (or, at least, the only technological minds within our reachable universe). He wrote:

> [T]he present theory of the expanding universe has made time, as know it, no longer infinite. If the entire universe was created in a single explosive instant a few billion years ago, there has not been a sufficient period for all things to occur even behind the star shoals of the outer galaxies. In the light of this fact it is now just conceivable that there may be nowhere in space a mind superior to our own ... . (Eiseley 1953)

This question started to be asked, independently, by thinkers of varying different fields during the 1940s: leading up, of course, to Enrico Fermi's famous articulation of it in 1950 (from which the so-called Fermi Paradox gains its name). Physicists, historians, and palaeontologists all asked the question—seemingly independently. What enabled this wave of intellectual convergence is most likely to have been the increasing awareness of the viability of the historicized view of the reachable universe, combined with the increasing popularity of the Space Age proposal that interstellar diaspora may one day genuinely be feasible. Indeed, a few years *before* Fermi asked the question, Arthur C. Clarke had already arrived at it during a speech given to the British Interplanetary Society at the White Horse Inn in London, on 5 October 1946. He asked whether humanity is 'the only intelligence in the Universe',

> or there are other, perhaps far higher, forms of life elsewhere? There can be few questions more important than this ... Planets are far commoner than we have [previously] believed: there may be thousands of millions in this Galaxy alone.

Clarke then went on to cite the sentiment, unlocked by radioactivity in the first decades of the 1900s, that we are ourselves only at the very beginning of civilization's potential history and development. Given this, he then asked:

> But if the Universe does hold species so greatly in advance of our own, then why have they never visited Earth?

> (Clarke 1946)

If you hold that life can spread but simply hasn't spread yet, and ask for the reason why no-one has done it before, then two answers seem to have immediately suggested themselves to the people who first asked these questions:

- It hasn't been done yet because we are among the first, or the others simply haven't had the time to get here yet.
- It hasn't been done yet because some limiting event reliably stops it from happening.

In a world that had recently been ushered into its atomic age, this latter was the answer historian Karl Jaspers arrived at when he (most likely independently) asked the same question as Clarke, three years later, in 1949.

> We can pose a singular question. Our history has lasted some six thousand years only. Why should this history occur just now, after the immeasurable ages of the universe and of the earth that have preceded it? Do not humans, or at all events, rational beings exist anywhere else in the universe? Is it not the natural development of the spirit to extend its operations into the universe? Why have we not long since had news, through radiations, from the universe? Communications from rational beings infinitely further advanced in technological development than ourselves? Can it be because all high technological development has so far led to the point at which the beings have brought about the destruction of their planet with the atomic bomb? Can some of the *novae* be end-effects of the activities of technological rational beings?

> (Jaspers 1953, 209)

This lurid conjecture, that *novae* are the tell-tale signs of other civilizations unlocking the atom and self-destructing, had floated around since at least 1918 (and continued to linger well after the Second World War). However, here it was ingeniously used as impetus for a gothic answer to the question of why, in a universe with finite yet enormous age, do we not already live in the ruins, tutelage, or dominion of other, more advanced intellects? Jaspers concluded that the silence of the skies may prove that premature extinction is 'inescapable'. But even if the fateful and unstable phase of history needs to be passed through 'a thousand times', the denizens of that world must treat 'each fresh instance'—every replaying of the tape—'afresh [as] the task of preventing the catastrophe, and that by means of every conceivable direct measure' (Jaspers 1953, 210). This is because accepting this form of explanation, we may be living in a universe characterized by a few glimmerings of light that reliably snuff themselves out, through cosmocataclysmic folly, but this doesn't negate the potential—no matter how slight—of being the first intelligent birth to do something more positive, less suicidal, with its astronomically unlikely endowment of life and mind.

Clarke, too, had acknowledged that there was potentially a vast cosmic future ahead of humanity but that all of this seemed to be gated on the results of the present, atomic epoch. In his speech, he acknowledged that 'atomic energy' will be the means of pushing humanity out of its 'planetary goldfish bowl', and further afield, across the 'gulf still lying between [humanity] and the stars'. But he acknowledged that this was in jeopardy: 'fifty years from now', this we may have ourselves destroyed this potential and its ruins will be decaying in 'radio-active wilderness'. Putting the point differently, Clarke explained that we now have the ability to 'destroy our civilization and slay the future before its birth' (Clarke 1946). The rocket may be 'one of the destroyers of civilisation', or one of its saviours, he concluded portentously.

With tragic irony, the early 1930s conversation that inspired Leó Szilárd to study nuclear physics, thus to invent the chain reaction behind the bomb, was one wherein a friend convinced him that unlocking the atom, and achieving interplanetary communication through atomic means, was the key to securing humanity's longer-term future from risks such as global wars (Szilárd 1979, 738).

Within a universe subsumed within history's horizon—being that arena wherein practical concepts of 'consequence', 'impact', and 'missed opportunity' have their bite—it becomes apparent just what *might* rest upon our attempt to spill intelligence and ethical agency beyond Earth, thus bringing life to the currently barren and dead expanses of wider space. This isn't about humans; nor even about our common terrestrial heritage; it may be about life, impartially, as an astrobiological whole. In a universe with a history, it is clear that it may be possible, yet simply unaccomplished. No-one else has yet 'woken up' the firmament, but it may still be achievable. Because, in a universe with a past, it becomes evident that we are either the first capable of understanding this potential, or, elsewise, the task of accomplishing is extremely hard such that none have succeeded before. Either way, it becomes obvious that there is an ethical reason behind arguments for future space settlement: whether we are merely amongst the first intelligences or others have tried and serially failed, accomplishing the spread of biology would allow our descendants to perhaps truly make an outsized positive impact on the aggregate amount of 'good'—of life, of complexity, of positive experience—that this universe will have manifested across its total history.

Otherwise, we cannot rely on the belief that there will be an infinitude of opportunities for others to fulfil this task, because genuine history allows for opportunities squandered and opportunities missed, and these days we must countenance the fact that the wider universe is itself thoroughly historical.

11. Conclusion

To sum up the story of this chapter, the historical emergence of ethical arguments for space settlement arose from the slow recognition that the wider universe is probably not invariantly full of life and intelligence. Assuming a directionless history for these 'ultimate entities', there was either no sense that going to space could be feasible—because all that has is achievable has already been achieved—nor that it could have a meaningful ethical impact on the history of the whole. But it was slowly, stage by stage, revealed that both the universe appears to have a history—within which opportunities and potentials emerge and can be lost—and, moreover, that life and intelligence have themselves had clear directional histories in the past: there was once *none*, and now there is *some*. From this it became possible to speculate upon how this directionality has not already exhausted its possible mileage: there was once *none*, now there is *some*, perhaps in some distant tomorrow there may *a lot more*.

The universe itself has also gained a beginning, which puts a backward stop on the amount of time within which prior intelligences might have already existed and achieved everything achievable. This makes it at least philosophically plausible that we are living in an epoch *before* intelligence or life becomes a widescale factor at meaningful scales. It also meant that it is also at least plausible that our descendants can consequentially affect the total amount of ethical value that will have been realized in the affectable cosmos, no matter how piecemeal or profound.

One might thus think of intelligence as a kind of nucleation of a crystal from a solution: a primary seed, tiny and fragile at first, but charged with the potential to tip the entire system into a dramatically new state. Perhaps it could be the spark, or catalyst, that triggers the shift from an abiotic to a biotic universe, from a sterile to a verdant one?

Ultimately, we have come to find ourselves in a universe that seemingly could house vastly more life, but also which, at the present moment, simply does not. In my eyes, there can be no good reason—no moral case—for this state of affairs. It is nothing but a brute fact and waste of opportunity. As with other arbitrary precedents of nature—from infant mortality to gravity—I hope our descendants, or others endowed with the ability to recognize this squander, liberate this universe of its abiotic default. Nature itself, passively or by panspermia, is not going reach those maxima without moral agents, in the far future, pushing for it: the way of independent nature is triumphant opportunism, and most of its best potentials are never realized by its random walks.

Future people could change that: making the universe a place that isn't defined by *wasted opportunity* but has manifested the *fullness* of all positive lifeways and forms of meaning, in all their likely staggering diversity, breadth, and intensity. Accidents aside, the time is there to explore this space prudently and preparedly, and to avoid the worse pitfalls that will undoubtedly litter the way. There is a grandeur in terrestrial evolution, with its several powers, but it can be so much more beyond. To close, a quotation from Olaf Stapledon's philosophical treatise, *A Modern Theory of Ethics*, from 1929:

> ... we are on solid ground in holding that the essence that is meant by 'good' is fulfilment, and in deriving the remote and the practical ideals from the needs of organisms of all ranks. Every actual organism which comes into existence claims fulfilment; and its claim must be taken into account in the ideal. But also every organism which might come into existence must be taken into account. And those must be brought to pass which, directly or indirectly, will afford the most complete fulfilment to the latent capacities of the active substance which is the cosmos.
>
> (Stapledon 1929)

Whether you detect here total view utilitarianism or Spinozism, the argument is powerful. Around a decade later, the astronomer Harlow Shapley—who, whilst studying distant galaxies, had given himself much time to think about humanity's position and potential within the wider universe—declared the following to an auditorium in Cleveland, Ohio:

> I may be obsessed, or suffering from anthropocentric delusions, but I cannot escape the feeling that the human mind and human curiosity are significant in this world— even perhaps in the cosmos of geological time and intergalactic space. With this impression (or illusion) that the human mind is the best of us, and the best of biological evolution, I cannot escape (and neither can you!) the feeling of responsibility to glorify the human mind, take it seriously, even dream about its ultimate flowering into something far beyond the primitive muscle-grinder and sensation-recorder with which we started.
>
> (Shapley 1963, 199–200)

You may not be convinced of the ethical arguments for pursuing this most flourishing, being happy with that muscle-grinding and sensation-recording legacy we have received from the blind, blood-soaked meandering of Darwinian process. Fair enough, but what about the *fun* of it all? Shapley concluded:

> We could, of course, ... deliberately refuse to grow, and go turtling through the ages, dull and static ... But it is better cosmic sportsmanship, and more exciting, to go to the top, to the limit of our abilities and aspirations, for there may be something at the rainbows' end that will make even the galaxies look trivial.
>
> (Shapley 1963, 193)

Notes

1. Copernicus believed the 'fixed stars' were not independent star systems, but part of an outer orb circumscribing our solar system, which thus makes up the 'whole universe'.

2. Copernicus believed the 'fixed stars' were not independent star systems but part of an outer orb circumscribing our solar system, which thus makes up the 'whole universe'.

3. In 1933, E. W. Barnes (1933, 435–6) was reasoning that recession means that we fortuitously live in that happy epoch wherein other galaxies are observable to us and radioactive decay of terrestrial elements remains ongoing: if humanity had 'first appeared' many billions of years hence, he concluded, 'we should neither have known radio-active bodies nor would our telescopes have revealed the great nebulae', such that it might be impossible to gain evidence of the scale of the cosmos in both time and space (given that galaxies would have receded beyond sight and radioelements terminally decayed). Here is the realization that it not only a question of history as to when we arrive at a certain piece of knowledge, but it is also a historical question as to whether we even can. (Barnes, indeed, speculates as to what other evidence—perhaps 'equally important'—may already have 'vanished' because of cosmic historicity.)

4. Otherwise, if you still maintain, against observable evidence, that we live in an eternal universe within which large-scale settlement is nonetheless possible then you must explain why, despite the fact it must have already happened, it remains somehow unobservable or invisible to us. Tsiolkovsky is an example of a thinker who reasoned himself into this latter position: he reasoned that the extraterrestrials are everywhere but are so much more advanced than us that they hide themselves from us. Hoyle, too, was led to similar conclusions.

References

'1/8 Inch of Ozone Alone Saves Life' *The New York Times* 30 October 1933, 19.

'A Radium Clock' (1905) 93(23) *Scientific American* 531.

Adams, F. and G. Laughlin. *The Five Ages of the Universe: Inside the Physics of Eternity* (New York, NY: Simon & Schuster, 1999).

'Alarming Situation' *Durham Globe*, Monday 16 May 1892.

Arrhenius, S. *Worlds in the Making: The Evolution of the Universe* (New York, NY: London, 1908).

Augustine. *Sancti Augustini Opera*, Vol 50 (Brussels: Brepolis, 1854).

Ball, R. S. 'How Long Can The Earth Sustain Life?' (1892) 51 *Fortnightly Review* 478–90.

Barnes, E. W. 'Contributions to a British Association Discussion on the Evolution of the Universe' (1931) 128 *Nature* 719–22.

Barnes, E. W. *Scientific Theory and Religion* (Cambridge: Cambridge University Press, 1933).

de Maillet, B. *Telliamed*, trans. A. V. Carozzi (Chicago, IL: University of Illinois Press, 1968).

Bernal, J. D. *The World, The Flesh, and the Devil* (London: Verso, 2017).

Blanqui, LA. *L'éternité par les astres: hypothèse astronomique* (Paris: Bailliere, 1872).

Bonaventure. *Doctoris seraphici: Bonaventurae Opera Omnia*, Vol 2 of 10 (Florence: Quaracchi, 1864).

Browne, T. *Selected Writings* (Oxford: Oxford University Press, 2014).

Bruno, G. 'On the Infinite Universe and Worlds' in D. W. Singer (ed.), *Giordano Bruno: His Life and Thought* (New York, NY: Greenwood Press, 1968) 225–380.

Caritat, J.-A.-N. *Outlines of an Historical View of the Progress of the Human Mind* (London: Johnson, 1795).

Chambers, R. *Vestiges of the Natural History of Creation* (London: Churchill, 1884).

Chamberlin, T. C. 'Soil Wastage' (1910) 73 *Popular Science Monthly* 5–12.

Chamberlin, T. C. 'Letter to Edwin E. Slosson, 17 January'. Box I, f.16, TCC Papers.

Chamberlin, T. C. 'The Future Habitability of the Earth' (1911) 2031 *Smithsonian Report, Publication* 371–89.

Clarke, A. C. 'The Challenge of the Spaceship: Astronautics and its Impact upon Human Society' (1946) 6(3) *Journal of the British Interplanetary Society* 66–81.

Crowe, M. *The Extraterrestrial Life Debate, 1750–1900: The Idea of a Plurality of Worlds from Kant to Lowell* (Cambridge: Cambridge University Press, 1988).

Cusanus, N. *Of Learned Ignorance*, trans. G. Heron (Eugene, OR: Wipf & Stock, 2007).

Darwin, Charles. 'Letter to J. D. Hooker, 9 Feb'. <https://www.darwinproject.ac.uk/letter/DCP-LETT-4769.xml> accessed 13 October 2021.

Darwin, C. 'Letter to J.D. Hooker, 9 Feb'. <https://www.darwinproject.ac.uk/letter/DCP-LETT-7471.xml> accessed 14 October 2021.

Darwin, G. H. 'Radio-activity and the Age of the Sun' (1903) 68 *Nature* 496.

de Sales, J.B. *De La Philosophie De La Nature, ou Traité De Morale Pour L'Espece Humaine*, Vol 1 of 6 (Paris, 1777).

Dick, T. *The Sidereal Heavens* (London, 1840).

du Prel, K. *The Philosophy of Mysticism*, Vol 2, trans C. C. Massey (London: Redway, 1889).

Eddington, A. 'The Internal Constitution of Stars' (1920) 11(4) *The Scientific Monthly* 297–303.

Eddington, A. 'The End of the World from the Standpoint of Mathematical Physics' (1931) 127 *Nature* 447–53.

Eddington, A. *The Expanding Universe* (London: Macmillan, 1933).

Eiseley, L. 'Is Man Alone in Space?' (1953) 189 *Scientific American* 80–7.

Esnault-Pelterie, R. 'Considérations sur les résultats d'un allégement indéfini des moteurs' (1931) 3(1) *Journal de physique théorique et appliqué* 218–30.

Etzler, J. A. *The Paradise Within Reach of All Men* (Pittsburgh, 1833).

Fontenelle, B. *Conversations on the Plurality of Worlds*, trans. A. Behn (London, 1688).

Galileo, G. *Dialogue Concerning the Two Chief World Systems*, trans. S. Drake (Berkeley, CA: University of California Press, 1967).

Gamow, G. *Birth and Death of the Sun* (New York, NY: Viking, 1940).

Gamow, G. *Biography of the Earth* (New York, NY: Mentor, 1948).

Goddard, R. *The Papers of Robert H. Goddard: 1898–1924* (New York, NY: Mc-Graw-Hill, 1970).

Goddard, R. 'The Final Migration' (1978) 31 *Beyond Reality* 20–23 & 60.

Gregory, W. *Handbook of Organic Chemistry* (New York, NY: Barnes, 1857).

Haldane, J. B. S. *The Last Judgement.* (New York, NY: Harper, 1927).

Haldane, J. B. S. 'The Origin of Life' 1929 148 *Rationalist Annual* 3–10.

Haldane, J. B. S. 'A New Theory of the Past' (1945) 33(3) *American Scientist* 129–45.

Hardy, T. *Human Shows, Far Phantasies, Songs, and Trifles* (London: MacMillan, 1925).

Helmholtz, H von. *Science and Culture: Popular and Philosophical Essays* (Chicago, IL: University of Chicago Press, 1995).

Heuer, K. *Men of Other Planets* (London: Victor Gollancz, 1953).

Huxley, J. *Evolution, the Modern Synthesis* (New York, NY: Harper, 1942).

Huxley, T. H. 'Evolution and Ethics II' (1893) 44 *Popular Science Monthly* 178–91.

Huygens, C. *The Celestial Worlds Discovered* (London: Timothy Childe, 1698).

Jaspers, K. *The Origin and Goal of History* (London: Routledge, 1953).

Jeans, J. 'The Wider Aspects of Cosmogony' (1928) 121 *Nature* 463–70.

Jeans, J. *The Universe Around Us* (Cambridge: Cambridge University Press, 1930).

Joly, J. 'Uranium and Geology' (1908) 28(725) *Science* 697–713.

Kandinsky, W. *Kandinsky: 1901–1913* (Berlin: Der Sturm, 1913).

Kepler, J. *Dissertatio cum nuncio sidereo nuper ad mortales misso à Galilaeo Galilaeo* (Prague: Sedesanus, 1610).

Klein, L. M. 'Huxley's Evolution and Ethics' (1893) 13 *Dublin Review* 589–98.

Kragh, H. *Cosmology and Controversy: The Historical Development of Two Theories of the Universe* (Princeton, NJ: Princeton University Press, 1996).

Lucretius. *On the Nature of Things*, trans. M. F. Smith (London: Hackett, 2001).

Maeterlinck, M. *The Double Garden*, trans. A. Teixeira de Mattos (New York, NY: Dodd, Mead, & Co., 1920).

Malthus, T. *Essay on the Principle of Population* (Cambridge: Cambridge University Press, 1992).

Manning, L. 'The Living Galaxy' (1934) 6(4) *Wonder Stories* 436–44.

Mather, K. 'A Geologic Forecast of the Future Opportunities of our Race' (1910) 83(2106) *Nature* 50–4.

Mather, K. 'Geologic Factors in Organic Evolution' (1924) 26(3) *The Ohio Journal of Science* 117–45.

McCabe, J. *The End of the World* (London: Dutton, 1921).

Milikan, R. 'Contributions to a British Association Discussion on the Evolution of the Universe' (1931) 129 *Nature* 709–15.

Morris, R. *Time's Arrows: Scientific Attitudes Toward Time* (New York, NY: Simon & Schuster, 1986).

Moynihan, T. *X-Risk: How Humanity Discovered Its Own Extinction* (Falmouth: Urbanomic, 2020).

Moynihan, T. 'Summons of the Silent Universe: The Relationship between Existential Risk and Cosmic Silence' in I. Crawford (ed.), *Expanding Worldviews: Astrobiology, Big History, and Cosmic Perspectives* (New York, NY: Springer, 2021a) 65–90.

Moynihan, T. 'The Interesting Case of Prince Vladimir Odoevsky And His Speculations on Civilization's Trajectories' (2021b) 129 (4) *Futures* doi:10.1016/j.futures.2021.102732.

Newcomb, S. *Popular Astronomy* (New York, NY: Harper, 1879).

O'Malley-James, J., Greaves, J. Raven, J., et al. 'Swansong Biospheres: The Biosignatures of Inhabited Earth-like Planets Nearing the End of Their Habitable Lifetimes' (2013) 8(S299) *Proceedings of the International Astronomical Union* 378–9.

Ord, T. *The Precipice* (London: Bloomsbury, 2020).

Robertson, H. P. 'On Relativistic Cosmology' (1933) 5(31) *Philosophical Magazine & Journal of Science* 835–48.

Russell, B. *Religion and Science* (Oxford: Oxford University Press, 1949).

Russell, H. N. 'Is the Universe Running Down?' (1925) 133(6) *Scientific American* 369.

Russell, H. N. 'The Highest Known Velocity' (1929) 140(6) *Scientific American* 504–5.

Rutherford, E. 'Some Cosmical Aspects of Radioactivity' (1907) 1 *Journal of the Royal Astronomical Society of Canada* 145–65.

Saleeby, C. W. 'Is the Human Race Mortal?' (1906) 112(672) *Harper's Magazine* 949–52.

Shapley, H. 'The Universe and Life' (1922) 146 *Harper's Magazine* 716–22.

Shapley, H. *The View from a Distant Star* (New York, NY: Basic Books, 1963).

Shaw, G. B. *Back to Methuselah* (New York, NY: Brentanos, 1921).

Shelley, P. B. *Essays, Letters from Abroad, Translations and Fragments*, Vol 1 (London: Edward Moxon, 1848).

Smith, R. *The Expanding Universe: Astronomy's 'Great Debate', 1900–1931* (Cambridge: Cambridge University Press, 1982).

Smith, A. *The Theory of Moral Sentiments* (London: Penguin, 2009).

Soddy, F. *The Interpretation of Radium* (London: Murray, 1909).

Soddy, F. *Radio-activity: An Elementary Treatise* (London: Electrician, 1904).

Soddy, F. *Science and Life: Aberdeen Adresses* (London: Dutton, 1920).

Stapledon, O. *A Modern Theory of Ethics* (London: Methuen, 1929).

Stapledon, O. *Last and First Men/Starmaker* (London: Dover, 1968).

Stout, R. *Evolution and the Origin of Life* (Napier: Dinwiddie, 1914).

Swagman, A. 'Ingle-Hook' *The Western Star*, Wednesday 19 June 1892.

Szilárd, L. 'Some Links in the Chain' *New Scientist*, Thursday 31 May 1979.

Tesla, N. 'The Problem of Increasing Human Energy' (1900) 60 *Century Magazine* 175–211.

Thompson, W. 'Presidential Address' (1871) 41 *BAAS Report* 200–2.

Thomson, W. 'On the Mechanical Energies of the Solar System' (1854) 21 *Transactions of the Royal Society of Edinburgh* 63.

Thomson, W. 'Physical Considerations Regarding the Possible Age of the Sun's Heat' (1861) 2 *British Association Report* 27-28.

Thoreau, H. D. 'Paradise (to be) Regained' (1848) 13(45) *The US Magazine and Democratic Review* 451–63.

Tipler, F. 'Extraterrestrial Intelligent Beings Do Not Exist' (1980) 21 *Royal Astronomical Society Quarterly* 267–81.

Tolman, R. 'On the Possible Line Elements for the Universe' (1029) 15(4) *Proceedings of the National Academy of Sciences* 297–304.

Tsiolkovsky, K. *Issledovaniye mirovykh prostranstv reaktivnymi priborami* (Moscow: Mash-inostroenie, 1967).

Voltaire. *Micromégas and Other Short Fictions*, trans Theo Cuffe(London: Penguin, 2002).

Wells, H. G. *The War of the Worlds* (London: Heinemann, 1898).

Wilson, W. E. 'Radium and Solar Energy' (1903) 68 *Nature* 222.

Winchell, A. *Sketches of Creation* (New York, NY: Harper, 1870).

Young, C. A. *The Sun* (New York, NY: Appleton, 1881).

19

Only a paper moon

The Artemis Accords and future human settlements

Christopher J. Newman and William Ralston

1. Introduction

The Apollo programme that saw humans land on the Moon was the culmination of an intense effort and vast expenditure by the United States. Despite the expenditure, the nature of the technology meant that any presence on the lunar surface would be temporary and not sustainable over the longer term (see Chaikin 1998). The remaining decades of the twentieth century were spent consolidating space activity in Earth orbit (Amos 2010). By the start of the twenty-first century, however, countries once again started to look again at the Moon. China, India, and the European Space Agency all promulgated plans for sustained lunar exploration (e.g. ESA 2016). It was, however, the US' announcement of Project Artemis which drew questions of lunar activity back into contemplation by the international community (Pearlman 2019). Project Artemis was announced in 2017. It plans to send women and men back to the Moon. Through this project, the United States is seeking to create an enduring and sustainable presence on the Moon in partnership between NASA, the private sector, and international collaborators (Berger 2019).

Any lunar settlement inevitably begs questions about the overarching legal framework, especially in the contentious area of resource extraction and utilization (Larsen 2021). The result of the dramatic change in the lunar exploration policy of the United States was an equally dramatic diplomatic initiative known as the Artemis Accords (NASA 2020a). The Accords, whilst not international treaties, are agreements negotiated between NASA and those international partners who seek to be involved in the Artemis programme. The announcement of the Artemis Accords at the virtual 71st International Astronautical Congress was the result of a concerted diplomatic effort (NASA 2020b).

Indeed, they have now been signed by 15 Partner States (Si-soo 2022). As such, the Accords are seeking to codify the way in which a sustained presence on the Moon would comport within established international agreements.

As international space law has developed, there has been discussion as to how settlements on other celestial bodies might be governed and what the role of law might be when humans establish a permanent presence on another planet (see e.g. Froehlich 2021). Foundational principles for regulating the activities of states in outer space were laid down in the Outer Space Treaty 1967. But the Treaty contains no real detail of how humans might establish a society in space beyond

Christopher J. Newman and William Ralston, *Only a paper moon*. In: *The Institutions of Extraterrestrial Liberty*. Edited by Charles S. Cockell, Oxford University Press. © Christopher J. Newman and William Ralston (2022). DOI: 10.1093/oso/9780192897985.003.0020 a prohibition on State appropriation of outer space and celestial bodies (Article II), forbidding military bases (Article IV), and making states internationally responsible for their national space activities, including non-governmental entities (Article VI). The Moon Agreement of 1979 attempted to codify provisions relating to exploration and the management of resources, but this was largely repudiated by the international community, including China, Russia, and the United States (Newman 2015, 32).

There has been considerable academic discussion of the way in which the Accords interact with the existing international law (see e.g. Johnson 2020; Wright Nelson 2020a; Deplano 2021). This discussion will touch upon that, whilst seeking to address the broader question about the evolution of space law. The question of the significance of the Accords will be addressed giving consideration as to the role that the Accords could have in shaping the way in which future human settlements might be governed beyond the terms of existing space treaties. Robinson (2004) identified that there was a key problem when considering any form of off-world governance: the lack of an underpinning shared value system upon which to base any subsequent regulatory or governance structure. As much as the Accords may represent a new epoch in space law, it is timely to ask whether they are another step towards identifying areas of commonality around which new space settlements can coalesce.

2. Lunar Governance and the Artemis Accords

The Accords themselves capture the core principles on which the partner states who wish to engage in Project Artemis have agreed to adhere. It is tempting to point to a superficial similarity to the way in which governance of the International Space Station (ISS) was managed in the 1990s. The ISS, however, was a partnership. Project Artemis is directed and funded by NASA and the United States. As von der Dunk (2020) states, legal coverage of the ISS project was by means of 'a single all-encompassing international treaty'. The Accords are not intended to be binding in the way that international treaties bind but they do clearly represent the 'price of admission' to the Artemis programme (Newman 2020) as they are negotiated by the United States with the partners bilaterally. Those states that wish to enjoy the opportunities, and presumably the lucrative contracts afforded through involvement in Project Artemis, are obliged to adhere to the Accords with the danger of exclusion and loss of opportunities serving as the incentive for compliance.

If they are not binding international law, the question then arises about the nature and characteristics of the Accords within the realm of space governance (Taichman 2021). The preamble expressly affirms the compliance with existing space treaties and bilateral agreements. The Accords, therefore, are political agreements which highlight shared understanding of substantive matters of existing space law. Compliance with the international space treaties will be the responsibility of individual partner nations (Larsen 2021, 37), but the Accords also go on to develop and codify existing norms of behaviour. To that extent, they can be said to reflect certain core values of the current international governance regime. But the Accords also clearly have been prepared in anticipation of humans becoming permanent settlers in outer space, with the Moon the first stop on a journey to Mars (Johnson 2020a).

An examination of the principles contained within the body of the Accords shows a fascinating mixture of provisions that Deplano (2021, 801) has been identified as falling within three broad categories. The first are those provisions which restate existing international space law in the context of lunar exploration. These principles, such as registration requirements, the rescuing of astronauts, and the provision of assistance will undoubtedly form the bedrock of any future legal framework for settlements. The Accords are, however, silent on the troublesome question of liability and fault in space operations. Perhaps, as the Accords stress that Project Artemis will be conducted within the ambit of the existing space treaties, liability will fall within the purview of the Liability Convention.

The second category of provisions are those which take elements of existing international law and add an additional interpretation to the rights and obligations contained within the Treaties. These provisions cover such themes as resource extraction and the interrelated idea of deconfliction of activities as well as lunar heritage protection. Wright-Nelson (2020b, 2) identifies the controversial nature of the interpretations of these areas within the Accords. Given that resource extraction will be crucial for the survival of any extraterrestrial settlement and the fabled 'trillion dollar space economy' is largely predicated on utilizing the vast mineral resources of the solar system, an examination of these criticisms will provide a crucial insight as to whether the Artemis Accords will provide a template for future settlements or whether they represent a dangerous distraction from the building of international consensus involving all members of the international community (Mosteshar 2020, 602–3).

The final categorization of principles identified by Deplano are those within the Accords that introduce concepts which are new to international law. It is worth noting that the principles contained within the Accords which fall into this this last category could be said to represent an identification of existing norms of behaviour rather than being entirely novel. The question of the legal status of historic lunar artefacts is one that has been the subject of significant discussion and lobbying.[1] Also, whilst not explicitly covered in the international space treaties, the management of space debris is something that the United Nations Committee on the Peaceful Uses of Outer Space (UNCOPUOS) has discussed in some detail (Martinez 2021) and falls within the remit of the UN Guidelines for the Long-Term Sustainability of Outer Space Activities.

3. Compatibility of the Artemis Accords with Existing Space Law

The assertion that the Accords are compliant with extant international space law is crucial for their widespread international acceptance. From the perspective of embedding themselves as a normative instrument within the governance of human space activity, being able to trace their lineage back to the foundational space treaties endows the Accords with legitimacy. When considering the handful of instruments of international law that exist to govern space activity, the foundational document is the Outer Space Treaty of 1967, which sets out the basic principles of space law. It is very much the product of the Cold War of the 1950s and 1960s, when geopolitical tensions between the United States and the Soviet Union first began and the development of competing space technology occurred at a frantic pace (Gabrynowicz 2004).

The ideological tensions between the United States and the Soviet Union gave rise to obvious anxieties about one of those two superpowers gaining dominance in outer space (and, thereby, gaining a significant strategic military advantage). In this context, it was clear that any legal vacuum could lead to conflict and tension. States came to the realization that some form of international cooperation was required to ensure that human activity in outer space was conducted for peaceful purposes only. Thus, in UNCOPUOS was created. It was within the legal sub-committee of this forum that space law was negotiated, in the shadow of the race to the Moon, and always with one eye on security considerations (Blount 2010–2011)

The Outer Space Treaty was adopted by the General Assembly of the UN on 19 December 1966, opened for signature on 27 January 1967 in London, Moscow, and Washington, DC and came into force on 10 October 1967. At the time of writing 111 nations have ratified the Outer Space Treaty, with a further 23 having signed it. The widespread acceptance of the 1967 Treaty has provided a bedrock of certainty to the regulation of human space activity and the Treaty itself is rightly regarded as the centrepiece of international law relating to space activity (Lyall and Larsen 2020, 51). Any future human settlements will almost certainly take account of the provisions of the Treaty into their new legal framework.

Central to the idea of establishing a lunar base is the idea that states are free to use and explore outer space, the Moon, and other celestial bodies. Article I of the Outer Space Treaty sets out the general parameters of the lawful uses of outer space:

> The exploration and use of outer space, including the Moon and other celestial bodies, shall be carried out for the benefit and in the interests of all countries ... and shall be the province of all mankind.

Article I goes on to provide that '[o]uter space, including the Moon and other celestial bodies, shall be free for exploration by all States without discrimination of any kind, on a basis of equality and in accordance with international law, and there shall be free access to all celestial bodies'. It goes on to guarantee 'freedom of scientific investigation in outer space, including the Moon and other celestial bodies'.

As Hobe (2009, 27) identifies, having defined the use of outer space in relatively wide terms, the subsequent Treaty Articles then place specific limitations on the use of outer space by States Parties. All this fits directly onto the way in which the Accords are structured. The Accords do not mandate the uses of any Project Artemis installation, rather they amplify and codify the limitations on unfettered usage. At the heart of the Outer Space Treaty is the notion (articulated in the preamble) that the exploration and use of outer space should be for peaceful purposes. Section 3 of the Accords explicitly affirms that activities of Project Artemis will be exclusively for peaceful purposes.

Indeed, the Accords (at Section 3) follow Article III of the 1967 Treaty and embed the entire extant framework of international law within Project Artemis by stating that not only will the activities be conducted exclusively for peaceful purposes but also in accordance with relevant international law. This is a powerful provision as it tacitly acknowledges that the Outer Space Treaty has created new rules of international law (so-called *leges spaciales*) but that it also exists within the continuum of the relevant rules of international law which apply to international relations *wherever they take place* (Ribblelink 2009, 65).

The inclusion of peaceful purposes and the acknowledgement of wider responsibilities under international law was a welcome addition to the Accords, following the increasingly bellicose rhetoric coming from the Trump administration during the later years of the 2010s (Smith 2017). Any future lunar settlement will present a tempting prospect for potential military expansion, especially given the creation of Space Forces. Within the Outer Space Treaty, Article IV provides *inter alia* that '[t]he establishment of military bases, installations and fortifications, the testing of any type of weapons and the conduct of military manoeuvres on celestial bodies shall be forbidden', and so a distinct Space Force/military base is unlikely, although that does not discount some form of defensive or supporting presence (Raymond 2021).

Although 'peaceful purposes' permeates the 1967 Treaty, military activity and support for terrestrially based military applications are now ubiquitous in Earth's orbital space. Despite persuasive academic arguments to the contrary, *peaceful purposes* is a term recognized by most space-active nations as meaning 'non-aggressive' (Jakhu, Chen, and Goswami 2020, 27). It seems unlikely that the United States will want to spend large amounts of money on lunar infrastructure without the means to protect these assets from the risk of attack from 'non-likeminded'[2] states. If the Accords are to provide a template for future space settlements, the acknowledgement of the international rule of law comes with the implicit recognition that *peaceful purposes* do not preclude a military presence. The possibility of conflict exists, not least when considering the provision of resources for establishing an off-world settlement.

4. The Utilization of Resources

The question of the legality of space mining and resource extraction is crucial to the establishment of a sustainable off-world settlement, be it on the Moon or beyond. To become self-sustaining, the settlements cannot simply rely on resources being brought from Earth. At the most basic level, settlers will need a breathable atmosphere, heating, water, and food. The Artemis Accords recognize the importance of the use of resources to survive and 'live off the land'. This is known as In Situ Resource Utilisation (ISRU) and Section 10(1) of the Accords explicitly recognized that space resources benefit humankind by providing critical support for safe and sustainable operations.

When considering ISRU in a lunar context, Elvis, Milligan, and Krolikowski (2016) have shown that both water and sunlight are not distributed equally across the Moon, with pockets of resource-rich territory containing ice-water and the 'peaks of eternal light'—highland regions near the lunar poles that receive almost permanent sunlight.

These are ideal places to place habitats as sunlight can provide a near constant source of solar power and may become highly prized pieces of lunar real estate. The Accords themselves make no reference to fairness or a commitment to equitable distribution of ISRU-rich sites, which whilst understandable, is something that would be desirable as human settlements further into the solar system will have even less opportunity to rely on resources shipped from Earth. Concentration of ISRU-rich sites in the hands of one state, company, or other grouping with access to space is foreseeable. Equitable sharing of *in situ* resources should be expected and required as part of a responsible and sustainable settlement framework and may well come under the notion of 'corresponding interests' under Article IX of the Outer Space Treaty (Marchisio 2009, 176)

There is also the broader question of mining the Moon for possible precious metals and other such commercial considerations. With it comes a whole host of questions, such as the ability to control sites that are being mined in on a celestial body and the right to trade any minerals that might have been mined. Whilst the creation of the necessary infrastructure to enable widespread and efficient commercial mining of the Moon remains in the future, the question of the legality of commercial space mining is one that continues to vex the space law community and discussions continue in UNCOPUOS as to how to manage commercially motivated resource extraction in space.[3]

The ambiguity in whether it is lawful to extract resources from celestial bodies arises from Article II of the Outer Space Treaty 1967, a provision which seeks to restrict the ability of states to lay claim to the exclusive possession of any part of outer space:

> Outer space, including the Moon and other celestial bodies, is not subject to national appropriation by claim of sovereignty, by means of use or occupation, or by any other means.

The competing arguments can be distilled into two fundamental interpretations of the position at international law. Some have taken Article II to be an explicit prohibition on state ownership and trading of lunar resources, advocating instead a multilateral, international legal order to administer and distribute (Hobe 2019, 158–65). This is a position that several countries hold—that space resources and 'outer space' are indivisible aspects of each other, therefore appropriation of any sorts would violate international law.

The second position is that Article II is not relevant as no claim of ownership being made in respect of the territory. Instead, resource utilization is a *usage* of the celestial body under Article I of the 1967 Treaty. Furthermore, under Article VI of the Outer Space Treaty, State Parties to the Treaty are internationally responsible for the authorization and ongoing supervision of non-governmental entities (Cheney and Newman 2018, 266). It is this interpretation that is represented within the Accords. Article 10(2) provides an explicit statement that the signatories of the accords affirm that the mining of space resources is in accordance with international law. This follows on from the passing of the US Space Act 2015, which put the right to use and trade space resources into US domestic law, complete with an assertion that this is not a claim of sovereignty under international law (see Masson-Zwaan and Sundhal 2021).

The blanket acceptance of this second interpretation has been criticized by several academics. Delpano (2021, 807) states that whilst the signatories to the Accords affirm that resource extraction does not violate Article II, they provide no clarity or specificity as to *how* it does not violate the non-appropriation principle. Furthermore, simply declaring that resource extraction is compliant with international law does not necessarily make it so. Accordingly, as the Accords are bilaterally negotiated between the United States and the partner states, they have the potential to undermine the whole Outer Space Treaty framework and the decision-making of UNCOPUOS (Mosteshar 2020, 602).

5. Deconfliction and 'Safety Zones'

It is not only the question of resource utilization where the compatibility with the existing international treaty regime has been questioned. Section 11 of the Accords provides for the avoidance of intentional activities that may cause harmful interference, a commitment to provide Artemis partners with necessary information regarding the location and nature of space-based activities under these Accords if they believe the activities might cause harmful interference or pose a safety risk. Perhaps most controversially, Section 11(7) of the Accords signals the intention to create 'safety zones'. These are 'buffer' areas, where activities that may cause harmful interference are likely to occur.

The idea is that by creating these zones, other users of the lunar environment will be notified of potential hazards and, by imposing certain conditions, the risks from these hazards will be minimized. There is little by way of detail as to how these zones would manifest themselves, but Wright Nelson (2020a, 605) suggests that measures within a safety zone could include such things as the requirement of filing of advanced travel plans within the zones, maximum speeds for rovers, suggested cordons around areas of high risk, together with details of possible radio interference and possible limits on use of rocket engines.

The suggested use of safety zones to deconflict space activities has raised some questions regarding compatibility with the Outer Space Treaty in two key areas. The first is that creating these zones in which conditions can be imposed is a breach of Article II and amounts to a de facto appropriation. The second is that, even if the creation of a safety zone does not amount to appropriation, it does restrict the freedom of access to all areas of the Moon, as guaranteed by Article I of the Outer Space Treaty. In both cases, this is unlikely. Whilst there may be a limitation on the use of the zoned areas, such limitations would not breach Article I of Article II when read alongside Article IX.

Article XI of the Outer Space Treaty is a broad-ranging provision of the Treaty which contains several distinct elements. It starts by exhorting State Parties to be guided by principles of co-operation and mutual assistance. Crucially, for the purposes of the Accords it requires that all states conduct all their activities in outer space, including the Moon and other celestial bodies with due regard to the corresponding interests of all other States Parties to the Treaty.

The notion of 'due regard' is a principle that is well recognized in international law, even if it has not been developed in respect of space activity. It imposes a duty on those conducting space activities, requiring that an act in space should be performed to 'a certain standard of care, attention or observance' (Marchisio 2009, 175). Article IX of the Outer Space Treaty also obliges States Parties pursuing exploration in outer space to have due regard to the environment, both in outer space and on Earth, 'so as to avoid their harmful contamination'. Finally, Article IX requires states to consult if they believe their space activities will result in 'harmful interference' (a term not defined in the Treaty).

It is the 'operationalization' of Article IX that the deconfliction element of the Accords are seeking to achieve. Section 11 starts with a clear affirmation of the existing international treaty law. Section 11(11) provides a commitment to respect the principle of free access to all areas of celestial bodies and all other provisions of the Outer Space Treaty in their use of safety zones. Indeed, the Accords go on to provide further commitment to 'adjust their usage of safety zones over time based on mutual experiences and consultations with each other and the international community'. If safety zones are operated in an open and transparent fashion, with consultation at the heart of the process, the contours of Article IX will start to come to into sharper focus and space will be safer for all users. The notification of activities that could harmfully interfere with another state's operations in space is something that any future off-Earth settlement will need to consider, and early transparency will help build confidence amongst the international community (Wright Nelson 2020a, 621).

Ultimately, the validity of safety zones will depend on the lunar activities that occur, and the hazards created by such activities. This will determine the way in which the safety zones manifest themselves. Disproportionate or inappropriate use of safety zones will undoubtedly cause tension and lead to conflict between Artemis and non-Artemis users of the Moon, just as the hoarding of off-world resources will lead to disputes about the legality of mining. Nonetheless, the interpretation of these provisions of the Outer Space Treaty within the Artemis Accords has been made in the context of an imminent mission to the Moon. This is understandable given that the Accords were negotiated as part of Project Artemis. If the broader role of the Accords is in shaping normative behaviour for off-world settlements is to be considered, it is necessary to look beyond the Moon and examine the broader legal framework for the settlement of humans in deep space.

6. An International Treaty—Return to the Moon Agreement?

The question then arises, if the Accords are not acceptable, or if they are only appropriate for the scope of Project Artemis, what is the alternative? The current geopolitical situation means that any treaty negotiation will have formidable hurdles to overcome if consensus is to be achieved (Danilenko 2016, 183). It should also be noted that there have been significant international efforts to resolve the specific legal questions regarding space mining, such as the Hague International Space Resources Governance Working Group and the subsequent promulgation of the building blocks (Bittencourt Neto, Masson-Zwaan, and Hofmann 2020). Despite this, and the positive developments in establishing a working group within UNCOPUOS, progress on the issue has been slow. Given that the Artemis Accords are open only to 'like-minded partners' an internationally agreed template for human settlements seems some way off.

One possible alternative, posited by a range of academics is the potential revivification of the Moon Agreement 1979 (e.g. Hobe 2010). Opened for signature on 18 December 1979, the Moon Agreement is the last of the 'children' of the Outer Space Treaty (OST), following on from the Rescue Agreement (1968), the Liability Convention (1972), and the Registration Convention (1976). Taken together with the OST, the provisions of which the Moon Agreement purports to 'define and develop', the 1979 Agreement was intended to be the international blueprint for the establishment and regulation of extraterrestrial settlements (Tronchetti 2017, 782).

Indeed, many of the Articles of Moon Agreement are essentially restatements of core principles of international law and of the Outer Space Treaty in particular. Thus, of the Moon Agreement, Article 2 provides that all activities on the Moon are to be conducted in accordance with international law and the Charter of the United Nations in particular. Article 3 carries forward the provisions of Article IV of the OST, restating that the Moon is to be used for exclusively peaceful purposes and prohibiting the militarization of the Moon and its orbit.

Article 6 provides that there shall be freedom of scientific investigation on the Moon by all States Parties (essentially rehearsing the second paragraph of Article I OST). However, Article 6(2) elaborates upon the OST by providing that, whilst conducting scientific investigations, States Parties have the right to collect and remove samples of minerals from the Moon, though such states should also 'have regard to the desirability of making a portion of such samples available to other interested States Parties and the international scientific community for scientific investigation'. Article 6(2) further provides that

> [s]uch samples shall remain at the disposal of those States Parties which caused them to be collected and may be used by them for scientific purposes ... States Parties may in the course of scientific investigations also use mineral and other substances of the Moon in quantities appropriate for the support of their missions.

In the Articles cited above, there could be nothing of any concern to any State Party to the OST. Indeed, Article 6(2) seems to provide a sensible and practical solution to questions concerning whether the collection and retention of minerals from celestial bodies for scientific purposes (and use of them in support of scientific missions) breached Article II of the OST (the prohibition of national appropriation of any part of outer space).

Despite these similarities with the OST and, indeed, the synergies with the provisions of the Artemis Accords, the chances of the Moon Agreement being revived at the current time are small. Article 7 obliges States Parties to 'take measures to prevent the disruption of the existing balance of [the celestial body's] environment, whether by introducing adverse changes in that environment, by its harmful contamination through the introduction of extra-environmental matter or otherwise'. This wording could place a significant economic burden on those seeking to establish settlements with a view to resource extraction (Newman 2015, 33).

It is, however, the provisions contained in Article 11 of the Moon Agreement which have ultimately limited wider adoption of this Treaty. Article 11(1) holds that '[t]he Moon and its natural resources are the common heritage of mankind'. Article 11 then goes on to give some notion of the implications of this concept. In Article 11(5), the Agreement provides that States Parties undertake to 'establish an international regime ... to govern the exploitation of the natural resources of the Moon'. Then, Article 11(7) spells out what that international regime is mandated to achieve, amongst which is Article 11(7)(d):

> An equitable sharing by all States Parties in the benefits derived from [the Moon's] resources, whereby the interests and needs of the developing countries, as well as the efforts of those countries which have contributed either directly or indirectly to the exploration of the Moon, shall be given special consideration.

Ironically, in trying to achieve a coordinated and cooperative approach to the utilization of space resources, Article 11 has essentially achieved the opposite. The incorporation of 'common heritage of mankind' (and its precise meaning) was a contentious issue during the negotiations of the Agreement and has remained so even after the Agreement was reached (Tronchetti 2010, 506). The difficulty arises from the inclusion of 'equitable' in Article 11(7)(d).

As Tronchetti (2010, 511) points out, there are two competing interpretations of the meaning of 'an equitable sharing ... in the benefits derived from those resources'. On the one hand, it could be taken to mean that resources and benefits derived from exploitation of celestial bodies are to be shared equally with all states irrespective of whether they have played any role in such exploitation (the interpretation contended for by developing countries). The counterpoint is that it could be interpreted as 'equal' in the literal sense of that word, meaning that those States who are actively engaged in the exploitation activities have a greater say in how the benefits are shared (the definition favoured by developed countries).

Tronchetti (2010, 511) also notes that, in addition to the contested definition of 'equitable' there is also much uncertainty over what is meant by the 'benefits' that are to be equitably shared. Are those benefits the profits of the activities? Or does it mean the sharing of the actual resources themselves? Does the term 'benefits' extend to the technology developed to extract resources? The vagaries attached to the Common Heritage Principle (Newman 2015, 34) and the ambiguities regarding the extent of resource sharing, coupled with the indeterminate nature of the body that will oversee such distribution means that states will be unlikely to commit to future exploration where such crucial concepts are unclear.

The Moon Agreement potentially provides a framework in circumstances where exploration of celestial bodies and extraction of celestial resources is being conducted by States Parties who are prepared to work together in a coordinated manner and harmonious spirit. However, those were not the prevailing circumstances at the time of the Moon Agreement, and they are not the circumstances that prevail now. The current geopolitical reality means that celestial resource extraction is likely to be extremely competitive. Furthermore, it is likely that the conduct of such activity will require the investment of significant amounts of private capital. The uncertainties created by the Moon Agreement do not support a commercially coherent case for such investment.

The Moon Agreement has been ratified by only 18 states and signed by an additional four states. Neither China nor Russia have ratified or signed the Agreement and in 2020 President Donald Trump rejected the notion that the Moon Agreement governs 'the promotion of commercial participation in the long-term exploration, scientific discovery, and use of the Moon, Mars, or other celestial bodies'. This repudiation expressly signalled to the world that the United States does not consider the Moon Agreement to reflect international law regarding lunar exploration.

7. Agreement and Consensus for Future Settlements

The competing opinions over the compatibility of provisions of the Artemis Accords with the Outer Space Treaty and the apparent repudiation of the Moon Agreement should not convey the impression that a solution for the governance of non-terrestrial settlements is unobtainable. There are a great many areas of shared agreement. Indeed, it is worth re-emphasizing that existing elements of international space law will still apply.

The Outer Space Treaty permits states to undertake the exploration, use of outer space, including the Moon and other celestial bodies. It also enshrines freedom of scientific access and freedom of access to space without discrimination. Non-appropriation of celestial bodies and the incorporation of international law have already been discussed. Article IV of the Treaty prohibits the stationing of nuclear weapons and weapons of mass destruction in outer space or on celestial bodies. Whilst Article V *inter alia* lauds astronauts as 'the envoys of mankind' (a figure of speech intended to emphasize that early space travellers who landed in hostile territory should not be used as political prisoners). More significantly Article V creates the humanitarian duty (reflected in Section 6 of the Artemis Accords) to take all reasonable efforts to offer all possible assistance to astronauts in distress (Von der Dunk and Goh, 2009, 97–8).

Within the Accords themselves, there is a great deal that has been welcomed. The notion of interoperability is already well-established as a desirable element of collaborative space ventures. Defined as 'the ability of a system to work with or use the parts or equipment of another system' (Salmeri 2020) interoperability has the potential to greatly reduce the expense for new state actors to embed existing, proven technologies within their infrastructure at a greatly reduced development cost. The establishment of common standards across a range of space-based applications has considerable safety benefits, allowing the creation of standard, universal docking mechanisms and other connectors. There are some potential problems associated with interoperability, such as the potential for commercial operators to use the standard systems to try and leverage commercial advantage, but as can be seen from the International Space Station this has not come to pass and the advantages far outweigh potential concerns. Future, off-world settlements would benefit greatly if all the design of all systems started from the premise of being interoperable with other the systems of other settlers.

The potential for scientific discovery by non-terrestrial settlements on other celestial bodies is perhaps one of the most exciting aspects of human exploration beyond Earth and the Moon. Given that NASA has exhibited a long-standing commitment to the sharing of data (Johnson 2020b) it is not surprising that there is a stand-alone commitment under Section 8 of the Accords to the open sharing of scientific data. Any discovery, including the potential discovery of life, will be subject to rigorous scrutiny and scepticism, the immediate dissemination of scientific data would seem to be a logical starting point for any human settlement. It is significant that the duty to share data will not be extended to private companies, ensuring that there will not be any conflict with potential intellectual property rights.

The principle of transparency is crucial to demonstrating the good faith of those involved in the Artemis Accords. Article IX of the OST places the principles of cooperation and mutual assistance at the heart of how space should be used and explored. From both a security and a scientific standpoint, openness of activity will go a long way towards dispelling some of the tensions that have been outlined. Whilst active cooperation cannot be imposed on states, an entrenched commitment to transparency will ensure that the potential for misunderstandings and conflicts is significantly reduced. A commitment to transparency should be embedded at the very heart of any future legal framework governing off-world settlements.

Notwithstanding these areas of consensus there are some omissions from the Accords that a sustainable off-world settlement would do well to consider. First, the environmental protections afforded by the Accords are limited to the prevention of lunar orbital debris. There is no distinct planetary protection policy. This is explicable as the Moon is largely regarded as lifeless, and there have been numerous sample return missions without any contamination. Additionally, there is a declaration at the start, acknowledging the benefits of coordination via multilateral forums. Whilst not explicitly mentioning it, the Committee on Space Research (COSPAR) would be likely to come within such fora and it has well established planetary protection guidelines (Cheney et al. 2020).

More broadly, any underpinning framework for a human settlement will need to consider the relationship of the settlement to the surrounding environment. From both an ecological and a scientific perspective, the way in which human settlers interact with another celestial body will be crucial. It is suggested that the embedding of measures to protect the environment of another celestial body is a fundamental concept that needs to be agreed at the outset of any mission. Mining is an inherently invasive and disruptive process to any environment. To be more than scavengers of the natural resources, the environmental impact of the settlement must be mitigated as far as possible (Newman 2016, 233).

Finally, there are wider political arrangements that will need to be considered as humans move beyond the Earth and the Moon. The distances between the Earth and the Moon mean that nation-states and the role of geopolitics will still be the determining factor in the overall governance of lunar settlements. The Artemis Accords, for example, make no reference to human rights—and this is perhaps understandable given the 'like-minded partners' that are signing up. But as settlements drift further away from the influence of the Earth, real-time communication will not be possible and the reliance on supplies from Earth diminishes, and then settlers will be forced to consider the relationship with Earth and with the nation-states that comprise our civilization.

Language is not neutral and in analysing a governance framework for off-world excursions, this discussion has been careful to use the word *settlement* as opposed to *colony*. The notion of colonialism is inextricably tied up with historical repression and conquest of Indigenous people (Ferrando 2016, 138). Those looking to settle on other celestial bodies must contemplate exactly what type of venture they are engaged upon when establishing their settlement. Robinson (2006) recognized that there may come a time when human settlers may wish to disestablish themselves from the ties to nations and geopolitics. This is foreseeable and should be within the contemplation of those establishing a blueprint for a settlement on a celestial body away from the influence of the Earth but is clearly not within the remit of the Artemis Accords.

Establishing the granular details of governance for an off-world settlement will largely depend on the infrastructure that emerges and is well beyond the purview of this discussion. There have been some attempts to identify what kind of political system might emerge once a settlement becomes established and has moved beyond the initial stages of simply surviving (Schmidt and Bohacek 2021). Indeed, Robinson (2006) has attempted to identify what a constitutional arrangement for a settlement on a celestial body away from Earth might look like. This is where we reach the boundary between the predictable developments that will occur because of Project Artemis and the longer-term journey of humanity as a spacefaring civilization.

8. Conclusion

With the Moon Agreement seemingly beyond the pale, the Artemis Accords are a tacit recognition that securing a meaningful, binding treaty is not practical given the current geopolitical landscape and the number of states now actively involved in UNCOPUOS. Once the United States decided to return to the Moon, the lack of consensus meant that there was little alternative but to try and establish a series of normative principles upon which to proceed. Notwithstanding the repeated claims of adherence to existing space treaties, the decision of the United States to open the Accords only to 'like-minded partners' could see a fracturing of consensus and compromise and open the floodgates to bilateral law-making.

The danger of a 'fractured' Moon, with different legal and political agreements governing activity, poses a risk to safety and threatens wider stability and security (Wright Nelson 2020b, 5). In addition, the move away from consensus and compromise means that the negotiating position of different states could become asymmetrical. It will, after all, be the United States that determines the qualification as a 'like-minded partner' for Project Artemis. As has been pointed out, the benefits of dialogue and cooperation between non-like-minded states was clearly demonstrated in the Apollo-Soyuz Test Project of the 1970s (Mosteshar 2020, 603). Human space exploration would be ineffably weakened if such projects were to become impossible. Discussions and negotiations in UNCOPUOS are a way that nations can keep lines of communication open.

There are the basic principles upon which agreement was reach amongst the international community at the time the Outer Space Treaty was promulgated. There is still broad consensus that the Treaty and the law-making process within UNCOPUOS are the ways in which the regulation of space activity is the preferred way to develop international law regarding outer space activities. Advocates of the Accords point to Section 10(4) which commits to ongoing discussions at the UNCOPUOS as to how the legal framework on resource utilization will develop. Critics of the Accords have countered that UNCOPUOS is the proper forum for interpreting the provisions of the OST and that the Accords serves to undermine inclusive international discussions (Mosteshar 2020, 602), but both at least recognize that UNCOPUOS must play a significant role in the development of international space law from now on into the future.

Merely signing the Accords does not breach the OST. Concern has been expressed over the mining of space resources and the potential implications of deconfliction activities, but in truth the technology and infrastructure to enable commercial lunar mining is still some way in the future. The establishment of functioning, independent settlements on celestial bodies away from the Earth and Moon is yet further into future. These tentative footsteps away from Earth heralded by Project Artemis represent a chance to explore shared values held by even the most implacably opposed States. By embedding the OST within the fabric of the Artemis Accords, the United States and partner states have acknowledged the importance of the rule of law. Perhaps it is that recognition coupled with the acknowledgement that cooperation and guarantees of mutual assistance are fundamental for safe and sustainable activities in space that provides the most significant legacy for future settlements.

Notes

1. See e.g. the work undertaken by 'For all Moonkind' <https://www.forallmoonkind.org/> accessed 5 July 2022.
2. The term 'like-minded partners' has been used extensively by the NASA to describe those states which might wish to sign up to the Artemis Accords. See e.g. <https://2017-2021.state.gov/dipnote-u-s-department-of-state-official-blog/space-exploration-and-the-artemis-accords/index.html> accessed 29 January 2022.
3. See for further details 'Working group established under the Legal Subcommittee agenda item "General exchange of views on potential legal models for activities in the exploration, exploitation, and utilization of space resources", available online at <https://www.unoosa.org/oosa/en/ourwork/copuos/lsc/space-resources/index.html> accessed 5 July 2022.

References

Amos, J. 'Obama Signs NASA Up to New Future' *BBC News* (2010) <https://www.bbc.co.uk/news/science-environment-11518049> accessed 30 January 2022.

Berger, E. NASA's Full Artemis Plan Revealed: 37 Launches and a Lunar Outpost. *Ars Technica* (2019) <https://arstechnica.com/science/2019/05/nasas-full-artemis-plan-revealed-37-launches-and-a-lunar-outpost/> accessed 30 January 2022.

Bittencourt Neto, O., Masson-Zwaan, T., and Hofmann, M. (eds). *Building Blocks for the Development of an International Framework for the Governance of Space Resource Activities. A Commentary* (Utrecht: Eleven International, 2020).

Blount, P. J. 'Renovating Space: The Future of International Space Law' (2011) 40(1) *Denver Journal of International Law and Policy* 515–32.

Chaikin, A. *A Man on the Moon: The Voyages of the Apollo Astronauts* (London: Penguin, 1998).

Cheney, T., Newman, C. J., Olsson-Francis, K. et al. 'Planetary Protection in the New Space Era: Science and Governance' (2020) 7 *Frontiers in Astronomy and Space Sciences* 589817.

Cheney, T. and Newman, C. J. –'Managing the Resource Revolution: Space Law in the New Space Age' in R. J. Wilman and C. J. Newman (eds), *Frontiers of Space Risk: Natural Cosmic Hazards & Societal Challenges* (Boca Raton, FL: CRC Press, 2018) 245–70.

Danilenko, G. M. 'International Law-making for Outer Space' (2016) 37 *Space Policy* 179–83.

Deplano, R. 'The Artemis Accords: Evolution or Revolution in International Space Law?' [2021] 70 *International and Comparative Law Quarterly*, 799–819.

Elvis, M., Milligan, T., and Krolikowski, A. 'The Peaks of Eternal Light: A Near-Term Property Issue on the Moon' (2016) 38 *Space Policy* 30–8.

ESA. 'Moon Village' *ESA* (2016) <https://www.esa.int/About_Us/Ministerial_Council_2016/Moon_Village> accessed 30 January 2022.

Ferrando, F. 'Why Space Migration Must be Posthuman' in S. J. Schwartz and T. Milligan (eds), *The Ethics of Space Exploration* (Cham: Springer, 2016) 137–52.

Froehlich, A. (ed). *Assessing a Mars Agreement including Human Settlements* (Cham: Springer, 2021).

Gabrynowicz, J. 'Space Law: Its Cold War Origins and Challenges in the Era of Globalization' (2004) 37 *Suffolk University Law Review* 1041–65.

Hobe, S. (2009). 'Article I' in S. Hobe, B. Schmidt-Tedd, K. Schrogl (eds), *Cologne Commentary on Space Law: Vol I Outer Space Treaty* (Cologne: Carl Heymanns Verlag, 2009) 25–43.

Hobe, S. 'The Moon Agreement—Let's Use the Chance!' (2010) 59 *German Journal of Air and Space Law* 372–81.

Hobe, S. *Space Law* (Nomos/Hart, 2019).

Jakhu, R. S., Chen, K. -W., and Goswami, B. 'Threats to Peaceful Purposes of Outer Space: Politics and Law' (2020) 18(1) *Astropolitics* 22–50.

Johnson, C. D. 'SpaceWatchGL Feature: The Space Law Context of the Artemis Accords (Part 1)' *SpaceWatch.Global* (2020a) <https://spacewatch.global/2020/05/spacewatchgl-feature-the-space-law-context-of-the-artemis-accords-part-1//> accessed 30 January 2022.

Johnson, C. D. 'SpaceWatchGL Feature: The Space Law Context of the Artemis Accords (Part 2)' *SpaceWatch.Global* (2020b) <https://spacewatch.global/2020/05/spacewatchgl-feature-the-space-law-context-of-the-artemis-accords-part-2/> accessed 30 January 2022.

Larsen, P. B. 'Is There a Legal Path to Commercial Mining on the Moon?' (2021) 83(1) *University of Pittsburgh Law Review* 6–49.

Lyall, F. and Larsen, P. B. *Space Law: A Treatise* (London: Routledge, 2020).

Marchisio, S. 'Article IX' in S. Hobe, B. Schmidt-Tedd, K. Schrogl (eds), *Cologne Commentary on Space Law: Vol I Outer Space Treaty* (Cologne: Carl Heymanns Verlag, 2009) 169–82.

Martinez, P. 'The UN COPUOS Guidelines for the Long-Term Sustainability of Outer Space Activities' (2021) 8(1) *Journal of Space Safety Engineering* 98–107.

Masson-Zwaan, T. and Sundhal, M. J. 'The Lunar Legal Landscape: Challenges and Opportunities' (2021) 46(1) *Air & Space Law* 29–56.

Mosteshar, S. 'Artemis: The Discordant Accords' (2020) 42(2) *Journal of Space Law* 591–603.

NASA. 'The Artemis Accords: Principles for Cooperation in the Civil Exploration and Use of the Moon, Mars, Comets and Asteroids for Peaceful Purposes' *NASA—The Artemis Accords* (2020a) <https://www.nasa.gov/specials/artemis-accords/index.html/> accessed 30 January 2022.

NASA. 'International Partners Advance Cooperation with First Signings of Artemis Accords'. *NASA.gov* (2020b) <https://www.nasa.gov/press-release/nasa-international-partners-advance-cooperation-with-first-signings-of-artemis-accords/> accessed 30 January 2022.

Newman, C. J. 'Seeking Tranquillity: Embedding Sustainability in Lunar Exploration Policy' (2015) 33 *Space Policy* 29–37.

Newman, C. J. '"The Way to Eden": Environmental, Legal and Ethical Values in Interplanetary Space Flight' in S. J. Schwartz and T. Milligan (eds), *The Ethics of Space Exploration* (Cham: Springer, 2016) 221–38.

Newman, C. J. 'Artemis Accords: Why Many Countries Are Refusing to Sign Moon Exploration Agreement'. *The Conversation* (2020) <https://theconversation.com/artemis-accords-why-many-countries-are-refusing-to-sign-moon-exploration-agreement-148134/> accessed 29 January 2022.

Pearlman, R. Z. 'NASA Names New Moon Landing Program Artemis after Apollo's Sister' *Space.com* (2019) <https://www.space.com/nasa-names-moon-landing-program-artemis.html/> accessed 30 January 2022.

Potter, S. 'New Zealand signs Artemis Accords' *NASA* (2021) <https://www.nasa.gov/feature/new-zealand-signs-artemis-accords/> accessed 30 January 2022.

Raymond, J. W. 'Opinion | How the U.S. Space Force is Trying to Bring Order to Increasingly Messy Outer Space' *The Washington Post* (2021) <https://www.washingtonpost.com/opinions/2021/11/29/space-activity-its-debris-increases-us-works-establish-international-norms-rules//> accessed 30 January 2022.

Ribbelink, O. 'Article III' in S. Hobe, B. Schmidt-Tedd, K. Schrogl (eds), *Cologne Commentary on Space Law: Vol I Outer Space Treaty* (Carl Heymanns Verlag, 2009) 64–9.

Robinson, G. S. 'No Space Colonies: Creating a Space Civilization and the Need for a Defining Constitution' (2004) 30(1) *Journal of Space Law* 169–79.

Robinson, G. S. 'Transcending to a Space Civilization: The Next Three Steps Toward a Defining Constitution' (2006) 32 *Journal of Space Law* 147–75.

Salmeri, A. '#spacewatchgl Opinion: One Size to Fit Them All: Interoperability, the Artemis Accords and the Future of Space Exploration' *SpaceWatch.Global*. <https://spacewatch.global/2020/11/spacewatchgl-opinion-one-size-to-fit-them-all-interoperability-the-artemis-accords-and-the-future-of-space-exploration//> accessed 30 January 2022.

Schmidt, N. and Bohacek, P. 'First Space Colony: What Political System Could We Expect?' (2021) 56 *Space Policy* 101426.

Si-soo, P. 'Israel Becomes 15th Nation to join Artemis Accords' *SpaceNews*. <https://spacenews.com/israel-becomes-15th-nation-to-join-artemis-accords/> accessed 30 January 2022.

Smith, M. 'Top Air Force Officials: Space Now is a Warfighting Domain' *News*. <https://spacepolicyonline.com/news/top-air-force-officials-space-now-is-a-warfighting-domain/> accessed 30 January 2022.

Taichman, E.A. 'The Artemis Accords: Employing Space Diplomacy to De-Escalate a National Security Threat and Promote Space Commercialization' (2021) 11(2) *American University National Security Law Brief* 112–46.

Tronchetti, F. 'The Moon Agreement in the 21st Century: Addressing its Potential Role in the Era of Commercial Exploitation of the Natural Resources of the Moon and Other Celestial Bodies' (2010) 36(2) *Journal of Space Law* 498–524.

Tronchetti, F. 'Legal aspects of space resource utilization' in F. von der Dunk and F. Tronchetti (eds), *Handbook of Space Law* (Cheltenham: Edward Elgar Publishing, 2017) 769–813.

Von der Dunk, F. 'The Artemis Accords and the Law: Is the Moon "Back in Business"?' University of Auckland Blog. <https://www.thebigq.org/2020/06/02/the-artemis-accords-and-the-law-is-the-moon-back-in-business/> accessed 30 January 2022.

von der Dunk, F. and Goh, G. 'Article III' in S. Hobe, B. Schmidt-Tedd, and K. Schrogl (eds), *Cologne Commentary on Space Law: Vol I Outer Space Treaty* (Cologne: Carl Heymanns Verlag, 2009) 94–102.

Wright Nelson, J. 'Safety Zones: A Near-Term Legal Issue on the Moon' (2020a) 44(2) *Journal of Space Law* 604–24.

Wright Nelson, J. 'The Artemis Accords and the Future of International Space Law' (2020b) 24(31) *Insights*.

20

Enlightenment beyond Earth

Anthony Pagden

For us today, in the twenty-first century, the importance of the legacy of the intellectual movement known broadly as the Enlightenment is everywhere apparent—although few of us may be immediately aware of it

It is so for two reasons.

The first is obvious enough. If we set aside for the moment the lingering controversy over the reality of the 'Scientific Revolution' of the seventeenth century, one thing still remains: that without the Enlightenment, modern science as we understand it today would not have taken the form it has here in the West. It might have happened in China or even in India, but not here where it did happen, in France, in Britain, in Germany, in Italy, in Spain, in Scandinavia, and in the Americas, both north and south. And without that science, and the technologies to which it has given rise, we would not today even be contemplating the possibilities and the challenges of any form of extraterrestrial ways of life.

The second reason is less obvious but equally pertinent.

In the recent past, what we now call the 'West' has become increasingly divided, intellectually, morally, and politically, into two broad groups who have sometimes radically opposing views about the worlds they inhabit. There are of course many shades in between but the dichotomy is there, and ultimately it derives from an acceptance or denial of the arguments put forward by the writers, philosophers, poets, dramatists, novelists, jurists, scientists, men and women of letters (for it was a multifaceted enterprise), who made up what *they* called—for this was a highly self-conscious movement—*the* Enlightenment (see Pagden 2013).

On the one side, we have the champions of enlightenment. They are those who, whilst they are prepared to acknowledge human frailty and the human capacity for doing harm, believe that it is possible to improve the human condition, through knowledge and science. And because they believe this, they also believe that there exists a 'human nature'—although few today would employ such a term—which is much the same everywhere. They hold, that is, that although matters such as cultures and religion are important, and that difference must be respected, this can only be so when such things conform to some set of values, some standard of behaviour, by those cultures towards all *their* own members, which every rational being of *whatever* culture, religion, or political system could be brought to understand and respect, if not necessarily to share.

To put it differently, they believe that the means by which humans order their lives cannot be anything other than human, consensual, intelligible, and changeable.

We might say that they give predominance to individual agency—to autonomy—and thus to individual freedom.

Anthony Pagden, *Enlightenment beyond Earth*. In: *The Institutions of Extraterrestrial Liberty*. Edited by Charles S. Cockell, Oxford University Press. © Anthony Pagden (2022). DOI: 10.1093/oso/9780192897985.003.0021

On the other side we have those who believe that cultures cannot be questioned or changed by those outside them, that there exists no such thing as a single 'rationality' or a single human identity, and therefore no single rule of justice, but only what the American philosopher Michael Walzer has called 'spheres of justice'—overlapping certainly but always distinct and always sovereign. Outside the community there can exist no grounds for making any kind of moral, social, anthropological, or religious judgement about anything that may be found within that community. The Indian caste system, for instance, so apparently and blatantly unjust to contemporary secular sensibilities, would, argues Walzer, be entirely just if it were accepted as such by all those who are affected by it. In fact, it is not—or so he says—so the question does not arise. All Walzer will accept as possibly universal criteria would be a list of 'negative injunctions against oppression and tyranny'—and these would inevitably be remarkably short—particularly if a system of caste were not to be included amongst them (Walzer 1983, 314–15; Benhabib 2011, 60–1).

For the members of this group—call them 'communitarians'—'the Enlightenment' in general, and Enlightenment universalism in particular, is nothing other than an attempt to impose what are, in fact, simply one set of norms on the whole of humankind. Today these are predominately Western, tomorrow they may well be Chinese. Enlightenment, therefore, necessarily seeks to destroy the power of any collective agency which opposes it, and is also, more often than not, 'imperialist', and, thus, ultimately tyrannical. Enlightenment, in the words of the English philosopher John Gray, is little more than an 'assault on cultural difference' and is the 'embodiment of Western cultural imperialism as the project of a universal civilization' (Gray 1997, vii).

On one thing both groups are broadly in agreement: 'enlightenment' supposes the existence of a single human nature—however basic—and the possibility of an, at least mutually comprehensible, set of cultural and moral values and the acceptance of a single understanding of what constitutes an order of justice, even if not of a specific legal or political system.

For those who lived through, and created it, enlightenment—*Aufklärung* in German, *Lumières* in French, *Ilustración* in Spanish, *Illuminismo* in Italian—meant very simply casting light where there had once been darkness. It meant demolishing prejudice—a favourite term of abuse described by Charles de Secondat, baron de Montesquieu, perhaps the most lastingly influential of the political and legal philosophers of the Enlightenment, as 'not what makes one unaware of certain things, but what make one unaware of *oneself*' (Montesquieu 1951, II, 230). By so doing, prejudice limits the necessary freedom of action which the individual must be allowed to exercise if she or he is to be able grasp the world as it is. Similarly, Immanuel Kant—one of the two most lastingly influential philosophers of the Enlightenment—in his famous attempt to answer the question 'What is enlightenment'—*was ist Aufklärung*—argued that enlightenment was precisely what allowed the individual to exit from what he called, his 'self-incurred minority', and by 'minority', Kant meant the 'inability to make use of one's own understanding without the guidance of another'. To do this the individual had to cast aside all those 'dogmas and formulas, those mechanical instruments for rational use (or rather misuse) of [mankind's] natural endowments' which Kant described as 'the ball and chain of his permanent minority'.

If I am to be morally responsible for my own actions, then I have to live, as Kant put it, without the aid, of 'a book that has understanding for me, a pastor who has a conscience for me, a doctor who judges my diet for me', and so forth. I need, he concluded, borrowing a famous line from the Roman poet Horace, 'to dare to think'—*sapere aude* ('An Answer to the Question: 'What is Enlightenment?' in Kant 1996, 17–18).

This is the necessary condition of freedom.

What this might eventually come to mean was perhaps best captured in the words of one of Kant's contemporaries (although he seems never to have read him), Marie Jean Antoine Nicolas de Carita, marquis de Condorcet. Condorcet was one of the creators of differential calculus and the first person to attempt to predict the possible outcome of human decision-making by using mathematics, a champion of equal rights for women and for all peoples of all races, and an abolitionist who devised the world's first state education system. (He was probably also the first person to use the term the 'West' in the sense it is broadly understood today, to describe not a region of the world but a group of peoples: the Europeans and all those of European descent now scattered across the globe.) In his *Sketch for a Historical Picture of the Progress of the Human Mind*, written in 1795 whilst in hiding from the forces of the Jacobin Terror, in a tiny room in the rue Servandoni in Paris, he described a future condition in which he believed that all mankind would acquire

> the necessary enlightenment to conduct themselves in accordance with their own reason in the common affairs of life, and to maintain them, free of prejudices, so that they might know their rights and be at liberty to exercise them according to their own opinion and their conscience, where all might, through the development of their faculties, obtain the certain means to provide for their needs.
>
> (Condorcet 1988, 264–9)

Condorcet's *Sketch*, and Kant's essay have often been taken to sum up all that was most important about the Enlightenment: namely, to put it very simply, that it was, in John Gray's sternly derogatory description, a movement 'promoting autonomous human reason and according to science a privileged status in relation to all other forms of understanding' (Gray 1997, 145).

And many contemporaries did, indeed, see it that way. They were all, after all, the self-styled heirs of the great intellectual movers and shakers of the seventeenth century, Descartes, Hobbes, Locke, Leibnitz, Spinoza Bacon, Newton, whose lasting achievement—quite apart from whatever else they might have done—had been, as Locke phrased it, to sweep away all 'the Rubbish that lies in the way to Knowledge'— most of it placed there by the Aristotelian, Thomist theologians of the previous two centuries (Locke 1975, 10). It was to the 'most enlightened man' (*L'homme le plus eclairé*) that Diderot dedicated one of the most characteristic projects of the Enlightenment, *L'Encyclopédie*, the sum, or as close as anyone could come to it, of all human knowledge.

This is one view of the Enlightenment. It is certainly the most familiar. But science—natural science, that is—was certainly not the whole of it. There was also what came to be called the 'sciences of man' and it is on this that I wish to focus here.

And I wish to begin with the other great philosopher of the Enlightenment, reading whose work, Kant claimed, had 'roused him from his dogmatic slumber': David Hume.

The two men could not have been more unalike. Kant was thin, dry, ascetic, sarcastic but humourless, and never left his native town of Königsberg. Hume was plump, witty, jovial, and travelled widely. 'You belong to all nations', Diderot wrote to him in 1768, 'I flatter myself that I am, like you, a citizen of that great city, the world'(Diderot 1875–7, V, 218). And so he was, and the compliment, as we shall see, was significant.

Hume took a rather different view of the place of reason in human understanding, best summed up perhaps in one of his most famous, and contentious, remarks that 'reason is and ought to be a slave of the passions and can never pretend to any other office but to serve and obey them' (Hume 1896, 415). By this, what Hume meant was that without our passions we would have no reason for reasoning nor for *acting* at all. It is our passions which motivate us to reason, and it is they that then activate our will. Linked to the passions are what the philosophers of the Enlightenment called 'sentiments', a now debased word, but one which for Hume meant those emotions which arise primarily from the passions of pride and humility, on the one hand, and from love and hatred on the other. There are a great many of them, but the most important for what I want to say here is what in the eighteenth century was called 'sympathy', and what we today might call 'empathy'. ('Sympathy' has, however, a far richer range of connotations than 'empathy'.)

This is perhaps best explained in *The Theory of Moral Sentiments* of 1759, the first major work of another key figure in any understanding of the Enlightenment, Adam Smith. When, wrote Smith, we see someone being tortured, then as long as 'we ourselves are at our ease, our senses will never inform us of what he suffers'. That is, we can never actually experience the physical pain of another human being, but we can, to some degree imagine it, and what prompts us to do that is a recognition of the humanity which we share with the victim—our 'sympathy', as Smith puts it, for 'our brother who is upon the rack'. We, the spectators, do not need to know where that person comes from, or to which tribe they belong. Our sympathy derives solely from the knowledge that he—or she—is our 'brother' (or sister). We do, however, need to know the cause of their sufferings. Only then will I be able to know how I might feel if I were to find myself in the same predicament. In other words, to be at all effective, my initial act of unthinking, instinctual sympathy has to activate my reason. This is because the compassion of the spectator must arise, in Smith's words, 'altogether from the consideration of what he himself would feel if he was reduced to the same unhappy situation, *and ... was at the same time able to regard it with his present reason and judgment*' (emphasis added). It is important to stress that, since it is based upon an emotion and may thus be 'used to denote our fellow-feeling with any passion whatsoever', sympathy depends on no prior moral, religious, or cultural similarities between the spectator and the victim. It emanates from within each one of us as free autonomous individuals, and even—so Smith claimed to believe—'the greatest ruffian, the most hardened violator of the laws of society is ... not altogether without it' (Smith 1976a, 10–12). Someone who was indeed 'altogether without it' would simply not be human. What this offers is a minimum psychological principle on which to base a claim to human sociability, both within individual communities and, what is still more significant for my purposes, between them.

'No passion', wrote Hume, 'if well understood can be entirely indifferent to us; because there is none, of which every man has not, within him at least the seeds and first principles.' 'We seem,' he went on, 'to have a concern even for those whose lives are remote from ours.' Why else, he asked, would we find, as we seem to, 'the fates of states, provinces, or many individuals' so 'extremely interesting', even if we have nothing particular at stake in what happens to them? Why otherwise do we read newspapers? Why are we moved by reading poetry and by the lives of beings that are entirely imaginary? I may not be able, or willing, to do anything about the present victims of the Taliban or—since my sympathies can also be aroused by the sufferings of the dead for whom I can do nothing—the victims of the Holocaust for instance. (Hume gave as his example for the same sentiment the cruelty of the Roman emperors Nero and Tiberius.) But this does not mean that the *thought* of them does not have the capacity to trouble me, and because of that, to become the principle of my future actions. 'No quality of human nature is more remarkable,' wrote Hume, 'both in itself and in its consequences, than that propensity we have to sympathize with others, and to receive by communication their inclinations and sentiments, however different from, or even contrary, to our own' (Hume 1970, 222). Enlightenment was in this way, as Condorcet said of it, a 'disposition of minds' across time and space. From here it was only a short step to imagining the existence of a world made up of diverse peoples all united by this common sympathy.

This then is the basis of Enlightenment universalism. To believe that the Indian caste system would be acceptable if all agreed to it would be to assume that we should feel no 'sympathy' with the sufferings of the Untouchables—even if, for some perverse reason, they were to consent to their suffering. It is also, however, obviously the case that, although we are able to experience sympathy with every other human on the planet no matter how remote or alien, we are more likely to *act* on that passion if the person, or persons are near and in some degree familiar to us, than we are if they are so remote, that we, in Smith's words, 'can neither serve nor hurt' them (Smith 1976a, 229–30, 154–5).

The familiarity required to compel us to move from a simple feeling of discomfort or benevolence to action—or at any rate the possibility for familiarity—was believed, however, to be a condition not of human psychology but of human history. Most of those we associate with the Enlightenment believed that all human societies move inexorably through various stages, determined largely by the means of production, from hunter-gatherers to pastoralists to agriculturists, to finally what was known as 'the commercial society'. At each stage human beings became more capable of controlling, and exploiting, their environments, richer and more diverse.[1] They also gradually increased their contact with, and inevitably their dependence on one another, until they succeeded, as Smith phrased it, in uniting 'in some measure the most distant parts of the world', so that by his day there were, in fact, very few peoples who were so utterly remote from us, that we—whoever 'we' might be—felt ourselves to be in no position to 'serve nor hurt' them (Smith 1976b, 626).

This final stage, which had laid the grounds at least for some future global order, was created by commerce. 'Commerce', however, was taken to mean something far more than mere economic exchange, far more than the 'invisible hand' of capitalism. Commerce had not only 'made property mobile'. It had done the same for people. It had become the channel by which peoples, in terms of groups of people, encountered one another, and through encounter, it was hoped, came to know and understand, one another. It was responsible, as Smith put it, for the 'communication of knowledge and of all sorts of improvements'. In Montesquieu's famous formulation, 'Commerce has made known the customs of all the peoples of the world, and spread everywhere.' It was this, he went on, which gave it the power to cure 'destructive prejudices; and it is almost a general rule, that wherever the customs are gentle there is commerce; and wherever there is commerce, customs are gentle' (Montesquieu, *De l'esprit des lois* in 1951, II, 585). And if at some still remote time there were indeed to be respect between nations, an equilibrium of wealth and power, then Smith no less than Montesquieu was convinced that nothing was more likely to bring it about than that 'that mutual communication of knowledge and of all sorts of improvements which an extensive commerce from all countries to all countries naturally, or rather necessarily, carries with it' (Smith 1976b, 629).

For Kant it had indeed been this commerce which had resulted in a state in which 'the community of the nations of the earth has now gone so far that a violation of rights in one place of the earth is felt in all' (*Toward Perpetual Peace* in Kant 1996, 329–31). The key term here, however, is 'rights', for it is on the basis of this that the other great project of the Enlightenment was built: 'cosmopolitanism'. Cosmopolitanism is not, of course, a creation of the Enlightenment as such, any more than is modern science. But the form it took in the eighteenth century has shaped the modern conception of a unified world more lastingly than any previous form of universalism.

Cosmopolitanism has often come in for some severe criticism. The cosmopolite, said Jean Jacques Rousseau, who was very hostile to the whole idea, is one who is guilty of 'loving the entire world in order to enjoy the privilege of loving no one' (*Du Contrat social* (1st version) in Rousseau 1959–1995, III, 287). In the scathing words of the English philosopher, Roger Scruton, 'Cosmopolitanism is the belief in, and pursuit of, a style of life which … [shows] acquaintance with, and an ability to incorporate, the manners, habits, languages and social customs of cities throughout the world.… The cosmopolitan is a kind of parasite who depends upon the quotidian lives of others to create the various local flavours and identities in which he dabbles' (Scruton 1982, 100). Similar comments have come from Hitler, from Stalin, and more recently from any number of nationalists and 'populists' across Europe. The objection is always at bottom the same: the cosmopolite has no direct allegiances or attachments to, or roots in, any particular society or community.

This is not, however, how the philosophers of the Enlightenment saw it. For them it was not a matter of detachment or contraction, of, as Scruton put it, 'dabbling' in the 'quotidian lives' of others. Nor was it a bid for the kind of flattening and homogenization—the 'Coca-Cola-ization' and 'MacDonald-ization'—of the world of which contemporary 'globalization' is so often accused.

It was indeed the very opposite. It was a demand for expansion and inclusion. What the philosophers of the Enlightenment were asking for was, in effect, a vision of the human that required a degree of attachment which, although it began at home, did not, because of that, simply stop at the nearest frontier—social, cultural, religious, or natural. Of course, I care more for my family than I do for the displaced masses in Syria or Iraq. What is being asked of me is instead a vision of the world which is capable of encompassing both my immediate family, my local community, *and* the displaced masses *within* a broader perception of humanity itself.

This is how Montesquieu expressed it. Since, he wrote in 1745, he was 'a man before I am a Frenchman, or rather because I am of necessity a man but only a Frenchman by chance',

> If I knew something useful to me, and harmful to my family, I would reject it from my mind. If I knew something useful to my family, and not to my country, I would try to forget it. If I knew of something useful to my country, and harmful to Europe, or useful to Europe and harmful to Mankind, I would look upon it as a crime.
>
> (*Mes Pensees* 11 in Montesquieu 1951, I, 981)

Adam Smith arrived at much the same conclusion by a somewhat different route. If, he said, 'the wise and virtuous man' was always willing, as Smith assumed he must be, to sacrifice his own private interests to those of the 'public interests of his own particular order and society', and if that same man recognized that this should in turn be 'sacrificed to the greater interest of the state or sovereignty, of which it is only a subordinate part', then it followed that 'he should be equally willing that all those inferior interests should be sacrificed to the greater interests of the universe' (Smith 1976a, 235).

Such persons are the true 'cosmopolites'—what Kant dubbed the 'cosmotheroi' (students of the world). All that they are being asked to do is to keep their eyes fixed upon the wider horizon of humankind itself. It was what Kant, significantly, called a form of 'global patriotism' ('Kant on the Metaphysics of Morals: Vigilantius' lecture notes' in Kant 1997, 406).

Global patriotism, however, demands both a grasp of the 'global' and a conception of the 'patria'. The former follows, of course, from the kind of shared sympathy which for Kant constitutes a moral duty. Humanity, he wrote,

> means on the one hand the universal *feeling of participation,* and on the other the capacity for being able to *communicate* one's innermost self universally, which properties, taken together constitute the sociability that is appropriate to humankind, by means of which it distinguishes itself from the limitations of animals.
>
> (Kant 2000, 229)

These are the inescapable facts of the human condition. They provided a basic reason why we should—why as humans we could not fail to—develop interpersonal relations with others from widely different cultures from our own and extend our benevolence to one another. And this, in Kant's view, generates a *right*—not, he insists, a mere 'philanthropic principle'—that each people 'try to establish community with all and, to this end, to *visit* all regions of the world' (*Toward Perpetual Peace* in Kant 1996, 329–31).

This is what he calls the *ius cosmopoliticum.* In the first instance it is fairly restricted one. It gives the 'traveller' a right to travel and of 'temporary sojourn' in any part of the world. This is not, of course, a right acquired by treaty. It does not include the right to permanent residence. And it derives solely from the fact that because I inhabit a sphere which I cannot leave—or at least Kant could not—and no matter where I set off from if I go on long enough, I will come back to where I started, this confers upon me what he called a 'common possession' of its surface. Only under such conditions would it be 'possible for [strangers] to enter into relations with the native inhabitants', and since societies cannot flourish in isolation from one another, much less in contempt of one another, it is a right that arises from the general concern for the good of the species. It had been, Kant had argued, precisely the tendency of the Greeks to isolate themselves from the rest of humanity, whom they then lumped together as 'barbarians', that had been 'the prime cause contributing to the downfall of their states' ('Kant on the Metaphysics of Morals: Vigilantius' lecture notes' in Kant 1997, 406).[2]

This is not an entirely new idea. It harks back to an ancient conception of the 'universal right of hospitality' and the idea that travel the 'rite of passage' was a natural one and therefore part of the 'law and nature and of nations', and it still has a place in the debate over the rights of refugees. It was for most what was called an 'imperfect right' in that it could be overruled in the interest of public safety and the like, which is why Kant accepts that the Chinese and the Japanese, whom he generally reviled precisely for isolating themselves from the rest of humanity, were entitled to place severe limitations on the activities of the European trading companies since—as he put it—having 'given such guests a try', they knew that in this case 'visiting' was merely a synonym for 'conquering' (*Toward Perpetual Peace* in Kant 1996, 329–31).

The origins of the claims made on behalf of the 'rite of passage' may be ancient, but Kant's version and the implications for it of Smith and Hume's theory of moral sentiments are really quite different. In all previous iterations it was nature—or more precisely, the natural law—which had given us this right. Here in this world of autonomous free agents who, in Kant's words, are 'meant to produce everything out of [themselves]', it is we ourselves who claim it through our own actions (*The Idea for a Universal History with a Cosmopolitan Aim* in Kant 2007, 111). Once again, we might say—and Kant and Hume would say indeed say—that this is the condition of true freedom.

The cosmopolitan right is also, in the end, not merely a right to survive and prosper; it is right having to do with our evolution as a species. This 'universal *cosmopolitan condition*', Kant declared, was nothing less than the 'womb in which all the original predispositions of the human species will be developed' (*The Idea for a Universal History with a Cosmopolitan Aim* in Kant 2007, 116–18).[3] Since, however, it constituted a 'global patriotism' it had also to possess a political and legal order. The form that this would take, Kant imagined—indeed predicted—would eventually be some kind of world association of nations: what he variously called a 'league of peoples', 'an international state', a 'universal union of states', a federation, a confederation, a partnership. Kant is nowhere precise in his description of what each of these might entail (and is inconsistent even in his use of the same term).[4] In outline, however, they all broadly correspond to Ian Crawford's vision of a future extraterrestrial federation based upon the principle of subsidiarity (Crawford 2015, 199–218).[5]

Kant, of course, knew nothing of subsidiarity, which was first articulated in the manner in which it is used today in the late nineteenth century by the Catholic bishop of Mainz, Wilhelm Emmanuel von Ketteler, and in 1891, it was incorporated into Pope Leo XIII's encyclical *Rerum novarum* as a means of combatting what the pope saw as the rising threat of liberalism and industrialization. The basic idea—which might initially in fact appear to be a distinctly liberal one—is that all government should be conducted at the lowest level possible. In the case of a federation (or an empire, as its more remote source is a principle of the Roman common law), this means that only those actions which directly affect all the member states should be decided by the federal government.[6] Such a principle would, as Kant insisted any federal government should, leave the individual states separate and independent as far as anything concerning their culture, language, and religion is concerned. It approximates to what the contemporary Irish-American philosopher Philip Petit calls the 'idea of externally undominated states' (Petit 2012, 2016). Subsidiarity, indeed, determines the limits of EU law and bears some resemblance to the tenth amendment of the US Constitution.[7] It also, of course, places severe limitations on the development of any true federation and is the source of many of the complaints about the EU's 'democratic deficit'. But since any future extraterrestrial federation would not be uniting a pre-existing body of states (or even, as in the US case, semi-autonomous communities) it would, theoretically at least, be in a far better position to draft for itself a constitution that would avoid the difficulties which have beset both the European Union and the United States.

The question that had bedevilled all previous attempts to create a compelling (and binding) inter-state law was, however, where any body of law which was not a positive, civil, one—in other words, one arrived at by a legislative authority of some kind—was to come from in the first instance? As I have argued the enlightened conception of 'sympathy', broadly understood, had provided a reason to believe that all human beings possess a disposition to communicate, and thus to reach an understanding with each other, and a corresponding right to the possibility of interaction with each other. On the basis of this it might be possible to create some kind of interpersonal legal order. But the relationships between individuals were one thing, those between states quite another, for whereas individuals within states were recognized to be bound by the rule of law, states in relationship to one another were not. As Kant phrased it, 'states considered in external relations to one another are (like lawless savages) by nature in a non-rightful condition'. Furthermore, he believed this to be a condition in which the very concept of a law—or right—would seem to be so inherently meaningless that 'it is difficult even to form a concept of this or to think of law in this lawless state without contradicting oneself' (*The Metaphysics of Morals* in Kant 1996, 485).

So long as this remained the case, no universal law—no 'law of nature and nations'—was even conceivable. All the previous attempts by the great natural law theorists of the seventeenth and eighteenth centuries—Hugo Grotius, Samuel Pufendorf, Christian Wolff, Emer de Vattel, all of whom Kant dismissed as 'the sorry comforters of mankind'—to create such a law was, in Kant's opinion, merely an indication of good intentions, but could have no purchase as law, because it was not subject, and in the world as he knew it could never be subject, to any 'external coercive power'.

In other words, what was required was not the 'law of nations', which was inextricably bound in with the 'natural law'—a form of law imposed by nature on humans—and reliant ultimately upon a hypothetical 'consensus of all the peoples of the world', but indeed a form of positive law, created by the states themselves and imposed, by means of treaties, on them by themselves. For this, in 1783 Jeremy Bentham coined the term 'international law'. Anything else, any 'laws of nature, fancied and invented by poets, rhetoricians, and dealers in moral and intellectual poisons' were, he declared memorably, 'simple nonsense ... rhetorical nonsense—nonsense upon stilts' (Bentham 2002, 317–401).

But since all positive law requires a legislative body and that can only operate in the context of a real, not hypothetical, international consensus, the world would first need a common global order of justice, and this is precisely what Kant's *ius cosmopoliticum* was intended ultimately to provide.

Kant indeed had seen—unlike all the authors of all the previous attempts to dream up a world federation, of which there were many—that such a union would only work on any scale if all the parts of which it was made up shared a common political regime. And not just any political regime. For although external 'undomination'—to use Petit's term—is not incompatible with internal domination, it is very unlikely that any 'unfree' or dominated state would be willing to enter into a relationship with any other state, dominated or 'undominated', in the first place.

All the states of which any future world federation was made up, would, therefore, have to be what Kant called 'representative republics'. By this was meant a state in which the sovereign is constrained to 'give his law as if it *could* have arisen from the united will of an entire people' ('On the Common Saying: "This may be true in theory, but it is of no use in practice"' in Kant 1996. 296–7, emphasis added). This is the definition of what he calls 'external (rightful) freedom'. It did *not* mean that the law had to be *made* by the people. Indeed, Kant is explicit in saying that this is not a democracy. The counterfactual is crucial. This would be a system in which the peoples are not free in Petit's understanding of what 'republican non domination' must mean. To be that they would have to be actively engaged in the business of law-making. They are, nevertheless, citizens and thus subject to a law which is equal for them all and which thus makes of them, in Kant's view 'co-legislating members of a state'. Only when all the states of the world, or of some self-defining part of it, had created such a constitution for themselves would they be able jointly to leave the state of nature and create a 'common law' amongst themselves.[8]

This 'common law' would be what we today call 'international law'. Of course, modern international law is frequently brushed aside by the so-called realists, much as Kant had brushed aside the 'law of nations' of his own day, in the belief that the international world as he knew it was still in a state of nature or what today is sometimes called a 'law-free policy space' (Hurd 2017, 17). In reality, however, today, unlike the late eighteenth century it is hard to imagine any modern state, even China, operating for long entirely outside the rule of law. For modern international law is, as Kant understood his *ius cosmopoliticum* to be, more than a set of rules, more than law as regulation. It is precisely *ius*—that is, justice, an order which determines what is to be discussed, how and in what form and language any final decision will be made (Waldron 2000, 227–43). It, in effect, now creates what in the eighteenth century was known as a 'civilization' (a word which is of legal origin, and has a quite precise legal meaning).

The Enlightenment vision of a 'cosmopolitan order' has provided the fundamental inspiration for the creation of a number of admittedly imperfect institutions, in some respects not utterly unlike the kind that Kant had hoped for: first the League of Nations—which for some of its founders was an explicitly Kantian project—and subsequently the United Nations, the International Court of Justice, the International Criminal Court, and the International Labor Organization, not to mention still more amorphous, and still impermanent, but potentially no less consequential acts such as the Kyoto Protocol, the Paris Agreement, etc. It has also provided the theoretical foundations for such modern conceptions as 'international justice', 'geo-governance', 'global civil society', and the 'Responsibility to Protect'. Then of course we have had, since 1948, a 'Universal Declaration of Human Rights'. True, demands made by this are vague, and there are many—the right to healthcare and education, for instance—to which even the most conscientious and wealthy Western countries do not subscribe. But it has been perhaps the single most influential concept in shaping foreign policy in the West since at least the late 1970s. And the very idea of a 'human right' would have been unimaginable without some conception of a single human nature and of a global community bound together by a common order of justice—by, in fact, something very close to Kant's *ius cosmopoliticum*.

So, in the light of all this, what can we say of the 'conditions for freedom beyond Earth'? For, in extraterrestrial space, we have indeed, as Charles Cockell nicely phrases it, a new 'state of nature' more extreme indeed than any which our Earth-bound ancestors encountered in America, and which will thus compel its settlers to draw up a new species of social contract amongst themselves. In some ways they will find themselves in a position not unlike the early settlers in America. Like them, they will, initially, be entirely dependent technically and economically on their respective 'mother' institutions, be they states or private corporations. They will require, as Charles Cockell has pointed out, the very real necessities of life for whilst 'everywhere on earth one is free to breathe the air', in space oxygen has to be manufactured and is thus dependent upon technologies imported from Earth. This will also, as he has stressed, present opportunities for a form of extraterrestrial tyranny, far greater than anything that currently exists on earth (Cockell 2009, 139–57). And even if private individuals, the Musks and the Bezos of today, and those that follow them, are able to provide the initial resources necessary for interplanetary travel and settlement, independently of any terrestrial state, they are also likely, if their power and influence were to become too threatening, to end up like the Virginia Company, the Dutch East India Company, and the British East India Company by being absorbed by the state. This presupposes, of course, that the 'state' as we currently understand it will continue in more or less its present form into the foreseeable future. But that may well turn out not to be the case. All modern democratic states are already in practice 'disaggregates'—to use Anne-Marie Slaughter's term—in that the powers of government are divided up between a large number of different interconnected institutions, many of which share their authority with comparable institutions beyond the state (Slaughter 2004, 12–15).

And the process of disaggregation is accelerating. It is very probable that any future space colony will eventually be dependent not on any particular state on Earth but on some kind of inter-state agency. In time, if it becomes possible to overcome the technical conditions which at present must tie any extraterrestrial settlement of human beings closely to the Earth, if, even, as seems not impossible, a new speciesification occurs as a consequence of the new environments to which these space colonists will have to adapt, entirely new states may evolve, just as they did eventually in the Americas.

Unlike the early American settlers, however, the early settlers in extraterrestrial space will not be able to carry with them as easily as their ancestors did the social and political institutions, customs, and systems of law of their former homelands; they will not be able to build New Englands, New Spains, or New Amsterdams in outer space. They will, however—initially at least—be humans like us, and they will still possess that capacity for 'sympathy' which Hume and Smith had identified as the source of all true human sociability. And since the inhabitants of these extraterrestrial settlements will, like the early American colonists, eventually become self-governing, they will also need to replicate some version not of the modern nation, as the American colonists did, but rather of the terrestrial, legal, and political institutions of international governance. To do that they will be thrown back on the Enlightenment vision of a universal—now no longer merely global—order of justice, grounded upon our common human nature and our irrefutable capacity to 'sympathize' and thus enter into communication, with one another. For in space those are the only human attributes we are likely to have in common. And if all that comes to pass, it can only be some version in principle—no matter how remote in practice—of Kant's cosmopolitan order. This was, as Kant himself had seen in 1795, an idea, a condition of future time; but as with all such ideas, he insisted, it was like Plato's *Republic*, one that should not be abandoned, 'under the very wretched and harmful pretext of its impracticability' (Kant 1998, 397).

So long as we are free—morally and intellectually—we can only go on hoping that some kind of cosmopolitan world will be achieved. More pressingly, the very fact that we are able to imagine such a state makes it our duty 'to work toward this (not merely chimerical) end' (*Toward Perpetual Peace* in Kant 1996, 337).

And if not on this Earth, then perhaps beyond it.

Notes

1. That Kant assumed that this process would be led by the European nations of the world, who are always the more civilized, and that they 'will in all likelihood eventually give laws to all others', there can be no doubt. But although Kant assumes that in the process of cultural amelioration, all the peoples of the world will approximate to the European, he also fiercely rejected the idea that his *Völkerbund* was intended to become a world-state or a 'universal monarchy'—in other words, an empire—which would have made of it, the 'graveyard of freedom' and a 'soulless despotism'. For a strongly critical view of Kant's cosmopolitanism see James Tully, 'The Kantian Idea of Europe. Critical and Cosmopolitan perspectives' in Pagden (ed). 2002, 331–58.

2. See Benhabib 2006, 22–4.
3. See also Kleingeld 2012, 76.
4. On Kant's uses of these terms, see Hurrell 1990, 183–205. On Kant's shifting views on the desirability of the world state (*Völkerstaat*) as opposed to a world federation (*Völkerbund*), see Mori 2004, 103–14.
5. See Latimer 2018, 586–601.
6. On the origins subsidiarity see Pagden 2022, 252–3.
7. 'The Powers not delegated to the United States by the Constitution, nor prohibited by it to the States are reserved to the States respectively or to the peoples.' Even after the massive extension of federal power which has taken place over the past three centuries, it is still true that federal law is, as Francis G. Jacobs and Kenneth L. Kars point out, 'designed to accomplish limited objectives' and is 'built upon legal relationships established by the states', and applies only 'so far as necessary for special purposes' (F. G. Jacobs and K. L. Kars 1986, I, 169–243 at 240).
8. What Kant, however, fails to explain is how such a society would be created in the first place except through revolution—and this he rejected as the unwarranted interruption of what he called the 'law of continuity' on which the legitimacy of all states ultimately depended. See Pagden 2014.

References

Benhabib, S. *Dignity in Adversity. Human Rights in Troubled Times* (Cambridge: Polity Press, 2011).

Benhabib, S. and Post, R. *Another Cosmopolitanism*, ed. R. Post (Oxford: Oxford University Press, 2006).

Bentham, J. *Rights, Representation, and Reform: Nonsense upon Stilts and Other Writings on the French Revolution*, eds P. Schofield, C. Pease-Watkin, and C. Blamires (Oxford: Oxford University Press, 2002).

Cockell, C. 'Liberty and the Limits to the Extraterrestrial State' (2009) 62 *Journal of the British Interplanetary Society* 139–57.

Condorcet, Marie Jean Antoine Nicolas de Carita, Marquis de. *Esquisse d'un tableau historique des progrès de l'esprit humain*, ed. A. Pons (Flammarion, 1988).

Crawford, I. 'Interplanetary Federalism: Maximising the Chances of Extraterrestrial Peace, Diversity and Liberty' in C. S Cockell (ed), *The Meaning of Liberty Beyond Earth* (2015) 199–218.

Diderot, D. *Œuvres completes*, eds J. Assevat and M. Tourneaux (Paris, 1875–1877).

Gray, J. *Enlightenment's Wake: Politics and Culture at the Close of the Modern Age* (London: Routledge, 1997).

Hume, D. *A Treatise of Human Nature*. ed. L. A. Selby-Rigge (Oxford: Clarendon Press, 1896).

Hume, D. *Enquiries Concerning the Human Understanding and the Principles of Morals*, ed. L. A. Selby-Rigge (Oxford: Clarendon Press, 1970).

Hurd, I. *How to Do Things with International Law* (Princeton, NJ: Princeton University Press, 2017).

Hurrell, A. 'Kant and the Kantian Paradigm in International Relations' (1990) 16 *Review of International Studies* 183–205.

Jacobs, F. G. and Kars, K. L. 'The "Federal" Legal Order: The U.S.A. and Europe Compared' in M. Cappelletti, M. Seccombe, and J. Weiler (eds), *Integration through Law. Europe and the American Federal Experience* (Walter de Gruyter, 1986) 169–243.

Kant, I. *Critique of Pure Reason*, trans. and eds P. Guyer and A. W. Wood (Cambridge: Cambridge University Press, 1988).

Kant, I. *Practical Philosophy*, trans. and ed. M. J. Gregor (Cambridge: Cambridge University Press, 1996).

Kant, I. *Lectures on Ethics*, eds P. Heath and J. B. Schneewind (Cambridge: Cambridge University Press, 1997).

Kant, I. *Critique of the Power of Judgment* (trans and eds P. Guyer and E. Mathews. Cambridge University Press, 2000).

Kant, I. *Anthropology, History and Education*, eds G. Zöller and R. B. Louden (Cambridge: Cambridge University Press, 2007).

Kleingeld, P. *Kant and Cosmopolitanism. The Philosophical Ideal of World Citizenship* (Cambridge: Cambridge University Press, 2012).

Latimer, T. 'The Principle of Subsidiarity: A Democratic Reinterpretation' (2018) 25 *Constellations* 586–601.

Locke, J. *An Essay Concerning Human Understanding*, ed. P. H. Nidditch (Oxford: Clarendon Press, 1975).

Montesquieu, Barone de, Charles de Secondat. *Œuvres Complètes*, ed R. Caillois (Paris: Gallimard, 1951).

Mori, M. *La pace e la ragione. Kant e le relazioni internazionali: diritto, politica, storia* (Bologna: Il Mulino, 2004).

Pagden, A. (ed). *The Idea of Europe. From Antiquity to the European Union* (Cambridge: Cambridge University Press, 2002).

Pagden, A. *The Enlightenment and Why it Still Matters* (Oxford: Oxford University Press, 2013).

Pagden, A. *The Pursuit of Europe A History* (Oxford: Oxford University Press, 2022).

Petit, P. *On the People's Terms. A Republican Theory and Model of Democracy* (Cambridge: Cambridge University Press, 2012).

Petit, P. 'The Globalized Republican Ideal' (2016) 9 *Global Justice: Theory Practice Rhetoric* 47–68.

Rousseau, J-J. *Œuvres Complètes*, eds, B. Gagnebin and M. Raymond (Paris: Bibliothèque de la Pléiade, 1959–1995).

Scruton, R. *A Dictionary of Political Thought* (London: Macmillan, 1982).

Slaughter, A-M. *A New World Order* (Princeton, NJ: Princeton University Press, 2004).

Smith, A. *Theory of Moral Sentiments*, eds D. D. Raphael and A. L. Macfie (Oxford: Clarendon Press, 1976a).

Smith, A. *An Inquiry into the Nature and Causes of the Wealth of Nations*, eds R. H. Campbell and A. S. Skinner (Oxford: Oxford University Press, 1976b).

Waldron, J. 'What is Cosmopolitan?' (2000) 8 *The Journal of Political Philosophy* 227–43.

Walzer, M. *Spheres of Justice: A Defense of Pluralism and Equality* (New York, NY: Basic Books, 1983).

21

Sovereign states, private actors, and (national) space laws. A rapidly evolving landscape

Stefania Paladini and Ignazio Castellucci

1. Space Age 2.0. A Brave New World

With more than 80 countries that have officially joined the 'space club' launching their domestically made satellites and an industry growing to US$446.9 billion in 2020 (The Space Foundation 2021), it is safe to say that nothing has changed so dramatically as the space sector over recent decades. The growing number of private actors in what has traditionally been the domain of nation-states only has led observers to label the present moment as 'Space Age 2.0' (Sheetz 2017, citing a Bank of America's report) highlighting the substantial qualitative difference from the past.

Moreover, when the existing framework was created, only a handful of countries were space active and the whole sector was dominated by the then two superpowers.

The commercial dimension of space has radically changed the way those actors interact and the purposes for their interactions as well, creating potential for both (international) cooperation or conflict.

2. The Existing Legal Framework

The existing legal framework at an international level consists in the United Nation's five treaties regarding space, developed from the 1960s to the 1970s. Their application is overseen by the UN agency UNOOSA (United Nations Office for Outer Space Affairs), an organization first established at the beginning of the space age (1959) with the name of UNCOPUOS (the United Nations Committee on the Peaceful Use of Outer Space), Department of Political and Security Council Affairs.

In 1992, the Division became UNOOSA, based in Vienna and including UNCOPUOS's functions (UNOOSA 2021).

There are the five treaties, listed in order as follows by year.

2.1 The 'Outer Space Treaty' (OST) 1967

The Treaty on Principles Governing the Activities of States in the Exploration and Use of Outer Space, including the Moon and Other Celestial Bodies is without any doubt the most important of the five treaties, the one that paved the way for all successive legislation, also preventing overt military action in space (Freedland and Blake 2017). 'The year 1967 represented a milestone in space history with the entry into force of the foundational instrument of international space law: the Treaty on Principles Governing the Activities of States in the Exploration and Use of Outer Space, including the Moon and Other Celestial Bodies', Simonetta Di Pippo, Director of United Nations Office for Outer Space Affairs, declared (UN 2017).

The OST states the cardinal principles ('space as a province of mankind', 'free for everybody use'; Article 1) and directives for human activities in space and provides the basic framework for them. No state can claim sovereignty of outer space, by use, occupation, or other means (Article 2), and all must act towards maintaining international peace in outer space, without using weapons of mass destruction (WMDs) (Article 3), considered by many 'the single most important provision in the space treaty system that deals with demilitarization of outer space, thus posing a potential legal challenge to the deployment of ASATs' [Anti-Satellite Weapons] (Chatterjee 2014, 29).

States are responsible for all their national activities in outer space, no matter if a body other than that state's government that directly carries them out, and are directly liable for damages caused by their space objects. These responsibilities are clearly spelt out in the following treaties.

2.2 The 'Rescue Agreement' (RA) 1968

This treaty, the Agreement on the Rescue of Astronauts, the Return of Astronauts and the Return of Objects Launched into Outer Space, which clearly mirrors the existing laws on international waters, extends the same principles and regulates the rescue and return for men and objects launched into space.

2.3 The 'Liability Convention' (LC) 1972

The Convention on International Liability for Damage Caused by Space Objects regulates the possible controversies amongst countries due to damage caused by space objects, re-asserting states' liability of states already contained in Article 7 of the OST.

2.4 The 'Registration Convention' (RC) 1976

This treaty, the Convention on Registration of Objects Launched into Outer Space, regulates the mechanisms of notification of the launch of space object into the Earth's orbit or beyond and establishes an UN-maintained registry of space objects.

2.5 The 'Moon Agreement' (Moon) 1979

The Moon Agreement is a special case in this framework, and it has historically not had the same importance compared to others, due to the scarce number of countries that have actually signed it.

It is, therefore, considered a sort of failed agreement, even though it has recently come back to attention due to the US-sponsored Artemis Accords and the ongoing plans to get back to the Moon.

Moreover, the letter of the agreement, including its definition of the Moon as the 'Common Heritage of the Mankind' (CHM) creates difficulties in the now not-so-far-fetched perspective of the harvesting and exploitation of natural resources (Listner 2011).

Which leads to another discussion on about how relevant the existing regulatory framework still is in the twenty-first century.

3. An Outdated Framework?

For what strictly concerns the internationally applicable space law, the outlook in 2021 is still the same as it was in 1967. But a brief look at the treaties' content shows that they are no longer suitable for the Space Age 2.0, or, at the very least, have problematic application, which was not the case back in the 1970s.

Today, however, these treaties can create all sorts of issues, and not only because of the sheer numbers of countries that have managed to put satellites in space.

It is important to mention that one of the main issues with international space law is that the apart from the 1972 LC, it does not provide a state-to-state dispute settlement mechanism, with the exception made for damages caused by space objects and the rules for the definition of parties' liability.

The most problematic aspect here, however, is the fact that many of the active satellites belong to private companies, which are not covered by any treaty and are therefore virtually outside the law.

An example of what can happen in situations like this comes from the famous case of the collision between the privately owned satellite Iridium 33 and the Russian junk Cosmos 2251 in 2009.

4. The Case of Iridium 33 and Cosmos 2251 Collision

On 10 February 2009, the derelict but functioning, Russian satellite Cosmos 2251 crashed into the active and privately owned Iridium 33 satellite, destroying it.

Whilst this was not the first accident in LEO (Lower Earth Orbit; Paladini 2019), it was the first instance in which two satellites in orbit collided by accident.

According to the existing rules, the 1972 Liability Convention (LC) was the treaty supposed to handle similar cases, except that it was unable to address this one.

First of all, there was the issue of establishing the responsibility for the collision, according to the LC provisions. Easier said than done.

The Russians claimed, with some reason, that their satellite was a junk and there was nothing they could do to avoid the collision. Iridium, on the other hand, did not even attempted a corrective manoeuvre, basically because it did not expect it to happen (the satellite tracking system is far from perfect, and a possible collision was lower in the Celestrak's SOCRATES top 10 list at that time; Kelso 2012).

In these conditions, there was not sufficient evidence for the application of the fault liability as required by the LC (Baker, 1998) to apportion responsibility for the collision to either Russia or Iridium.

Even more importantly, Iridium was a private corporation and thus it fell outside the remit of the LC, leaving the case in a legal vacuum and highlighting the need for rethinking the entire framework of space law on a different and more reliable basis.

5. Space Law, Governance, and Regulation—Back to the Future?

As non-state space actors develop their activities at a pace unthinkable of only two or three decades ago, current developments suggest that the existing legal regime based on instruments of 'classic' international space law will prove less and less satisfactory, due to its lacunae, loopholes, intrinsic inability to reach non-state actors and activities, and to govern public, private, and public–private activities, situations, and incidents in an increasingly crowded extra-atmospheric space.

5.1 Space Law and Governance: a complex environment

This inadequacy, vis-à-vis the fast-growing, fragmented, and diversified nature of the interests at stake, is apparent both on the 'passive protection' side (e.g. in relation to liability for space activities, as demonstrated by the Iridium case) and in the 'active rights' side (e.g. in relation to private, commercial exploitation of space, and related activities)—not to mention the 'police', 'security', and/or 'military' dimensions.

Research is developing beyond 'classic' scholarship on space law to considering the different ramifications of space activities, and it has already produced a significant amount of literature (Jankowitsch 2015). Most scholarly writing, however, focuses on specific aspects of space law (international liability, private contracts, national security, and so forth). Some important books deserve to be praised, collecting together pieces of research, with individual chapters devoted to the different dimensions of space law, trying at least to present the different aspects of it in a manageable vehicle, providing a valuable tool for reflection on the entire area (von Der Dunk 2015; Achilleas 2020).

However, there still is a distinct feeling and a consensus on the absence of a general framework for all those normative dimensions, as well as on the lack of a general legal theory of space law, capable of presenting knowledge about this complex field.

5.2 The Westphalian paradigm

Space law and normative environment clearly challenge the 'Westphalian paradigm'—the idea of absolute sovereignty of nation-states within their political boundaries, conventionally associated with the Peace of Westphalia (1648), based on modern philosophies and political developments before and after that event (Bodin 1583). European modern, 'Westphalian' states exercise political, military, and legal pressure on one another; each of them, defined by its territorial boundaries, features its own jurisdiction and applicable national law as one of the most typical expressions of the state, to display/enforce sovereignty and national identity.

The Westphalian vision of the state and state law could thus be represented by a chessboard, with well-defined squares: no gradual transitions or nuances exist when moving from one square to the next; law would change in a discrete, black-and-white way, jumping over a hair-thin line from one square into another. Within the allotted territory of each square, the position and distance from the centre is theoretically irrelevant for the purposes of the applicable national law, as well as for the exclusion of foreign ones. In this conception of law, even commercial law suffers from attempts to domesticate it, as national commercial codes and laws enter into force. At the beginning of the nineteenth century, Fichte theorized about the 'closed commercial state': a water-tight entity in which all plans—political, economic, social, cultural, legal—have identical, overlapping geographic extension, dominated by national politics and national law (Fichte 1800).

Rules applicable to relations amongst modern states, meanwhile, have developed to form what is now international law, hailing from early modern European inter-state comity, consolidated practices, and treaties—perhaps justifying Schmitt's description of it as *ius publicum europae* (Schmitt 1997)—then expanded in the nineteenth century to cover all the world's states' international relations.

Such an environment is precisely at the roots of lawyers' unease, vis-à-vis the challenges posed by space activities and related normative issues.

5.3 The need for a post-Westphalian paradigm

A different, non-Westphalian theory may be needed to make sense of current developments of space law. The second half of the twentieth century has already been characterized by a marked departure from the Westphalian paradigm in trade and economic matters—in a reverse quest for uniformity of regulation in transnational economic activities, including private and public–private ones, rediscovering contemporary forms of pre-modern law merchant/*lex mercatoria* a few decades before the globalization process started (Rose 2000; Toth 2017).

It might be a good idea to emulate this process for space law, looking backwards to the state of affairs before the Westphalian process began.

The pre-modern *ius commune* environment was based on a quite liquid state of legal affairs, with the German Empire and the Catholic Church playing the role of universal powers, and the existence of a vast European normative space featuring generally acknowledged shared values, enshrined in the body of *ius commune* developed by legal scholars (Pihlajamaki, Dubber, and Godfrey 2018), and a lack of a notional distinction between 'national' laws and 'international' law, as we know it.

Kings, princes, lords, bishops, abbots would produce their own rules: they would all be intrinsically valid and potentially applicable, in a complex, pluralist environment. Other normative forces, of a public, private, or communitarian nature—religious congregations, guilds and chambers of commerce, large business enterprises, local communities of merchants or dwellers, etc.—would also play the game of normative production, whether doctrinal, stipulated, case-based, or use-based, in written or customary form. Global merchants of those times would develop their own customary *lex mercatoria* through practice, arbitration, compilation of trade usages.

Scope and reach of any set of norms would be associated with a community and the related interests of its members—according to the dynamics of societies and space identified by George Simmel (1992)—rather than to a territory: a clear vision of this different attitude would be visible in the reach of European powers' laws over their respective colonial empires and networks of commercial outposts, as well as over their military and commercial fleets reaching all parts of the globe. Each normative system extended its rule—in principle without limit and in practice as far as possible—subject to the variable effects of power negotiations between competing, often overlapping, orders and related institutional mechanism. Members of human communities, thus, normally found themselves subjected to a number of competing normative orders (Pihlajamaki et al. 2018).

There were no legal chessboards there.

The pre-modern legal environment may perhaps be represented as featuring many antenna masts, each broadcasting signals of normative power: much like FM radio transmitters operating in the same frequency band, in the same area. At the same time, the universal broadcast of *ius commune*, travelling on a different frequency band, is a fundamental part of the environment, coming from a distant past, or just being inherent in the universe—like waves emanating from the Big Bang, impossible to suppress entirely.

In such an environment, signals get easily scrambled. The proximity of receivers to normative broadcasting, interference of other transmitters, and power of emissions determine the different broadcasts' actual reach, influencing specific situations of life irrespective of the size of the plot of land where the mast is located: let's just think of the small size of the Roman Catholic Church state, vis-à-vis the global reach of Canon law, or that of medieval mercantile and port cities, the municipal statutes of which regulated a significant part of the Western world's commerce—a process nowadays being replicated by the growing 'legal power' of some cities (Davidson and Tewari 2019) or micro-jurisdictions having high global importance.

5.4 The case for global legal pluralism

This description seems to be applicable, *mutatis mutandis*, to the present-day global legal environment. Not by chance, frequent references are made, in contemporary scholarly debate on global law, to a neo-medieval model (Friedrichs 2001).

New attention is devoted to law and society studies and, especially, to legal pluralism—a description through which legal anthropologists, earlier in the twentieth century, would read the legal structure of colonial and post-colonial societies, identifying the existence of multiple normative layers, semi-autonomous vis-à-vis one another (Moore 1973; Griffiths 1986; Tamanaha 2008)—identifying the existence, today, of a 'global legal pluralism' (Snyder 2002) in which, again, a high number of normative orders play a game of negotiation, pressure and counter-pressure, mutual cross-infiltration, and, sometimes, eventual hybridization.

Public and private political, social, and economic geographies produce complex interactions amongst many different layers, patches, or clusters of normativity, administering institutions, adjudicative and other enforcement mechanisms, with innumerable operational scenarios. Interaction amongst the different normative orders takes place according to the individual circumstances of specific transactions or incidents, rather than according to a general normative framework coordinating them.

Global legal scholarship is at work to determine the viability of a framework and to develop a general theory to read together this multitude of different normative phenomena. As the notions of sovereignty, border, jurisdiction, and sources of law are challenged and get fuzzier, new importance is recognized in multi-shaped or amorphous normative entities, societal roots, and actual reach of normative forces.

Ideas and notions that seemed to belong to the past gain centre stage again, such as those of 'empire' and its relations with that of 'political boundaries', and lines defining the respective areas of influence amongst different imperial entities become blurred.

A novel interest is also displayed in relation with the notion of 'frontier', with new geopolitical frontiers appearing in non-territorial spaces, such as Antarctica, the Moon, extra-atmospheric space, or the digital one, by public and private actors.

The effectiveness of competing legal orders, or specifically identifiable normative eco-systems—in other words the capability of expressing an operational rule which is actually enforced by the relevant actors, which in turn implies the capacity to attract those actors—is, thus, rather than legitimacy and abstract validity of sources and norms, the key element to understand such a normative environment.

The 'success' of a normative system—its legal power, we might say—is not directly determined by its intrinsic formal contents, nor by the territorial or population size of the entity expressing it, which may be small, negligible, or just non-existent/not relevant. It is, rather, related to the importance of the subjects, and matters, attracted and actually governed.

This is the case, for example, with global giants of technology and their internal 'normative systems', enforceable irrespective of any state legal system—even, for example, against the President of the United States, as the world witnessed in January 2021, Sacco 1991, 1991a or with some present-day small and micro-jurisdictions able to punch well above their weight in banking, finance, IT services, etc.

5.5 A new paradigm for space law

Coming, finally, to space law: when dealing legally with the extra-atmospheric legal environment, 'space' is certainly different from 'place'—the latter normally being associated, in legal terms and in lawyers' parlance, to a territorial sovereign/jurisdictional entity and to the related local 'sources' of law; the former normally lacking the mentioned elements.

Not by accident, initial developments of space law have been conceptualized and implemented in the twentieth century as an expansion of the classic regulation of the high seas, binding on sovereign states, such as the OST 1967, with rules mimicking that of maritime international law, developed in its modern terms since Grotius.

This process has already taken place with air law, and more recently with laws regulating all spaces with no overall national jurisdiction, like Antarctica and the Moon. Mars, outer space, or the digital space could be the new frontiers for these kind of legal developments, including debate or clashes similar to those associated with the legal regime of the high seas—between visions of them as nobody's space, or everybody's (or the strongest's) person's or state's space, and visions of the same as frontiers which may, in some instances at least, be territorialized (Somos 2020, 2012).

At the same time, legal globalisation produced an environment characterized by a very complex normative pluralism: important non-state actors are active on the global scene, producing rules and normative (eco-)systems, which hardly qualify as (parts of any) national legal systems. Different normative systems of different natures, as a matter of fact, do interact freely with state laws and international laws, negotiating their respective areas of influence and effectiveness, based only on their respective de facto capacity of seeing their norms (reflecting their nature, agendas, inner mechanisms) attracting operators and prevail, being enforced in real life and on the economy. A good example here, which is both enlightening and of immediate understanding, are the 'legal systems' that social networks provide, which illustrate well these dynamics.

Dialectics between different normative orders will produce cross-fertilization and hybridization of rules, with the most important actors of both public and private nature making more or less stable alliances, cooperating to develop their legal instruments consistently, accommodating the needs of both, whilst competing, cooperating, or conflicting with other public–private blocs.

5.6 National states and territorial space laws

As private actors develop their activities at least partially in a legal vacuum, some national states are developing their space laws with a view to attracting and regulating them, laying the conditions to develop favourable legal eco-systems reflecting their own needs.

National norms will reflect the different space powers' approaches both to space and to the law in general, accommodating variable proportions of political input, state legislation and regulations, mechanisms interfacing them with international law, soft law, other state laws, rules, and best practices of the private sector, *lex mercatoria*, etc.—with a view to fostering the growth of a nationally characterized legal eco-system, able to provide a framework for the entire spectrum of activities related to the space industry—from research and development (R&D) to space travel ticketing.

This ability in developing an appropriate, user-friendly normative framework or eco-system—to be balanced with states' capacity to maintain control of sensitive areas and classified information to the extent necessary—will be critical in attracting space economic actors and operators—space mining companies, insurance companies, IT service providers, just to mention a few—by virtue of internal growth of the industry as well as due to flag- and forum shopping phenomena.

State jurisdictions' success in expansion in the new frontiers of space and the space economy will be determined by the 'importance'—whatever this may mean—of space activities regulated, irrespective of the same jurisdictions' 'terrestrial' size. And, of course, by those jurisdictions' relations with others on the Earth's surface: much like the importance of Swiss banking law, of Panamanian maritime law, or of many micro-jurisdictions' tax and financial laws, which have determined those countries' status as major (legal) players in their respective fields, irrespective of their geographic size.

6. Which National Space Law? A Comparative Approach

Some of the discussion points raised above are already apparent in a few national space laws, old and new.

And, if 'national space law' has been often used in the past by countries to define and regulate the remit of activities of their space agencies—and therefore, not inclusive of any private actor, such as companies—there is evidence things have changed in the recent addition or amendment in the existing national space laws, starting with the United States.

Whilst an in-depth treatment of these national laws is beyond the scope of this chapter, it is important to give them a cursory look to understand in what sense things have been evolving in recent decades. Some national laws look specifically at the (future) exploitation of space resources, others are devoted to space activities as a whole.

In a specific case (China), it is the lack of a space law itself that is worth of a closer look.

6.1 USA: U.S. Commercial Space Launch Competitiveness Act—Title 51 (Public Law 114–90 114th Congress, 25 November 2015)

Article 51 states 'a United States citizen engaged in commercial recovery of an asteroid resource or a space resource under this chapter shall be entitled to any asteroid resource or space resource obtained, including to possess, own, transport, use, and sell the asteroid resource or space resource obtained in accordance with applicable law, including the international obligations of the United States' (Article 51).

As a matter of the fact, the US law has been the first in the world to address openly the question of the ownership of a 'space resource' (e.g. minerals mined from celestial bodies), affirming, for the first time, that if the celestial bodies themselves belong to nobody (or everybody), the resources mined belong to the countries that mined them. This law filled a legal vacuum, providing an applicable regime without which the concept of resource mining in outer space was severely lacking and whose legitimacy could even be questioned altogether (Hobe 2007).

The formulation was therefore particularly important, also because the US law does not infringe the 1967 OST provisions in any way. As it stands, it integrates them, clarifying a potentially problematic point and paving the way for private initiatives to take part in the commercial exploitation of space resources.

6.2 Luxemburg: 'Loi du 20 juillet 2017 sur l'exploration et l'utilisation des ressources de l'espace' and following regulations

Luxembourg was the second country to legislate on the remit of the space resource, with a law(Loi du 20 juillet 2017) essentially mirroring the US text and allowing the appropriation of the resources to the company (or the national agencies) that extracts them.

The fact should not come as a surprise; with solid institutions, a strong banking sector, and privileged location of many international bodies, Luxembourg has long been a favourite for company holdings in the European Union. It has had a track record in the satellite business since the 1980s and initiated its space research and development plan back in 2008 (Space Resource Luxembourg 2017).

Compared to the United States, the Luxembourg law goes further, transitioning from the principle to the application of the framework itself and detailing procedures for obtaining authorization for the space activities under government supervision, both absent in the US legislation.

As a result, Luxembourg is now a thriving European place of space industry and research, attracting private companies and universities under the lead of the local space agency (Luxembourg Space Agency). The 2017 Law was modified and expanded in 2020 with the Loi du 15 décembre 2020 on space activities.

6.3 The UK—Space Industry Act 2018

The Space Industry Act 2018 is only the last of a series of regulations, from the Outer Space Act 1986 (OSA) to Chapter 38 of the Deregulation Act 2015, which is not surprising, considering the UK's early start in the space sector. But this law springs from different interests. In both US and Luxembourg cases, space mining and the physical acquisition of outer space resources are clearly at the origin of the legislation.

The UK Space Industry Act 2018 is quite different. It regulates commercial spaceflights and spaceports, in line with both future space tourism possibilities and suborbital flight activities, aimed at the growing market of small satellites.

> The [new] Space Industry Bill gives companies the ability to launch satellites from UK soil, putting us at the forefront of the new space race, and helping us to compete as the destination of choice for satellite companies worldwide.
>
> (The UK Transport Minister, Johnson 2018)

Regulations that implement that law (The Space Industry Regulations 2021, 792/21) have recently been enacted, even though it is too soon to assess their actual impact on British domestic industry.

7. China's (Missing) Space Law

Few (if any) doubt China is the emerging superpower in the space sector.

If anything, 2021, with the successes of the Mars landing and the Chinese ISS (Tiangong) completion in LEO, has only reinforced the claim. So, it is even more surprising to discover that China has no space law to regulate the sector, not even one like the Japanese Basic Space Law (L 43/2008) that regulates the government activities and the national space agency (JAXA) at a general level.

As a matter of fact, China has always treated the space sector more as the object of an administrative framework regulation, delegating the organization of space activities to government acts, apart from general guidance issued at central level, such as the White Papers for Space 2000, 2006, 2010, and the latest one to date and the first of Xi Jinping's era, White Papers for Space 2016 (State Council 2016).

This is probably due to the way the Chinese space sector was organized, and the unique characteristics of the CNSA (Chinese National Space Agency), which, albeit the name, is something completely different from NASA and the ESA and is more a publicity and external relations body than anything else.

The real activities are under the remit of CASC (China Aerospace Science and Technology Corporation) and CASIC (China Aerospace Science and Industry Corporation) the Chinese SOEs that dominate the sector, operating in an actual legislative void under the overall supervision of SASTIND (State Administration for Science, Technology and Industry for National Defence), a central government body.

> With the growth of the Chinese space program, such a lack of structured national space law is beginning to show its limits and to raise concerns about its negative impact on business opportunities and the ability of China to fully comply with international obligations. One should keep in mind that international space treaties (China is part to four international space law treaties) are not self-executing, thus requiring states to adopt domestic measures to ensure their effective implementation.
>
> (Tronchetti 2019)

What China has done thus far it to take a quite 'minimalistic' approach to the implementation of the international treaties, issuing two low-level administrative acts (i.e. ministerial regulations, sitting low in the hierarchy of Chinese legal system) that regulate the launch and registration of the space objects:

> 7.2. Measures for the Administration of Registration of Objects Launched into Outer Space 2001 (CHN) ('Registration Measures'); Interim Measures on the Administration of Licensing the Project of Launching Civil Space 2002 (CHN) (Licensing Measures).

In regulatory terms, a major innovation came in 2014, when the Chinese government issued the so-called Document 60 (namely, Guiding Opinions of the State Council on Innovating the Investment and Financing Mechanisms in Key Areas and Encouraging Social Investment; 国务院关于创新重点领域投融资机制鼓励社会投资的指导意见, drafted by the NDRC (National Development and Reform Commission; State Council, 2014). In this document, aimed at reforming strategic sectors of national security relevance, the government included civil space development as a key area of innovation, and, at the same time, allowing private investment in commercial space activities. This represented a major change with the previously state-dominated sector, encouraging the development of the private space sector (Mao 2017; Huang 2019) even though Chinese private initiative in civilian space sector began much earlier than that (Lingye, 2017).

More recently, the NDRC issued an updated industrial catalogue (全国鼓励外商投资产业目录; NDRC 2019) of permitted investment, included thereafter in the long-awaited 2019 law that reorganizes foreign investments in China (FIE).

The new law promotes foreign investment in a number of previously closed or semi-closed industries, including civil satellite design and manufacturing, payload, component manufacturing, on-board testing equipment manufacturing, satellite communication system equipment, and civil satellite application development).

This represented a major policy shift that could open the door for commercial space companies in China to contact foreign investors (and helping these new companies secure investments), but it also requires a more solid and heavily regulated national legal framework.

8. A Fast-evolving Landscape

The four cases (the United States, Luxembourg, the United Kingdom, China) are only a limited sample of existing domestic space law examples but they are sufficient to illustrate both the complexity and the challenges to any (foreseeable) harmonization amongst them.

It has been observed (Marboe 2015) that there are both advantages and disadvantages to national laws, and, if they are directly applicable and are enforceable, which is not necessarily the case with obligations of a public international law nature, they are also generally not applicable outside a given nation's territory. This will be the case, in general, for space law which might include subjects of space law that fall within the remit of national territories (e.g. spaceports, resources brought back to Earth, satellites) but not by definition outer space, which exists is beyond all national borders. Moreover, as we have seen, space laws are still limited in number, and different in the extent and activities covered.

> National space legislation is available in only a few states. There are sufficient and forward-looking legal regimes in some countries, but this is not the case in others. This is an obstacle to the creation of a level playing field in the space sector. If all the

actors were bound by similar and fair rules in the carrying out of space activities, this would be to the long-term benefit of all. By contrast, in a situation of many diverse national legal frameworks, the phenomena of 'nationality planning' or the choice of 'flags of convenience', as known from the law of the sea, are also potential problems relevant in the law of outer space.

(Marboe 2015, 45)

What is certain is that the presence of highly different national space laws can lead to contrasts amongst them, which can easily escalate to conflict, as seen in other trade activities and industries.

Would the growing Chinese private sector in space activities (a recent IDA report counted 78 companies, OneSpace, Landspace, iSpace, LinkSpace, ExPace, Galactic Energie, and Spacety, amongst them; IDA 2019) affect the enactment of commercial space laws (both at home and overseas) and eventually lead to a more conflict in outer space? It is certainly a question worth asking.

9. How We Deal with a Developing Space Normative Environment. Scenarios and Opportunities

In the increasingly complex landscape illustrated above, there are a few possible alternatives to suggest, all presenting opportunities and challenges, as well as objective requirements to make them work, as will be discussed now.

Empiric evidence in the early twenty-first century warrants going beyond the usual two-dimensional split of law into national and international.

Globalization multiplies opportunities for normative systems, which are well connected politically and logistically, and offer attractive normative products, to govern specific segments of economic activities and networks, and to define necessary 'landing spaces' or localization (e.g. for civil, tax, or criminal purposes) in a convenient setting, thus expanding their reach in across the global spaces.

Normative interactions and aggregations amongst different normative systems, of different nature, reproduce geo-political dynamics—geo-legal, may we say—including mutual indifference, cooperation, negotiation, competition, conflict, and warfare (Castellucci 2020).

Whilst it is difficult to make accurate forecasts, it can certainly be suggested that the normative landscape, 10 years from now, will be characterized by a growing number of private corporations developing their space-related internal rules. A reasonable number of national space (and EU) legislations and agencies will be increasingly involved in space business regulation and governance, providing legal frameworks to economic, non-economic, and dual space activities, such as telecommunications and IT, space mining, space travel, scientific research, etc.—cooperating with their private economic champions, and/or seeking to attract foreign ones within their normative eco-systems.

International law will most certainly be developed—customarily, through treaties, and through international administrative law developed within international organizations—to provide additional governance at inter-state level. Normative products of hard and soft law aimed at the private sector will most likely be produced by national and international bodies or agencies, along with contractual instruments and best practices of a private origin.

The sector's expansion will be characterized to some extent by 'frontier', or 'Wild West' logics, with active national powers playing the role of security providers to protect their national interest in their geo-legal expansion, at least partially coinciding with that of their private allies in business.

Private activities, however global, will still need on-Earth location for some purposes: to launch vectors, to develop studies, to hire and pay personnel, to localize and regulate space-related contracts; and they cannot avoid localization for others (tax, administrative, criminal)—much like in global finance and its 'landing' jurisdictions. We may, thus, see space-business-friendly terrestrial jurisdictions become preferred destinations for forum and flag shopping in relation to space activities, with commercial networks and supply chains, providing space-related goods and services, mostly localized in those business-friendly jurisdictions' respective normative eco-systems.

This possible scenario will warrant skilled people managing its nerve centres.

A characteristic of most pluralist legal eco-systems—except in cases of irreconcilable dissociation of factors leading to paralysis or disruption—is their long-term tension towards hybridization, initially produced in the main areas of overlapping of the different semi-autonomous legal fields (Moore 1973) characterizing them.

Hybridization, in turn, is typically determined by operators and actors being operational in more than one normative system—for example, with religious, or customary, or private economy leaders vesting significant positions in the political or administrative apparatus of the state—thus being able to deploy/transfer experience and develop inter-system solutions which are workable in all different, semi-autonomous, normative legal fields.

Discerning and interested state jurisdictions and private space actors will have to develop normative frameworks in constant dialogue amongst themselves, and with international law fora, endeavouring to promote conversant normative instruments despite their different types and natures, maximizing the overall efficiency of the general space legal environment they will contribute to shape; or, at least, that of a more limited normative eco-system occupied and managed by it and its allies and friends.

In legal terms, a class of legal professional in law-making, in international organizations, in academia and research, and in the business and professional communities will be needed, able of conversing with different public and private legal environments, and knowledgeable about the dynamic of normative hybridization, to develop the field fruitfully, in all directions—through research, scholarship, legislation, contract-drafting, arbitration, litigation, and public–private negotiations.

10. Conclusions

We are still at the beginning of Space Age 2.0 and we can only glimpse at future developments, from space tourism and orbiting habitats to Moon and asteroid mining.

What is certain is that all will require private actors and capital to thrive, and private enterprise will entail a far more complex legal framework both at international and national level than is currently available.

As we have seen, 'classic' international law is already no longer sufficient to deal with the present reality of space human activity, still less suffice in the foreseeable future.

If the recent past is any guide, we can assume that the number and level of disputes will increase steadily in the coming years, following the continuous growth of the economic value of space activities and the number of the public and private actors which take part in them. Cases like the Iridium collision and similar instances will become more common, with consequences in economic, legal, and environmental terms flowing from them.

In the absence of mechanisms of dispute settlement more robust than those previewed by the LC, the risk is of a progressive weakening of the applicability of space law as a whole, eventually resulting in a general uncertainty of the entire regulatory framework.

However, there is evidence that a deep reshaping of the legal framework of outer space has already started, even though with different velocities and characteristics depending on sectors and subjects of interest. If anything, there is an evident ongoing diversification and hybridization of the space law landscape as we used to know it.

A lot of effort will be needed to ensure this process proceeds in ways satisfactory for relevant actors, though.

Specific knowledge of the sector and its peculiarities, comparative law skills, a deep understanding of the dynamics of normative pluralism and hybridization will be critical to develop adequate, effective, useful legal studies in this arena, as well as effective operational tools.

The next 10 years will be key to determine whether the various national space laws will successfully interact with one another and eventually enable private and public actors to avoid conflicts of interests in the pursuit of the humankind's conquest of space.

Acknowledgements

Whilst this chapter is the result of a joint work, Stefania Paladini has especially contributed to the abstract and sections 1, 2, 3, 4, 6, 7, 8, 10 Ignazio Castellucci, to sections 5 and 9.

References

Achilleas, P. and Hobe, S. *Fifty Years of Space Law/Cinquante ans de droit de l'espace* (Leuven: Brill, 2020).

Baker, H. A. 'Liability for Damage Caused in Outer Space by Space Refuse' (1988) 8 *Annals of Air & Space Law* 183–227.

Bodin, J. (1583). *Les six livres de la Republique* (reprinted with Aalen Scientia 1977, Paris, 1583).

Castellucci I. (ed.). 'Geopolitics and the Law (SI)' (2020) 2 *Gnosis: Rivista italiana di intelligence.*

Chatterjee, P. Legality of Anti-Satellite Under the Space Law Regime (2014) 12 *Astropolitics* 27–45.

Davidson, N. M. and Tewari, G. (eds.), Global Perspectives in Urban Law: The Legal Power of Cities, Juris Diversitas Series (London: Routledge, 2020).

Von der Dunk F. (ed.). *Handbook of Space Law* (London: Edward Elgar, 2015).

Fichte, G. *Der Geschlossene Handelstaat* (New York, NY: Wentworth Press, 2018).

Freedland, S. and Blake, D. 'As the World Embraces Space, the 50 Year Old Outer Space Treaty Needs Adaptation' *The Conversation* (2017) <https://theconversation.com/as-the-world-embraces-space-the-50-year-old-outer-space-treaty-needs-adaptation-7983> accessed 3 September 2021.

Friedrichs, J. 'The Meaning of New Medievalism' (2001) VII(4) *European Journal of International Relations* 475–501.

Griffiths, J. 'What is Legal Pluralism?' (1986) *Legal Pluralism & Unofficial Law* 1–24.

Hobe, S. 'Adequacy of the Current Legal and Regulatory Framework Relating to the Extraction and Appropriation of Natural Resources' (2007) 32 *Annals of Air and Space Law* 114–35.

IDA. 'Evaluation of China's Commercial Space Sector' (2019) <https://www.ida.org/-/media/feature/publications/e/ev/evaluation-of-chinas-commercial-space-sector/d-10873.ashx> accessed 30 October 2021.

Jankowitsch, P. 'The Background and History of Space Law' in F. von der Dunk (ed.), *Handbook of Space Law* (London: Edward Elgar, 2015) 1–28.

the UK Transport Minister. https://www.gov.uk/government/news/new-laws-unlock-exciting-space-era-for-uk

Kelso, T. S. ('Analysis of the Iridium 33-Cosmos 2251 Collision' Center for Space Standards and Innovation (2012) <http://celestrak.com/events/collision/> accessed 31 October 2021.

Lingye, M. 'Development Status of Major Commercial Aerospace Enterprises in China (国内主要商业航天企业发展现状)' (2017) 16 November *Satellite Applications* (卫星应用) <http://www.taikongmedia.com/Item/Show.asp?m=1&d=25027> accessed 30 October 2021.

Listner, M. J. 'The Moon Treaty: Failed International Law or Waiting in the Shadows?' (2011) *The Space Review* <http://www.thespacereview.com/article/1954/1> accessed 31 October 2021.

Marboe, I. *National Space Law in Handbook of Space Law* (London: Elgar Elwin, 2015).

Moore, S. F. 'Law and Social Change: The Semi-autonomous Social Field as an Appropriate Subject of Study' (1973) 7(4–719) *Law & Society Review.*

National Development and Reform Commission. 'Catalogue of Encouraging Foreign Investment Industries' (2019) <http://wzs.ndrc.gov.cn/glwstzcyml20190201.pdf> accessed 30 October 2021.

Paladini. S *The New Frontiers of Space* (Macmillan, 2019).

Pihlajamaki, H., Dubber, M. D., and Godfrey, M. *The Oxford Handbook of European Legal History* (Oxford: Oxford University Press, 2018).

Rose F. (ed.). (*Lex Mercatoria: Essays on International commercial Law in Honour of Francis Reynolds* (London: Routledge, 2000).

Sacco, R. 'Legal Formants: A Dynamic Approach to Comparative Law' (1991) 39(1) *Winter American Journal of Comparative Law* 1–34 (Instalment I) -issue 2, Spring, 343–401 (Instalment II).

Sacco, R. 'Legal Formants: A Dynamic Approach to Comparative Law' (1991a) 39(2) *Spring American Journal of Comparative Law* 343–401.

Schmitt, C. *Der Nomos der Erde im Völkerrecht des Jus Publicum Europaeum* (Berlin: Duncker & Humblot, 1997).

Sheetz, M. 2017. The space industry will be worth nearly $3 trillion in 30 years, Bank of America predicts. https://finance.yahoo.com/news/space-industry-worth-nearly-3-180819064.html

Shiong, W. 'Investing in China's Space Dream' *PE Hub Network* (14 January 2019) <https://www.pehub.com/2019/01/investing-in-chinas-space-dream/> accessed 30 October 2021.

Simmel, G. *Soziologie* (Frankfurt: Suhrkamp Verlag KG, 1992).

Snyder, F. 'Governing Globalisation' in M. Likosky (ed.), *Transnational Legal Process: Globalisation and Power Disparities* (Cambridge: Cambridge University Press, 2002) 265.

Somos, M. 'Selden's Mare Clausum: The Secularisation of International Law and the Rise of Soft Imperialism' (2012) 14 *Journal of the History of International Law* 287–330.

Somos, M. 'Open and Closed Seas: The Grotius–Selden Dialogue at the Heart of Liberal Imperialism' in E. Cavanagh (ed.), *Empire and Legal Thought* (Leiden: Brill, 2020) 322–61.

The Space Foundation (2021). The Space Report 2020. https://www.thespacereport.org/uncategorized/global-space-economy-nears–447b [30 October 2021].

Space Resource Luxembourg. 'Luxembourg Space Capabilities—Luxembourg Cluster Catalogue 2017' (2017) <http://clustermembers.luxinnovation.lu/space/wp-content/uploads/sites/4/2016/10/07697_LUXINNOVATION_SpaceCapabilities_05-2017-Web.pdf> accessed 26 October 2021.

State Council Information Office of the People's Republic of China. 'Guiding Opinions of the State Council on Encouraging Social Investment in Investment and Financing Mechanisms in Key Innovation Areas (国务院关于创新重点领域投融资机制鼓励社会投资的指导意见) Document 60' 16 November 2014 <http://www.gov.cn/zhengce/content/2014-11/26/content_9260.htm> accessed 30 October 2021.

State Council Information Office of the People's Republic of China. 'Full text of white paper on China's space activities in 2016' (2016) <http://english.gov.cn/archive/white_paper/2016/12/28/content_281475527159496.htm> accessed 30 October 2021.

Tamanaha, B. 'Understanding Legal Pluralism: Past to Present, Local to Global' (2008) *Sydney Law Review* 375–30.

Toth, O. *The Lex Mercatoria in Theory and Practice* (Oxford: Oxford University Press, 2017).

Tronchetti, F. *The Exploitation of Natural Resources of the Moon and Other Celestial Bodies: A Proposal for a Legal Regime* (Leiden: Martinus Nijhoff Publishers, 2009).

Tronchetti F. 'Space Law and China' (2019) *Planetary Science* <https://doi.org/10.1093/acrefore/9780190647926.013.66> accessed 11 July 2022.

UN. International Space Law: United Nations Instruments (United Nations, 2017) <http://www.unoosa.org/res/oosadoc/data/documents/2017/stspace/stspace61rev_2_0_html/V1605998-ENGLISH.pdf> accessed 18 October 2021.

UNOOSA. <http://www.unoosa.org/> accessed 30 October 2021.

Xiaoyun, H. 'Private Commercial Satellite Companies Have Sprung Up, and "Changguang Satellite" has Completed 250 million yuan of Angel Financing (民营商业卫星企业异军突起，「长光卫星」完成 2.5 亿元天使轮融资)' (2019) <36kr.com/p/5174039.html> accessed 31 October 2021.

22

In space, nobody can copyright your scream

Burkhard Schafer

1. Countdown to the Chapter in 10, 9, 8 ...

This chapter explores intellectual property (IP) implications of a putative space colony populated initially only by autonomous, 'intelligent' and in varying degrees creative robots, serving as an advance party for possible human occupation (Smith and Plattard 2020; on the question of competition vs cooperation between robots and humans in space exploration see Launius and McCurdy 2008). It brings together for the first time the discussion surrounding IP rights in outer space with the bourgeoning discourse surrounding robot rights, and in particular IP rights for the creative outputs of machines. Such a setting, I will argue, is, unlike Earth-based use of robots, a much more plausible candidate for robot rights, which puts the position advanced here into contrast with the popular 'relational' theory of robot rights that derives such a recognition from their proximity to humans and the values attributed to them in close human-robot interactions. In exploring these ideas, we will also discuss more generally the role IP rights may or may not have to play to protect space liberty.

As creativity and artistic production are a key aspect of this chapter, I will first briefly set the scene for the discussion by drawing on the public imaginary, and in particular the collaborative works by Fritz Lang and Thea von Harbou—a collaboration that amongst other things gave us the countdown from 10 to zero as a launch sequence for space rockets. In section 2, I introduce briefly the type of IP rights that could be relevant to the scenario described above, and give a short account of the issues they raise under the existing IP regime when a discovery, innovation, or creation is made in space. The issue of space IP has been in the margins of space law, with only a very small number of studies conducted so far. What these studies have in common though is that they consider a human astronaut as inventor or creator. In section 3, I introduce the emerging discourse surrounding creations and inventions made by artificial intelligence (AI) and robots, a topic that has over the last few years gained considerable international traction and, unlike space IP, been subject to legislative consultations, reform proposals, and increasingly also court decisions. In the final section, these two issues are brought together, asking if a society of robots, far away from any direct human interference, would benefit from or be hindered by a future IP regime that accepts machines as creators and inventors of works.

2. When Friede Meets Maria

In 1929, *Die Frau im Mond* ('Woman in the Moon') premiered at the UFA-Palast am Zoo in Berlin, drawing a captivated audience of 2,000 visitors, amongst them one Albert Einstein. Directed by the great Fritz Lang, it was based on the novel *The Rocket to the Moon* by his collaborator and then wife, Thea von Harbou. Whilst by far not the first story, or for that matter the first film, to depict space travel (that honour probably falls to *Le Voyage dans la Lune* by Georges Méliès from 1902), it introduced for the first time 'serious' concepts of space exploration to a mass audience, including a multi-stage rocket. Lang insisted on scientific accuracy, hiring two pioneers of rocket technology as scientific advisors, Hermann Oberth and Rudolf Nebel, collaborator and mentor of a certain Wernher von Braun respectively (Freund 1993). Ironically maybe, giving this parentage that also explains why the rocket looked strikingly similar to the V2 weapon Braun and Oberth would later build for Nazi Germany, the name of the female protagonist and the spaceship named after her is *Friede*, 'Peace'. When at the end of the film, Friede, a capable scientist in her own right, decides to stay behind on the Moon with her lover, her action is also a declaration of freedom, defying conventional morality and gender roles of the time. Her drugged (and generally useless) husband by contrast is returned in the damaged rocket to Earth, having been unable to handle the challenges of the new environment.

Harbou is arguably more famous for writing the screenplay of *Metropolis*, the techno-dystopian film about the future of work in a world of robotics, anticipating also what we would call in modern parlance smart cities. Its protagonist, Maria, fights for justice for the workers, but is undermined in her effort by a robot that takes on her likeness and subverts her standing within her community. This makes *Metropolis* also a precursor of current discussions surrounding the legal and ethical implications of anthropomorphic robots, their potential for deception, and the danger of deep fakes. It is the legal implications that arise in the intersection between these two, the exploitation of human and robotic labour on the one hand, space travel and space exploration on the other, that this chapter is going to explore.

As the 'Woman in the Moon' shows, long before space explorations became a physical possibility, humans dreamt of the opportunities that the exploitation of the resources of the Moon and other space objects could bring. The film is as much a story of human avarice and greed as it is of technology and the dream of conquering space. It is the promise of gold that motivates the explorers, both the heroes and their antagonists, and the riches that can be gained by the first to build a workable transport system between Earth and Moon.

For obvious reasons, as a value proposition this idea would have been doomed. Not only because of the costs of transporting the gold, though these would have been substantial too. Gold acquires its value mainly from its scarcity, and having a steady and almost inexhaustible supply of gold (which needs not even be mined, as it is in the film in abundance on the very surface of the Moon) would quickly destroy its market value. It is only in the margins of the story and by implication that we encounter another type of value in the film, the value of ideas, concepts, and knowledge. When the antagonists try to gain access to the rocket, they break into the office of its inventor and steal the plans and drawings, using them as leverage for a place on the ship.

It is however these ideas, the intangible property, that arguably is of greater real value than the gold. In a case of life imitating art, the Gestapo seized all the drawings and plans for the 2m tall, functional rocket that Oberth and Nebel had planned to launch at the opening night, recognizing correctly where the true value was located.

Even though the problems with the business model of the protagonists in 'The Woman in the Moon' are rather obvious, the idea that space exploration will give access to unprecedented mineral wealth remained a main driving force behind real-world space exploration and its funding ever since. 'New goldrushes' are frequently announced by academics and the popular press alike (see e.g. Feinman 2013; Pelton 2016; Pearson 2018), though so far, commercially viable mining in space remains elusive.

This fascination with exploitation of physical resources also dominated the legal discourse. The international treatises that together form the current 'constitution' of space law aimed from the beginning to protect these resources from becoming the sole property of those nations with the most advanced space program.

The foundations of international space law were created in five international agreements (Jenks 1965):

- the Treaty on Principles Governing the Activities of States in the Exploration and Use of Outer Space, including the Moon and other Celestial Bodies (henceforth 1967 Outer Space Treaty or OST);
- the Agreement on the Rescue of Astronauts, the Return of Astronauts and the Return of Objects launched into Outer Space
- the Convention on International Liability for the Damage Caused by Space Objects
- the Convention on registration of Objects Launched into Outer Space
- the Agreement Governing the Activities of States on the Moon and Other Celestial Bodies

They establish as an underlying principle the idea of space as a commons. Article I of the OST in particular proscribes that space exploration is 'for the benefit and interests of all countries, irrespective of their degree of economic or scientific development, and shall be the province of all mankind'. In a similar vein, Article II provides for so-called non-appropriation of space, so that celestial bodies are not 'subject to national appropriation by claim of sovereignty, by means of use or occupation, or by any other means'. As one of the key functions of a property right in a tangible object is however to exclude others, this also creates limitations for the possibility to 'own' the gold the protagonist in 'Woman in The Moon' coveted—notably they all were private entrepreneurs driven by profit motives, not, as would later happen in real life, nation-states competing in an increasingly adversarial environment for reputation and strategic power.

The consensus on space as a commons has shifted in the aftermath of the enactment of the Commercial Space Launch Competitiveness Act by the United States in 2015. In a move that would have delighted both the heroic Wolf Helius—who despite his idealism is first and foremost an entrepreneur and industrialist—and the shady five 'Brains and Chequebooks' as his enemies, it permits US companies and citizens to 'engage in the commercial exploration and exploitation of space resources', which also includes water and minerals. Whether the Act is consistent with the OST is a subject of considerable academic debate (Tronchetti 2016; Koch 2018). We only note that seven years after its passing, there still hasn't been any commercially viable exploitation of space minerals of the type that the law tried to encourage.

A much more realistic value proposition for space exploration is a different type of goods, intangible goods in the form of data, knowledge, and ideas for new products, processes, and plants. The thieves who stole the plans for the rocket acquired something of much greater commercial value than the eventual expedition would generate. Ideas and knowledge do not take up precious cargo space. They can travel much faster than physical resources, limited only by the transmission speed of the communications equipment. And unlike physical resources, they are non-exhaustive, increasing in value the more people can benefit from them. Who would have owned the map that Helius and his friends drew of the landing site and its geological features? Who the patent in the invention that allowed them to repair their stricken craft? As we will see below, copyright and patent law are intimately linked to a concept of territory and jurisdiction, so 'where', legally speaking, these works were created, and which country has jurisdiction over them this becomes an important question. A future space colony may want to trade with Earth, also to assure its economic independence and sustainability, but the 'goods' most likely to be traded on their side will be intangible assets and the wealth of new information and knowledge that they will generate. The legal framework that governs 'ownership' over this type of asset is IP law.

Despite this importance of intangible assets as a value proposition for space exploration, there is hardly any mention of them in the international treaties that constitute space law.

One of the few exceptions is a short reference to IP rights in the Declaration by the United Nations Committee on the Peaceful Uses of Outer Space on International Cooperation in the Exploration and Use of the Outer Space for the Benefit and the Interest of All States form 1996. It states that:

> [s]tates are free to determine all aspects of their participation in international cooperation in the exploration and use of outer space on an equitable and mutually acceptable basis. Contractual terms in such cooperative ventures should be fair and reasonable and they should be in full compliance with the legitimate rights and interests of the parties concerned as, for example, with intellectual property rights.

However, the main target of this provision are inventions made on Earth, to create the technologies needed for space exploration, especially in complex international collaborations such as the International Space Station, not inventions made in outer space. WIPO, the World International Property Organization, had tried since the late 1990s to initiate a process of clarification and, if necessary, reform of the IP regime as it pertains to works created in outer space. Starting in 1997, WIPO conducted a series of studies and roundtables with a particular focus on inventions made in outer space (WIPO 2004, 8). However, none of these initiatives garnered sufficient support for legislative action.

There has been a small number of academic papers on the topic of IP rights for inventions and creations made in outer space (see e.g. Malagar and Magdoza-Malagar 1999; Cromer 2006; Lockridge 2006; Miles 2008; de la Durantay, Golla, and Kuschel 2014; Doldirina 2015; Leepuengtham 2017). A recurrent theme highlights the fragmented and unsatisfactory state of the current regime, so for instance Sterling's particularly compelling case for reform (Sterling 2008).

What these studies have in common is that first, they focus on creations and inventions that are taking place on vessels, such as the International Space Station and similar 'mobile systems', but (with the notable exception of Sterling) don't discuss the issues created by a putative space colony on a celestial body. As we will see, most of the solutions they advocate under current law break down in this scenario. Second, they focus on human inventors and creators, the archetypical subjects of IP law, but do not discuss innovation by autonomous machines.

Over the last few years, creative AI systems that compose music, write news articles, paint pictures, take photographs, or tell (not so funny) jokes brought the concept of computational creativity to a wider audience. Less publicly but potentially with a much higher economic impact, AI systems are also increasingly capable of making discoveries that, if had they been made by a human, would give rise to a patentable innovation. These technological breakthroughs also let to a lively debate about the role copyright and patent law should play in this new creative ecosystem, a discussion which is increasingly moving beyond academic speculation. Both the European Union and the United Kingdom have been exploring law reform to protect works by machines, whilst UK and Australian courts came to opposite conclusions regarding the status of an AI as inventor on a patent application.

Whilst there may not be an immediate business case to equip robots for space exploration with the ability to compose music (but then again, why not?), they will have to perform some of the activities that could give rise to a copyrightable products: making photographs, compiling data in a database, drawing maps for coordination purposes with other machines, or generating human readable reports.

Furthermore, to be able to survive under conditions that their human designers can only partially anticipate, they will have to invent solutions to problems that might be deemed patentable inventions. This could be the construction of a new antenna, or a new and more efficient way to communicate with other robots. They will need to be able to repair and modify themselves, making the outcome of their activity even less predictable for their designers, and hence closer to the type of novelty that also attracts legal protection.

As we will argue, the emergence of AI systems capable of creating and inventing is posing a further difficulty for a putative IP regime for creativity in outer space. To prepare the ground for this argument, the next section introduces briefly the relevant IP rights, develops some shared fact patterns that give rise to issues, and indicates why creativity in outer space is an IP issue. We then discuss in more detail the issues for patent and copyright law, leading to an analysis of the additional complexity that is generated when the inventor or creator is an autonomous machine. In the final section, we ask how to address the regulatory gaps that this discussion has demonstrated, and how our choices matter for the question of space liberty.

3. Notice and Takedown Requests from Ground Control to Mayor Tom

In this section, a number of relevant IP rights are introduced. Focus is on those features that raise potential issues in our scenario. First we set the scene and introduce a range of IP rights. We then discuss in particular copyright and patent law issues.

3.1 Setting the scene

In 2013, shortly before the end of his mission at the International Space Station (ISS) Commander Chris Hadfield released a video of himself performing David Bowie's 'Space Oddity' that was transmitted to Earth. There it was released as a YouTube video where it was viewed over 50 million times.[1] Whilst viewing the video is free, the traffic it generated for YouTube is in itself of economic value, and we can easily imagine how it could have been monetized through the site's advertising system. This then raises the question if Hadfield had rights in the recording of his performance that he could have used to generate revenue for himself, and also what type of permission, if any, he needed to get from David Bowie to perform Bowie's song in space. Hadfield, it should be noted, had indeed received permission from the rights holders and observed the limitations of this license so carefully that the video was temporarily removed form YouTube when the licence had temporarily lapsed.

This gives us the basic fact pattern for IP law questions in outer space:

- What are the rules regarding the use in space of a work that was created on Earth and is protected by an IP right?
- What are the rules regarding the use in space of a work that was created in space by someone else?
- What are the rules regarding the use of a work that was created in space by someone else on Earth?

Or put differently, we can ask who, if anyone, owns the rights to a work that was created in space (the question if a work is protects) and secondly, who, if anyone, is liable for the infringement of the rights of someone else if the potentially infringing act took place in space (the infringement question)

The field of law that regulates the ownership over the results of human ingenuity and creativity is IP law. Conceived in the seventeenth and early eighteenth century respectively (Sherman and Bently 1999, 207) and thus co-evolving with the emergence of the modern nation-state, it is unsurprisingly tied to concepts of jurisdiction and territory on Earth. It matters, at least for some rights, in which country the creator was located at the moment of creation, or with which national authority they register their work. Similarly, it matters where an alleged infringement of the right tool place, and which state has jurisdiction to hear the legal complaint. Outer space, legally speaking, is however extraterritorial, it 'belongs to nobody because it belongs to everybody' (Winkler 1999, 44). We will see that this is not categorically true, but for now it helps us to understand why there is an issue that needs addressing.

Whilst copyright and patents are the most well-known IP rights, there are a number of other types of rights recognized as well, though not all equally by all jurisdictions. In addition to patents and copyright, there are in particular industrial design rights, trademarks, plant varieties, trade dress, and trade secrets as distinct forms of rights over intellectual creations. Similar to copyright, there is also integrated circuit lay-out design protection that protects the three-dimensional topography of a chip of semiconductor devices. Some jurisdictions recognize 'utility models', a right similar to patent but with a simplified and less onerous protection for more incremental inventions. In the European Union, there is furthermore a *sui generis* right in some databases that protects the effort and the choices made when compiling information into a database.

Remember the setting from the introduction: we assume one or several settlements in outer space, initially populated solely by autonomous machines. Their activity and explorations create value, in the form of knowledge, data, and other intangible assets. Can we envisage a world where they trade these assets either with Earth, or with each other? Being able to trade in this way would constitute a first step towards self-sufficiency, increased independence from the nation or commercial entity that send and owns them, and with that, liberty. Conversely, such a colony would be inevitably in a precarious and vulnerable position, and incapable of matching the significant research capacities that we find on Earth. As with the colonies of old, exploitative systems could emerge where the colony trades mainly in raw data, but the 'value adding' analysis that turns the data into knowledge, or inventions, is carried out on Earth. By cutting a colony off from information and knowledge, its very survival could be threatened. Are therefore the colonies entitled to use inventions made on Earth without having to worry about patent licences and, conversely, can they copy and distribute news from Earth between them without compensating the creator of the news?

The focus of this chapter is on patents and copyright as the most relevant forms of IP rights for our setting, and also because the principles and problems that we will discuss apply to some of the lesser-known rights too. A few comments however on some of the other rights can serve as edge cases for the main argument that this chapter will develop.

Overall, the issues that these other rights face in space have attracted even less attention by academics, governments, or legislators, with the possible exception of database rights: satellites generate already now significant amount of data, some of it in the form of images, a type of work that is capable of copyright protection. Some of that data gets selected, curated, and organized into databases prior to transmission, on the satellite by an AI (for examples see e.g. Espositio et al. 2019). Satellite data therefore comprise one of the cases where already today, the question of space IP and IP in works generated by machines intersect. This then gives rise to the question of who owns the rights into both, the individual images taken by a satellite according to the commands of one of its algorithms, and the database that another algorithm created from them. Who owns these works when they were created 'nowhere', and literally by 'nobody'? Whilst copyright in photographic images is at least internationally recognized, the separate database right is only recognized within the European Union (Malgieri 2016)—does EU law extend to outer space?

Another more unusual IP right are plant breeders' rights that give the breeder of a new variety of plant the exclusive right over the propagating material (such as seeds) for 20 to 30 years, depending on the type of plant (see generally Koo, Nottenburg, and Pardey 2004; Llewelyn and Adcock 2006). They reward breeders for the intellectual effort invested in creating new and hopefully beneficial varieties of a plant, for instance one that is resistant to pests. For the survival of a human space colony, the ability to breed plants that survive under the conditions of a different planet would be decisive, but even short of such an ambitious goal, experimentation with plants and seeds has been an important aspect of space-based research, including the creation of new varieties (see e.g. Chengzhi 2011; Mohanta et al. 2021).

To be capable of protection, the new variety must amongst other things be *distinct*, that is it must differ from all other known varieties; these distinguishing characteristics must be *uniform*, that is expressed by all the plants of the new variety; it must be *stable*, that is the same from generation to generation. This indicates the main difficulty for plant breeder's rights: very unlike patents, where the validity can be determined immediately at the point of application, the new variety has to be grown for several seasons to determine whether it is not just distinct and uniform, but that the distinctiveness and uniformity is stable over time.

To gain registration, the breeder must therefore submit seeds to a national office —though increasingly, those countries that recognize plant breeder's rights exchange information about the outcome of tests to enable faster registration in multiple countries. The plant variety office then grows it for one or more seasons to evaluate its stability. Here we can see some of the IP problems that this right generates for space: first, we note, as for most IP rights, the close connection between a state's territory and the scope of the right: national testing and registration offices have to validate the application, and the protection is limited to the jurisdiction of that country. Second, a physical sample has to be sent to at least one of these national offices—which in the case of a space colony would create significant costs. Finally, the registration office grows the seed under Earth conditions, which may not be an appropriate or relevant test environment for this new variety. Maybe under Earth gravity, the plant does not survive, especially when its distinguishing characteristic is subject to selection pressure here.

Trademarks are another IP right that has, thus far, been little discussed in space law (one rare exception is Borovik 2019). They are mentioned here mainly for their entertainment value, as they recall the Cold War joke about Nixon who, on having heard that the Soviets had painted the moon red, then ordered NASA to paint 'Coca-Cola' in white over it.

There is a somewhat more serious side to the use of trademarks in our scenario. Their main function is to facilitate the decision-making process for human shoppers and to allow them to discriminate, or maybe better, discern between different products on offer (Schachter 1927). Trademarks, so at least one important theory justifying them argues (McKenna 2012), enable us to get what we really want, by preventing consumer confusion (Calboli and Farley 2012). Their ultimate aim is to promote the ability of consumers to make purchasing decisions based on accurate information in the market without excessive research costs. However, whilst humans have to make decisions using their heavily bounded rationality, AI systems and robots are not in the same way limited. This has led some authors to proclaim the death of trademark law, since AI shopping assistants won't need trademarks to decide between products (Grynberg 2019). If there were commerce between Earth and the robot-populated space colony (and trademark law only applies to commercial contexts) only one side would need trademarks as a crutch to facilitate their decision-making, with possible implications for key trademark concepts such as the likelihood of confusion. Trademarks protect consumers from confusing a valuable product from a cheap imitation, but what fools a human does not necessarily fool a machine.

On the other hand, the very fact that a product is made in space could be a selling point on Earth. In addition to trademarks, which allow buyers to identify the producer, this could also create interest in a more direct way to certify extraterrestrial origin, through yet another form of IP right, a 'Geographical Indication' or GI. GIs are labels on products which indicate from which specific geographical location the product originates. Famous terrestrial examples are 'Champagne' for sparkling wine form the champagne region in France, Parmegiano Reggiano from certain parts of Italy (but not just Parma), and, since 2021, Rooibos tea from South Africa.

According to Article 22.1 of the Agreement on Trade-Related Aspects of Intellectual Property (the TRIPS Agreement), GIs are 'indications which identify a good as originating in the territory of a Member State, or a region or locality in that territory, where a given quality, reputation or other characteristic of the good is essentially attributable to its geographical origin'. Obviously, that GIs must refer to the territory of a member state makes this provision not directly applicable to products from space colonies, though possibly, as we will see, from space vessels.

The final IP right to mention for our discussion before we analyse copyright and patents in more depth are trade secrets. It may be helpful to think of them as the last resort when patenting an invention has failed, was too expensive, or for some other reason undesirable. Patent law aims to strike a balance between secrecy and openness that ultimately benefits everybody, the inventor but also society. In exchange for a temporal monopoly, the patent holder discloses in the patent application the relevant features of the innovation. Once the term of protection is over, others can then build on the disclosed features. Alternatively, whilst an invention is still protected, the disclosed information can still facilitate the discovery of new, possibly better and non-infringing ways to achieve the same result. Once it is known how a problem can be solved in principle, finding new solutions often becomes easier. In the absence of such a protection, inventors would have to keep as many details as possible of their invention secret and undisclosed.

Trade secret law creates a limited protection for ideas that are only protected through this 'silent' form of IP, discouraging in this way some forms of industrial espionage. But a competitor who simply stumbles across the same solution to a problem would be within their rights to use their discovery. The possibility to simply keep new information, knowledge, and innovations secret is a necessary backdrop for this chapter: whilst it may be fanciful to discuss the question of whether a colony of robots needs IP rights, denying space-based innovation and creation by machines protection could incentivize the builder of these machines from sharing their information, merely though appropriate design features. This would harm the cooperation between robots of different owners/builders, and also the free exchange of ideas between the space colony and Earth, both highly undesirable consequences. We will come back to the default position towards the end of this chapter as it will play a crucial role in the argument to extend, despite some misgivings, the IP regime to space and space colonies.

3.2 Copyright and patent law in space

With this, we can now discuss two of the most important forms of IP in more depth, copyright and patents. Focus is on those aspects of the law that make them relevant or problematic for our discussion.

Copyright gives the creator of a creative work—typical examples are books, songs, news items, or photographs—a temporary monopoly to distribute, copy, and generally commercially exploit their work. Copyrights are 'territorial rights' in the sense that if a state grants an author copyright in their work (typically because the author is a national of the country, or the work was first published on the territory of the granting country), it does not extend beyond the territory of that specific jurisdiction. However, over the course of the last century, international conventions and treatises such as the Berne Convention and the TRIPS Agreement not only harmonized significant parts of the law, but also ensured that foreign works are regularly given the same protection as domestic work, provided the foreign country is also a signatory to the said convention or treatise.

Depending on jurisdiction, a broad range of creative outputs can be protected, ranging from poetry to motion pictures, from choreography and musical compositions to paintings and sculptures. Importantly in our context, photographs and computer software can also enjoy copyright protection.

Copyright arises in most jurisdictions automatically at the time a work is created, registration is not necessary. Protection lasts for a set number of years after the death of the author, typically 50 or 70 years. Reflecting the diverging traditions of the continental European and the Anglo-American approach to copyright, two different types of rights are conferred on the author. The first set are moral rights, a notion that originates in the continental European tradition. They protect the non-commercial interest of artists in their creation, which includes the right of attribution—that is, to be recognized and named as the creator—and rights to the integrity of the work. The latter may give the author for instance a right to object to distortions or mutilation of the work if these are 'prejudicial to the author's honour or reputation', potentially even after the artist relinquished ownership (and in some jurisdictions the author can't waive these 'natural' rights). Commercial exploitation rights by contrast focus on the right to license, sell, or otherwise alienate the work on a market, and in that way open the possibility for the creator to benefit financially from their work.

Infringement of a copyright requires an active act of copying, that is the infringer must be aware of the existence of the work that they are copying—so a creator who creates a work that happens to be identical or near-identical to one already in existence, but does so independently and without having been exposed to the older work, does not infringe the copyright of the creator of the older work.

The second important IP right, patents, protects the interest of inventors in 'useful innovations'. It bundles a set of rights that together create a temporary monopoly for the exploitation of an invention, and excludes during that period others from using the protected idea without a license from the patent holder. Unlike copyright, it is granted only after an application has been filed by the inventor, been evaluated by the patent office, and the patent subsequently registered. As noted above, this process of registering a patent with a description detailed enough to enable a 'person skilled in the art' to replicate the invention balances the interests of the inventor with those of society: in exchange for temporary protection, the new idea is disclosed. Unlike copyright, infringement occurs when an unauthorized person brings a product to the market that is based on the protected idea, whether or not the infringer knew about the registered patent. With other words, it is crucial that a would-be inventor searches the patent register first for possible prior art, which also means that it has to be possible to carry out such a search reasonably fast and without too much by way of costs.

Conversely, patentable ideas have to be 'novel', which means that if an invention has already been made public, whether protected by a patent or not, then it is not possible to patent that idea any longer. Some jurisdictions such as the United States recognize a grace period, under which the inventor can disclose the invention early without preventing them from later filing a patent application. Whilst copyright is created at the moment a work is created, patents require registration, which can lead to time delays between discovery and protection. In situations where more than one person made the same discovery at the same time, the first to file the patent application will be granted protection, regardless of the date of actual invention.

With these observations, we have identified all those aspects of copyright and patent law that can create issues in our scenario:

- Both are territorial in nature, that is they only protect against infringement on the territory of the state that granted the right.
- Nationality of the creator and place of the publication of a work determine in copyright law which country has jurisdiction.
- The main addressee of both systems are humans and their creative capabilities. In particular, the laws aim to incentivize creativity and inventiveness.
- In patent law in particular, it can matter *when* an invention was made, disclosed, or registered.

Questions of territory, time, and creator/inventor all become contested once we leave Earth. We began this chapter with Chris Hadfield and his video of performing 'Space Oddity'. Let us for the moment assume counterfactually he had not acquired a licence from Bowie. What is the legal situation?

The first issue we face is the territorial nature of copyright. As noted above, copyright is territorial, that is any infringing action has to take place within the territory of a country that has granted the right to the infringing work. Furthermore, we noted that space is extraterritorial. This would indicate that everything is *Hunky Dory* for Hadfield, so why did he in real life ask for, and obtain, a licence to perform the work?

According to Article VIII of the OST, just as with ships on the high sea, the state that registers an object that is launched into space retains jurisdiction over it. Space vessels are therefore 'quasi-territorial', which means for legal purposes whatever happens on a registered spaceship that does not stay in space is treated as if it had taken place on Earth, and on the territory of the registering state.

This means in particular that IP rights in an object or work that exist on Earth do not get suddenly extinguished once a vessel reaches outer space. The International Space Station (ISS) is a special case, but its legal regime nonetheless comes broadly to the same result. Article 21 of the Space Station Agreement is one of the rare provisions in space law that explicitly address expressly IP rights: 'for purposes of intellectual property law, an activity occurring in or on a Space Station flight element shall be deemed to have occurred only in the territory of the Partner State of that element's registry'. In other words, Bowie's copyright is protected on any vessel that is registered to a country that also on Earth recognizes his copyright, and with that every signatory state of the Berne Convention. Where did the infringement occur? This depends mainly on where the infringement takes effect. Hadfield transmitted his performance to Earth, where it was seen by millions. Under Article 8 of the WIPO Copyright Treaty, the effect of a transmission is where the transmission was received, and, as the WIPO Issue Paper (WIPO 2004, 6) clarifies: 'This Article is also applicable to transmission to and from a spacecraft.'

The situation is somewhat more complicated if the effect of the infringement had been on board the ISS itself, for instance a performance to all other astronauts on board at that time. This could have led to rather strange results, given how close to each other and continuous the various segments are. Assume Bowie had licensed Hadfield to perform his song 'everywhere in the United States'. Under quasi-territoriality, this also allows him to perform the song on the US segment of the ISS. But had Hadfield during his performance floated slowly from the Destiny module of the ISS, registered to the United States, into the Columbus module registered to the European Space Agency (ESA) (and with that the 20 ESA member states simultaneously), he would have crossed the border between two copyright regimes, and the moment he crossed the threshold between the modules, he would have violated Bowie's copyright.

What about Hadfield's IP right in the recording of his performance? To simplify things, assume he had recorded a song he wrote himself whilst in the ISS. Here the situation is even more straightforward: it is the nationality of the author rather than the place of creation that matters. As a Canadian, he is citizen of a member state to the Berne Convention, and therefore acquires copyright in the work that would be protected under the relevant national copyright laws. Anyone in the United Kingdom making a copy of his transmission and distributing it would breach his copyright.

Whilst this analysis relied on the OST alone, a number of states have enacted national laws that explicitly affirm the solution for IP laws. In Germany, Article 2, a law that implements Article 21(2) of the Space Station Agreement, provides that 'activity taking place in or on an element registered by the European Space Agency ... shall be deemed to have taken place within the scope of application of this Act as regards the protection of industrial property rights and copyright'. For patent law,

Section 105 of the US Patent Act similarly brings any 'invention made, used or sold in outer space on a space object or component thereof' under the scope of US patent law.

Whilst this analysis is straightforward and seems to indicate that space copyright may not be such a problem after all, it rests on the facts that (i) Hadfield was on board of a vessel registered by a state and (ii) the effect of his activity was on Earth or other parts of a registered vessel, and with that 'quasi-territorial'. A more problematic picture emerges when we vary these conditions.

What would the law say if Hadfield had performed the song not on the ISS but the Moon? As long as the performance is transmitted to Earth, not much changes. His copyright remains the same as it is linked to his nationality, not the place of creation. The infringement happens where the transmission is received, on Earth. Real differences however emerge in the situation where his audience too is on the Moon, comprising other astronauts rather than, *Earthlings*. Now the infringing act is extraterritorial, and as IP rights are territorial in nature as we learned, no infringement is taking place. In this situation, we face an asymmetry: inhabitants of a space colony can freely use works created on Earth, or inventions made on Earth, provided the effect of these actions stays within the confines of their colony, or the celestial body on which it is located. Conversely, people on Earth would not be permitted without licence to copy works created by the colonists (provided these are still citizens of a nation on Earth), or use their inventions if these have been registered at a national patent office on Earth.

This result might be desirable, for the same reasons that critics of our current terrestrial patent law advocate asymmetric patent laws for developed and developing countries. These would allow the latter more leeway to use without licence patents registered in developed countries, whilst giving innovations from their fledging research ecosystems full international protection (see e.g. Sterckx 2004; Brown 2012). Space colonies will inevitably be fragile, vulnerable, and in many ways dependent on the more powerful Earth. Unlike on Earth however, current law comes without the need of reform or intervention to an asymmetric allocation of rights and duties.

Unfortunately, the analogy with developing countries is misleading in one crucial aspect. Developing countries on Earth have IP regimes that encourage their own citizens to be inventors and creators—this is, after all, the rationale for IP rights. Whilst it may be desirable to allow them access to the fruits of intellectual labour of scientists and writers in rich and developed countries for free or through a mandatory, low-cost licence, their own citizens enjoy IP protection for their inventions and creations also against each other, and thus are incentivized to create and invent. But for the same reasons that the rights of an IP holder on Earth are not infringed when the copying only had effect on the celestial body, the same holds true in the relation between the colonists. Assume Hadfield had composed a song on the Moon. Another astronaut gets hold of a copy of his notes, likes what he sees, and before Hadfield had time to do so sells it under his name amongst the other members of the crew. For the same reason that Hadfield did not violate Bowie's copyright in the case variation above, the action of his crewmate, as morally reprehensible as we may find it, did not break any laws.

In other words, whilst some aspects of the current IP regime for outer space have potentially desirable side effects, they also create significant gaps in IP protection. This is why Sterling proposed his alternative Copyright Treaty for outer space, which would treat all of space as one territory, and apply terrestrial laws to them through analogous institutions and procedures. However, this solution takes the form of an 'all or nothing'—the additional protection that space colonies have under the current regime disappears, and the treaties creates a level playing field in law where in terms of power and capabilities, the field is anything but even.

Patent law largely follows the analysis for copyright, with a number of additional complications. As noted above, for patent law the moment in time when a patent is registered can be crucial. Furthermore, unlike with copyright, 'parallel invention' is also an infringement. Both factors together would make it even more problematic to extend Sterling's proposal to patent law. If inventions made in the space colony had to be registered at a national registry on Earth, the inevitable time delay that the long-range transmission of data would bring could expose inventors in the colonies to additional risks under the 'first to file' system, whilst the information about their invention is in transmission. The greater the distance between colony and Earth, and the more difficult the communication, the prohibition of parallel invention would create additional risks and burdens especially for the colonies: they would infringe an Earth patent even if they came up with the same idea, entirely independently, and prevented from finding out about the patent through the difficulties of communication. The burden to carry out patent searches on registries on Earth would lie entirely with them. If on the other hand there were to be corresponding local 'space territory patent offices' and registries, the 'first to file' system would require a single temporal reference frame to determine priority—and whilst arguable far-fetched, this can become an issue if the distances between Earth and colony become more substantial. In any case though, also a mutual requirement to carry out patent searches in registries held on other planets would be prohibitively burdensome.

3.3 Self-made machine man

As we have seen in the discussion so far, the nationality of the creator or inventor can also play an important role. Because Hadfield is a Canadian, his performance, albeit made in space, automatically creates copyright protection in Canada. Through international treatises, this is then extended for all practical purposes globally. As the objective of IP law is to incentivize innovation and creative endeavours, its targets are humans, whose needs and desires can be leveraged by the law to create incentive structures. Sterling (2008, 5) briefly discusses images taken by a robotic device, but as a limiting case only: as long as there is no human intervention at all, the raw images that the machine produces are incapable of IP protection, only through selection, editing, and transformation by a human can they become works for IP purposes.

The recent public interest in machines that can create and innovate has also resulted in a reinvigorated discussion about the IP implications (for examples see Guadamuz 2021). More than any other private law regime, IP law seems to centre around what makes us human, our ability to dream about new worlds and envision ways to reach them, to create works that appeal not just to our utilitarian instincts but to something else.

In an opinion piece for WebProNews, the novelist and technology writer Jason Lee Miller stated:

> Chess is one thing, but if we get to the point computers can best humans in the arts— those splendid, millennia-old expressions of the heart and soul of human existence— then why bother existing?

Research in computational creativity is, however, not a recent phenomenon. The possibility of true machine creativity was already discussed in Alan Turing's seminal paper from 1950 that gave us the Turing Test (Turing 1950). Long before the advent of modern computers, 'algorithmic' creation of musical works using nothing but a set of strict rewriting ruled together with a dice (as random generator) were a popular Victorian pastime (see for a historical analysis Schafer et al. 2015, 222ff). From the 1980s onwards, Margaret Boden and others transformed the research landscape and formally established computational creativity research as a legitimate academic endeavour (Boden 1998; Colton, De Mántaras, and Stock 2009). It was only in the last few years however that very public examples of machine creativity started a public debate about machine creativity. From AlphaGo's famous 'move 37' that wrongfooted a human Go master by its originality (Menick 2016) to the AI that generated blog posts that fooled thousands into believing they were written by a human (Hao 2020), more and more examples of AI systems challenge our conception of creativity by creating the type of work we previously considered unique to humans and the human mind. AI systems now create poetry (Oliviera, Mendes, and Boavida 2019), paintings (Colton et al. 2015), compose music (Lopez-Rincon, Starostenko, and Ayala-San Martín 2018) or even entire musicals (Jordanous 2017). Whilst these activities may be less relevant to the needs of space exploration, the automated composition and taking of photographs most certainly is, and already contributes significantly to our understanding of outer space (Vertesti 2015; on AI-generated photography see Lan and Sekiyama 2015). "Drawing maps" is another important skill to turn data into a useful format when exploring new geographies (Li and Hsu 2020), as is the ability to modify or entirely rewrite software code (Becker and Gottschlich 2021), all of which fall into the category of works protected by copyright as discussed above.

Robots are not only competing with humans in the arts, but also in the field of innovation. Genetic programming (GP) is a form of AI modelled after the process of biological evolution that systematically solves high-level problems by improving upon a set of candidate solutions of known performance (Koza and Koza 1992). Whilst human operators specify the seed solutions, fitness measures, and the termination criteria, there is usually no human intervention during the program's execution. GP has been used to independently recreate known patented inventions, generate non-infringing work-around solutions, and is responsible for the creation of at least one known patented invention that is known to have been created using GP (Koza 2010 265).

An example of particular relevance for our topic, the antenna for the miniature satellites used in NASA's Space Technology 5 mission, was designed using

GP algorithms (Hornby, Lohn, and Inden 2006). From a set of existing antenna designs, the AI evolved a set of novel antenna designs that met the mission's pre-defined requirements. The unusual structural designs produced in this way were radically different from traditional human designs and to human designers look non-intuitive, whilst having a number of notable functional advantages over any known antenna.

These breakthroughs have generated significant legal debate. What is the status of a work generated by a computer without human input, in a copyright and patent regime devised to incentivize human labour and reward human ingenuity? Should they be owned by the developer of the AI, its operator, the first person to identify the result as valuable, or should they have no protection at all and be part of the creative commons? (see e.g. Grimmelman 2015; Schafer et al. 2015; Guadamuz 2021). Unlike the examples of IP discussed above where international treatise have resulted in significant harmonization, the answers to these questions still differ significantly between countries. Section 9(3) of the UK Copyright Act, for instance, has an explicit provision on computer generated works that give copyright to the person 'by whom the arrangements necessary for the creation of the work are undertaken' (Lee 2021).

If a robot were to create these works on board a spaceship, there might be a complex team of engineers on Earth even though they have no direct input into the final creation, nor can they anticipate how the work will look. Concerns like this, that the connection between any human 'author candidate' and the eventual work is too indirect, lead many to argue that works that are created without direct human input are not protected, and should not be protected, by copyright (see e.g. Aplin, Schafer, and Li, 118). This also seems to be the position that for now most legal systems are taking, though they differ in the amount of human input that will be required in individual cases to turn the AI back into a mere tool of a human creator.

Similar disagreement and divergences can be observed in patent law. Within the space of just a few weeks, courts in Australia and the United Kingdom respectively affirmed or denied the registration of the DABUS AI as inventor of a patent (Abbot, Matulionyte, and Nolan 2021). As with the debate surrounding copyright, there has been a growing academic debate about the status of AI-generated innovations (see e.g. Fraser 2016; Fraser and Schafer 2017; Ravid and Liu 2017).

According to our hypothesis, the majority of works and innovation in space in the near future will be made by autonomous machines. So how does this debate intersect with the question of IP in outer space? What would have been the situation if we replace Hadfield by a machine?

For a scenario where the production of the work takes place in the ISS, the territoriality principle applies. To the extent that the applicable national legal system ties copyright to the nationality of the author (as most under the Berne Convention do) we face a first problem. The robot-Hadfield is a machine, and whilst currently machines in outer space will be owned by the launching nation, they do not have 'nationality' for the purpose of IP law; stunts such as the award of Saudi Arabian nationality to the Sophia robot—which was a mere publicity move without legal meaning—notwithstanding (Parviainen and Coeckelbergh 2021). This is also a problem when we imagine a fully autonomous robo-Hadfield to compose or preform a song on the Moon: we would have to extend the idea that objects in outer space are owned by and under the flag of the launching state, and argue that robots, for IP purposes at least, are also that state's "nationals". In a further complication, we can think of an entire spaceship as a robot, in line with Iain M. Banks science fiction stories about the 'Culture'. If that ship then takes images that are copyrightable, it would be at the same time 'territory' of the launching state and a 'national' (for IP purposes) of that state.

Assuming we can overcome this hurdle, other problems come to the fore. If robo-Hadfield moves from the Canadian part of the ISS to the European one, the applicable law now becomes a conglomerate of 20 different IP jurisdictions, some of which in turn disagree on the question how robo-generated works should be treated. A UK court, applying Article 9 cited above, will look for the person or persons who took all the necessary steps to allow robo-Hadfield to perform the song—it is likely that this will be a large group of people, scientists, engineers, administrators, from different jurisdictions. A German court may decide that the output was lacking an author, and for that reason has no copyright protection at all. Whilst in cross-jurisdiction litigation on Earth, the rules of international private law are meant to resolve issues like this, their suitability to settle similar issues under space law remains highly contentious, the main reason why Durantaye, Golla, and Kuschel (2014) argued for a major revision of space IP law.

Similar issues arise when our putative robo-Hadfield leaves the ISS and performs, in analogy to the scenarios above, on a planet, or makes an invention there. Whether the invention is patentable under its name depends on the jurisdiction of the chosen registry, and is less problematic. For any copyright law claim we have to decide whether the space law concept of 'national flag' and the IP law concept of 'nationality' are sufficiently aligned, and then in addition we have to align this assessment with the question whether or not a machine can be an author. Let us assume the robot is British, but the first transmission of its work is to the United States. The US court would then have to determine if the work was generated by a 'national' of a member state of the Berne Convention. Assuming this is answered in the affirmative, the court then has to decide if in this case the UK approach that treats computer-generated work as copyrightable applies, or the US approach that so far refuses this possibility.

Equally complex, if not more so, are issues surrounding the violation of copyright or patents by a machine, the 'upstream' question of the input, rather than the output, of creative machines (Schafer et al. 2015) All robots have sensors with which they can record material potentially owned by someone else, they all need training data as input. Whether or not using data without a licence for the training of an AI, which then creates works of its own with little or no similarity to the input data, is itself highly contested. It was for instance subject to a recent consultation by the UK Intellectual Property Office to which the author made submission (Aplin, Schafer, and Li 2022).

To illustrate this point, let us again go back to Hadfield's Bowie performance. We argued above that even on the ISS, this was a possible violation of Bowie's copyright, and a license was needed. Similarly, if a robot had performed that work, Bowie's rights would have been an issue, the guilty party probably the person who downloaded a version of the song to the robot. But what if the AI had simply scanned radio transmissions from Earth and had autonomously picked one song to perform on board? Even though there was no direct human input, the owner of the machine (in our case the Canadian government) remains liable.

So far, human and AI infringement follow similar lines. What if Hadfield had not performed a specific Bowie song but his own composition, made after listening non-stop to hundreds of Bowie recordings, and as a result creating something that sounds to the expert ear very much in the 'style' of Bowie? From an IP perspective, as long as that song is sufficiently different from any individual Bowie composition, this is unproblematic. Protected are specific expressions, not the underlying ideas or personal styles, and learning from the masters is generally permitted. For an AI, the answer may differ, because it has to store, if temporarily, a copy of the data, images, songs, etc. on which it is trained, and this alone may be sufficient for a copyright violation. This holds at least in Europe that does not benefit from the more flexible fair use and non-expressive use doctrines of the United States.

If the ISS had a copyright licence to receive radio from Earth and play it to the crew, could the AI then legally use this music to also write its own, in a similar style, for the astronauts' entertainment? Again, it depends—quite possibly if the AI is in the US segment and the transmission restricted to it, quite possibly not in the European part of the ISS as under European (and UK) law, the right to read or listen to a protected work only exceptionally also entails the right to data mine it. What if the AI-generated song is not broadcast internally on the ISS, but exclusively to the astronauts when they are outside the vessel, maybe to help them psychologically? This may be safer, legally, as now the infringement takes place outside any state's territory.

In a final reiteration of our analysis above, what if the interaction is not between humans and machines but exclusively between machines? Consider a group of robots from nation A on the surface of a planet taking images of their surroundings and transmitting these to each other for coordination purposes. Robots from nation B, a friendly competitor, intercept the data in transmission, analyse the data faster, and as a result save considerable energy, a very scarce resource. Is this an IP issue? Under space law, there is an overriding obligation to make information of this type available 'eventually', but this could be through a license rather than allowing from the beginning free use for everyone. Whatever the answer to this question may be, as the effect of any infringement is again purely experienced in outer space there is no copyright remedy for nation A.

In summary, we have seen how a space IP regime that was already fractured and beset with difficulties becomes even more complex, and also more fractured, once works are created and innovations are made by machines. Here, the significant divergences between jurisdictions that exists on Earth puts additional strain on the much less detailed space law. Similarly, as Sterling noted, current space IP law leaves, intentionally or unintentionally, for better or worse, a significant amount of creative and innovative activity in space less protected than its counterpart on Earth. Whilst this issue is less pressing whilst vessels and astronauts leave Earth only temporarily, (and most of the information is send back to receivers on Earth) it would become a real issue for any long-term space settlement. There these type of data, information and works are mainly of benefit to other space-dwellers and not Earthlings.

4. The State is Needed Even by a Society of Robots

We can now draw the different strands of the discussion together. As we have seen, our IP regime would require considerable changes to accommodate extraterrestrial production of work once the creators of these works leave the confines of a space vessel and build colonies on other celestial bodies. It would require even more adjustments if the invention or work is done by in an autonomous machine.

This leaves us with a range of policy options:

(1) IP minimalism: the territorial principle remains restricted to works created by humans on board of registered space vessels. The use of ideas, copying or distribution of works and inventions of someone else are not infringing when their effect remains limited to the celestial body. This is the status quo.

(2) territorial extension with equality: space becomes a single territory for legal purposes. Rules of terrestrial IP law apply without discrimination, machines remain excluded from being holders of IP rights, the question if anyone else has IP rights in them needs to be resolved by global consensus. This is in essence an extension of the proposal by Sterling.

(3) territorial extensions with inequality: space becomes a single territory, inhabitants of space settlements create IP that is effective against other settlers and against humans on Earth, but they remain free to use terrestrial works and inventions provided that use has not impact on Earth. This accounts for the vulnerability and dependency of space colonies.

(4) As with (3), but the principle is extended to colonies inhabited exclusively by autonomous machines, who can now 'trade' or 'exchange' their outputs between each other.

From a liberty perspective, solution (1) has superficial appeal. Whilst libertarians in particular emphasize the importance of property for freedom, even they often exclude IP rights from this commitment (Kinsella 2009; McElroy 2011). IP rights create artificial scarcity, and state-backed monopolies, that come at significant cost to other liberties such as freedom of speech in particular (see e.g. Samuelson 2002). Whilst an IP regime rewards and incentivizes creativity for some, it can also prevent creative endeavours by others, especially if it is enforced badly (McLeod 2005). In the context of space, *not* extending these monopolies unnecessarily has particular appeal and resonates with the underlying philosophy of space law that sees space as a common, not to be exploited or owned by individual states or corporations. Indeed, some academic commentators have argued that the OST also prohibits the attribution of IP rights to things like images of celestial bodies taken in space, though this argument is ultimately unconvincing (for a discussion see Durantaye et al. 2014, 4). Cyberlibertarians in turn argued for a cyberspace free of IP (Lobato 2011), where 'data wants to be free' (Wagner 2003), and whilst that fight was lost, it can still serve as a blueprint to argue against enforcing IP rights if the effect of the violation 'stays in space'.

However, this stance faces two problems. First, it does not account for the power imbalance between Earth and a putative space colony. As argued above, the main 'tradeable asset' of such a colony is likely to be knowledge. Leaving it unprotected in space, whilst keeping tight IP regulations on Earth, would exacerbate this problem. It is not by coincidence that on Earth powerful developed nations that keenly enforce the IP generated by their own industries are quite open to the idea that 'Indigenous knowledge' forms part of a 'commons' and are as such unprotected. This gives them ample opportunity to leverage their superior research capacities to make use of knowledge and ideas shared traditionally openly within Indigenous communities in developing countries, for instance for drug development, without having to pay for access to this information (Drahos and Frankel 2012).

The second problem links back to our discussion of trade secrets above. Patent law in particular tries to strike a balance between protection and openness. In the absence of patent protection, inventors will take additional steps to keep their idea for themselves disclosing only what is needed to build a machine or produce a drug. Technological countermeasures that make backwards engineering more difficult become a prized good. In a hostile and dangerous environment such as space, sharing vital information rather than keeping it hidden is even more important, so creating incentives for sharing vital. "Hard-baking" inflexible access restrictions into the equipment that can't be overridden would be a dangerous design solution that should be discouraged.

Option 2 we briefly discussed, and option 3 extends it to robots. Treating space as a single territory, with its *sui generis* laws modelled on Earth as proposed by Sterling minimizes the fragmentation and eliminates many of the difficulties that otherwise would be an obstacle for efficient trade in ideas between Earth and colonies. Option 2 is simpler, but it does not address the power imbalances, with a possibly significant cost for liberty in space. Option 3 is the one preferred in this chapter. It allows colonies to leverage their main asset, their new ideas, knowledge, and information in trade with Earth, and protects them there. It also protects IP in intra-colony interactions, acting as an incentive for innovation and also as an enabler for liberty. By providing a business model for artists and innovators, IP contributes to the ability of a society to differentiate and specialize in tasks. An artist can choose to be an artist because others chose to be shepherds and the IP regime enables the trade of songs for sheep. But option 3 continues to treat space as 'extraterritorial' as far as Earth-generated IP is concerned. Consequently there is no infringement if its effects are only in space. In practical terms, an innovator on Mars need not worry if their innovation has already been registered on Earth—something also practically impossible to ascertain- as long as it is used exclusively in space. The innovator on Mars would have to consider the issue though if they then wanted to license it on Earth, where any local patent with earlier registration date would prevail.

The myriad new issues that are created by AI systems and robots as innovators and creators would have to be decided in this option uniformly for this new territory. The answer we give to these questions may be very different from those for an Earth context, especially if under our initial scenario the colony is populated entirely or mainly by machines, and the information exchange is mainly between them.

Why would we even consider AIs or robots as authors or inventors for IP purposes, let alone do so in space, where the effects of their actions are limited to other machines? IP, as we noted, tries to create incentives, and allows humans to be creative whilst meeting their need of food, shelter, and the appreciation of their peers. Surely,

AI systems are not driven by these desires, the incentive rationale fails, and with that the need for an IP regime that as we argued comes at a cost to liberty too?

This question connects our debate to the recent discussion (or rather controversy) surrounding robot rights (on this discussion see e.g. Richardson 2016; Gunkel 2018; Bennett and Daly 2020; Birhane and van Dijk 2020).

In the IP sphere, creative or inventive AI is often seen as a threat to human creativity, a competition we can't win. This is the basis of the recent EU proposal for IP reform that would strictly limit protection for computer-generated works. For others, the very fact that machines are now able to compete with humans in this most humane of endeavours shows they have moved beyond mechanical devices and acquired a new quality that deserved protection (see in particular Cedillo-Lazcano 2020).

With that, the issue also becomes an important argument in an even wider debate surrounding the recognition of robots as legal persons. Surely, entities that can move our hearts through the works they create are deserving some type of protective legal status? IP thus becomes one battle line in the fight for robot rights, and with that a discussion that intersects even more directly with the theme of liberty that is at the heart of this book. For the detractors of robo-rights such as Birhane (Birhane and van Dijk 2020) they are part of the dystopia of Habou's *Metropolis*, a corporate ploy to increase control over the workforce and restrict their freedoms. In a similar vein, Bryson (2010), in an expression she now regrets even though she stands firmly by the underlying idea, argues that robots should and remain slaves. Rights are by the people and for the people only (Bryson, Diamantis, and Grant 2017).

By contrast, advocates of robot rights regularly see the debate an essential extension of a much longer 'quest for freedom' that has to overcome its anthropocentric shackles. The relational theory of robot rights as advocated by Gellers, Coeckelbergh, or Gunkel in particular takes the social meaning that human imbue in (some) machines as their starting point.(see eg. Gellers 2020; Gunkel 2018, Coeckelbergh 2010). Robots become possible rightsholders precisely because we relate to them in specific ways, give them meaning and values. From a virtue ethics perspective, someone who mistreats an anthropomorphic robot like Sophia, especially if it is capable of expressing (mimicking the expression of) pain or discomfort, is behaving in an 'unhealthy' way, psychologically and morally. Attributing rights to robots then also elevates humans, and allows us to form better, and healthier, relationships with our environment. In a society that recognizes slavery, nobody can, ultimately, be free. For Gellers in particular, the quest for robot rights then becomes the logical next step the historical trajectory of enhancing liberty by widening the scope of rights, to whatever 'other' our society may encounter—humans of a gender other than that of the rulers, a different skin pigmentation, or not humans at all, but animals, and now robots.

From a legal perspective, I remain sceptical of this approach. As long as humans are around, it is always possible to find a much easier, more conventional way to reach a desired legal outcome than treating robots as holders of rights. In the IP context, this means in particular that as long as robots produce works primarily for a human audience, we can shift the point of creation to the moment where a human selects, from all the outcomes that the machine generates, the one that meets their objectives most. This was also the answer we gave to the UK IPO consultation on that very question (Aplin et al. 2022).

However, this answer assumes that there are still humans around, and human interests are at least indirectly involved. In the scenario of our space colony populated by autonomous machines, even this last connection to a human mind gets severed, and with that also most of the legal and moral objections against robot right. But would they 'want', or need them? As they don't feel pain, have no desires or needs, what would their purpose be?

In his work on *Eternal Peace*, Immanuel Kant argued that even a 'community of devils' is capable to create a state under the rule of law, and not only that, the laws of reason would compel them to do so (Kant 1983, Vol 7, 223). In the *Anthropologie*, he further developed the analogy between the laws of nature, the laws of reason, and the laws of humans, by arguing that the state is in essence the '[m]achine-being, provided by reason' (Kant 1983, Vol 10, 686 my translation). This notion of the state as a gigantic mechanical machine, driven by reason, brings us back to *Metropolis* with which we began our chapter.

If reason alone enables even devils to form a state and, as Höffe and other Kantians argue (Höffe 1988; Pawlik 2006), not just enables but necessitates it, then a society of robots might be in need of a state-like constitution and the concept of rights.

One function of rights is to solve coordination problems. To say that driver A (or indeed, 'the car') has right of way over driver B simply means that they know what to do when arriving at a junction, and what sanctions follow if something happens. Robots on a celestial body will face numerous coordination problems they will have to solve without real-time human input, and a language of rights may enable this, just as it does with humans. In the scenario briefly described above, two robot collectives have a conflict regarding a piece of information which the first group acquired, the second one may want to use, but in doing this is benefitting from the other group's efforts. A language of rights could for instance create a reciprocal duty next time round when the roles are conversed.

Coordination and cooperation issues like this will be ubiquitous in a robot-populated space colony (Martinez and Leitner 2010). Deciding the choices that Option 3 generates in favour of a recognition of robots as holders of IP rights, as authors and inventors, could thus provide an experimental environment for the exploration of space liberty on a large scale. In this view, the best reason for robot rights would therefore be the opposite of what the relational theory envisages. As far away from human control as possible, and from any harm humans may suffer, robots will, so predicts Kant, develop a system of rights appropriate *for them*. Computer modelling in artificial society research confirms this prediction at least in part (Flentge, Polani, and Uthmann 2001; Kurahashi and Terano 2005).

Liberty then is for machines to develop their own system of rights—from the robots for the robots, and without necessarily tying them to a rigid human-constructed system of rights. Just as AI systems have, on a memorable occasion, already developed their own language to talk to each other and, in particular, to negotiate (Lewis et al. 2017), an appropriate rights regime could naturally emerge through the mere interaction of robots on their colony. The main contribution that the issue of space IP could thus bring for the question of space liberty might be an incentive to cut loose the ties that bind our machines to us, and be a first step recognizing them as suitable holders of (some) rights.

Acknowledgments

The work on this paper was supported by AH/S002782/1 "Creative Informatics" and EP/V026607/1:"Trustworthy Autonomous systems"

Notes

1. <https://www.youtube.com/watch?v=KaOC9danxNo> accessed 6 July 2022.

References

Abbott, R., Matulionyte, R., and Nolan, P. 'September. A Brief Analysis of DABUS, Artificial Intelligence, and the Future of Patent Law' (2021) 125 *Intellectual Property Forum* 10–16.

Aplin, T. F. and Schafer, B. and Li, P. 'Response to UK IPO Open Consultation on AI and IP: Copyright and Patents' January 7, (2022). Available at SSRN: https://ssrn.com/abstract=4003626 or http://dx.doi.org/10.2139/ssrn.4003626.

Becker, K. and Gottschlich, J. 'AI Programmer: Autonomously Creating Software Programs Using Genetic Algorithms' (2021) July *Proceedings of the Genetic and Evolutionary Computation Conference Companion* 1513–21).

Bennett, B. and Daly, A. 'Recognising Rights for Robots: Can We? Will We? Should We?' (2020) 12(1) *Law, Innovation and Technology* 60–80.

Birhane, A. and van Dijk, J. 'Robot rights? Let's talk about human welfare instead.' Proceedings of the AAAI/ACM Conference on AI, Ethics, and Society. New York (2020) https://doi.org/10.1145/3375627.3375855

Boden M. A. 'Creativity and artificial intelligence.' (1998) 103 *Artificial Intelligence.* 347–356.

Borovik, B. 'Zombie (Trademarks) From Outer Space' (2019) 24 *Intellectual Property and Technology Law Journal* 323.

Brown, A. E. L. *Intellectual Property, Human Rights and Competition: Access to Essential Innovation and Technology* (Cheltenham: Edward Elgar Publishing, 2012).

Bryson, J. J. 'Robots Should Be Slaves' (2010) 8 *Close Engagements with Artificial Companions: Key Social, Psychological, Ethical and Design Issues* 63–74.

Bryson, J. J., Diamantis, M. E. and Grant, T. D. 'Of, For, and By the People: The Legal Lacuna of Synthetic Persons' (2017) 25(3) *Artificial Intelligence and Law* 273–91.

Calboli, I. and Farley, C. H. 'The Trademark Provisions in the TRIPS Agreement' in Carlos M. Correa ed. *Intellectual Property and International Trade: TRIPS Agreement,* (Wolters Kluwer, 2016) 159–60.

Cedillo-Lazcano, I. AI© R. (2020) 34(2) *International Review of Law, Computers & Technology* 201–13.

Chengzhi, L. 'Agronomy in Space—China's Crop Breeding Program' (2011) 27(3) *Space Policy* 157–64.

Coeckelbergh M. 'Robot rights? Towards a social-relational justification of moral consideration.' (2010)12(3) *Ethics and information technology* 209–221.

Colton, S., De Mántaras, R. L., and Stock, O. 'Computational Creativity: Coming of Age' (2009) 30(3) *AI Magazine* 11.

Colton, S., Halskov, J., Ventura, D., et al. 'The Painting Fool Sees! New Projects with the Automated Painter' (2015) Proceedings of the Sixth International Conference on Computational Creativity, Park City 189–196.

Cromer, J. D. 'How on Earth Terrestrial Laws Can Protect Geospatial Data' (2006) 32 *Journal of Space Law* 253.

de la Durantaye, K., Golla, S. J., and Kuschel, L. '"Space Oddities": Copyright Law and Conflict of Laws in Outer Space' (2014) 9(6) *Journal of Intellectual Property Law & Practice* 521–30.

Doldirina C. 'IP Rights in the Context of Space Activities' in F. G. von der Dunk. and Tronchetti, F. (eds), *Handbook of Space Law* (Cheltenham: Edward Elgar Publishing, 2015) 953.

Drahos, P. and Frankel, S. (eds) 'Indigenous Peoples' Innovation and Intellectual Property: The Issues' in *Indigenous Peoples' Innovation: Intellectual Property Pathways to Development* (ANUU Press Canberra, 2012) 1–28.

Esposito, M., Conticello, S. S., Pastena, M., et al. 'In-orbit Demonstration of Artificial Intelligence Applied to Hyperspectral and Thermal Sensing From Space' (2019) 11131 *CubeSats and SmallSats for Remote Sensing III* 111310C.

Feinman, M. 'Mining the Final Frontier: Keeping Earth's Asteroid Mining Ventures from Becoming the Next Gold Rush' (2013) 14 *Pittsburgh Journal of Technology Law and Policy* 202.

Flentge, F., Polani, D., and Uthmann, T. 'Modelling the Emergence of Possession Norms Using Memes' (2001) *Journal of Artificial Societies and Social Simulation.*

Fraser, E. 'Computers as Inventors—Legal and Policy Implications of Artificial Intelligence on Patent Law' (2016) 13 *SCRIPTed* 305.

Freund, R. 'Frau im Mond' in G. Dahlke and G. Karl (eds), *Deutsche Spielfilme von den Anfängen bis 1933. Ein Filmführer.* 2. Auflage (Berlin: Henschel-Verlag, 1993) 193–195.

Gellers, Joshua C. *Rights for Robots: Artificial Intelligence, Animal and Environmental Law* (Milton Park: Routledge, 2020)

Grimmelmann, J. 'There's No Such Thing as a Computer-Authored Work—And It's a Good Thing, Too' (2015) 39 *Columbia Journal of Law & Arts* 403.

Grynberg, M. 'AI and the "Death of Trademark"' (2019) 108(2) *Kentucky Law Journal* 199–238, 215.

Guadamuz, A. 'Do Androids Dream of Electric Copyright? Comparative Analysis of Originality in Artificial Intelligence Generated Works' (2021) *Intellectual Property Quarterly* 147–176.

Gunkel, D. J. *Robot Rights* (Cambridge: MIT Press, 2018).

Hao K. 'A college kid's fake, AI-generated blog fooled tens of thousands. This is how he made it.' *Technology Review* (2020) https://www.technologyreview.com/2020/08/14/1006780/aigpt-3-fake-blog-reached-top-of-hacker-news. Accessed 25.08.2022

Höffe, O. *Der Staat braucht selbst ein Volk von Teufeln* (Ditzingen: Reclam 1988)

Hornby, G., Globus, A., Linden, D., et al. 'Automated Antenna Design with Evolutionary Algorithms' (2006) *Proceedings of Space 2006, San Jose* 7242–7250

Jenks, C. W. *Space Law* (London: Stevens, 1965).

Jordanous, A. 'Has Computational Creativity Successfully Made It "Beyond the Fence" in Musical Theatre?' (2017) 29(4) *Connection Science* 350–86.

Kant, I. *Werke in Zehn Bänden*, edited by Wilhelm Weischedel. (Darmstadt: Wissenschaftlichen Buchgesellschaft 1983).

Kinsella, S. 'Intellectual Property and Libertarianism' (2009) 17 November *Mises Daily*. https://mises.org/library/intellectual-property-and-libertarianism.

Koch, J. S. 'Institutional Framework for the Province of all Mankind: Lessons from the International Seabed Authority for the Governance of Commercial Space Mining' (2018) 16(1) *Astropolitics* 1–27.

Koo, B., Nottenburg, C., and Pardey, P. G. 'Plants and Intellectual Property: An International Appraisal' (2004) 206(5700) *Science* 1295–7.

Koza, J. R. 'Human-competitive Results Produced by Genetic Programming' (2010) 11 (3–4) *Genetic Programming and Evolvable Machines* 251–84.

Koza, J. R. *Genetic Programming: On the Programming of Computers by Means of Natural Selection*, Vol. 1 (Cambridge, MA: MIT Press, 1992).

Kurahashi, S. and Terano, T. 'Analyzing Norm Emergence in Communal Sharing via Agent-Based Simulation' (2005) 36(6) *Systems and Computers in Japan* 102–12.

Lan, K. and Sekiyama, K. 'Autonomous Viewpoint Selection of Robots Based on Aesthetic Composition Evaluation of a Photo' in *2015 IEEE Symposium Series on Computational Intelligence* (IEEE, 2015) 295–300.

Launius, R. D., McCurdy, H. E. *Robots in Space: Technology, Evolution, and Interplanetary Travel* (Baltimore, MD: JHU Press, 2008).

Leal Martinez, D. and Leitner, J. 'On Cooperation in a Multi Robot Society for Space Exploration' in *Proceedings of the 10th International Symposium on Artificial Intelligence, Robotics and Automation in Space, i-SAIRAS 2010* (European Space Agency, Sappuro 29.08.2010) 834–9.

Lee, J-A. 'Computer-generated Works under the CDPA 1988' in J-A. Lee, R. Hilty, and K-C. Liu (eds). *Artificial Intelligence and Intellectual Property* (Oxford: Oxford University Press, 2021) 177–95.

Leepuengtham, T. *The Protection of Intellectual Property Rights in Outer Space Activities* (Cheltenham: Edward Elgar Publishing, 2017).

Lewis, M. et al. 'Deal or no deal? Training AI bots to negotiate' (2017) https://engineering.fb.com/2017/06/14/ml-applications/deal-or-no-deal-training-ai-bots-to-negotiate/ accessed 25.08.22

Li, W. and Hsu, C. Y. 'Automated Terrain Feature Identification from Remote Sensing Imagery: A Deep Learning Approach' (2020) 34(4) *International Journal of Geographical Information Science* 637–60.

Llewelyn, M. and Adcock, M. *European Plant Intellectual Property* (London: Bloomsbury Publishing, 2006).

Lobato, R. 'Constructing the Pirate Audience: On Popular Copyright Critique, Free Culture and Cyber-Libertarianism' (2011) 139(1) *Media International Australia* 113–23.

Lockridge, L. A. W. 'Intellectual Property in Outer Space: International Law, National Jurisdiction, and Exclusive Rights in Geospatial Data and Databases' (2006) 32 *Journal of Space Law* 319.

Lopez-Rincon, O., Starostenko, O., and Ayala-San Martín, G. 'Algorithmic Music Composition Based on Artificial Intelligence: A Survey' in *2018 International Conference on Electronics, Communications and Computers (CONIELECOMP)* (IEEE, Cholula, 21–23 February 2018) 187–93.

Malagar, L. B. and Magdoza-Malagar, M. A. 'International Law of Outer Space and the Protection of Intellectual Property Rights' (1999) 17 *Boston University International Law Journal* 311.

Malgieri, G. '"Ownership" of Customer (Big) Data in the European Union: Quasi-Property as Comparative Solution?' (2016) 20(5) *Journal of Internet Law* 2–17.

McElroy, W. 'Contra Copyright, Again' (2011) 3 *Libertarian Papers* 1.

McKenna, M. P. 'A Consumer Decision-making Theory of Trademark Law' (2012) 98 *Virginia Law Review* 67.

McLeod, K. *Freedom of Expression®: Overzealous Copyright Bozos and Other Enemies of Creativity* (London: Doubleday, 2005).

Menick, J. 'Move 37: Artificial Intelligence, Randomness, and Creativity' (2016) 16 *Mousse Magazine*. 2020.

Miles, C. 'Assessing the Need for an International Patent Regime for Inventions in Outer Space' (2008) 11 *Tulane Journal of Technology and Intellectual Property* 59.

Mohanta, T. K., Mishra, A. K., Mohanta, Y. K., et al. 'Space Breeding: The Next-Generation Crops' (2021) 12 *Frontiers in Plant Science* 771985.

Oliveira, H.G., Mendes, T., Boavida, A., et al. 'Co-PoeTryMe: Interactive Poetry Generation' (2019) 54 *Cognitive Systems Research* 199–216.

Parviainen, J. and Coeckelbergh, M. 'The Political Choreography of the Sophia Robot: Beyond Robot Rights and Citizenship to Political Performances for the Social Robotics Market' (2021) 36(3) *AI & Society* 715–24.

Pawlik, M. 'Kants Volk von Teufeln und sein Staat' (2006) 14 *Jahrbuch fuer Recht und Ethik* 269.

Pearson, E. 'Space Mining: The New Goldrush' *ScienceFocus* (2018) <https://www.sciencefocus.com/space/space-mining-the-new-goldrush/> accessed 6 July 2022.

Pelton, J. N. *The New Gold Rush: The Riches of Space Beckon!* (Berlin: Springer, 2016).

Ravid S.Y., Liu X. 'When artificial intelligence systems produce inventions: An alternative model for patent law at the 3a era.' (2017) 39 *Cardozo Law Review* 2215

Richardson, K. 'Sex Robot Matters: Slavery, the Prostituted, and the Rights of Machines' (2016) 35(2) *IEEE Technology and Society Magazine* 46–53.

Samuelson, P. 'Copyright and Freedom of Expression in Historical Perspective' (2002) 10 *Journal of Intellectual Property Law* 319

Schafer, B. and Fraser, E. 'Self-Made (Machine) Men: IP Implications of Inventions by Robots' (2017) *Jusletter IT*. https://jusletter-it.weblaw.ch/issues/2017/23-November-2017/self-made-(machine)-_e60f32e0bf.html__ONCE&login=false.

Schafer, B., Komuves, D., Zatarain, J. M. N., et al. 'A Fourth Law of Robotics? Copyright and the Law and Ethics of Machine Co-production' (2015) 23(3) *Artificial Intelligence and Law* 217–40.

Schechter, F. 'Rational Basis for Trademark Protection' (1927) 40(6) *Harvard Law Review* 813–83.

Sherman, B., Bently, L. *The Making of Modern Intellectual Property Law*, Vol. 1 (Cambridge: Cambridge University Press, 1999).

Smith, A. and Plattard, S. 'A Roadmap for the Robotic Facilitation of Off-World Living' (2020) 73 *Journal of the British Interplanetary Society* 97–102.

Sterckx, S. 'Patents and Access to Drugs in Developing Countries: An Ethical Analysis' (2004) 4(1) *Developing World Bioethics* 58–75.

Sterling, J. A. 'Space Copyright Law: The New Dimension A Preliminary Survey and Proposals' (2008) 45 *Journal of the Copyright Society of the USA* 345.

Tronchetti, F. 'IV–Space Resource Exploration and Utilization of the US Commercial Space Launch Competitiveness Act: A Legal and Political Assessment' 2016 41(2) *Air and Space Law* 143–56.

Turing, A. 'Computing Machinery and Intelligence' (1950) 59(236) *Mind* 433.

Vertesi, J. *Seeing Like a Rover: How Robots, Teams, and Images Craft Knowledge of Mars* (Chicago, IL: University of Chicago Press, 2015).

Wagner, R. P. 'Information Wants to Be Free: Intellectual Property and the Mythologies of Control' (2003) 103(5) *Columbia Law Review* 995–1034.

Winkler, G. 'Raum und Recht-Dogmatische und theoretische Perspektiven eines empirisch-rationalen Rechtsdenkens' in *Forschungen aus Staat und Recht*. Wien-(New York: Springer, 1999) 120.

WIPO. 'Intellectual Property and Space Activities Issue paper prepared by the International Bureau of WIPO' (2014) <https://www.wipo.int/export/sites/www/patent-law/en/developments/pdf/ip_space.pdf> accessed 6 July 2022.

23

Justice in space

Demanding political philosophy for demanding environments

James S. J. Schwartz

1. Introduction

> ... it is fraudulent to comfort people with promises of liberties that they cannot actually enjoy because necessary constituents of the enjoyment, like protection from physical safety, are lacking. It is fraudulent, in other words, to promise liberties in the absence of security, subsistence, and any other basic rights.
>
> —Henry Shue, *Basic Rights*

According to an intuitive application of 'ought-implies-can' reasoning, societies which exist in conditions that do not enable the protection of the basic rights of their citizens (because of extreme resource scarcity, for instance) do not thereby have an obligation to protect the basic rights of their citizens. Whilst it might be morally regrettable that basic rights go unprotected in these societies, this does not constitute a miscarriage of justice, given that justice cannot demand that anyone perform duties which are impossible to conduct. For ease of expression, let us call such societies 'burdened societies', and let us call the environmental conditions which give rise to burdened societies 'demanding environments'.

Extracting from this, we might say that, generally, in burdened societies—in societies that exist in very demanding environments—citizens should not expect social guarantees protecting the exercise of their basic rights. Rather, we should approach burdened societies with what I will call a *lenient* political philosophy. A *lenient* political philosophy would concede that burdened societies, due to the (possibly permanent) very demanding nature of the environments in which they exist, bear little to no obligation to ensure that citizens' basic rights are protected. Whilst this need not foreclose the pursuit of progressive *reforms* within burdened societies (e.g. a lenient political philosophy still might direct burdened states to, where possible, develop their capacities to protect citizens' basic rights), it nevertheless allows that extreme vulnerability is to be expected as an unavoidable (even if regrettable) fact of life for citizens in burdened societies. In other words, justice is only possible to a lesser degree in demanding environments.

Space societies will exist in environments that are more demanding than any environment housing terrestrial societies. If the reasoning of the preceding paragraphs is correct, then it would appear to follow that space societies will also bear little to no

James S. J. Schwartz, *Justice in space*. In: *The Institutions of Extraterrestrial Liberty*. Edited by Charles S. Cockell, Oxford University Press. © James S. J. Schwartz (2022). DOI: 10.1093/oso/9780192897985.003.0024 obligation to protect the exercise of citizens' basic rights. Moreover, since space environments are comparatively more demanding than terrestrial environments, space societies have correspondingly *weaker* duties to protect basic rights compared to any terrestrial society. Therefore, political philosophy for space polities should be *lenient* as well. We should not expect much from space societies, given that they will experience unique struggles simply to come into and maintain their existence. If citizens of space societies suffer and die routinely from starvation, asphyxiation, poverty, radiation exposure, or decompression, this would be regrettable but it would, unfortunately, be the best that can be expected in the harsh, unforgiving environments of space.

I believe the conclusions of the previous paragraph should be enthusiastically rejected. What is more, I maintain that the exact opposite conclusions are warranted: that political philosophy for space polities should be *demanding* rather than *lenient*. We should expect *more* from space societies than we generally expect from burdened terrestrial states.

The crux of my argument, which is presented in section 2, is that from the demanding nature of the potentially habitable environments in space, we *cannot* infer that any society which existed in those environments would be *burdened* in the relevant sense. The degree to which a space society might become burdened will depend on many factors, including especially the outcomes of decisions about how well to provision and support its founding. We can reduce the chances of burdened space societies coming into existence in the first place in part by setting a *demanding* standard. In this chapter, I shall argue that one necessary (but probably not sufficient) component of this standard is the provision of guaranteed life support. We should not view it as morally permissible to found space societies that are not capable of guaranteeing that their citizens are able to access to adequate life support.

My argument is broadly egalitarian in nature and borrows from Henry Shue's defence of subsistence rights (Shue 1996). Thus, I shall assume (rather than argue for) the existence of 'positive' rights (rights to the provision of goods or services) in addition to the more routinely recognized 'negative' rights (rights to non-interference).[1] Shue argues that we cannot claim that basic rights have been upheld if there are not also protections against 'standard threats' (i.e. predictable and remediable environmental and social barriers) to the enjoyment of these rights, and Shue includes threats to subsistence amongst the standard threats faced by terrestrial states.[2] Thus, rights-protecting societies have obligations to provide reasonable guarantees that citizens are protected against standard threats. I shall argue that the same principle applies to space societies, and that the demanding nature of the space environment (in that it poses some especially extreme standard threats) does not mitigate the obligation to protect citizens against standard threats.

The very demanding nature of what I am proposing is likely to invite the objection that I am engaged in some kind of practically unmoored, idealistic, or utopian speculation about what is owed to citizens of space societies. Might the *demanding* approach to political philosophy be guilty of the same sins that various ideal theories in political philosophy have been accused of committing—for example, of failing to be action guiding, or failing to specify feasibly implementable principles, or failing to acknowledge the regrettable shortcomings of human motivation? In section 3 I argue that the intuitive force of these kinds of objections is defused in the case of space societies. Whilst these considerations might constitute compelling objections to what can be hoped for the *reformation* of already-existing burdened societies, they do not apply to the case of the *founding* of new space societies, because it is within our power to refrain from instigating space settlements that are incapable of meeting demanding requirements. The existence of burdened space societies is not an assumption of this discussion. Rather, it a reprehensible possible future state of affairs that we should aim to avoid. Nevertheless, I concede to the critic that pessimism about human motivation (in particular, the motivation required for building and maintaining just societal institutions in space) is a potentially significant source of concern; however, I shall argue that this does not present a compelling obstacle to the pursuit of demanding political standards for space societies.

2. The Demanding Space Environment

Space environments are demanding in the sense that they are instantaneously lethal to humans and moreover few contain all the necessary ingredients, and in adequate concentrations, for producing or accessing the most basic of human necessities: food, water, and air (Cockell 2015). In these respects, space environments are more demanding than any environment on Earth, including the Antarctic and the deep ocean. From a basic life support perspective, then, space environments are extremely impoverished and insecure places to inhabit. Life support insecurity, radiation poisoning, and depressurization will be amongst the standard threats that must be confronted in space. Let me group all of these concerns together under the heading of 'life support security/insecurity'.[3] Since threats to life support security are threats to basic human survival, they are thereby threats to the possibility of enjoying any kind of human life at all in space.

In such a situation, it might be thought that we should reduce our expectations regarding justice for persons living in space societies: the demanding nature of the space environment means there can be no guarantees of life support security. Whilst this might be morally regrettable, nothing better can be expected for those living in space. Thus, it cannot be sensibly maintained that citizens of space societies have *rights or entitlements* to adequate life support or protection against radiation and, correspondingly, it cannot be sensibly maintained that the societal institutions of space societies have duties or obligations to guarantee secure access to life support.[4] Again, whilst this might be morally regrettable, nothing better can be expected for or demanded by those living in space.

The implications of this position are severe. Since the exercise of *any basic human rights at all* depends on access to adequate life support (at the very least!), the above argument is destructive of the very idea of rights for citizens of space societies. However, if we follow Henry Shue's (1996) account of basic rights, then we have a path for resisting this pessimistic conclusion.

As Shue argues, an important motivation for identifying and circumscribing human rights is 'to prevent, or to eliminate, in so far as possible the degree of vulnerability that leaves people at the mercy of others' (Shue 1996, 30). On this account, what is needed, in order for an individual to enjoy a right, is that this individual also enjoys 'protection against the threats that could ordinarily be expected to prevent, or hinder to a major degree, the enjoyment of the initial right' (Shue 1996, 32). On the part of those who share an obligation to protect this individual's rights, this involves providing 'social guarantees ... that would prevent the most common and serious threats from preventing or acutely hindering the enjoyment of the substance of the right', and this, in turn, 'places especially heavy emphasis upon preventing standard threats' (Shue 1996, 32).

Shue characterizes standard threats as 'ordinary and serious but remediable threats' (Shue 1996, 32); in addition, standard threats must be predictable (Shue 1996, 33).[5] Thus, Shue is not arguing for the protection of rights against all conceivable threats but only against standard threats that can be anticipated and defended against. Moreover, Shue is comfortable acknowledging that which threats count as 'standard' varies across contexts—indeed, this is a feature of his position, rather than a bug:

> What is, for example, eradicable changes, of course, over time. Today, we have very little excuse for allowing so many poor people to die of malaria and more excuse probably for allowing people to die of cancer. Later perhaps we will have equally little excuse to allow deaths by many kinds of cancer, or perhaps not. In any case, the measure is a realistic, not a utopian, one, and what is realistic can change.
>
> (Shue 1996, 33)

The question before us, then, is whether threats to life support security present as standard threats—as ordinary, predictable, serious, and remediable threats—for space societies. If so, then we can leverage Shue's position to posit the existence of a right to life support security (since no other rights could be exercised in the absence of life support). In space, after all, a right to subsistence would imply a right to life support security, since life support is constitutive of subsistence. More generally, and extracting from this, we could argue that the demanding nature of the space environment implies that in space there usually will be more standard threats to confront, meaning more expansive social guarantees will usually be needed in order to protect even the most basic rights against these threats. In other words, we need correspondingly demanding political standards for polities in demanding space environments.

Nevertheless, a critic might offer that space societies are, in general, unlikely to be in positions where it is possible to guarantee life support security. The demanding nature of space environments is the very source of this threat, and it hardly seems remediable, much less eradicable. Indeed, the critic might continue, space societies are likely to constitute *extreme* examples of what John Rawls (1999) calls *burdened* societies, those which 'lack political and cultural traditions, the human capital and know-how, and often, the material and technological resources needed to be well-ordered' (Rawls 1990, 90).

Of particular salience to life support security is the availability of breathable air and water, which in space must be manufactured from *in situ* resources (presuming the necessary resource inputs are even available *in situ*). Thus, scarcity of life support inputs might be a fixture of space societies, depending on where they are located and on what means these societies employ to provide life support. Here a critic could point to what has been posited by Katherine Eddy, *viz.*, that rights do not exist when conditions make their protection too demanding. As Eddy argues, in situations in which 'resource deficits are long-standing and unalterable', and thus that 'the resultant scarcity [is] a feature of the normal case', we cannot coherently maintain that there exist rights to the access or enjoyment of these resources (Eddy 2008, 470). To demonstrate how this impacts the status of rights, Eddy asks the reader to consider two worlds—one in which HIV treatment is easily accessible and inexpensive, and one in which HIV treatment is scarce, very costly, and difficult to administer:

> I think it is fairly uncontroversial to assert that in the first world, the state would have a duty to provide the treatment to those in need; the relevant factors contributing to the existence of the duty are both the HIV-sufferer's hugely important interest in the drug (she might die without it), and the ease with which it could be provided. But the possibility (even the future possibility) of a situation like the first world is not enough to establish the existence of a duty if we happen to find ourselves in a world like the second. If our world is like the second world where the drug is scarce and expensive, then it is not the case that the HIV sufferer has the right to the drug. She does not have the right because her world is one where it remains too burdensome for the state to fulfill the duty.
>
> (Eddy 2008, 471)[6]

Continuing from there, the critic could construct a parallel argument to Eddy's, focusing on space societies, in which we are to imagine two space societies—one where life support security is cheap and accessible, and another where life support security is expensive and frequently unavailable. Here the critic would conclude that only in the first is society obliged to provide life support to citizens in need, owing to citizens' hugely important interest in life support (they would die without it), and in how easily it could be provided. In the second space society, where life support is scarce and expensive, there is no right to life support security, because it would be too burdensome for the society to fulfil the duty.[7]

The intuitive force of this criticism involves a critical element that is present in the case of terrestrial societies but not (as yet) present in the case of space societies (of which there are none): terrestrial societies are not free to choose their starting points in time, their locations, their cultural makeups, their technological capacities, etc. Rather, societies' developmental pathways are inextricably tied together with their histories and their relationships with other societies, as well as with their physical environments. These histories and relationships affect, to varying degrees, how feasible it is for societies to protect the basic rights of their citizens and to provide them with essential services. Societies do not become burdened by choice; no justice-loving individuals would seek to *create* burdened societies if there were less-burdensome alternatives available. For much the same reason, burdened societies cannot become unburdened simply by fiat; demanding, unfavourable conditions cannot be wished away. Likewise, societal institutions capable of protecting against standard threats to the exercise of basic rights cannot be wished into existence.

However, it is well within our power to determine the starting point in time, location, cultural makeup, and technological capacities of early space societies. It is well within our power to research and learn what must be researched and learned in order to create space societies that *begin life* protecting against all of the standard threats that will be faced wherever in space these societies are founded. After all, these societies' access to resources, and thus their capacities to respond to threats—standard and non-standard alike—largely will be determined by *our* prior decisions about where and when to found space societies, on what resource supply chains we have pursued during their construction, and so forth. The ability of space societies to do something as fundamental to survival as reconnoitre for life support resources could be deeply constrained or greatly enhanced based on how supportive of a start the founders provided.

Thus, founders' decisions will have a significant impact on the developmental pathways open to space societies. We have a responsibility to avoid initiating space settlement when it would force the resulting society onto developmental pathways that we know are likely to generate conditions of life support insecurity. As Shue argues, 'it is unreasonable to arrange human institutions in a way that threatens vital interests when they can be arranged in a way that does not threaten vital interests' (Shue 1996, 127). And if the only options seemingly available to us involve creating burdened space societies, we should at that point reconsider the wisdom of founding any space societies at all, at least until our epistemic situation improves.[8] It would be an *expansive* moral failure on our part if we knowingly and avoidably created burdened societies in space.

There will not exist any burdened societies in space unless we deliberately set out to create them, or unless we create them inadvertently by engaging in negligent, over hasty, or premature space settlement efforts (whether publicly funded or privately funded). Whether a space society begins life with guaranteed life support security is, I submit, a minimal criterion for evaluating our 'readiness' to initiate space settlement. If we do not know how to do this, we simply do not know enough to pursue the creation of societies in space.

It is easy to see why such a fundamental consideration could be so easily and frequently overlooked, especially in the context of theorizing about justice. Typically, our engagement with politics and political philosophy takes place in *reform* contexts, in other words in (or in support of) attempts to *respond* to already-existing injustices in already existing polities. Because Earth is so extensively settled there is seldom need to consider the question of what should be said about the *creation* of new societies in wholly unsettled and unclaimed territories. But what space ethics and political philosophy require in the case of space settlement are not values, principles, theories, and institutions of *reform* but rather values, principles, theories, and institutions of *founding*. At the foundational stage, with very few developmental pathways closed off, feasibility principles such as 'ought implies can' cannot issue much guidance about what is and is not feasible for space societies. For the same reasons, our terrestrially moderated intuitions about what is and is not politically feasible cannot be regarded as authoritative. I submit, then, that instead of calling for a *lenient* approach to political philosophy, these foundational situations call for *demanding* standards.

Suppose, for instance, that a previously uninhabited territory was discovered on Earth, one that fell beyond the claim of any existing state, and one moreover that was mostly salt flats and not easily inhabited by humans. Suppose that, nevertheless, the international community came together with the goal of creating a new society in this territory (perhaps it is uniquely insulated from the effects of climate change), despite the serious and endemic famine risk. Suppose further that the international community decided in the process to ignore all available evidence about the impact of famine conditions on human communities, and dedicated no resources to ensuring that those living in the territory would have reliable access to food and water. We would rightly claim that this would be a morally reprehensible method of founding a new society, and that it would be a method that we should not pursue in the first place, especially when much better alternatives are open to us (or at least when so little work has been done ruling out better alternatives). It is not obvious why space settlement proposals should warrant a different judgment.[9]

The principle I am driving at here is that the founders of *new* societies have an obligation to utilize state-of-the-art knowledge from the social and natural sciences as well as ethics and political philosophy. This obligation cannot be satisfied by token appeals to speculations from the seventeenth and eighteenth centuries about the content of human nature (or its purported fixity). It cannot be satisfied by superstitious and historically dubious appeals to notions of manifest destiny or frontierism. What is known about the conditions that are generative of poverty, starvation, discrimination, extreme inequality, and so forth—especially what has been revealed by the last several decades of academic research into these topics—cannot permissibly be ignored in the contemplation and construction of new polities. We should demand more from new societies precisely because, in general, they will be founded in circumstances of greater epistemic (and often material and technological) privilege compared to their progenitors and predecessors. This holds whether we are considering some entirely new terrestrial society (*per impossibile*) or a new space settlement. Founders would not be morally 'off the hook' for the extensive human suffering that would be likely to arise in any space societies they created that did not guarantee life support security. They would be clearly and identifiably responsible (at least to some positive and blameworthy degree) instead.[10]

3. Epicycles of Ideal Theory

Any entreaty that political or social institutions should meet demanding standards is liable to be met with the reply that this is an unrealistic and/or utopian demand without any practical import. Thus, whilst discussions about demanding standards of justice might constitute an interesting intellectual exercise, they are of scant relevance to political philosophy (which should confine itself to discussions of feasibly implementable principles and recommendations). Arguably, the preceding discussion constitutes an exercise of the oft-reviled 'ideal theory' approach to theories of justice, in so far as one of the many ways of understanding the ideal/non-ideal divide is that 'the debate on ideal and non-ideal theory focuses on the question of whether feasibility considerations should constrain normative political theorizing and, if so, what sorts of feasibility constraints should matter' (Valentini 2012, 654). I have, after all, just offered reasons for holding that feasibility constraints of a certain variety (the demanding nature of the space environment) ought not to constrain normative political theorizing about space societies, at least when it comes to protections against standard threats to life support. Does that leave my view open to the same purported deficiencies as other utopian/ideal theories in political philosophy?

In one sense this kind of objection is entirely beside the point. There are no already existing space societal institutions that provide a real-world vantage point from which to judge new proposals as realistic or utopian (especially at a time when the idea of a space society in any form seems an implausible fantasy in its own right). Since there are no space societies in need (or not) of reform, there are no fixed circumstances constraining the possibilities of reform for space societies. Space societal institutions which fail to guarantee life support are, at the time of writing, just as real, and just as fantastically utopian as space societal institutions which guarantee life support.

However, there is an element of the realist/non-ideal critique of ideal theory that appears to apply non-vacuously to theorizing about justice for as-yet-non-existent space societies, and these would come from *motivational* critiques of ideal/utopian theories. It is unrealistic to expect humans to be (or become) motivated to implement the recommendations of utopian theories, especially very demanding ones. The insinuation is that such theorizing is thereby pointless since it sits at such a great distance from real-world applicability.

In application to space settlement, the claim would be that it is foolishly optimistic to think that humanity could be motivated to create a space society capable of providing guaranteed life support and, therefore, there is no good reason to engage in theorizing about space when it issues this kind of ideal or utopian requirement. The empirical claim about human motivation that opens this paragraph is sufficiently dubious that it should not be treated as a background assumption, but for the sake of argument I am willing to grant to critics that it is true in the following way: that, *at the time of writing*, it is foolishly optimistic to think that *those parties working hardest to bring space settlements about* (space billionaires and space settlement advocacy groups) could be motivated to create a space society capable of providing guaranteed life support.

Note that this latter claim is fully compatible with it being *true* that it is *perfectly reasonable* to think that *humanity* could be motivated to create a space society capable of providing guaranteed life support. And even if it is not true that humanity is currently so motivated, human motivation is subject to change over time (as are the motivations of space billionaires and space advocates). This suggests a response to both kinds of concerns (pessimism about the motivations of the particular humans and groups most likely to settle space in the short term, and pessimism about the motivations of humans generally). Because human motivation is subject to change, so too is human motivation related to space settlement. As Valentini appreciates, 'non-conclusively-justified pessimism about human nature' should no more guide our political theorizing than non-conclusively justified optimism about human nature (Valentini 2017, 25). Ideal/utopian theorizing about justice, in turn, plays a conceptually foundational role in inspiring changes in the ends of human motivation, or at least in helping to give direction to ways human motivation can be bent towards greater justice over the long term. For instance, Pablo Gilabert has argued that '[a]mbitious political pictures can inspire political action, setting long-term agendas for dramatic improvements of social life' (Gilabert 2012, 48), a feature also emphasized by (Erman and Möller 2013) and (Estlund 2020). Theorizing about demanding standards of justice for space societies could, then, contribute to shifting human motivation towards endorsing and applying these standards.[11]

Thus, even if constraints on human motivation provide compelling reasons for rejecting terrestrially oriented ideal theories, this would not imply that these constraints refute demanding standards for (as yet non-existent) space polities. Given that we are dealing with such a prenatal arena of political philosophy, it is methodologically counterproductive to insist on narrowly 'realistic' constraints on political theorizing about space societies.[12] Political philosophy for space societies, then, constitutes a subject that is ripe for the hypostasizing and theorizing about political ideals of many kinds, including ones that appear very demanding to contemporary political theorists.

4. Conclusion

I have argued that there are good reasons for requiring the founders of space societies to ensure that their societies begin life and continue on developmental pathways that guarantee life support security. In particular, I have argued that, because it is within our collective power to refrain from founding space societies until we have learned how to guarantee life support security, we cannot justifiably found new polities in space that do not (at least) guarantee life support security. Therefore, we ought to opt for *demanding* as opposed to *lenient* approaches to political philosophy for space societies.

I have not, however, said much about what demanding requirements would actually entail for the founders of a space society. That is, I have not offered any specific institutional prescriptions, other than the prescription that there exist institutions that effectively provide life support security. Such an omission is deliberate, because any details of implementation will vary dramatically based on the time and location of the founding of the settlement, and of the availability of enabling technologies, resources, and infrastructure.[13] It is worth pointing out that demanding requirements exist, nevertheless, because the more widely this is appreciated, the more likely it will be that, when the time comes, actual, adequate steps are taken to ensure that we do not found a space society that does not protect against standard threats to life support security.

We have no business *creating* societies where human beings will be forced to toil without promise of water or even breathing air. We can do better, and we must do better. How could we otherwise found stable and sustainable space societies?[14]

Notes

1. For arguments supporting the existence of positive rights, see Shue 1996 as well as Holmes and Sunstein 1999.
2. After all, one can no more exercise one's rights when one has starved to death than one can exercise one's rights when unjustly confined by hostile actors.
3. Below I will use expressions such as 'guaranteed access to life support' and 'guaranteed life support security' synonymously. Thus, nothing distinguishes the obligation to guarantee *access* to life support from the obligation to guarantee life support *security*; these obligations are one and the same.
4. Throughout I shall roam freely between talk of basic *rights* and basic *entitlements*; however, I do not aim to endorse any particular account of the nature of rights or entitlements. My aim is only to discuss the *content* of the basic rights or entitlements for space polities. For those who are less than enamoured with rights talk, feel free to imagine that my arguments concern the content of the duties or obligations of space polities and societal institutions.
5. See Reeves 2015 for a more detailed discussion of the 'standard threats' approach to rights.
6. Eddy's intended targets are ideal theories of political justice that invoke what she calls 'ideal rights', so the reader should be aware that I am extending her position somewhat beyond its primary scope of application.
7. In my response I will circumvent rather than refute this objection. However, an alternative response would be to adopt a conditional or transitional approach to duties and rights, e.g. Meyers 1981: Ashford 2008; Gilabert 2008; Fuller 2012. Such views admit of the possibility of rights violations in burdened societies, whilst prescribing as duties only those reforms which are realistically capable of helping this society reform into one that (eventually) is burdened no longer. This provides the basis for a similar response with respect to space societies. Though space societies might be forced to begin life as burdened societies, we can still offer certain progressive prescriptions, such as that these societies must prioritize the construction of institutions capable of protecting citizens' basic rights, even if the process of construction is tragically long and arduous. The perspective I defend below does not view the creation of a burdened space society as morally permissible in the first place.
8. This is not a vague entreaty that humans 'learn our lessons' on Earth before venturing into space, but, as I make clear above and below, an entreaty that humans support space settlement only if they are committed to identifying and addressing any space settlement knowledge gaps using all relevant state-of-the art knowledge and research methods.
9. It could be argued that, in both cases, individuals who are prepared to 'pay their own way' and to take on any risks involved should be permitted to do so, regardless of how ill-prepared their efforts appear to others. This argument neglects consideration of those born into space societies, who might find themselves living within inescapable and perilous circumstances beyond their power to change. No one can be justified in *creating* conditions of avoidable hardship and also in subjecting future generations to these conditions, *especially* in the absence of an exercisable right to emigrate (Schwartz 2021).
10. All I have technically argued for here is that the provision of life support security is a necessary condition on a minimally just space society. I leave it open for the time being whether more is required, though I have elsewhere argued for the need for space societies to protect further rights or entitlements as a matter of course; see Greenall-Sharp et al. 2021 and Schwartz 2018.
11. This suggests another form of reply, one that is less rooted in egalitarian political philosophy and more compatible with libertarian approaches: That it would be *to the benefit of*

space billionaires and space settlement advocates to support only the founding of those space societies which provide guaranteed life support. In other words, if these individuals *became* motivated to meet *demanding* requirements, then they would be more likely to succeed in their space settlement ambitions. After all, without life support, there can be no functioning space society. Without a functioning space society, none of the major space settlement goals of space billionaires and space settlement advocates can be sustainably upheld. There is no such thing as a 'backup planet' that does not provide its people with life support.

12. For a related argument supporting a similar 'open' methodological approach to space *ethics*, see (Schwartz 2016).
13. This is compatible with it still being always impermissible to attempt to found space societies in locations where it is simply *not possible* to guarantee life support.
14. I thank Jonathan Trerise for helpful discussion and comments.

References

Ashford, E. 'The Duties Imposed by the Human Right to Basic Necessities' in T. Pogge (ed.), *Freedom from Poverty as a Human Right: Who Owes What to the Very Poor?* (Oxford: Oxford University Press, 2008) 183–218.

Cockell, C. 'Freedom in a Box: Paradoxes in the Structure of Extraterrestrial Liberty' in C. Cockell (ed.), *The Meaning of Liberty Beyond Earth* (New York, NY: Springer, 2015) 47–68.

Eddy, K. 'Against Ideal Rights' (2008) 34 *Social Theory and Practice* 463–81.

Erman, E. and Möller, N. 'Three Failed Charges Against Ideal Theory' (2013) 39 *Social Theory and Practice* 19–44.

Estlund, D. *Utopophobia: On the Limits (if any) of Political Philosophy* (Princeton, NJ: Princeton University Press, 2020).

Fuller, L. 'Burdened Societies and Transitional Justice' (2012) 15 *Ethical Theory and Moral Practice* 369–86.

Gilabert, P. 'Global Justice and Poverty Relief in Nonideal Circumstances' (2008) 34 *Social Theory and Practice* 411–38.

Gilabert, P. 'Comparative Assessment of Justice, Political Feasibility, and Ideal Theory' (2012) 15 *Ethical Theory and Moral Practice* 39–56.

Greenall-Sharp, R., Kobza, D., Houston, C., et al. 'A Space Settler's Bill of Rights' in O. Chon Torres, T. Peters, J. Seckbach, et al. (eds), *Astrobiology: Science, Ethics, and Public Policy* (Beverly, MA: Wiley-Scrivener, 2021) 377–88.

Holmes, S. and Sunstein, C. *The Cost of Rights: Why Liberty Depends on Taxes* (New York, NY: W.W. Norton & Company, 1999).

Meyers, D. 'Human Rights in Pre-Affluent Societies' (1981) 31 *The Philosophical Quarterly* 139–44.

Rawls, J. *The Law of Peoples* (Cambridge, MA: Harvard University Press, 1999).

Reeves, A. 'Standard Threats: How to Violate Basic Human Rights' (2015) 41 *Social Theory and Practice* 403–34.

Schwartz, J. 'On the Methodology of Space Ethics' in J. Schwartz and T. Milligan (eds), *The Ethics of Space Exploration* (New York, NY: Springer 2016) 93–107.

Schwartz, J. 'Worldship Ethics: Obligations to the Crew' (2018) 71 *Journal of the British Interplanetary Society* 53–64.

Schwartz, J. 'A Right to Return to Earth? Emigration Policy for the Lunar State' in M. Rappaport and K. Szocik (eds), *The Human Factor in Settlement of the Moon* (New York, NY: Springer, 2021) 193–205.

Shue, H. *Basic Rights: Subsistence, Affluence, and U.S. Foreign Policy* (2nd edn, Princeton, NJ: Princeton University Press, 1996).

Valentini, L. 'Ideal vs. Non-ideal Theory: A Conceptual Map' (2012) 7(9) *Philosophy Compass* 654–64.

Valentini, L. 'On the Messy "Utopophobia vs. Factophobia" Controversey: A Systematization and Assessment' in M. Weber and K. Vallier (eds), *Political Utopias: Contemporary Debates* (Oxford: Oxford University Press, 2017) 11–33.

24

Extraterrestrial governance

Why the constitutions of planets should be grounded in the constitution of their inhabitants

Michael Shermer

> But what is government itself, but the greatest of all reflections on human nature? If men were angels, no government would be necessary. If angels were to govern men, neither external nor internal controls on government would be necessary. In framing a government which is to be administered by men over men, the great difficulty lies in this: you must first enable the government to control the governed; and in the next place oblige it to control itself.
>
> —James Madison, Federalist Paper No. 51

We are about to establish settlements on the Earth's moon and on Mars in the next few years or decades, and it's not even clear what food these astronaut explorers should take with them for optimal nutrition, much less what governing ideas and documents they should bring along for political sustenance.[1] Given that the twenty-first-century solar system settlers and—let's project optimistically, the twenty-third-century interstellar explorers—will have similar biological natures as our Paleolithic colonists had when they undertook their exploration and settlement of Earth, it would be prudent to start thinking about how best to govern an extraterrestrial civilization, starting with what we have learned already from the scientific study of governance on our planet. What have we discovered?

I started thinking about this issue when Elon Musk announced in September of 2017 his intention to set up a Martian colony before the end of the 2020s, explaining his motivation to a South by Southwest (SXSW) conference audience in March of 2018 thus: 'If there's a third world war we want to make sure there's enough of a seed of human civilization somewhere else to bring it back and shorten the length of the dark ages', adding that the endeavour will be 'difficult, dangerous, a good chance you'll die'.[2] Curious to know his thoughts on the subject of extraterrestrial governance, on 16 June 2018 I whimsically tweeted at the SpaceX CEO:

Michael Shermer, *Extraterrestrial governance*. In: *The Institutions of Extraterrestrial Liberty.*
Edited by Charles S. Cockell, Oxford University Press. © Michael Shermer (2022).
DOI: 10.1093/oso/9780192897985.003.0025

Michael Shermer ✓ @michaelshermer · Jun 16   ⌄

When you start the 1st Mars colony @elonmusk what documents would you recommend using to establish a governing system? U.S. Constitution/Bill of Rights? Universal Declaration of Human Rights? Humanist Manifesto? Atlas Shrugged? Against the State: An Anarcho-Capitalist Manifesto?

💬 158   ⟲ 277   ♡ 2.9K   ⅲ

Minutes later, I received this reply from Musk:

Elon Musk ✓ @elonmusk · Jun 16   ⌄

Replying to @michaelshermer

Direct democracy by the people. Laws must be short, as there is trickery in length. Automatic expiration of rules to prevent death by bureaucracy. Any rule can be removed by 40% of people to overcome inertia. Freedom.

💬 1.9K   ⟲ 2.5K   ♡ 18K   ✉

Given that Twitter may not be the ideal platform for fleshing out complex thoughts like establishing a new governing system on another planet, I thought Musk's additional comments at the SXSW conference—in response to a question about what type of government he envisions for the first Mars colony—were revealing:

> Most likely, the form of government on Mars would be somewhat of a direct democracy where people vote directly on issues instead of going through representative government. When the United States was formed representative government was the only thing that was logistically feasible. There was no way for people to communicate instantly. A lot of people didn't have access to mailboxes, the post office was primitive. A lot of people couldn't write. So you had to have some form of representative democracy or things just wouldn't work at all. On Mars, everyone votes on every issue and that's how it goes. There are a few things I'd recommend, which is keep laws short. Long laws ... that's like something suspicious going on if there's long laws.[3]

That certainly sounds reasonable. Who wouldn't prefer fewer laws and more freedom? Unfortunately, most people on both the political left and the political right, both parties of which have all reliably increased the size the government in their charge. And it is why libertarian sentiments, so eloquently expressed in the title of Matt Kibbe's libertarian manifesto, *Don't Hurt Other People and Don't Take Their Stuff*,[4] are embraced by so few people (the Libertarian party is lucky to ever get even 1 per cent of the vote in national elections). Even religious moral sentiments, such as the Golden Rule to 'do unto others as you would have them do unto you', along with its silver derivative 'don't do to others what you don't want done to yourself', have to be expanded upon with holy books, commandments, commentaries, gospels, torahs, psalms, scrolls, midrashes, and the like.

The reason is to be found in the nature of human interaction and conflict resolution, which grow exponentially as small colonies, bands, and tribes coalesce into larger chiefdoms, states, and empires. Musk's direct democracy could be effective with just a few dozen inhabitants, which is roughly the size of hunter-gatherer bands. But look what happens as population numbers increase through both fecundity and immigration—both of which are likely in an initial extraterrestrial founding colony—as worked out by the UCLA geographer Jared Diamond from his studies of the hunter-gatherer peoples of Papua New Guinea: a small band of 20 people generates 190 possible dyads, or two-person interactions ($20 \times 19 \div 2$), small enough for informal conflict resolution. But increase that 20 to 2,000 and you're facing 1,999,000 possible dyads ($2000 \times 1999 \div 2$). Here a 100-fold population increase produces a 10,000-fold dyadic rise. Scale that up to cities of 200,000 or 2,000,000 and the potential for conflict multiplies beyond comprehension, along with it laws and regulations needed to insure relative harmony and efficiency. As Diamond explains in his book, *Guns, Germs, and Steel*:

> Once the threshold of 'several hundred,' below which everyone can know everyone else, has been crossed, increasing numbers of dyads become pairs of unrelated strangers. Hence, a large society that continues to leave conflict resolution to all of its members is guaranteed to blow up. That factor alone would explain why societies of thousands can exist only if they develop centralized authority to monopolize force and resolve conflict.[5]

Let's apply this analysis to what Musk has in mind when he says that 'the threshold for a self-sustaining city on Mars or a civilization would be a million people'. That number generates nearly 500 billion dyadic combinatorial possibilities ($1,000,000 \times 999,999 \div 2 = 499,999,500,000$), meaning any hoped-for manual of 'short laws' would soon become volumes of bureaucratic regulations and laws, which as we know from history results in complex political and legal systems of suffocating regulations, entangling restrictions, government overreach, and suppression of individual freedom and autonomy. Extraterrestrial explorers and settlers will need to figure out how to prevent a bureaucracy from expanding in response to the accelerating dyadic combinations as their population increases, which has happened in every government on Earth as if it were a law of nature.

Thus, the general challenge for Martians is the same as it has been for Earthlings for millennia: to strike the right balance between freedom and security. It would be interesting for Martian colonists to experiment with and design new systems of governance never tried on Earth, but these Martians will be Earthlings with all the inner demons that come bundled with the better angels of our nature.

To many people, the idea of humans settling down on an extraterrestrial body sounds overly optimistic, highly improbable, and even hallucinatory. 'The idea of Elon Musk to have a million people settle on Mars is a dangerous delusion' said the Astronomer Royal Lord Martin Rees. 'Living on Mars is no better than living on the South Pole or the tip of Mount Everest.'[6] Indeed, but that is surely what it must have seemed to our Paleolithic ancestors who—barely eking out a living in the nooks and crannies of sub-Saharan Africa and Ice Age Europe in the evolutionary bottlenecks that nearly exterminated our species some tens of thousands of years ago[7]—nevertheless over those long gone millennia spread out across all of Earth's continents and, eventually, island-hopped their way across the vast Pacific Ocean, which must have seemed as daunting to those early explorers as interstellar space appears to us today.

Given our species' exploratory nature and its propensity to widen our horizons by branching out from our home territories to put down stakes and settle new lands, it is not completely crazy to imagine how an interstellar and even galactic civilization might develop. Founder populations on the moon and Mars could eventually spread themselves throughout the solar system, establishing communities on the moons of Jupiter and Saturn and, in the fullness of time and with new propulsion technologies (we're talking centuries, if not millennia), cross the vast distances of interstellar space and over millions of years star-hop their way around the galaxy.[8]

What might happen with our moral and political nature? Assuming that we are not going to genetically engineer greed, avarice, competitiveness, aggression, and violence out of our nature—given that these characteristics are essential to who we are as a species and all have an evolutionary logic to them—instead, as we branch out across the galaxy, there will not be one civilization but many, not one hominin species but multiple. Given the distances and timescales involved, there is likely to be many species of spacefaring hominins in which each colonized planet will act like a new 'founder' population from which a new species evolves, reproductively isolated from other such populations (the very definition of a species).[9] These civilizations will vary even more than nations on Earth varied before globalization, with dozens, hundreds, possibly even thousands of different civilizations in which sentient beings may flourish—a vast array of peaks on the galactic political landscape.

If this were to happen, sentience itself would become immortal, inasmuch as there is no known mechanism—short of the end of the universe itself trillions of years from now in our accelerating, expanding cosmos[10]—to cause the extinction of all planetary and solar systems with all their different sentient hominin species at once.[11] In the far future, civilizations may become sufficiently advanced enough to colonize entire galaxies, genetically engineer new life forms, terraform planets, and even trigger the birth of stars and new planetary solar systems through massive engineering projects.[12] Civilizations this advanced would have so much knowledge and power as to be essentially omniscient and omnipotent. (I call this extrapolation Shermer's Last Law: Any sufficiently advanced extraterrestrial intelligence is indistinguishable from God.)[13] If this also sounds delusional, recall that as long ago as 1960, the physicist Freeman Dyson showed how planets, moons, and asteroids could be torn apart and reconstructed into a giant sphere or ring surrounding the Sun to capture enough solar radiation to provide limitless free energy.[14] Such a device—now called a Dyson Sphere—would enable a sub-Type I civilization like ours to transition into a Type II civilization. And there are today astronomers and organizations searching for signs of such extraterrestrial technologies in our galactic neighbourhood, such as Dyson spheres or ETI probes.[15] What do these types represent?

In a 1964 article on searching for extraterrestrial civilizations, the Soviet astronomer Nikolai Kardashev suggested using radio telescopes to try to detect energy signals from other solar systems in which he suggested we might find three different types of civilizations: a Type I civilization can harvest all of the energy resources of its planet; a Type II civilization can master the power from its sun; and a Type III civilization can harness the energy from its entire galaxy.[16] Each of these types would presumably give off a different energy signature indicating intelligent life (much like astronomers elsewhere in our galaxy might detect the signature of oxygen, methane, and other gases in our atmosphere commensurate with living organisms).[17]

In 1973 the astronomer Carl Sagan suggested another civilization classification system based on information storage, concluding that we are a Type 0.7 civilization.[18] The physicist Michio Kaku estimates that it will take humanity 100–200 years to reach Type I, a few thousand years to achieve Type II, and between 100,000 and 1 million years to attain Type III.[19] He also speculates on a Type IV civilization that would utilize the dark energy of the entire universe, and even a Type V civilization that would be able to tap into the energy of multiple universes.[20] The Mars Society founder and aerospace engineer Robert Zubrin has his own classification system and defines his civilizational types by the extent to which its inhabitants have dispersed: Type I across its home planet, Type II across its home solar system, Type III across its home galaxy.[21] A Type IV, then, would be one that inhabits the entire known universe, and Type V the multiverse.

Whatever classificatory system one prefers, my larger point in this thought experiment is that any such civilization could not achieve this level of development without also being advanced politically, economically, and morally as well. But what if we evolve to be evil? What if our species becomes even more vainglorious, power-obsessed, egocentric, and megalomaniacal?

Given how easily the Spanish conquered the Aztecs with only a few thousand-year cultural and technological differences between them, might human exploration of the cosmos and settlement of other planets prove equally catastrophic for whoever is less advanced whom we might encounter?[22]

Many prominent scientists think it would.[23] Stephen Hawking, for example, says:

> We only have to look at ourselves to see how intelligent life might develop into something we wouldn't want to meet. I imagine they might exist in massive ships, having used up all the resources from their home planet. Such advanced aliens would perhaps become nomads, looking to conquer and colonize whatever planets they can reach.

Given the history of encounters between earthly civilizations where the more advanced enslave or destroy the less developed, Hawking concluded: 'If aliens ever visit us, I think the outcome would be much as when Christopher Columbus first landed in America, which didn't turn out very well for the Native Americans.'[24] Jared Diamond echoed these sentiments in denouncing as suicidal folly a 1974 message sent from the Arecibo radio telescope into space: 'If there really are any radio civilizations within listening distance of us then for heaven's sake, let's turn off our transmitters and try to escape detection, or we're doomed.'[25]

The moral character of extraterrestrials has long been a staple of science fiction, which often reflects our own ethical dilemmas.[26] Many stories portray aliens as evil marauding conquerors intent on enslaving or exterminating humans. Examples include H. G. Wells' *The War of the Worlds* (in which invading Martians are unwittingly defeated, not by humans but by a fortuitous and deadly virus), H. P. Lovecraft's *The Call of Cthulhu* (Cthulhu is a monstrous entity who lies 'dead but dreaming' in the city of R'lyeh, a place of non-Euclidean madness sunken below the depths of the Pacific Ocean), and *Independence Day* (in which steely-eyed air force pilots gun down enemy alien spaceships). Even with all his sanguine hopes for the future, Gene Roddenberry still peppered his twenty-third-century *Star Trek* galaxy with martial species like Klingons and Romulans, and assimilating entities like the Borg.

Less common but, in my assessment, more realistic are themes in which aliens are not only technologically more advanced than us but morally superior as well. Examples include *Doctor Who* (an alien from an extraterrestrial race called the Time Lords, who acts as a custodian of humanity), and the alien race in Carl Sagan's *Contact* (which provides humans with the plans for building a worm-hole travelling spaceship). The classic film in this genre is the 1951 *The Day the Earth Stood Still*, a Christian allegory in which an alien named Klaatu (who takes the name 'Mr Carpenter' whilst visiting Earth) admonishes humans for threatening nuclear annihilation and insists that they will not be allowed to join the planetary community as long as they retain nuclear weapons. A giant robot with a deadly ray beam named Gort stands at the ready next to the spaceship, the enforcer of peace. Klaatu wants to deliver his message to all the leaders of Earth's nations but is refused, so (like Jesus) he mingles amongst the common people, is tracked down, and killed by government authorities, and put into a tomb-like morgue. Gort retrieves Klaatu and resurrects him,[27] after which he addresses a group of scientists, warning them that if humanity's morality does not advance to keep pace with its destructive technology they will not be allowed to survive. 'There must be security for all—or no one is secure', Klaatu lectures his now-enthralled audience.

> This does not mean giving up any freedom except the freedom to act irresponsibly. Your ancestors knew this when they made laws to govern themselves—and hired policemen to enforce them. For our policemen, we created a race of robots. Their function is to patrol the planets ... and preserve the peace. The result is that we live in peace, without arms or armies, secure in the knowledge that we are free from aggression and war—free to pursue more profitable enterprises.[28]

His message of redemption complete, Klaatu's heavenly ascension completes the allegory.

From science fiction to science conjecture, a reasonable hypothesis is that any extraterrestrial species with self-replicating molecules like DNA, and that reproduces sexually to produce genetic variation, will most likely have evolved by means of natural selection. If so, then they will probably have evolved something like the moral emotions we have, in which case they will cooperate or compete based on the same game theoretic models (like the Prisoner's Dilemma) that generate an array of responses to other sentient beings. Thus, the odds are that ETIs will have a suite of moral emotions not unlike ours. The difference is that they would be morally advanced by virtue of being chronologically further along in their social evolution, the arc of their moral universe having already bent to justice.

Now, let's reverse the arrow and valance of the thought experiment in which we are the exploring, expanding, settling intelligent species encountering either empty planets or those with less advanced civilizations. Although we only have a sample size of one, and our species does have an unenviable track record of first contact between civilizations, the data trends for the past half millennium are encouraging: colonialism and slavery are abolished and outlawed, the percentage of populations that perish in wars has dramatically declined, crime and violence are way down, civil liberties are way up, and across the globe the desire for representative electoral democracies is sky-rocketing, along with literacy, education, science, and technology. These trends have made our civilization more inclusive and less exploitative. Extrapolate that 500-year trend out for 5,000 years or 500,000 years and we get a sense of what an extraterrestrial intelligence might be like. Any civilization capable of extensive space travel will have moved far beyond exploitative colonialism and unsustainable energy sources such as fossil fuels.[29] (NASA already has plans to move beyond chemical rockets to alternative means of powering space exploration. To think that a super-advanced alien spaceship would arrive on Earth in order to exploit our oil and gas seems ludicrous.) Enslaving the natives and harvesting their resources may be profitable in the short term for terrestrial civilizations, but such a strategy would be unsustainable in the long run, which is the time frame of the thousands of years needed for interstellar space travel.[30]

Back to our first baby steps off our planet: if we do establish a base and then a colony on Mars, how shall it be governed? Elon Musk thinks it should be governed by a *direct democracy*. This sounds good in theory, but in practice such a system can easily slide into a tyranny of the majority, which is why direct democracies are historically rare. In recent centuries we have learned that some sort of representational system appears to be necessary, the *representational democracy* of the sort practised in the US Constitutional Republic works fairly well, although as is all too apparent representatives can be heavily influenced by special interest groups. Perhaps a variation of *delegative democracy* might be tried, in which voters have an option to delegate their vote to others, although that could quickly degenerate into trading or selling votes. In *Federalist* No. 10, Madison outlined the problem with competing factions in a direct democracy ('a landed interest, a manufacturing interest, a mercantile interest, a moneyed interest'):

> [A] pure democracy, by which I mean a society consisting of a small number of citizens, who assemble and administer the government in person, can admit no cure for the mischiefs of faction. A common passion or interest will be felt by a majority, and there is nothing to check the inducements to sacrifice the weaker party. Hence it is, that democracies have ever been found incompatible with personal security or the rights of property; and have, in general, been as short in their lives as they have been violent in their deaths.

A number of scientists and science fiction writers, reflecting on the establishment of colonies on other worlds, have made the analogy of Europeans colonizing the Americas, but this only goes so far given the fact that those incipient settlers at least had air to breath, water to drink, and plenty of potential food on the hoof, in the ground, and in oceans, lakes, and rivers. The lack of these basic commodities generates additional problems for the political governance of Mars—there's no air, food, or (that we know of yet) water there! As well, the 1967 Outer Space Treaty that the United States signed prohibits anyone from 'owning' Mars. What would be the incentive to colonize Mars if there's no guarantee that the work you do to live there would result in any type of ownership? Although working the land and air to produce resources is not directly proscribed by the treaty, doing so in a manner that doesn't lead to tyranny is another matter entirely.

These and related problems were addressed by the University of Edinburgh astrophysicist Charles S. Cockell in a series of meetings with scientists and scholars from varied fields in two conference proceedings titled *Human Governance Beyond Earth* and *The Meaning of Liberty Beyond Earth*. To learn more about what to do when the most basic necessities of life—oxygen, water, and food—are under the control of one company (SpaceX?) or one government (the United States of Mars?), I spoke with Dr Cockell by Skype,[31] starting with the observation that Earthlings colonizing Mars will be nothing like Europeans colonizing North America. 'Space is an inherently tyranny-prone environment', Cockell told me. 'You are living in an environment where the oxygen you breathe is being produced by a machine.' On Earth, he notes, governments can rob their people of food and water, 'but they can't take away your air, so you can run off into a forest and plan revolution, and you can get your friends together and you can try to overthrow a government'.

In habitats on the Moon or Mars where oxygen production is controlled by a single entity, there must be some guarantee that the air supply cannot be cut off to citizens. It would seem, then, that common ownership of the air and the machines that produce it through a single entity would follow, and I suggested as much to Cockell. But to my surprise he responded: 'I would go in completely the opposite direction. I would fragment as much as possible. I would try to create plurality in the means of production and great competition and have many people able to produce oxygen. So what you're trying to do is decentralize', he and his team concluded after studying Earth-bound systems, because 'centrally planned governments generally end up as not very good experiments'.

What about corporations that capture a majority market share and become so dominant that they can monopolize the market and turn tyrannical, I inquire? 'Well, these things happen on the Earth', Cockell historicized, noting that documents like the Bill of Rights are designed to keep tyranny in check. True, but not without violence, revolutions, and wars, I rejoin. 'It's hard work to get people to believe in freedom and to fight for it', Cockell admitted, 'and I think in space it's going to be even more hard work. So you've got to give people freedom of movement, freedom of information—it's really no different from the Earth, it's just more expanded and more vigorous of what we need to do on the Earth to maintain freedom.'

If he were to recommend to the first Mars colonists what documents they should take with them to help design their new society, Cockell unhesitatingly offered 'the US Constitution, Bill of Rights and the Declaration of Independence', adding that the latter 'is not just about independence; it's also about the ideals of free governance'. Here the analogue is fitting, given that on a planet in which nearly every square metre of land (and much of the sea) is already under some form of governmental legislation, there are few opportunities to try anew, so the US experiment in governance is one well worth emulating. 'One of the moments of genius of the founding fathers ... was they recognized that human beings can never be made perfect. Ambition must be made to counteract ambition. You can't make people perfect. You must take a cynical view and assume they are very imperfect and create the checks and balances to hold those slightly more negative aspects of humans in check.'

Amongst the new ideas Cockell and his colleagues came up with is modularity, literally incorporating liberty into architecture. 'You could modularize a settlement so that there's lots of oxygen production machines, lots of food production machines, such that the failure of any one of them does not threaten the whole settlement.' Decentralization 'allows people to do their own thing. The disasters happen when you try to artificially construct societies that are wholly controlled from the centre or where there's no organization and you create an anarchic society. The best forms of society have always been ones that are flexible, and modify themselves over time as fashions and ideas change, and that's why, I think, Western democracies are reasonably successful at keeping people happy to the maximum extent you can try and do that. And I think in space there is going to be nothing different there.'

Robert Zubrin, aerospace engineer, President of the Mars Society, author of *The Case for Mars* and *The Case for Space*, and one of the most visionary scientists I have ever known, told me when I queried him on this matter, 'I'm not going to specify a government for Mars. There will be groups of people that go to Mars and they'll have different ideas on what the best government should be and what form of government will maximize human potential and opportunity. In fact, I think this will be a major driver for the colonization of Mars—there will be groups of people who have novel ideas in these respects and will not be popular and they'll need a place to go where they can give these ideas a spin.'[32]

Invoking Darwin's idea of natural selection, Zubin went on to suggest that those with governance ideas that work will succeed and those that don't, won't. This, he suggests, is not unlike what happened with the founding of the United States. The liberal ideas from the Enlightenment that the founders evoked were not unknown in Europe, but the long-established power structures prevented them from flourishing there. 'Mars won't be utopia', Zubrin added, invoking instead the US founders' description of America as a grand experiment in governance. 'It will be a lab. It will be a place where experiments are done.' But what if a Martian tyrant turns off the air of the people in order to control them, I inquire? 'Specialization leads to empowerment', Zubrin countered, explaining that autocrats could control peasants in medieval Europe because they were living in such a simple society that tyrants could do away with them if they didn't obey. On Mars, everyone will be critical to everyone else's survival, so 'extraterrestrial tyranny is impossible'.

Perhaps these extraterrestrial settlers can run different governing experiments and see what works, a fluid and dynamic system Thomas Jefferson envisioned in 1804: 'No experiment can be more interesting than that we are now trying, and which we trust will end in establishing the fact, that man may be governed by reason and truth.'[33]

There are, as well, a variety of social experiments in setting up new societies here on Earth that differ from nations and states from which we may glean insights into extraterrestrial governance. These come in the form of *unintentional communities*, such as shipwrecked sailors stranded on an island; *intentional communities*, such as the communes established in America in the nineteenth century and Kibbutzim founded in Israel in the twentieth century; and *artificial communities*, such as online gaming communities developed in the twenty-first century. These natural experiments in living that differ from those I've been discussing are deeply explored by the evolutionary sociologist Nicholas Christakis in his 2019 book *Blueprint: The Evolutionary Origins of a Good Society*.[34] Let's look at a few examples from unintentional communities and see what light might be shown on what the first Martians should do to set up a good society.

First, says Christakis, at the core of all good societies is a suite of eight social characteristics, including: (i) the capacity to have and recognize individual identity; (ii) love for partners and offspring; (iii) friendship; (iv) social networks; (v) cooperation; (vi) preference for one's own group (i.e. 'in-group bias'); (vii) mild hierarchy (i.e. relative egalitarianism); and (viii) social learning and teaching. Whatever the right balance of these characteristics will be for the first Martians remains to be seen, but the overall balance to be sought is between individualism and group living—individual autonomy balanced with commitment to the community.

Unintentional communities are natural experiments that have struggled to find this balance, a type of 'forbidden experiment' that would never get the approval of a research IRB (Institutional Review Board). Being stranded in a remote place is one such natural experiment, and, believe it or not, there's a database of such forbidden experiments in the form of shipwrecks with survivors, or in the subtitle of an 1813 work in this genre, 'A Collection of Interesting Accounts of Naval Disasters with Many Particulars of the Extraordinary Adventures and Sufferings of the Crews of Vessels Wrecked at Sea, and of Their Treatment on Distant Shores'. Christakis includes a table of 24 such small-scale shipwreck societies over a 400-year span from 1500 to 1900, with initial survival colony populations ranging from four to 500, with a mean of 119 (2,870/24 = 119.5), but with much smaller numbers of rescued survivors, ranging from three to 289, with a mean of 59 (1,422/24 = 59.25), reflecting their success or failure at striking the right balance. The duration of these unplanned societies ranged from two months to 15 years, with a mean of 20 months (461.5/23 = 20.06; one group was rescued after 13 days so I didn't count them).

Some of the survivors killed and ate each other (murder and cannibalism), whilst others survived and flourished and were eventually rescued. What made the difference? 'The groups that typically fared best were those that had good leadership in the form of mild hierarchy (without any brutality), friendships amongst the survivors, and evidence of cooperation and altruism', Christakis concludes. The successful shipwreck societies shared food equitably, took care of the sick and injured survivors, and worked together digging wells, burying the dead, building fires, and building escape boats. There was little hierarchy—for example, whilst on board their ships, officers and enlisted men were separated, but on land successful castaways integrated everyone in a cooperative, egalitarian, and more horizontal structure, putting aside prior hierarchical class differences in the interest of survival. Camaraderie emerged and friendships across such barriers were formed.

The closest thing to a control experiment in this category was when two ships (the *Invercauld* and the *Grafton*) wrecked on the same island (Auckland) at the same time in 1864. The island is 41.8 kilometres long and 27.5 kilometres wide and lies 466.7 kilometres south of New Zealand, truly isolated. The two surviving groups were unaware of one another, and their outcomes were starkly different. For the *Invercauld*, 19 out of 25 crew members made it to the island but only three survived when rescued a year later, whereas all five of the *Grafton* crew made it to land and all five were rescued two years later. 'The differential survival of the two groups may be ascribed to differences in initial salvage and differences in leadership, but it was also due to differences in social arrangements', Christakis explains. 'Among the *Invercauld* crews, there was an "every man for himself" attitude, whereas the men of the *Grafton* were cooperators. They shared food equitably, worked together towards common goals (like repairing the dinghy), voted democratically for a leader who could be replaced by a new vote, dedicated themselves to their mutual survival, and treated one another as equals'. Take note future Martians.

In between unintentional and intentional communities is arguably the most famous of these forbidden experiments, which began in the early morning hours of 28 April 1789, when Fletcher Christian, the Master's Mate of the HMS *Bounty*—a modest-size merchant ship crewed by 45 men and loaded with over 1,000 pots of breadfruit trees bound for the Caribbean islands where they would be delivered as cheap slave fodder for the British colony there—seized control of the ship from Captain William Bligh, released the captors into a 7 metres launch, and sailed the *Bounty* into the mists of history where they were not found until 1808, when only one member of the original crew, John Adams, survived to tell the tale of what happened. I researched and wrote about this event in my book *Science Friction*, employing evolutionary psychology as a deeper explanation of what happened and why.[35] It is a valuable lesson for future Martians on how not to set up a new society.

The *Bounty* departed Portsmouth on 23 December 1787. Ten months and 27,010 miles later it arrived in Matavai Bay, Tahiti, where it was anchored for five months, plenty of time for the young male crew members to become romantically involved with the native women, including Fletcher Christian. After departing Tahiti, a number of the smitten men grew restless without their love interests that, when coupled to the normal (at the time) draconian discipline imposed on them by Bligh for relatively minor offenses, led to the explosive response that fateful April morning in 1789. After the mutiny, Bligh and his loyal followers sailed the little launch 5822 kilometres to safety—one of the greatest open-ocean voyages in recorded history—affording him the chance to tell the world what happened and why. Here is Bligh's explanation provided in his published account:

> It however may very naturally be asked what could be the reason for such a revolt, in answer to which I can only conjecture that they have Idealy assured themselves of a more happy life among the Otaheitians than they could possibly have in England, which joined to some Female connections has most likely been the leading cause of the whole business.

After seizing the ship, Christian and his followers returned to Tahiti, and on 23 September 1789, they left the island for good, knowing the Royal Navy would track them down and execute them, carrying a total of nine male mutineers, six Tahitian men, and eleven Tahitian women. On 15 January 1790, they arrived at one of the remotest rocks in the Pacific, Pitcairn Island, unloaded the ship, and a week later torched it in a final gesture of commitment to their new life on this far-flung outpost. The seeds for the failure of governing a long-term society were already apparent in the sex ratio: fifteen men to eleven women; not good for domestic tranquillity, much less the general welfare. After three years, the woman living with mutineer John Williams died, so he took the wife of one of the Polynesian men, leading to jealousy, violence, and, on 20 September 1793, retribution, when five of the mutineers were killed, including Fletcher Christian, along with all of the Polynesian men. In subsequent years after the massacre, one mutineer committed suicide, another was killed by one of the other mutineers, and still another died of asthma. By 1800, the only male survivor was John Adams, who lived until 1829, long enough to tell the tale of the fate of the *Bounty's* mutineers and why the Pitcairn society failed.

In my analysis of the mutiny on the *Bounty* and the settlement on Pitcairn island, I made the distinction between proximate causes (e.g. imbalanced sex ratio) and ultimate causes (e.g. attachment and jealousy) for the failure of both Captain Bligh on the *Bounty* and of Fletcher Christian on Pitcairn, starting with certain basic facts about human nature: we are a hierarchical social primate species and one of the most sexual of all primates, where males are more openly and intensely obsessed with obtaining sexual unions and equally preoccupied with protecting those unions from threats from other males (mate guarding). The evolutionary foundation for the mutineer's heightened state of emotions is born out in modern neuroscience, which shows that the attachment bonds between men and women are powerful forms of chemical addiction, especially in the early stages of a relationship. The neurotransmitter dopamine produces emotional attachments so strong that they are literally addictive, stimulating the same region of the brain that is active in drug addictions. The mutineers were literally going through withdrawal when they seized the *Bounty* and returned to Tahiti.

On Pitcairn, the mutineers failed to find the right balance between hierarchy and egalitarianism. Both Bligh and Christian were products of the Royal Navy, very much steeped in hierarchy and the accompanying status (or lack thereof). Anthropologists tell us that most social mammals, all social primates, and every human community ever studied, shows some form of hierarchy and social status. On the *Bounty,* the hierarchical tensions between Bligh and Christian led to the mutiny, and on Pitcairn

Island the hierarchical tensions between Christian and the mutineers, and between the mutineers and the male Tahitians (with the racist attitudes of the former towards the latter, common for the day), when coupled with the sex ratio imbalance and accompanying jealousy and violence, led to its demise.

The lessons for Martian colonists from Pitcairn mutineers are clear enough: start off with a balanced sex ratio, structure a political system more horizontal than hierarchical, eliminate racist and misogynist attitudes, and accentuate cooperation and attenuate competition.

The long-term historical trend in moral progress that I document in *The Moral Arc*,[36] points to the fact that we may, in fact, need a Leviathan—an overarching government of some form—to attenuate our inner demons and to accentuate our better angels. But let's not restrict ourselves by the categorical thinking such labels as nation-state or city-state impose, as these are just ways of describing a linear process of governing groups of people of varying sizes. Such systems have taken many forms over the centuries, but in time they have narrowed their scope to include most of the following characteristics that the majority of Western peoples enjoy today (call them the Justice and Freedom Dozen):

(1) A liberal democracy in which the franchise is granted to all adult citizens.
(2) The rule of law defined by a constitution that is subject to change only under extraordinary circumstances and by judicial proceedings.
(3) A viable legislative system for establishing fair and just laws that apply equally and fairly to all citizens regardless of race, religion, gender, or sexual orientation.
(4) An effective judicial system for the equitable enforcement of those fair and just laws that employs both retributive and restorative justice.
(5) Protection of civil rights and civil liberties for all citizens regardless of race, religion, gender, or sexual orientation.
(6) A potent police for protection from attacks by other people within the state.
(7) A robust military for protection of our liberties from attacks by other states.
(8) Property rights and the freedom to trade with other citizens and companies both domestic and foreign.
(9) Economic stability through a secure and trustworthy banking and monetary system.
(10) A reliable infrastructure and the freedom to travel and move.
(11) Freedom of speech, the press, and association.
(12) Mass education, critical thinking, scientific reasoning, and knowledge available and accessible for all.

Martian political, economic, and legal institutions are likely to vary from earthly ones according to the most basic needs of the first colonists, but any such variation will be necessarily bounded by human nature. Perhaps SpaceX or NASA will create a division of Social Engineers—a team of legal scholars, political scientists, economists, social psychologists, and conflict resolution scholars—and they'll come up with a wholly different system of governance.

Or maybe the Martian colonists themselves will think of something new as they experiment with different solutions to social problems. Who knows what that space seed might sprout in the coming centuries and millennia? The prospects of a new form of government being discovered on Mars and exported back to Earth makes any such exploratory mission to the red planet worthwhile. Either way, extraterrestrial governance is a useful analogue for terrestrial governance, inasmuch as we do not have the final answers here in order to ensure the survival of our own civilization. Whatever we do, we had better get it right, because, as the political commentator Charles Krauthammer wrote in his aptly titled book *Things that Matter*:

> Politics, the crooked timber of our communal lives, dominates everything because, in the end, everything—high and low and, most especially, high—lives or dies by politics. You can have the most advanced and efflorescent of cultures. Get your politics wrong, however, and everything stands to be swept away.[37]

Notes

1. The nutrition reference is to the fact that the US Government recently overturned a half century of nutritional advise. See e.g. M. J. Stampfer and W. C. Willett. 'Rebuilding the Food Pyramid' (2006) 1 December *Scientific American* <https://bit.ly/3fdisCC>.
2. E. Musk. 'Elon Musk Answers Your Questions' SXSW conference, March 2018 <https://bit.ly/2tD8zsx>.
3. Ibid.
4. M. Kibbe. *Don't Hurt People and Don't Take Their Stuff: A Libertarian Manifesto* (New York, NY: William Morrow, 2014).
5. J. Diamond. *Guns, Germs, and Steel* (New York, NY: W. W. Norton, 1996) 268.
6. R. Baibhawi. 'Elon Musk's Idea of Colonising Mars is "dangerous delusion" Says Astrophysicist' (2021) 14 March *Republicworld.com* <https://bit.ly/3x7bqWg>.
7. Doron M. Behar, Richard Villems, Himla Soodyall, et al. 'The Dawn of Human Matrilineal Diversity' (2008) 82(5) *American Journal of Human Genetics* 1130–40; S. H. Ambrose. 'Late Pleistocene Human Population Bottlenecks, Volcanic Winter, and Differentiation of Modern Humans' (1998) 34(6) *Journal of Human Evolution* 623–51; D. Whitehouse. 'When Humans Faced Extinction' *BBC News* 9 June 2003 <https://bbc.in/2Wr8JSG>.
8. M. Kaku. *Physics of the Future: How Science Will Shape Human Destiny and Our Daily Lives by the Year 2100* (New York, NY: Doubleday, 2011).
9. The evolutionary biologist Ernst Mayr defined a species as 'a group of actually or potentially interbreeding natural populations reproductively isolated from other such populations.' E. Mayr. 'Species Concepts and Definitions' in *The Species Problem* (American Association for the Advancement of Science, Publication No. 50, 1957). Mayr offers this expanded definition: 'A species consists of a group of populations which replace each other geographically or ecologically and of which the neighboring ones intergrade or hybridize wherever they are in contact or which are potentially capable of doing so (with one or more of the populations) in those cases where contact is prevented by geographical or ecological barriers.' See also: E. Mayr. *Evolution and the Diversity of Life*

(Harvard University Press, 1876); E. Mayr. *Toward a New Philosophy of Biology* (Cambridge, MA: Harvard University Press, 1988).

10. L. Smolin. *The Life of the Cosmos* (New York, NY: Oxford University Press, 1997); A. Liddle and J. Loveday. *The Oxford Companion to Cosmology* (Oxford University Press, 2009); S. Weinberg. *Cosmology* (New York, NY: Oxford University Press, 2008).

11. F. Dyson. 'Time Without End: Physics and Biology in an Open Universe' (1979) 51(3) July *Reviews of Modern Physics* <http://www.aleph.se/Trans/Global/Omega/dyson.txt>.

12. J. Pollack and C. Sagan. 'Planetary Engineering' in J. Lewis, M. Matthews, and M. Guerreri (eds), *Resources of Near Earth Space* (Tuscon, AZ: University of Arizona Press, 1993); L. Niven. *Ringworld* (New York, NY: Ballentine, 1990); O. Stapleton. *The Starmaker* (New York, NY: Dover, 1968).

13. M. Shermer. 'Shermer's Last Law' (2002) January *Scientific American* 33.

14. F. J. Dyson. 'Search for Artificial Stellar Sources of Infra-Red Radiation' (1960) 1131(3414) *Science* 1667–8.

15. See e.g. A. Loeb. *Extraterrestrial: The First Sign of Intelligent Life Beyond Earth* (New York, NY: Houghton Mifflin Harcourt, 2021).

16. N. Kardashev. 'Transmission of Information by Extraterrestrial Civilizations' (1964) 8 *Soviet Astronomy* 217. Kardashev computed the energy levels of the three types to be: Type I ($\sim$4 × 10^{19} ergs/second), Type II ($\sim$4 × 10^{33} ergs/second), and Type III ($\sim$4 × 10^{44} ergs/second).

17. J. Heidmann, . *Extraterrestrial Intelligence* (New York, NY: Cambridge University Press, 1992) 210–12.

18. C. Sagan. *The Cosmic Connection: An Extraterrestrial Perspective.* (New York, NY: Anchor Books, 1973) 233–4. Sagan's information storage capacity metric increases one order of magnitude at each step, starting with A = 10^6 bits of information, B = 10^7 bits ... Z = 10^{31} bits. Sagan estimated we were at 10^{13} bits in 1973, making us a Type 0.7H civilization. My own calculations lead me to conclude that in 2014 we were at 10^{21} bits—a zettabyte— thus we are presently a Type 0.7P civilization. I based my estimate on the 2010 figure cited earlier from Diamandis that at the end of 2010 we were producing 912 exabytes of information. An exabyte is one quintillion bytes. A quintillion is 10^{18}. I estimate that we have exceeded 1,000 exabytes by 2014. A thousand exabytes is a zettabyte, or 10^{21} or a 1 followed by 21 zeros. As the Kardashev scale is logarithmic—where each increase in power consumption on the scale requires a huge leap in production—we have a ways to go. Fossil fuels won't get us there. Renewable sources such as solar, wind, and geothermal are a good start, but in order to achieve Type 1.0 status in this typology we need to go nuclear, for example, fusing 1,000 kg of hydrogen into helium per second, a rate of 3 × 10^{10} kg/year. As a cubic kilometre of water contains about 10^{11} kg of hydrogen, and our oceans contain about 1.3 × 10^9 km of water, this would give us ample time to make the transition to the next level.

19. Kaku (2011) note 8. See also: M. Kaku. 'The Physics of Interstellar Travel: To One Day Reach the Stars' (2010) <http://mkaku.org/home/articles/the-physics-of-interstellar-travel/>.

20. M. Kaku. *Parallel Worlds: The Science of Alternative Universes and Our Future in the Cosmos* (New York, NY: Doubleday, 2005) 317.

21. R. Zubrin. *Entering Space: Creating a Spacefaring Civilization* (New York, NY: Putnam, 2000) x.

22. C. Maccone. 'Evolution and History in a New "Mathematical SETI" Model' (2014) 93 *Acta Astronautica* 317–44.

23. D. Brin. 'Shouting at the Cosmos ... Or How SETI has Taken a Worrisome Turn into Dangerous Territory' (2006) September <https://lifeboat.com/ex/shouting.at.the.cosmos>.

24. S. Hawking. *Into the Universe with Stephen Hawking* (Discovery Channel, 2010); J. Leake, 'Don't Talk to Aliens, Warns Stephen Hawking' *The Sunday Times* (London), 25 April 2010. See also S. Shostak. *Sharing the Universe: Perspectives on Extraterrestial Life* (Berkeley, CA: Berkeley Hills Books, 1998).

25. J. M. Diamond. *The Third Chimpanzee: The Evolution and Future of the Human Animal* (New York, NY: Harper Perennial, 1991) 214. For an overview on the risks of contact with extraterrestrial intelligences from the perspective of risk management assessment see M. Neal. 'Preparing for Extraterrestrial Contact' (2014) 16(2) *Risk Management* 6387.

26. For more on this P. Davies *Are We Alone?* (New York, NY: Basic Books, 1995); P. Davies. *The Eerie Silence: Renewing Our Search for Alien Intelligence* (New York, NY: Houghton Mifflin Harcourt, 2010); S. C. Morris. *Life's Solution: Inevitable Humans in a Lonely Universe* (Cambridge: Cambridge University Press, 2003).

27. Gort is instructed to go get the deceased Klaatu after the Mary Magdalene-like character who has developed a friendship with Klaatu instructs Gort 'Klaatu barada nikto', one of the most famous lines in all science fiction. In the original script, after Klaatu is resurrected he explains to the astonished onlooker that this is the future power of science and technology, but the Motional Picture Association of America's censor for the film, Joseph Breen, determined that this was unacceptable and forced the producer to add the line 'that power is reserved to the Almighty Spirit'. See J. Blaustein, R. Wise, P. Neal, et al. 'Making the Earth Stand Still' DVD Extra, 20th Century Fox Home Entertainment, 1995.

28 E. H. North. Script for *The Day the Earth Stood Still*. 21 February 1951 <https://bit.ly/3fIGVjH>.

29. G. Michael. 'Extraterrestrial Aliens: Friends, Foes, or Just Curious?' (2011) 16(3) *Skeptic* 46–53.

30. A. A. Harrison. 'The Relative Stability of Belligerent and Peaceful Societies: Implications for SETI' (2000) 46(10–12) *Acta Astronautica* 707–12; D. Brin. 'The Dangers of First Contact: The Moral Nature of Extraterrestrial Intelligence and a Contrarian Perspective on Altruism' (2010) 15(3) *Skeptic* 28–35.

31. Personal correspondence, 30 June 2018 <https://bit.ly/2KruElT>.

32. Personal correspondence, 17 June 2019 <https://bit.ly/2J181SN>.

33. T. Jefferson. 'Letter to Judge John Tyler Washington' 28 June 1804 <https://bit.ly/292vEbR>.

34. N. Christikas. *Blueprint: The Evolutionary Origins of a Good Society* (New York, NY: Little, Brown, 2019).

35. M. Shermer. *Science Friction: Where the Known Meets the Unknown.* (New York, NY: Times Books, 2005) 111–29.

36. M. Shermer. *The Moral Arc: How Science and Reason Lead Humanity Toward Truth, Justice, and Freedom* (New York, NY: Henry Holt, 2015).

37. C. Krauthammer. *Things That Matter: Three Decades of Passions, Pastimes, and Politics* (New York, NY: Crown, 2013) 2.

25

Global legal pluralism and outer space law

The Association of Autonomous Astronauts as a socio-legal community

Saskia Vermeylen

1. Introduction

In all likelihood, private corporations and not governments will fund the first settlements on Mars. This raises some interesting questions about the legal authority in Mars settlements. Corporations are usually not bound by the same rules and principles as national governments. There is thus a danger that the first settlements may find themselves in a legal vacuum unless our earthly legal traditions get propelled into space. As Yuk Chi Chan argues, initially, space travellers will be mainly bound by existing Space Law which consists of a mix of public international law and domestic legislations and regulations (Chan 2021, 183). The legal rules on Mars fall under the auspices of the 1967 *Treaty on Principles Governing the Activities of States in the Exploration and Use of Outer Space, including the Moon and Other Celestial Bodies* (OST) which sets out the core principles of non-appropriation and non-militarization when governing Mars. Furthermore, Article VI of the OST asserts that national governments are responsible for the activities of non-governmental entities, including private companies. But whilst Article II prohibits national states to own any surface on Mars by claim of sovereignty, Articles VII and VIII provide state parties with jurisdiction and control over any space objects launched and registered under their authority. This means that national governments in the initial phase still carry a sufficient amount of legal power. However, there is no legal clarification as to how this legal regime that still relies heavily on the jurisdiction and rule of law of the nation-state will evolve in the longer term.

This scenario gets even more complicated against the background of recently passed national legislation, such as the US Space Resource Exploration and Utilization Act of 2015 (Space Act) which in Title IV allows US citizens and commercial entities to own mined resources in space. As more countries will be looking into passing national legislation to govern space activities conducted by their citizens, the legal situation on Mars could become very complex with overlapping jurisdictions making claims over resources. International space law is mostly silent about the laws ruling settlements and the closest example we currently have available is the legal framework of the International Space Station (ISS) (Farsaris 2021). It can be assumed that the first Mars settlements will follow the lead of the ISS which already functions as an elementary form of space society. As such it provides already insights into initial legal principles that will govern space settlements. The Intergovernmental Agreement (IGA) of the ISS foresees that every person in the ISS continues to be the subject of their home country's national laws. In the longer term, it is not unforeseeable that Martians will start contesting being controlled and governed by the states that were responsible for the first settlers. As Chan summarizes concisely: 'it would seem that the present space law regime is inadequate in governing a Mars colony. A separate agreement, or perhaps an entirely new legal system, will have to be created to rise to the challenge' (Chan 2021, 183).

The IGA of the ISS is analysed in this chapter using the theoretical concept of legal pluralism in order to assess how space law needs to evolve so it can become a source of law that is no longer deeply entangled in state sovereignty. As sovereignty-based jurisdiction in outer space is as good as non-existent, legal pluralism both as a praxis but also as a normative framing can open the possibility to think more progressively and speculatively about the future legal norms that could govern a human presence in space and on Mars.

Just for clarity, I would like to add here that I am not championing Mars settlements or the development of extractive industries in space. This chapter is more like a thought experiment and seeks to offer an alternative idea about how to govern space in the future. Space settlements, even just as a theoretical experiment, provide the perfect scenario and background to research how the inevitable intermingling of normative legal systems can lead to a more inclusive version of space law. As the role of the sovereign state as the ultimate legislator is already reduced in space law, it seems like a logistical step to embrace more enthusiastically the idea that legal pluralists have come to endorse: that societies always consist of multiple overlapping communities which can include state-based formal and governmental communities, but can also encompass non-state-based communities such as non-governmental organizations, pressure groups, activists, local communities, and many more.

With the exception of the work done by Eduard van Asten (2011), this chapter is one of the very few studies examining space law from a global legal pluralist perspective. This is remarkable because global legal pluralism provides a solid theoretical platform conducive to describe and analyse the opportunities that are emerging in the new global legal system which is characterized by hybrid legal spaces that consist of a 'web of jurisdictional assertions by state, international and non-state normative communities' (Berman 2018, 151). I argue that space law fits the characteristics of global legal pluralism and by extension global law perfectly. Within the context of this chapter, global law can be best described as a transformative legal project that seeks to change the way in which the legal community (in the broadest sense) understands law, law-making, legitimacy, and the validity of legal sources. As succinctly summarized by Kulovesi, Mehling, and Morgera (2019, 408), with globalization the organization of the legal universe has become multi-layered and pluriverse. Multiple actors, ranging from states, governmental and non-governmental organizations, law firms, corporations, financial markets, and even academic communities, are shaping the creation of new legal norms that are no longer state-centric or universal, the latter referring to international law. I argue that space law—both in its current arrangement through the relationship between national and international law and in its future form with multinational corporations potentially playing an increasingly normative role in law making—is a textbook example of a global legal system.

Importantly, referring to Günther Teubner's (1997) work on global legal pluralism, according to Kulovesi, Mehling, and Morgera (2019, 418), global law also offers the opportunity to question critically the validity of legal norms and the ontology of law. Traditional theorists of global law, such as Neil Walker (2015), consider highly elitist and specialized groups of experts, most notably academic lawyers, as communities with legal authority to 'fashion and shape global law' (Walker 2015, 31). I extend this category of legal experts to artistic and political communities. I use the example of the Association of Autonomous Astronauts (AAA) as a counterexample to the ISS as a social community in space shaping the normative framing of space law. Whilst global legal pluralism does indeed provide the possibility to include non-legal communities as norm-setting entities, it is not common to incorporate artistic communities. As I have argued elsewhere (Vermeylen 2021) space law is entangled with colonial practices of law-making, therefore, I also seek to decolonize space law and I find the dialogue between law and arts a useful conduit from which to reform the normative framing of space law. This can be best illustrated by an example.

Origins of the Seven Stars (2010) is a small, framed, text piece by the artist Carey Young, exposing how outer space travel is entangled in a colonial archive. Young comments that the artwork 'intersperses a native American Myth from the Wyandot Nation (as transcribed by the anthropologist Marius Barbeau) with a U.S. Presidential Directive from 2004, a legal text which intends to advance US interests via space exploration.'[1] Young's work addresses, and I quote from an exhibition catalogue, the monolithic power of legal systems. The artist examines law as a conceptual and abstract space in which power, rights, and authority are played out through varying forms of performance and language. Young's work 'call[s] law's authority into question and create[s] slippages in the law by playfully adopting as well as disrupting its forms and methods and by highlighting its gaps, ambiguities, and subjectivities' (Young 2013).[2]

I will follow a similar approach. I will first discuss legal pluralism and apply the sociological and anthropological approach of legal pluralism to the ISS. This is followed by a systems approach to legal pluralism which is applied to the arts collective, the AAA. The chapter will end with a short discussion how the views of the AAA can influence law-making and become a source of global law that can reform space law into a more inclusive body of law that can protect space from exploitation and appropriation by commercial entities.

2. Empirical-Positivist Legal Pluralism and the International Space Station

There is not one theory of legal pluralism. Its definition is also part of a wider and very broad enquiry about jurisprudence or the question of what counts as law. Although some die-hard legal positivist may still disagree, it is increasingly accepted that the conception of law confined to state law, which also includes municipal and international law, is far too narrow. Legal pluralism is thus the opposite of state-centrism and treats non-state law as a legitimate source of law and norma-tive order. Emmanuel Mellisaris' distinction between the empiricist-positivist strand and systems-theoretical strand of legal pluralism is a useful classification (Mellisaris 2004, 57). Whilst empirical legal pluralism incorporates anthropological and socio-logical legal pluralism, systems-theoretical legal pluralism focuses more on the idea of dispersed legalities and engages with Teubner's systems-theoretical legal plural-ism. Within the context of this chapter, I will draw loosely upon the two different approaches to make the point that the future generation of space law should also include other normative orders than state law.

Early examples of empiricist legal pluralism study the law's responsiveness to the community and their needs and they focus in their enquiry on the non-state legal orders and their relationship with state legal orders (Melissaris 2004, 59). The work of Eugen Ehrlich and his idea of living law have been very influential. Ehrlich was a professor of Roman law at the University of Csernowitz in Bukowina which was part of the Austrian Empire (an area that is now part of Ukraine). In 1936, Ehrlich published his book, *Fundamental Principles of the Sociology of Law*, in which he puts forward the idea that the centre of gravity of legal development does not lie in jurisprudence, judicial decisions, or legislation, but that it is society itself that decides on the law question. According to Ehrlich the functioning of law should not just be regulated through positivist law and the development of state law and court rules. As daily life takes place mostly far away from the places of author-ity such as the state and the courts, the operation of law is therefore also to be found in the everyday social life of communities. For Ehrlich, living law is 'the law which dominates life itself even though it has not been posited in legal propositions' (Ehrlich 1936, 493).

Although it is too early to see the contours of how a living law system might look like for settlements on Mars, it is very likely that over time space communi-ties will increasingly cut ties with the legal structure and jurisdiction that applies on Earth. However, as I will illustrate below, taking the example of the ISS as the first space community, the legal structure and normative values are more in tune with an anthropological form of legal pluralism. Astronauts and crew on the ISS have the potential to become over time a semi-independent social society that is temporary suspended in space, and disconnected from Earth. However, in reality, the centre of law on the ISS is still ruled by a gravitational pull downwards towards the sovereign state. How this fits with a classical and anthropological reading of legal pluralism is explained below.

Another but related form of empiricist legal pluralism has its roots indeed in anthropology. For example, Bronislaw Malinowski (1926) studied the hierarchical coexistence of perceivably separate legal systems (Berman 2007, 1170). Different forms of social interaction were studied to distil the legal rules that were operating in 'primitive' societies (Hamilton 2009, 8). Anthropologists recognized the existence of multiple normative orders and studied the dialectical interaction between and amongst these normative orders (Berman 2007, 1171). In the early days of colonial-ism, indigenous legal systems were left untouched by the colonial powers as it was not in their economic or political interest to extend their influence into the 'native' legal system. This changed, however, in the late eighteenth century when, in the British colonies through indirect rule, the colonial powers had to rely on pre-existing political authorities which implied that through the use of indigenous leaders or chiefs the colonial powers were interpreting customary norms in the native courts (Tamanaha 2008, 383–5). British and native legal systems existed side by side, sometimes overlapping, often conflicting, and both having an influence on daily life. Customary law's jurisdiction was usually in the areas of family law (such as marriage, inheritance, and familial property rights) and customary or religious conflicts. After decolonization, customary law became even more popular as state legal institutions were often poorly developed, underfunded, and had little capacity (Tamanaha 2008, 385).

However, as Brian Tamanaha shows, legal pluralism has a much longer history that goes further back in time than the colonial period. Early variations of legal pluralism can be found in medieval times when it was quite common to have an eclectic mix of different sorts of law and institutions, each occupying the same space, sometimes conflicting or complementary, but typically lacking any overarching hierarchy or organization (Tamanaha 2008, 377). Tamanaha illustrates further that these forms of law included local customs, Germanic customary law, feudal law, commercial law (*lex mercatoria*), canon law, and Roman law. Some of the laws were unwritten; others were in written code form. Different types of courts existed, depending on the law. They included manorial, municipal, merchant, guild, church, and royal courts. Each court had a different type of judge: barons or lords of the manor, burghers (leading city residents), merchants, guild members, bishops, and kings, or their appointees. Jurisdictional rules varied for each court, equally for the laws to be applied related to the person involved—their status, descent, citizenship, occupation, or religion—as well as to the subject matter at issue (Tamanaha 2008, 377).

The coexistence of more than one body of legal norms was a common phenomenon in Europe until the Treaty of Westphalia in 1648 divided Europe into separate, secular territories under the authorities of a sovereign who could, according to the Treaty of Augsburg (1555), decide about the religion of his/her citizens. Together, these Treaties changed the political and legal landscape; territorial states would become the central political and legal unit of Western Europe (Tamanaha 2008, 379–80). In the process customary law became part of the state legal system and legal professionals started to interpret and make decisions about what could be deemed as customary, but this did not always correspond with the actual customs. The monopolization of law by states in Western Europe reduced legal pluralism at home just as a new wave of legal pluralism emerged as a by-product of colonialism (Tamanaha 2008, 380–1).

In this empirical and positivist version of legal pluralism, the state still assumes legal authority over territories and populations. As we know, in international space law territorial sovereignty is forbidden, due to the provisions in Articles I and II of the OST conceptualizing space as a *res communis*, meaning that space is free to be used by all states without discrimination of any kind. This means that national laws cannot be applied in space on the basis of territorial jurisdiction. However, national jurisdiction is still applicable in space through personal jurisdiction in the context of states being responsible for the control over activities that are conducted by its nationals (see Article VI of the OST). A quasi-territorial jurisdiction exists in relation to Articles VII and VIII and the liability and jurisdiction of the nation-state in space for space objects that are launched from its territory or facility. Some of these principles have later been reinforced in other arrangements such as the Convention on International Liability for Damage Caused by Space Objects (1972).

This rather ambiguous and complex legal regime of the OST also applies as the first order of law onto the ISS and private partners, (e.g. companies providing material or crew). The second layer of laws applicable to the ISS are multilateral and bilateral agreements that are linked and are part of the ISS's history. When former US President Ronald Reagan initiated the development of the ISS, he invited other countries to participate in the development of a permanently inhabited civil space station located in Earth's orbit. After many years of negotiations, the Intergovernmental Agreement (IGA) of 1988 was signed but later amended in 1998, when, after the fall of the Soviet Union Russia joined the cooperation. The IGA of 1998 is signed by 15 partner states and five Cooperating Agencies. Although consistent with the existing body of international space law, not all legal issues that are being faced by the ISS, such as criminal law and intellectual property rights, are represented in the five existing outer space regulations at the international level, hence the promulgation of the IGA which acts as the constitution of the ISS (van Asten 2011, 123). In addition to the IGA, there are also four Memoranda of Understanding (MoU) between various states' agencies, including the Crew Code of Conduct for the ISS crew. This is a soft law (non-binding) instrument dealing with the governing of human behaviour on the ISS.

Some of the legal questions that are applicable for the ISS, are possibly also applicable for the first Mars settlement(s). In terms of which jurisdiction is applicable in the Mars settlement, in the short term, Article 5 of the IGA stipulates that the jurisdictional principles provided by the Outer Space Treaty and the Registration Convention are also applicable to the ISS. This means in practice that a nation which registers a space object pursuant to Article II of the Registration Convention, can then, in line with Article VIII of the OST, retain jurisdiction over such object or, in the case for the ISS, flight element (Farsaris 2021). Article 5 paragraph 1 of the IGA is thus a typical example of the state's quasi-territorial jurisdiction being extended to the flight elements. Importantly, Article 5 paragraph 2 also refers to the nationality principle as it specifies that each Partner to the IGA retain control over personnel in or on the Space Station. Articles 21 and 22 repeat this form of jurisdiction based on nationality with respect to intellectual property rights and criminal law. Furthermore, as space law expert Frans von der Dunk remarks, within the ISS, the principle of quasi-territorial jurisdiction prevails over personal or national jurisdiction, except when it concerns criminal law (von der Dunk 2006, 23–4). Concretely this means that within each compartment of the ISS, in the first instance, the jurisdiction of the registered state predominates over the jurisdiction of personnel (van Asten 2011, 125).

Finally, to finish this section regarding which laws are applicable to the ISS, the third layer of laws concerns national laws and regulations. These consists of national laws that von der Dunk and van Asten describe as dealing with operational legal fields such as security, safety, and liability (von der Dunk 1998, 107). They mainly deal with space objects and issues relating to licensing requirements, insurance issues, and technical aspects (van Asten 2011, 125). Another body of national laws consists of general laws that are applicable within a state to regulate the activities and behaviour of private actors. For example, the US Patent Act 35 USC § 105 (2003) states that space objects in outer space should be considered an extension of US territory in relation to inventions made, used, or sold on a space object registered by the United States (van Asten 2011, 126). In other words, national states can declare an extension of their jurisdiction in outer space through national laws. A more recent example that could potentially be an important game changer for the private sector is the US Space Act which provides that 'any asteroid resources obtained in outer space are the property of the entity that obtained such resources, which shall be entitled to all property rights thereto' (Space Act).

Based on the analysis of the laws that are applicable to the ISS, it can be argued that in the first instance, Mars settlements will be ruled and governed through extending national jurisdiction into space. From a legal pluralist perspective, it is too soon to know to what extent non-state and informal legal norms and principles have already emerged on board the ISS. This would require more socio-legal and anthropological research but it is not unlikely that in everyday practices the crew on board of the ISS are developing their own rules and regulations according to their own daily rhythms on the ISS. Initially, these are very informal rules but they could over time develop into more formal ones.

However, for now it seems safe to conclude that the laws that regulate the ISS include some elements of Ehrlich's living law. Although a straightforward territorial jurisdiction based on sovereignty is rejected in space law, the notion of state law still applies and resembles some of Ehrlich's ideas which replaced the principle of territoriality with the older principle of personality. During medieval times when distinct groups often shared a specific territory, each had their own tribal laws which travelled with them when they moved to other territories. Each territory thus had multiple jurisdictions. But with modernity and Enlightenment ideas, citizenship was increasingly linked to territorial absolutism which eventually led to the formation of an 'absolute state as a compulsory institution' (Mohr 2003, 61). A good example of how the ISS may incorporate some of the earlier ideas around multiple jurisdictions existing coincidently can be found in Article 22 second paragraph of the IGA which broadens the possible criminal jurisdiction on board of the space station and offers the choice between passive nationality and territoriality principles. When a criminal act is committed by a national of a Partner state, that Partner state has the primary criminal jurisdiction over its national who is the offender. However, the Partner state of the victim, or on whose flight element the crime was conducted, that Partner state may exercise their respective prosecutorial interest, and after consultation within a limited time frame, the Partner state of the offender may agree that the crime will be prosecuted according to the jurisdiction of the victim's Partner state.

But despite these examples, all three layers of law applicable to the ISS are still based on state-centred traditions which Ehrlich argued can be challenged for the following three reasons. As summarized by Singh (2010), for Ehrlich, states emanate only second-order norms in the form of legal propositions which are hierarchically inferior to social norms produced in normative orderings. Secondly, Ehrlich also broadens the notion of legal norm to include social norms, thereby undermining the traditional legal norm from the state. Thirdly, state law must be made inferior to societal law in order to avoid state monopolization resulting in the homogenization of legal norms across associational lines. Or in other words, the state must remain subordinate in order to protect plural values in the law. Whether the law is and can be responsive to such a request is discussed in the next section.

3. Systems-Theoretical Legal Pluralism and the Association of Autonomous Astronauts

Following the first wave of early legal pluralists, a new body of scholars emerged in the 1970s and 1980s, questioning the hierarchical modelling of one legal system and the centrality of the state in law-making (Berman 2007, 1171). The changes have been summarized by Berman as follows: first, plural systems are semi-autonomous, they operate together with other legal fields but are not governed by other legal fields; second, legal systems interact and influence each other; and third, law is broadly defined so that many types of non-official normative ordering could be treated as law. These concepts open up the possibility of introducing the concept of legal pluralism beyond colonial societies and apply it in industrialized settings (Berman 2007, 1171–2). Influenced by John Griffiths' work, the ideology of legal centralism was rejected (Griffiths 1986, 3).

Particularly, the latter point has provoked a lot of debate. Opponents of legal pluralism labour the point that there are conceptual differences between state law and other normative orders. They argue that state law is so successfully dominant and its application in courts so specific that it would be erroneous to consider other normative norms which are more flexible, vague and negotiable in comparison to state law as law (von Benda-Beckmann and von Benda-Beckmann 2006, 16). According to the ethnocentricity argument, using law for non-state normative orders is imposing a Western Eurocentric concept of law, and other normative ideas are just 'jammed' into a pre-conceived idea of what is law and could, in the process, become distorted (von Benda-Beckmann and von Benda-Beckmann 2006, 14). Finally, the 'all is law' argument is seen as a concept that is too wide, whereby everything could be perceived as law and consequently crucial differences between various normative systems can be broken down (von Benda-Beckmann and von Benda-Beckmann 2006, 17). Given that the ISS model is based on the extension of quasi-territorial and national jurisdiction in space, it is highly likely that the legal space community will initially make similar objections to the idea that Mars communities could develop their (own) space laws and space settlements could be governed through a legal plural system that consists of terrestrial and Martian laws.

However, as I already mentioned in the introduction, legal pluralism is also part of a global legal system. The global legal system consists of 'an interlocking web of jurisdictional assertions by state, international and non-state normative communities' (Berman 2007, 1159). Increasingly, states must share jurisdiction over the same activity or they must share their legal authority with multiple courts (e.g. at international, national, and regional levels). Or they may even share authority with non-state legal or quasi-legal norms. In these sites of hybrid legality, the overlapping legal authorities may clash and cause conflict and confusion. Different options exist to resolve such conflicts, the most common ones are imposing or reinstating the primacy of territorial or state governed legal systems or seeking universal harmonization of legal systems and norms (Berman 2007, 1162). An alternative option that is endorsed by global legal pluralists is to embrace legal pluralism and actively 'seek to create or preserve spaces for conflict among multiple, overlapping legal systems' (Berman 2007, 1164). This version of global legal pluralism is a strong reaction against an overly state-centric or universalist approach. Pluralism is thus not just normative but is also a descriptive framework or a juris generative model. This acknowledges that different legal communities pursue different norms and global legal pluralism studies the interaction between the different actors without suggesting a hierarchical ordering between them. Instead, it focuses on the plural procedural mechanisms, institutions, and practices that provide a platform for the communication between plural voices (Berman 2007, 1166).

One of the leading scholars engaging with global legal pluralism is Teubner, who asks questions about what counts as distinctively legal and how state law and other laws are interrelated. He studies these questions not from an empirical position but instead focuses on the communication between different laws (Melissaris 2004, 62). Melissaris summarizes Teubner's main ideas as follows (Melissaris 2004, 62). Teubner rejects the idea that normativity would be the defining criteria when deciding on the recognition of a legal order. Teubner argues that theories of normativity are too heavily focused on the distinction between legal and non-legal phenomena, thereby neglecting the dynamic and process-led character of legal pluralism. Teubner equally discards functionality theory which promotes social control as the ultimate arbiter and consequently is not helpful when differentiating between the legal and the non-legal as these rules are usually too inclusive and every social relationship or function could be seen as legal. Instead, Teubner, inspired by a linguistic turn in legal theory, argues that legal pluralism is 'a multiplicity of diverse communicative processes that observe social action under the binary code legal/illegal' (Teubner 1983, 1451).

What is particularly interesting in Teubner's work is the level of abstraction he achieves through systems theory which helps to move beyond the sometimes emotive debate about what counts as law when linked to specific legal orders. For Teubner, studying legal pluralism through the prism of systems theory provides 'a basis for studying what is legal within society without either conflating all law with the official law of the state or losing its ability to separate what is legal from the rest of society' (Nobles and Schiff 2012, 267–8).

Whilst systems theory is terribly complex, in order to stay concise, I will just outline some of its major points in relation to Teubner's work on legal pluralism and I draw mainly upon the work of Nobles and Schiff to explain systems theory and its relevance for legal pluralism. According to systems theory, modern society consists of different communication subsystems. There could be, for example, an economic system, a legal system, a cultural system, and so forth. Each system is autopoietic, meaning that it forms its communication from itself, so each subsystem's communication applies a binary code. In the case of law, the code that is applied is legal/illegal (Nobles and Schiff 2012, 270). Because systems theory concentrates on coding it has more potential to have a more inclusive view on what is law whilst distinguishing between the centres and peripheries of the legal system (Nobles and Schiff 2012, 270). In the case of the law, the centre is the courts with their responsibility to make a decision about what is law. In order to do so, they have developed a level of systematic complexity that is beyond reach of most laypeople to use and understand (Nobles and Schiff 2012, 276). Therefore, Luhmann refers lay legal communications to the periphery. However, what is important to note is that this distinction between centre and periphery does not indicate a hierarchical relationship between the two. All legal communications are just communications, and the communication of the courts are not necessarily more legal than the communications of lay peoples (Nobles and Schiff 2012, 267). Legal meaning only develops when the system is applying the binary code legal/illegal, implicitly or explicitly, when constructing itself and its environment (Nobles and Schiff 2012, 267).

From the point of view of systems theory of legal pluralism, it could be envisaged that in the longer-term, Mars settlements' legal system may at the centre still be ruled by the national jurisdictions of the registered objects and people living on Mars. However, at the periphery, Mars' legal system would also accommodate the new legal rules that emerge on Mars through the interaction with other systems in the space settlements. In systems theory, the main issue is not about finding out whether a particular society has a legal system or not. What systems theory does is to understand society and the different interactions that are taking place within society (Nobles and Schiff 2012, 291). Systems theory and how it is used to analyse the law starts from the basic belief that modern society consists of functionally differentiated social systems. Functional differentiation fractures modern society so that the options of communication (which is needed in order to create meaningful actions) become structured separately within each system (Nobles and Schiff 2012, 291).

When it comes to Martian communities, systems theory goes against the long-held belief in classical legal pluralism that these communities are closed systems. Systems theory shows that Martian communities are part of a wider environment and society, including the nation-states that will, because of technological, logistical, and socio-legal dependency, have a lengthy presence in Martian societies. But what systems theory can cater for is the creation of an environment wherein communities show willingness to be part of a common set of discursive forms and push forward the agenda of procedural pluralism. One way of achieving this is through focusing on a technique that is widely used in anthropology: thick descriptions of the ways in which various procedural mechanisms, institutions, and practices actually operate as sites of contestation and creative innovation that eventually can form new laws.

Whilst it is beyond the possible to study Martian communities and their communicative strategies, as a thought experiment I am introducing in the final section of this chapter the Association of the Autonomous Astronauts (AAA). The AAA as a case study offers insights regarding how an intentional community of autonomous astronauts used space, technology, and the wider contextual environment of space travel as an opportunity to imagine new forms of social interaction (Sammler and Lynch 2021, 715). The reason why I use the AAA as an example is twofold. First, its communicative tactics can illustrate how global legal pluralism can work in action. Second, it also allows me to include a more critical angle in the study of global legal pluralism and space travel. It builds on the wider call in studies on global law that communities of practice could also become normative setting. Whilst, for example, in the area of global environmental law, Indigenous and local communities are being considered as experts in law-making through their customary law-making practices (see e.g. Vermeylen 2015), I argue that a similar approach could also be adopted for space law. Indigenous legal practices as sources of global law (i.e. norm-setting) are even recognized in international law, of which the Convention on Biological Diversity (CBD) is a good example.

Interestingly, Indigenous law as a communicative device that disrupts and changes law-making procedures has also been expressed through artistic methods and processes of which the *Ngurrara Canvas* is a good example (Vermeylen 2021, 273–7). Therefore, the unease that the space law community may experience about the idea that the practices of the AAA could be interpreted as legal procedures is a reflection of the deeply conservative approaches in space law. Ironically, whilst space law prohibits the extension of territorial jurisdiction into outer space by claims of sovereignty, it is still a body of law that operationalizes its rule of law through extending the state's quasi or national jurisdiction into space through international and national laws, and multilateral and bilateral agreements. Space law is one of the very few areas in law that has not yet embraced idea of global law and the inclusion of non-state and non-governmental communities of practice into law-making. Paradoxically, it is very likely that it is also an area that in the future may face some unprecedented challenges to expand its horizons when space tourism and the first Mars settlements become a possibility.

The AAA was formed in 1995 and was active for five years until it disappeared in 2000. The AAA was a network of community space programmes including 30 chapters across the United Kingdom, Europe, North and South America, and Oceania. They drew inspiration from the Situationist International (SI), a group of artists that experimented with radial activism in order to overturn existing political establishments. The SI developed a new politics of resistance that was based on mocking, reusing, or distorting existing forms of politics and governance in order to undermine and expose the socio-economic conditions that enabled exploitation (Bonnett 1999, 25). Although it was not its main site for contention, the SI expressed as early as the 1960s that space travel was an activity that needed subversion (Bonnett 1999, 25). For example, in 1969 Eduardo Rothe wrote in the magazine, *Internationale Situationniste*:

The first men to go beyond the atmosphere are the stars of the spectacle that hangs over our heads day and night … As an example of survival in its highest manifestation, the astronauts make an unintentional critique of the earth: condemned to an orbital trajectory—in order to avoid dying of cold and hunger—they submissively (technically) accept the boredom and poverty of being satellites.[3]

Rothe concludes by predicting that

[m]an will enter space to make the universe the playground of the last revolt … We will enter space not as employees of an astronautic administration or as volunteers of a state project, but as masters without slaves reviewing their domains: the entire universe pillaged for the workers' councils.[4]

The AAA was influenced both by the subversive techniques that were used by the SI to criticize and change the status quo as well as the message that space travel was an activity that could be changed, or literally in the language of the SI turned around (*détournement*). Both the SI and the AAA wanted to mutate technology so it could be repurposed for different political ends, which was the liberation of the working class (Bonnett 1999, 26).

Launched in 1995, the AAA created a worldwide network of local, community-based groups dedicated to building their own space ships. Refashioning the SI slogan 'Below the Paving Stones: The Beach!', the AAA proclaims 'Above the Paving Stones: The Stars!', explaining that 'the AAA has formed an approach to technology that is primarily concerned with investigating how a specific technology is used and who gets to use that technology ... Autonomous Astronauts create a complex interactive project that anyone can participate in, and which completely changes existing notions of space travel.'

Remarkably, whilst space law is entangled with technological advancements, for the AAA, technological utopianism is more like a dystopia. The AAA is therefore also clear in its message that technological advancements should not propel space exploration. Instead, space travel should be instead be led by the possibilities of social conduct and experimental relationships (Bonnett 1999, 26). This makes the AAA the world's only space programme, notes AAA's astronaut, Skeet, which makes technological issues secondary to the social relationship of space communities (Skeet 1996).

As can be read in the First Annual Report of the Association of Autonomous Astronauts, 'Here Comes Everybody!', the AAA were dismayed how space travel programmes missed the opportunity to do things differently in space:

> What new perspectives are available to an independent, community-based space exploration program? Autonomous Astronauts create a continuous revolt based on their own selfish desires for adventure. What else could bring these pioneers together? Each human is, in their own way, a complete universe of thought and being—so how could any Autonomous Astronaut possibly endorse government exploration programs which are connected, by their military associations, to the wilful destruction of entire universes? In zero-gravity the present-day capitalist economic framework will be transcended, as the technology required to sustain life in space will also be harnessed to guarantee a life of endless pleasure and zero-work (work in the sense of capitalist wage slavery). Autonomous Astronauts construct their own continuous games, which they alone have chosen to play, whilst zero-gravity will dissolve the competitive urges that are so discouraging to projects based on planet earth.
>
> (Inner City AAA 1996, 6)

It is through this desire to do things differently in space that the AAA used tools such as psychogeography, art, music, performance, and pamphlets to construct and show-case other ways of knowing space than state- and corporate-sponsored knowledge that has so far been serving the space endeavour (Sammler and Lynch 2021, 714).

Whilst arguably the biggest production of knowledge was focused on subverting technological advancements for other—more peaceful and non-capitalist—purposes, the
AAA also challenged space law of which the following extracts are good examples
(these examples are not exhaustive).

First, it is critical of the existing laws, treaties, and international tribunals whose
purpose is to regulate the vertical extension of capitalist production modes into space.
In 'Space is the Place', the AAA critiques that space is used as 'the planetary hope' to
build an extension of a terrestrial economic system in space so that 'a new territory'
comes into being that can accommodate new producers and consumers. Space is thus
reduced to a 'province of accumulation' (Inner City AAA 1996, 17).

Second, in its Second Annual Report of the Association of Autonomous Astronauts, 'Dreamtime is Upon Us!', the AAA rejects the rationale of government space
programmes that are dominated by the views of engineers, and distance themselves
from terraforming projects on Mars as they associate human settlements on Mars
as actions of a capitalist system that is 'completely out of control, has exhausted the
earth's resources and requires another planet to devour' (Inner City AAA 1997, 5–6).
This statement is also accompanied by denouncing space programmes that are still
intertwined with Earth-based notions of national borders and state controls (Inner
City AAA 1997, 7). In its Second Annual Report, the AAA directly opposes space
tourism and private entrepreneurship in space, as these are continuations of the
capitalist myth of free markets (Inner City AAA 1997, 7).

Third, the Fourth Annual Report of the Association of Autonomous Astronauts,
'Space Travel by Any Means Necessary!', includes a Treaty on Use of Outer Space,
Including the Moon and Other Celestial Bodies. The Treaty is the manifesto of
Nomadic Individuals who are addressing the Governments of the Member States
of the Joint Space Treaty Agreement, pointing out explicitly the ambiguity and tension I highlighted earlier in the OST. The OST declares that outer space shall be
free for exploration and use by all states but it also 'dictates' that the activities of
non-governmental entities in outer space shall require authorization and continuing supervision by the appropriate state party. The Nomad AAA states, therefore, in
Article 1 of its Treaty that:

> [t]he exploration of space shall be carried out for the benefit of all peoples, individ-
> uals and species of the cosmos, irrespective of their location in space and time. This
> treaty does not recognise any notion of country or state and, further, seeks to abolish
> such notions as an intrinsic part of this agreement.
>
> (Inner City AAA 1999, 22).

In this quotation there is a clear detachment of the current space programmes and
the laws that accompany and facilitate space travel. In their reflection as about the
actions and the pamphlets of the AAA, Sammler and Lynch remark that the AAA
as a futurist community reflects how the political organization and labour organizations in space could look like (Sammler and Lynch 2021, 116). In that vision there is
a strong critique against the military and state-led control of outer space. Pointedly,
already in the 1990s, the AAA was already signalling its concerns about the privatization of outer space through the capitalist and imperialist notions of property that

would guide future access into space. In its Fourth Report, the AAA reflects upon the private sector plans and how they seek to promote the utilization and material resources to benefit the expansion of capitalism in space (Inner City AAA 1999, 48). They compare the colonial expansion in space with the actions of the colonists in America (Inner City AAA 1999, 48). It is in these outspoken criticisms that other normative values about law-making get traction in their manifesto.

4. Conclusion

As a way of conclusion, as Sammler and Lynch clarify, the AAA criticizes how space travel has been wrapped up with specific onto-epistemologies that present off-world activities as particular forms of technological and scientific achievements (Sammler and Lynch 2021, 715–16). This leaves, according to the AAA, no room for other ways of knowing and being in outer space. I argue that a similar point can also be made about space law. As outer space is increasingly opened up as a new frontier for commercialization and extractive entrepreneurship, the role of the nation-state in regulating off-planetary access may shift. Judging from the initial changes, it could well be that the role of the nation state could actually become bigger rather than smaller. As more recent developments, such as the Artemis Accords and the US Space Act, seem to suggest, the United Nations and international space law may become subordinate to national laws or bilateral agreements that create a legal environment that is conducive for the further development of the commercial space sector. I argue that there is a sense of urgency for space law to learn from other global legal practices. International biodiversity law is a good example where non-state actors are given a platform from which they can influence and promote their living law practices. Ultimately, future commercial space travel opens up a new exciting enquiry for the law. Part of global legal pluralism is to query the legality and legitimacy of the law. Instead of falling back onto state-centric and universal approaches to law-making, space travel could open up a new imaginative practice for the law.

Legal pluralism is all about creating an inclusive legal environment so that a diverse set of legal manifestations can become equal partners in lawmaking processes. As Melissaris argues, law from a legal pluralist perspective is about opening up the space for political deliberation, transparency, and publicity (Melissaris 2013, 174). I argue that the vision of the AAA is providing the first seeds for nurturing an inclusive dialogue about the legal meaning of space being for the benefit of humankind at the dawn of NewSpace. Surely this should also include a participatory approach towards law-making. If space is for the benefit of humankind, by extension it seems only fair that humankind also has a voice in the legal processes that decide how space should be governed in the future.

Admittedly, legal pluralism may for some space lawyers just be a theoretical exercise. But as Melissaris points out, any legal theory always reaches its limits (Melissaris 2013, 180). In order to implement foundational law, it needs politics. Translating legal theory into substantive matter is not just the job of the lawyer, it needs a collective, or, in the words of Ehrlich, it needs living law. As I have shown in the chapter, living law subordinates state law. So after all, there may be hope that one day the AAA's Treaty finds traction beyond the earthly astronaut and travels upwards, as the AAA would like to see it, on a self-made space rocket.

Notes

1. C. Young. 'Origins of the Seven Stars' (2010) Archival injekt print, 49.8 × 29.8 cm <http://www.careyyoung.com/works#/origin> accessed 13 December 2020.
2. C. Young. 'Legal Fictions' at Migros Museum für Gegenwartskunst, Zurich, text in *Mousse Magazine* Paula Cooper Gallery, 9 September 2013 <http://moussemagazine.it/carey-young-migros/> accessed 13 January 2020.
3. <http://www.bopsecrets.org/SI/12.space.htm> accessed 22 December 2022.
4. <http://www.bopsecrets.org/SI/12.space.htm> accessed 22 December 2022.

References

Berman, P. 'Global Legal Pluralism' (2007) 80 *Southern California Law Review* 1155–238.

Bonnett, A. 'Situationist Strategies and Mutant Technologies' (1999) 4 *Angelaki: Journal of the Theoretical Humanities* 25–32.

Chan, Y. C. Protecting the Million-Year Picnic: The Importance of Importing the Rule of Law to Mars' in A. Froehlich (ed.), *Assessing a Mars Agreement Including Human Settlements* (Cham: Springer, 2021) 181–91.

Ehrlich, E. *Fundamental Principles of the Sociology of Law* (Cambridge, MA: Harvard University Press, 1936).

Farsaris, A. E. 'The International Space Station (ISS) Intergovernmental Agreement as a Precedent for Regulating the First Human Settlements on Mars' in A. Froehlich (ed.), *Assessing a Mars Agreement Including Human Settlements* (Cham: Springer, 2021) 63–74.

Griffiths, J. 'What is Legal Pluralism?' (1986) 18 *Journal of Legal Pluralism and Unofficial Law* 1–55.

Hamilton, J. A. *Indigeneity in the Courtroom: Law, Culture, and the Production of Difference in North American Courts* (New York, NY: Routledge, 2009).

Inner City AAA 'Here Comes Everybody—The First Annual Report of the Association of Autonomous Astronauts' (London: Inner City AAA, 1996).

Inner City AAA 'Space Travel by Any Means Necessary—The Fourth Annual Report of the Association of Autonomous Astronauts' (London: Inner City AAA, 1999).

Kulovesi, K., Mehling, M., and Morgera, E. 'Global Environmental Law: Theory, Challenge and Promise' (2019) 8 *Transnational Environmental Law* 405–35.

Malinowski, B. *Crime and Custom in Savage Society* (Littlefield: Adams & Co., 1926).

Melissaris, E. 'The More the Merrier? A New Take on Legal Pluralism' (2004) 13 *Social & Legal Studies* 57–79.

Melissaris, E. 'From Legal Pluralism to Public Justification' (2013) 3/4 *Erasmus Law Review* 173–80.

Mohr, R. 'Law and Identity in Spatial Contests' (2003) 5 *National Identities* 53–66.

Nobles, R. and Schiff, D. 'Using Systems Theory to Study Legal Pluralism: What Could be Gained?' (2012) 46 *Law & Society Review* 265–96.

Sammler, K. G. and Lynch, C. R. 'Spaceport America: Contested Offworld Access and the Everyman Astronaut' (2021) 26 *Geopolitics* 704–28.

Singh, S. 'Eugen Ehrlich's 'Living Law' and its Legacy for Legal Pluralism' (2010) Available at SSRN: <https://papers.ssrn.com/sol3/papers.cfm?abstract_id=1660606>.

Skeet, J. 'Dreamtime is Upon Us—The First Annual Report of the Association of Autonomous Astronauts' (London: Inner City AAA, 1996).

Tamanaha, B. Z. 'Understanding Legal Pluralism: Past to Present, Local to Global' (2008) 30 *Sydney Law Review* 375–411

Teubner, G. 'Global Bukowina: Legal Pluralism in the World Society' in G. Teubner (ed.), *Global Law without a State* (Aldershot: Dartmouth Publishing Co Ltd, 1997) 3–18.

Teubner, T. 'Substantive and Reflexive Elements in Modern Law' (1983) 17 *Law and Society Review* 1443–62.

United Nations Office for Outer Space Affairs *Treaty on Principles Governing the Activities of States in the Exploration and Use of Outer Space, including the Moon and Other Celestial Bodies* (1967) <http://www.unoosa.org/oosa/en/ourwork/spacelaw/treaties/introouterspacetreaty.html> accessed 10 July 2022.

van Asten, E. 'Legal Pluralism in Outer Space' in Nandasiri Jasentuliyana Keynote Lecture on Space Law & 2nd Young Scholars Session. *New Perspectives on Space Law.* Proceedings of the 53rd IISL Colloquium on the Law of Outer Space, 2010 The International Institute of Space Law, Paris. 67–80

Vermeylen, S. 'Comparative Environmental Law and Orientalism: Reading beyond the "Text" of Traditional Knowledge Protection' (2015) 24 *Review of Environmental Comparative International and European Law* 304–17.

Vermeylen, S. 'Space Art as a Critique of Space Law' (2021) 54 *Leonardo* 115–24.

Vermeylen, S. 'Canvases as Legal Maps in Native Title Claims' in U. Dieckmann (ed.), *Mapping the Unmappable? Cartographic Explorations with Indigenous Peoples in Africa* (Bielefeld: Transcript Verlag, 2021) 261–90.

von Benda-Beckmann, F. and von Benda-Beckmann, K. 'The Dynamics of Change and Continuity in Plural Legal Orders' (2006) 53–54 *Journal of Legal Pluralism* 1–43.

von der Dunk, F. G. *Private Enterprise and Public Interest in the European Spacescape* (International Institute of Air and Space Law, Faculty of Law, Leiden University, 1998).

von der Dunk, F. G. 'The International Legal Framework for European Activities on Board the ISS' in F. G. von der Dunk and M. M. T. A Brus (eds), *The International Space Station, Commercial Utilisation from a European Legal Perspective* (Leiden: Martinus Nijhoff Publishers, 2006) 15–32.

Walker, N. *Intimations of Global Law* (Cambridge: Cambridge University Press, 2015).

Young, C. '"Legal Fictions" at Migros Museum für Gegenwartskunst, Zurich' *Mousse Magazine Paula Cooper Gallery*, 9 September 2013 <http://moussemagazine.it/carey-young-migros/> accessed 10 July 2022.

26

On libertarian communities in/around outer space

Is ecology an antithesis to liberty?

Matjaz Vidmar

1. Ecosystem as Analytical Framework for Governance

Ecosystem is defined as the relationship between the totality of an environment (biotop) and all living entities within (biocenosis). As such, the mutual dependency between the living space and its inhabitants creates an interlinked web of (balanced) constraints and affordances, that both make life possible and at the same time constrain its form and abundance. Hence, an ecological framework by necessity limits absolute liberty of living beings, since their life cannot exist outside a favourable ecosystem. In other words, our (mutual) dependence on ecosystemic services of other living organisms and our common reliance on the non-living environment forces us to behave in a considerate way, or face system-wide disruption that may threaten our survival and that of our cohabitants.

The expansion of our living space into outer space asserts a two-way pressure on the ecosystem. On the one hand, living organisms need to adapt to the environmental conditions and realign strategic partnerships with co-dependents. On the other hand, the new environment gets somewhat reconfigured and modified to suit the requirements of the living organisms through their activities upon and with the non-living matter. Furthermore, the 'expansion' of ecosystemic framework allows for new perspectives on the existing conditions of life and a re-evaluation of our responses to the existing and new affordances and constraints.

In particular, when looking at the extreme conditions of outer space, two critical issues emerge. The environmental constraints placed upon living organisms are so severe that only a technological intervention allows for their survival. However, with technology-assisted survival comes dependence on the source of technology, which is a social conditioning that can be used for the control and deprivation of liberty. Most often, survival strategies in extreme environments require knowledge and skills that extend beyond any individual. Hence, communities form as ways to develop and retain a stable configuration of relationships between organisms and their embeddedness in the biotop. Rule-based systems of behaviour such as hierarchical command structure or value systems for resources are both at the core of enabling the success of community organization(s) to distribute risks to individuals and benefits to all. This is linked to the need to aggregate and understand complex systems in an actionable, practical way.

Matjaz Vidmar, *On libertarian communities in/around outer space*. In: *The Institutions of Extraterrestrial Liberty*. Edited by Charles S. Cockell, Oxford University Press. © Matjaz Vidmar (2022). DOI: 10.1093/oso/9780192897985.003.0027

The second issue is that the critical environment in outer space lends itself particularly strongly to a re-evaluation of the preciousness of the existing ecosystem. It is well documented from the early days of space programmes that a transformative cognitive shift occurred in astronauts when they left our planet's atmosphere as they all found looking 'back' towards the Earth much more awe-inspiring that looking 'forward' into the blackness of outer space (White 1987). Moreover, iconic 'backward-looking' imagery, such as the Apollo 8's *Earthrise*—the Earth appearing over the lunar horizon—captured the public imagination in ways transcending the fascination with the stars and other celestial bodies.

This chapter will examine the interlinking of these two issues with a view to understand the governance framework conditions in outer space. It will start by reviewing the critical contribution of Social Studies of Outer Space to the understanding of the 'expansion' of the ecological framework into outer space through an examination of sociotechnical imaginaries—a type of future visions shaping collective behaviour (Jasanoff and Kim 2009). Furthermore, it will expose a core tension between how the introduction of technology to challenging environments established a (libertarian) community of interest within and beyond Earth's orbits whilst at the same time changing the conditions of the environment to force new constraints upon the liberty of the participants. Finally, the chapter concludes with an attempt to resolve this tension through an acceptance of 'techno-ecological' co-dependency in a way that recognizes the need to work within the 'extended' environment (more) independently from the original ecosystemic framework as a way to establish a new framework of liberty within the constraints of the ecological reality.

2. Expansion of Ecological Framework to Outer Space

The near-Earth outer space is getting increasingly 'occupied' by technological means: the introduction of artificial satellites to receive and transmit information and in some cases harbour life (e.g. space stations). This has led to the establishment of business interests and civic visions for the utilization of space resources and led to outer space 'invading' everyday life through the emergence of an Earth–space economy. Coupling these developments with the broad liberalization of social structures and an increase in public participation in governance and innovation of technology domains, a critical need has arisen for social science to analyse and (more actively) engage in the emerging discussion about techno-scientific futures and societal impacts of life with and in outer space. At the moment, social science is somewhat lagging behind these aims, hence an integrated programme of research is emerging in Social Studies of Outer Space (Alvarez et al. 2019). It has been particularly recognized that due to the multi-dimensional proliferation of this field, a need arose to examine a more systemic view of space exploration and industry, focusing on systemic and ecological dimensions and perspectives (Vidmar 2020).

More specifically, the Social Studies of Outer Space have in recent years focused on a mixed enquiry of the cultural framework for space exploration and industry, especially anthropological studies of the way space is 'constructed' through discourse and practice (Messeri 2016). This is further enhanced by the direct analysis of sociotechnical imaginaries (Tutton 2018) as well as how these imaginaries interlink with framework(s) of (risk) governance (Vidmar 2019) and (neo-)colonization (Alvarez 2020). Particular interest is being paid to the contention between the public discourse surrounding these visions, their 'ground truth' within the socio-political and economic context, and the social structures they (re)produce. Three techno-scientific (eco)systems are especially prominent: the expansion of human settlement to/in outer space, the appropriation of outer space and its resources, and the jarring juxtaposition to the state of the Earth's environmental and social reality and future outlook.

Lead protagonists, both in the private and the public sphere, are advancing utopian narratives of the seamless expansion of our ecosystem into outer space as a/the 'solution' to the earthly challenges as well as an opportunity to reset or at least reconfigure the broken environmental and social relationships. Especially strong narratives have formed around Martian and lunar environments as settings for future cooperative 'villages' (Moon Village Association n.d.) and 'settlements' (The Mars Society n.d.). Similarly attractive seems to be the space in between, despite its lack of planetary affordances such as firm ground, or perhaps precisely because of the absence of land-based constraints (Asgardia—The Space Nation n.d.). Using well-defined terrestrial issues, such as the sustainability crisis, geopolitical inequality and infighting, and potential cosmic existential threats (i.e. catastrophic meteorite impact), prominent individuals, community groups, as well as popular culture productions all call for investment in technology that would enable relocation of people into the imaginary (places). It is important to note that these narratives are far more prevalent in the Western cultural sphere, where a post-humanist version of settler colonialism to right the historic wrongs in the form of astrofuturism (Kilgore 2003) is both echoing the Western approach to colonization as well as promising a new dawn of a more integrated humanity. Exactly how such a transformative process is to be achieved is often left undefined.

Hence, more than ever before a social scientific study of the technical systems underpinning these expansionist narratives is vital. Especially important to address is the seeming re-emergence of practices of place-making that ignore or minimize the impact of environmental constraints on the establishment and evolution of social relationships and structures. At a time of emerging technical capability to enact these imaginaries, this is particularly worrying. In the absence of robust critique, naïve metaphors are emerging as a way to describe solutions to concerns around the implications of space exploration and settlement. For example, it is increasingly apparent that space exploration since 1950 has employed unsustainable practices, such as discarding disused material and equipment not only in 'wastelands' but also in or near valuable locales, such as the low and the mid-Earth orbits. The problems caused by 'space junk' are now increasingly apparent, in particular in conjunction with some brash attempts at demonstrating space 'power' through deliberate shooting-down' and, hence, disintegration of assets, creating even more numerous and harder to control debris. The 'remote' location of such places and incidents does not inspire the same urgency to address these issues as would environmental pollution more central to our living space (e.g. oil spills). Nonetheless, the proposed solution is the same: the practice of 'cleaning'. Through such an everyday metaphor, the problem appears to be 'solved', even though the application of an actual 'cleaning' solution is still in the earliest stages of development, it is very likely to be inadequate to deal with the totality of the issue, and the criticality and immediacy of the danger is severe. Noting such examples of a very acute lack of any understanding of the transformative process the emerging space imaginaries represent, or indeed lack of any evidence of them being properly considered at all, makes a social scientific study of the interrelation between ecological and technological expansion into outer space essential.

3. Libertarian Communes in Outer Space

In order to improve our understanding of the ecological constraints and affordances within the outer space context, the imaginaries giving rise to them need to be looked at in more detail. Especially interesting are the references given within these narratives linking the understanding of the earthly ecosystem to the social structures emerging in the new expanse. We turn to two particular sociotechnical imaginaries: Elon Musk's ambition for 1 million-strong society on Mars by 2050 and the Space Nation—Asgardia. In essence, both these projects offer a sort of utopian libertarian communes with self-referential ecology and governance, free from earthly constraints. Though approaching the imaginary future in space from different perspectives, Musk from technology side and Asgardia from the social dimension, they centre on an age-old premise that the new environment will give rise to a new (better) community organization.

The more publicly visible amongst these plans is Musk's stated intent to develop a large-scale transportation system which would enable 1 million people to live on Mars in a matter of decades (Drake 2016), though he initially started off with a more modest goal of 80,000 (Coppinger 2012). His focus is primarily technological: his first priority is to develop rockets big enough to carry payloads required for such an operation. Interestingly, the objective to settle Mars has been his stated goal all along, the SpaceX's more commercial programme of satellite launches and support for human spaceflight programme are only means to achieve that end. Communicating his vision in a series of keynote presentations, interviews, and tweets, Musk promises all the ingredients for a prosperous settlement. Echoing the narratives of the 'American Dream', he claims Mars 'will be the planet of opportunity', noting that 'there will be a lot of jobs on Mars!' Important to note is that outer space environment constraints mean that the journey to Mars is on a one-way ticket, it being impractical, in terms of energy, to stage any return missions in the first phase of the settlement. In a similarly structured project called Mars One, thousands of people applied for a one-way ticket to a 'Big Bother'-type reality TV show set on Mars (Tutton 2018). Though the project ultimately failed (Foust 2019), it demonstrated that without considering the actual ecological reality, the vision of an underspecified libertarian space commune is very much appealing.

Speaking of the settlement itself, Musk is far vaguer: whilst noting humans will have to live in 'glass domes at first', his future plans revolve around terraforming the red planet, including detonating nuclear explosions over Martian poles to extract carbon dioxide and water (Walls 2019). Either way, two clear visions are set for the

Martian settlement in this imaginary: the environment is to be occupied by technological means and the commune/ity established in its wake is to be libertarian in its outlook. This exposes the core naiveté or lack of consideration of the need to shape the (eco)systemic evolution of the expansion into the new ecological framework. Matters of environmental challenges—such as high energy radiation—are often brushed off as 'technical problems' or even a matter of chance. In an interview, Musk noted that 'a bunch of people will probably die at the beginning' (Gorgan 2021), commenting at other times that one of the settlers' characteristics should be the spirit of adventure, individuals unafraid to place themselves in peril. However, based on the public reaction to such statements, the constraints of the journey to Mars and settling there seems to represent an infringement on people's liberty to a degree largely unacceptable to most, though not all. The dependence on a set of available technologies is likely to be conditioned by trust in their performance and a degree of dependence, both of which include a system of social relationships. Here, the implication is clear: you have to believe in Musk's vision and his technology in order to accept the risk to place your life in the hands of his engineers. Moreover, you seemingly have little recourse to move outside such ecological governance framework once in outer space, if you do lose faith. In short, under the 'technology first' view of outer space settlement, your liberty is the price to pay for the 'opportunity'.

However, not all imaginaries of future space communes are so technocentric. Perhaps the most well known and most developed social imaginary of the past decade is Asgardia, a proposal for a non-territorial space settlement, offering supra-national (or 'transnational') libertarian governance and a 'citizenship' model based on membership application, with tiers. At the moment, this includes free 'followers' as well as fee-paying 'residents'. The promise is that earthly squabbles of the old kind are banned, singling out religion and politics as two such contentious issues. In the words of its founder and 'Head of Nation', Igor Ashurbeyli: 'Asgardia can offer the entire Earth civilization an alternative way'. The group's stated aims are to create 'a fully fledged independent digital state recognized by earthly nations' and 'ensure peaceful space exploration, protect our home planet from cosmic threats and lead the development of new Space Law to eliminate militarization of space' (Mission n.d.). This extends to 'facilitating the first human childbirth in space', which would lead towards human 'immortality as a species'—an echo of the visions of transplanetary settlements as guarantees against existential threats on Earth.

Claiming over 1 million 'Asgardians' have joined the project, it certainly seems an appealing proposition to leave behind past conflicts and constraints of earthly relationships and start building a new social organization of equals. The plan is that one day, Asgardians could do this physically—move to live in a massive orbital space station ('space arc') or even live on the surface of the Moon or Mars. For the moment, though, the only Asgardian physical presence in space is in the form of a small (2.7 kilogram) satellite bearing its name and containing digitally stored photos of its members, state symbols and documents, and two particle detectors to examine the space environment. The mission's aims are twofold: to test the durability of solid-state drive storage under radiation conditions in the low Earth orbit as well as serving as Asgardian 'extraterrestrial territory', which they would like to get recognized as a state by the United Nations—an unlikely prospect, according to legal analysts (Alshamsit, Ballestett, and Hanlonftt 2018).

However, the challenge of this particular (social) expansion into outer space is not only reflecting on the limitations of the existing political organization on Earth but also its almost inevitable reproduction in the new expanse. Though claiming to be 'totally democratic', Asgardia is described as 'constitutional monarchy' due to the 'kingdom' having 'magical' meaning, and invoking religious connotations and correspondingly. Almost absolutist executive power is vested in the Head of Nation with little transparency as to the decisions that Head of Nation takes (Harby 2018). Curiously, similarly to Musk, Ashurbeyli is a billionaire, retired from his original business (in defence manufacturing) and uses his profile and wealth to forward this imaginary. This illustrates that even when the primary objective is a more holistic approach to place-making beyond the Earth, the realities of practically implementing a framework of governance fall well short of the stated libertarian ambitions of organizations 'beyond' the state.

Though perhaps lacking in the realization of their forward-facing libertarian (re)configuration, these socio-technical entities are already ecologically disruptive. Though Musk's SpaceX has not yet developed a settlement, it has established a small foothold in the off-Earth ecosystem by introducing satellites, rocketed bodies, and other material to it. That expansion has also yielded a reflection on the existing (earthly) ecosystem, for instance by championing technological/material reusability and redefining commercial business models for the space–Earth economy. It has also highlighted the ethical and practical challenges of expansion into outer space, especially in a libertarian way. The test launch of SpaceX's Falcon Heavy rocket in 2018 was a means of promoting a car (Musk's Tesla Roadster) launched into space as a sort of publicity stunt, making the ultimate example of 'space junk'.

Whilst aesthetically perhaps more pleasing than blocks of concrete used in many prior test launches, there are significant socio-ecological implications. Besides a nearly indefensible waste of a car at a time of rampant global inequality and challenges of sustainability (Robinson 2018), the addition of objects into outer space underlines the fact there is already quite a lot of stuff out there, in particular in orbits close to the Earth. Though the functional objects are creating an 'environment' which provides referential interaction back to Earth through services like remote sensing and limited biological habitation (chiefly on the International Space Station and similar habitats), the discarded material is increasingly seen as a threat. Due to the high energy motion of these objects and their uncontrollability, they are at risk of colliding and extensively damaging 'useful' artefacts, such as satellites, as well as potentially scattering debris that further endangers others. This is described as Kessler Syndrome: an exponential effect that could create a massive field of debris, making traversing certain orbits extremely risky and thus effectively halting our ecological expansion anywhere into outer space.

It seems the danger of human-introduced 'space junk' could be potentially placing an existential constraint on our liberty to pursue the expansion of our ecosystem. Alternatively, our desire for ecological expansion places a constraint on our liberty to produce more space waste. Either way, even in the purely material and technological expansion of the ecosystem (without living organisms being involved directly in the *in situ* interaction) is a challenge to the framework of (absolute) liberty. Hence, it seems likely that with further development of various communes and communities in outer space, a techno-ecological perspective will be the reference frame for their organization, whether their protagonists want it or not!

4. Towards 'Techno-ecology'

Once the frame of analysis of the ecosystemic expansion into outer space shifts towards a combination of technological and ecological considerations, a wider array of concepts becomes available to understand its relationship to individual and communal liberty. Of crucial importance are both the relationship between communal criticality of the environmental co-dependence and the individual's ability for decision-making. A techno-ecological perspective also allows for understanding how any of the emerging outer space communes—since these societies are primarily though to be self-referential—interact with the existing ecosystem on Earth, including its social structures. Ultimately, the techno-ecology is a way to reimagine constraints and affordances of the off-Earth environments and allows for both recognition of ways in which ecological considerations limit liberty *in situ* as well as how they allow for (partial) liberation from existing/past governance frameworks.

As outlined at the beginning of this chapter, the most fundamental ecological principle is the intrinsic relationship between living organisms and their environment. One critical consideration for expanding terrestrial life into outer space is then the structure and the affordance of the new environment. For instance, anthropological studies have been looking at 'waste' as 'matter out of place' (Douglas 1966), a relational perspective that enables the understanding that 'waste' is a matter of context and can be re-evaluated as non-waste. Hence, 'space waste' could also be seen as a 'space resource'—tonnes of metal, silicates, and the like, which can be repurposed for future missions and projects. The recognition that shifting our (social) perspective on resources and functions of materials and artefacts can lead to a certain amount of independence from the originating ecosystem allows to start the process of ecosystemic (re)configuration within/of the new environment.

Conversely, an ecological perspective on life on the surface of Mars or at an Asgardian 'space arc' needs to take into consideration the context of the *in situ* social relationships. Of particular concern here is that it is the ecological interdependence which frames the critical fragility of life's existence beyond established ecosystemic boundaries. When moving to a techno-ecology, the technological intervention which enables the much more limited ecosystem to (continue) to exist in an otherwise hostile environment becomes the governing force within such an ecosystem. In addition, if the technology itself is somehow dependant on the physical and/or social affordances of the originating ecosystem (i.e. Earth), that is a particularly acute limitation. That liberty under such circumstances will be severely limited goes without saying, the question is for how long? Is independence (or more crucially, self-sustainability) ever possible? In spite of the optimistic sociotechnical imaginaries of thriving independent civilizations in outer space, the future is uncertain. It is possible that eventual emancipation is achievable, however with its pursuit, the early settlers are condemned to a long period of severely limited liberty. Is that process sustainable or will it lead to (self-) destruction of the community? Even if the society survives, its initial of deprivation of self-governance may leave it incapable of envisioning, let alone enacting, any libertarian ideals currently put forward.

The bottom line of this experimental exploration of the interplay between ecology and liberty is that they are indeed in conflict when extending the ecosystemic framework. On the one hand, past ecological constraints and affordances have shaped our systems of thought and governance in a way that frames liberty as a function of our ecosystem. Since reaching out to space is an extension (and not the establishment) of the ecosystemic framework, we are projecting our past/current perspectives into the new/imaginary as well. This is exacerbated when a techno-ecological reality dictates that a strong referential dependency remains in place. On the other hand, as the ecosystemic framework is stretched, a more systemic re-evaluation of the existing social relationships occurs. This has been mentioned previously in the context of SpaceX's disruption to the satellite and human spaceflight market without direct reference to the grander vision of settlements on Mars. Therein lies perhaps a glimmer of hope that techno-ecology is not antithesis to liberty—as the pursuit of the ecosystemic expansion once realized may bring about an entire (new) governance framework, both at the origin (i.e. on Earth) as well as in the expanse (i.e. outer space).

Hence, a degree of techno-eco optimism is perhaps not unwarranted, though extreme caution should be advised. Historiographical studies of past colonizations clearly show that the application of past governance frameworks to expanses of (the Western) social sphere on Earth caused untold suffering. The adoption of a 'business-as-usual' attitude or dismissal of ecological constraints as 'teething problems' do not bode well. Similarly problematic are replicating everyday metaphors—such as cleaning outer space of space junk—or social structures like a 'constitutional monarchy'. Whilst these may or may not survive the contact with the real techno-ecology of outer space, the damage they can do in the process of attempted application may make it unviable for any real sense of liberty to emerge. As such, perhaps ecology is not antithesis to liberty but a biotop to its biocenosis: neither can exist without the other and they are mutually interdependent.

References

Alshamsit, H., Ballestett, R., and Hanlonftt, M. L. D. 'Space Station Asgardia 2117: From Theoretical Science to a New Nation in Outer Space' (2018) 16 *St. Claire Journal of International Law* 37.

Alvarez, T. 'The Eighth Continent: An Ethnography of Twenty-First Century Euro-American Plans to Settle the Moon' The New School ProQuest Dissertations Thesis. https://www.proquest.com/openview/95e2de8e8c8e9fb9397908dcb3ff7d08/.

Alvarez, T., Clormann, M., Jones, C., et al. 'Social Studies of Outer Space' (2019) In Innovating STS Digital Exhibit, curated by Aalok Khandekar and Kim Fortun. Society for Social Studies of Science. August 2019.

Asgardia. 'The Space Nation' <https://asgardia.space/en/> accessed 27 January 2022.

Coppinger, R. 'Musk Outlines Mars Colonization Vision' SpaceNews.com (2012).

Douglas, P. M. *Purity and Danger: An Analysis of Concepts of Pollution and Taboo* (London: Routledge, 1966).

Drake, N. 'Elon Musk: A Million Humans Could Live on Mars by the 2060s' (2016) National Geographic Magazine. https://www.nationalgeographic.com/science/article/elon-musk-spacex-exploring-mars-planets-space-science.

Foust, J. Mars One Company Goes Bankrupt' SpaceNews.com (2019).

Gorgan, E. '"A Bunch of People Will Die" Going to Mars, Elon Musk Says' *autoevolution* (2021) <https://www.autoevolution.com/news/a-bunch-of-people-will-die-going-to-mars-elon-musk-says-160013.html> accessed 28 January 2022.

Harby, B. 'Asgardia: The Problems in Building a Space Society' *BBC Future* <https://www.bbc.com/future/article/20180803-asgardia-the-problems-in-building-a-space-society> accessed 28 January 2022.

Jasanoff, S. and Kim, S. H. 'Containing the Atom: Sociotechnical Imaginaries and Nuclear Power in the United States and South Korea' (2009) 472(47) *Minerva* 119–46 <https://doi.org/10.1007/S11024-009-9124-4> accessed 9 July 2022.

Kilgore, D. W. D. *Astrofuturism: Science, Race, and Visions of Utopia in Space* (Pittsburgh, PA: University of Pennsylvania Press, 2003).

The Mars Society. <https://www.marssociety.org/> accessed 27 January 2022.

Messeri, L. *Placing Outer Space* (Durham, NC: Duke University Press, 2016).

Mission. <https://asgardia.space/en/pages/mission> accessed 28 January 2022.

Moon Village Association. <https://moonvillageassociation.org/> accessed 27 January 2022.

Robinson, N. 'Why Elon Musk's SpaceX Launch is Utterly Depressing' *The Guardian*, 7 February 2018 <https://www.theguardian.com/commentisfree/2018/feb/07/elon-musk-spacex-launch-utterly-depressing> accessed 28 January 2022.

Tutton, R. 'Multiplanetary Imaginaries and Utopia' (2018) 43 *Science, Technology, and Human Values* 518–39. https://doi.org/10.1177/0162243917737366

Vidmar, M. 'On the Practices of Risk Re-Normalisation: "Knowing" the Known Unknowns in Public Discourse on Outer Space Exploration' (2019) 56 *Teorija in Praksa* 814–35.

Vidmar, M. 'Transplanetary Ecologies: A New Chapter in Social Studies of Outer Space?' (2020) 39(2) *EASST Review* 57–60.

Walls, M. 'Looks Like Elon Musk Is Serious About Nuking Mars' Space.com (2019).

White, F. *The Overview Effect: Space Exploration and Human Evolution* (3rd edn, American Institute for Areonautics and Astronautics (AIAA): 1987) https://doi.org/10.2514/4.103223.

27

Law and liberty on the Moon

Frans G. von der Dunk

1. Introduction

As 'law' is, in a manner of speaking, a form of mathematics working with words and punctuation marks instead of numbers and mathematical symbols, but essentially requiring a similar level of precision if meaningful discussion is not to be replaced by shouting matches, it is important to define at the outset the key terms and concepts involved in a particular subject matter before engaging in further legal analysis. That is certainly no different for a topic as exciting and thought-provoking as the issue of law and liberty on the Moon.

Of the three key concepts thus making up the title of this contribution, 'Moon' is easiest to define. The 'Earth's only natural satellite',[1] it is recognized by the 1967 Outer Space Treaty,[2] the seminal international treaty providing for the legal framework for all space activities and respected as such by all spacefaring nations of the world,[3] as one of the 'celestial bodies'[4] subject to particular attention in this context. A later treaty, the 1979 Moon Agreement,[5] would present an effort to further detail the Outer Space Treaty's regime specifically with respect to the Moon and other celestial bodies, but as it has only been ratified by a limited number of signatories,[6] one should be very careful not to attribute too much value to it from a legal perspective.[7]

As for 'liberty', often viewed as a synonym for the word 'freedom', it has been variously defined as 'the ability to do as one pleases', and more focused on *political* liberty, as 'the state of being free within society from control or oppressive restrictions imposed by authority on one's way of life, behaviour, or political views'.[8] The fundamental dichotomy here is between the freedom of individuals, considered an innate and naturally just state of affairs, and the threat of governmental authority to take away those freedoms.

At the same time, government *has* to play a fundamental role in society also for those freedoms to 'function' as desired: given the realities of human fallibility, human interaction, and human societies, the freedom of one person easily impacts upon or even infringes the freedom of another, and this is where state authority would have to step in. A proper balance has to be achieved as between the freedoms of the one and the freedoms of the other by some overarching authority. That is also where 'the law' comes in, to define the precise extent and scope of any individual liberties *vis-à-vis* the accepted need for governmental authority to protect other individual's liberties—and beyond that, to protect the common interests of society at large as well.

'Law' after all (not including here the laws of physics) is very much a societal tool, a social construct to address human interaction and activity precisely to try and achieve

Frans G. von der Dunk, *Law and liberty on the Moon*. In: *The Institutions of Extraterrestrial Liberty.*
Edited by Charles S. Cockell, Oxford University Press. © Frans G. von der Dunk (2022).
DOI: 10.1093/oso/9780192897985.003.0028

the proper balance between the various freedoms of the various constituent elements of society—those same fallible humans—as well as *vis-à-vis* the public common good at large.

Most people would immediately understand that the way the law tries to achieve that is through reflecting a general sense of justice and fairness—this is the ethical aspect of the law. The general sense of fairness and justice normally would be shifting in any society over time, but even if the law is usually, by its very nature, lagging behind, it is supposed to reflect at least the main tenets of what at any given time the society in question considers to be just and fair.[9]

In real life, however, it is probably true to say that in many cases the practical aspect of the law is more important or at least more to the point: the effort to establish some level of predictability in human behaviour for the common good of all concerned. There is nothing inherently more just or fair in driving on the left-hand side of the road as opposed to driving on the right-hand side, yet a binding choice clearly has to be made either way if traffic is not to be replaced by total and continuing chaos and, quite regularly, disaster. That is also, by the way, why the law is apt to lag behind societal developments with respect to what justice and fairness are supposed to be: it is its inherent sturdiness, which allows it to escape, to some extent, the heat of the moment, the hype of the day, therefore provides at least a modicum of predictability and foreseeability.[10]

2. The Role of Law of Outer Space

All this is in principle no different for outer space, albeit that the special nature of outer space and human activities therein raises some further complexities. Symbolically, that already begins with the fact that 'left' and 'right', as convenient tools for practical arrangements on Earth, as such lose their significance in the three-dimensional and in principle infinite realm of outer space and would have to be replaced by more sophisticated concepts.

First, when discussing 'space activities' as the category of human activity that 'space law' would essentially need to address,[11] far more even than in other realms of international law, the relevant activities are usually not conducted by individual humans as such but by individual or groups of humans formally acting on behalf of so-called legal fictions such as commercial companies and sovereign states. These legal fictions mean that the actions of certain *humans*—making a satellite broadcast a commercial TV show or politically tinged messages, by fiddling with controls in a ground station—are for legal purposes first and foremost attributed to the company of which they are employees or the state of which they are propaganda officials. It would be the company paying any fines for violating nationally agreed and enforced rules of frequency usage; it would be the state having to repair any international violation of the rights of (an)other state(s).

Second, it should be realized that most law presumes a certain 'unity of location' of three constitutive elements: the activity (of which law determines the extent of legality), the actor undertaking that activity (which is held accountable for it by law), and the target of the activity (which law allows to legally challenge the activity) are usually present in the same state, allowing a single national law logically to address all three elements in one coherent go.[12]

From this angle, space law is—again—very special in that, from the perspective of 'location', one then needs to discern three fundamentally different categories of space activities each requiring its own approach to dictating, limiting, and/or protecting whatever liberties might be at issue. Only for a relatively exceptional category of human spaceflight activities will the human actors (to be) targeted by the law find themselves in the same legal realm as the activity itself and the result thereof: in outer space.

Even here, whilst normally the 'unity of location' makes it easy to apply and enforce law—since all constituent elements are situated in the same state—in this case the 'global commons' nature of outer space means that it does not and cannot geographically form part of *any* state.[13]

As a consequence, of the two universally recognized standard approaches for states to assert and enforce law they can *not* apply jurisdiction on the basis of territory to assert law in that realm, only jurisdiction on the basis of nationality—since astronauts do not lose their nationality merely on account of being present in outer space. In addition, space law offers states a somewhat limited additional tool: by registering space objects launched into outer space, states can exercise quasi-territorial jurisdiction over such space objects and its personnel even if temporarily present outside of the space object.[14]

The overwhelming category of activities in outer space, however, by contrast is actually and essentially 'remote-controlled', meaning that the actors are somewhere down on Earth, in other words usually on some state or other's territory, and thereby subject to its territorial jurisdiction, whilst the result of their activities plays out somewhere in outer space, achieved by way of radio signals or similar technologies. Of course, states can still *also* exercise nationality-based jurisdiction over the actors and/or quasi-territorial jurisdiction over space objects involved as relevant.

In addition, a third major category of space activities concerns activities actually, at least for a major part, not even resulting *itself* in activities in outer space: the activities related to the launch of objects into outer space, activities starting on Earth and then traversing (usually national) airspace. Here also, states can avail themselves of territorial, nationality-based, and/or quasi-territorial jurisdiction as relevant.

3. International Space Law and Liberty

Given the above, space law became a substantive, almost tangible issue with the first human activity in outer space: the launch of Sputnik-1 by the Soviet Union in 1957. The first legal issue immediately arose: given that air law had recognized the fundamental sovereignty of underlying states over their respective airspaces,[15] could the altitude at which Sputnik-1 overflew many other countries—including, interestingly, the Soviet's Cold War adversary, the United States—be considered still to be within airspace, allowing the countries overflown to prohibit such overflight or subject it to such conditions as they might see fit? Or would it need to be addressed as being subject to a different regime, because at that altitude it was no longer in any state's airspace but in something to be viewed as, legally speaking, fundamentally different, being labelled 'outer space'?[16]

The result of this, and similar, legal questions was the establishment in 1958 of a special United Nations Committee on the Peaceful Uses of Outer Space (UNCOP-UOS)[17] under the *aegis* of the United Nations (UN) General Assembly (GA) to address precisely such questions in its Legal Subcommittee. Its activities resulted first in a UNGA Declaration setting out the main legal principles applicable to outer space and human activities there,[18] almost all of which were then transformed into legally binding obligations under the Outer Space Treaty.[19]

From the perspective of law and liberty on the Moon, ultimately five key clauses of the Outer Space Treaty stand out as providing relevant parameters and pointers thereto.

Perhaps the most fundamental clause concerns Article II, which provides: 'Outer space, including the Moon and other celestial bodies, is *not subject to national appropriation* by claim of *sovereignty*, by means of use or occupation, or by any other means.'[20] This essentially means that no single state can treat outer space or any part thereof as part of national territory (unlike the airspace as discussed above in the context of Sputnik-1) and thereby apply its national legal system without further ado—including any laws prohibiting alternatively promoting or allowing for certain liberties.

A clause almost logically following from Article II is Article I, which *inter alia* states: 'Outer space, including the Moon and other celestial bodies, shall be free for exploration and use by all States without discrimination of any kind, on a basis of equality and in accordance with international law, and there shall be free access to all areas of celestial bodies.' Note, that this freedom as such is only directly enjoyable by *states*; whether and to what extent private entities can benefit from it depends on the state under whose jurisdiction they reside.[21]

At the same time, the clause reflects the famous Lotus Principle, stating:

> International law governs relations between independent States. The rules of law binding upon States therefor emanate from their own free will as expressed in conventions or by usages generally accepted as expressing principles of law and established in order to regulate the relations between these co-existing independent communities or with a view to the achievement of common aims. Restrictions upon the independence of States cannot therefore be presumed.[22]

This is a clear manifestation of the baseline freedom of states to act (or to allow their private sector to act) in outer space: limitations to that freedom are and can only be provided by *international* law, notably through such treaties as the Outer Space Treaty itself. Here, prohibitions come to mind regarding the use of outer space for orbiting weapons of mass-destruction,[23] the use of celestial bodies such as the Moon for military purposes,[24] or the use of outer space without due regard (whatever its precise contents might be) for the legitimate activities of other states.[25]

As it were as the flipside of the baseline freedom for states, Article VI holds states internationally responsible not only for compliance with international law for their own, 'public' activities in outer space but uniquely also for the activities of private actors as long as qualifying as 'national activities in outer space'[26] of the state in question. As a follow-up, Article VI then requires at least the 'appropriate State' to exercise 'authorization and continuing supervision' over such activities, a clause usually given effect by adopting national space legislation including systems for authorization of private sector space activities.[27]

A fourth key clause to a limited extent makes up for the impossibility to exercise territorial jurisdiction in and over (parts of) outer space, since as already discussed above in regard to Article VIII states can register space objects and thereby can continue to exercise jurisdiction over it and over anyone on board thereof on a so-called quasi-territorial basis. In other words: states *are* entitled to determine to what extent liberties under national law apply to activities on board of space objects registered by them, read to anyone on board thereof.

Finally, Article III as a kind of fall-back clause ensures that general international law continues to apply to outer space—as long as not specifically otherwise provided for by space law. Though the reference is specifically to the UN Charter[28] and 'the interest of maintaining international peace and security and promoting international cooperation and understanding',[29] it reflects the general understanding that state and human activities in or with respect to outer space are not so special that they fall, in principle, outside of the ambit of international law.

This is, effectively, the most substantial principle of the Outer Space Treaty in terms of liberty, beyond the freedom of states as such, as it does allow analysis now to turn to the extent in which general international law has honoured the concepts of freedoms and liberties as briefly pronounced earlier in the context of humans themselves.

4. General International Law and Liberty

The 1948 Universal Declaration of Human Rights[30] is generally acknowledged to be the original source of rights of individual humans as these currently form part and parcel of international law, by addressing states and urging (if not exactly requiring) them to honour such rights with as little reluctance and exceptionalism as possible.

Most importantly, Article 3 establishes the 'right to life, *liberty* and the security of person',[31] whilst Article 4 backs this right to principled personal freedom up with the prohibition on slavery and servitude, and further supportive clauses require non-discrimination, due process and fair trial, freedom of movement, expression, and suchlike, as essentially major elements of those same personal freedoms.[32]

Whilst the Universal Declaration on Human Rights, for instance by way of its reference to 'Articles' instead of principles or paragraphs, suggests it has the nature of a treaty, directly binding upon states parties, as a GA Declaration strictly speaking it is *not*, and therefore not legally binding as black-letter law in and of itself. As the Preamble itself states: the GA proclaims the Declaration 'as a *common standard of achievement* for all peoples and all nations ... to *promote respect* for these rights and freedoms and ... secure their universal and effective recognition and observance'.[33]

To a large extent, that drawback has been corrected by the establishment of the UN Covenants, which are binding as treaties at least upon the states that ratified them. Thus, the 1966 International Covenant on Civil and Political Rights[34] aims to ensure 'that any person whose rights or freedoms as herein recognized are violated shall have an effective remedy, notwithstanding that the violation has been committed by persons acting in an official capacity'[35] and proceeds with referencing many of the freedoms or freedom-related human rights already espoused by the Universal Declaration on Human Rights.[36]

At the same time, the possibility of abrogating several of the very freedoms espoused by the International Covenant on Civil and Political Rights is specifically mentioned:

> In time of public emergency which threatens the life of the nation and the existence of which is officially proclaimed, the States Parties to the present Covenant may take measures derogating from their obligations under the present Covenant to the extent strictly required by the exigencies of the situation, provided that such measures are not inconsistent with their other obligations under international law and do not involve discrimination solely on the ground of race, colour, sex, language, religion or social origin.[37]

Ultimately, therefore, whilst the relevant principles on individual liberties are well articulated also as *legal* principles, the practice of states in their sovereignty all too often using the exceptions and loopholes or simply bending the international canon of human rights to their particular liking cautions for easy transposition of any of those human rights to the uniquely special environment of outer space.

The same also goes for the concurrent International Covenant on Economic, Social and Cultural Rights[38] which provides for a number of fundamental group rights, such as those related to the self-determination of peoples and the effective protection of minority rights, which also generally underpin individual liberties.[39]

This means, in turn, effectively that the application of any individual or minority group liberties in a lunar context depends on the implementation by states, in other words, requires of the relevant states to use their national legal regime to apply, implement, and enforce those.

5. Future Implementation of Human Liberties on the Moon?

All of the above, of course, will have to occur in a context where the Outer Space Treaty allows the application of individual states' jurisdictions, beyond the limited context of a space object duly registered, only on the basis of the nationality of the persons involved, and within that context on the basis of either nationality of persons involved or quasi-territorial jurisdiction over the space object.[40]

It bears noting that Article VIII of the Outer Space Treaty refers to jurisdiction over the registered space object 'and personnel thereof', technically also applying to such 'personnel' if and when operating outside the space object, properly speaking. This clause had envisaged such short-duration Extra-Vehicular Activities (EVAs) as space walks or Moon walks, which raises questions as to its applicability in the context of long-duration settlements on the Moon or other celestial bodies.

The term 'personnel of a spacecraft', after all, assumes the person in question to play a key role in the operation of that spacecraft. When, however, humans travel on board a spacecraft to a celestial body to settle there for the long term, and at best occasionally but shortly revisit the spacecraft having brought them there, such an assumption surely would lose its relevance—and thereby the applicability of the jurisdiction of the state of registry of the spacecraft over the person at issue.

Bearing this in mind, it is possible to distinguish at least three different level scenarios of how states might wish to apply whatever domestic law they have on issues of individual or minority group liberties to settlements on the Moon or other celestial bodies.

Scenario 1 would be the simple scenario. In this scenario, the spacecraft taking humans to the celestial body where the settlement is to be established is registered by the same state of which all humans on board are nationals. In this scenario, it is simply a matter of that same state applying its laws on liberties, preferably both to the spacecraft as a piece of quasi-territory and to the humans on the basis precisely of their nationality.

As long as the spacecraft is used as (part of) the habitat being established, the issue discussed above regarding the extent to which the national law of the state in question applies to 'personnel' becomes effectively solved, as that state can still apply its laws to the settlement based on the unfirm composition of the inhabitants in terms of their nationality being of that same state. Even if (parts of) the habitat would be built from locally extracted materials, questioning the applicability of Article VIII of the Outer Space Treaty and the Registration Convention as these were supposed to apply to objects *launched* into outer space as opposed to *constructed* in outer space, as long as the inhabitants are all nationals of that same state this result would not change.

Over time some of the fundamental underlying assumptions might change. Once the habitat has been up and running for years, or even better, decades—once the humans living there have started to become fully self-sufficient, have conceived children, and may well have lost any interest in ever coming back to Earth—the original (sense of) connection with the motherland is likely to fade into the distance. Living on the Moon would most probably require not only different habits and activities but, as a logical consequence thereof, also laws different from what makes sense down on Earth, and the urge to develop their own, Moon-oriented legal regimes will mature to a point that the lunar establishment would wish to tear loose from the motherland.

It can only be hoped that such 'decolonization' would not generate the same violence that it has only too often generated down on Earth, and that the motherland will see the reasonableness in allowing the habitat to stand on its own feet, legally too, and become independent. Even so, it would call for a paradigm change in international law to the extent that an 'extraterrestrial' state should now be recognized basically on a par with all states that have existed so far on Earth only.

Whether and to what extent such a new state will in its own national legal order allow for individual and other liberties, whether by inheriting them from the motherland (likely perhaps at least as a point of departure) or otherwise, will then be entirely up to the inhabitants of such a new state.

Scenario 2 is a less simple one. Here, the—presumably considerable—number of spacecraft bringing the new settlers to the Moon are still registered by one single state, but the people carried on board may come from different nationalities. As long as the journeys last, of course, the national laws of the state of registry of those spacecraft would still dictate and determine the existence of any liberties. As a default, that would also apply to the extent the spacecraft themselves will serve as the (primary) habitat, although the more the actual journey recedes into the past, the less likely those settlers *not* from the state of registry will consider themselves bound to such a national law and the less likely their states of nationality will treat them as subject basically to a foreign state's jurisdiction only.

Already prior to any profound move towards independence, legal disputes may thus arise as to which national laws determine the extent of liberties to be enjoyed by the various settlers, and once independence begins to evolve, the resulting regime will probably be a new mix of the various original national laws concerned. It also means that, for such independence to come about peacefully, not only the state of registration of the spacecraft involved but also all states of nationality of the settlers involved would have to accept that their 'strong arm of the law' should no longer extend to the Moon. In other words, the likelihood of *some* serious conflict arising regarding the future legal status of such a settlement, including its rights to determine its own take on liberties, would be substantially larger.

The development of habitats from resources extracted from or found on the Moon itself, further to the principled lack of applicability of the registration regime to such constructions, would only further hasten such a process of loosening any ties with the original terrestrial motherlands which are after all still engaged through quasi-territorial and/or nationality-based jurisdiction.

Scenario 3 is the really complex one: here the various spacecraft involved in developing the settlement will come from—read, be registered by—different states, each therefore entitled to exercise to that extent quasi-territorial jurisdiction, as well as carrying humans from different nationalities, also potentially throwing a further number of states entitled to exercise their nationality-based jurisdiction into the mix. Needless to say, the problems of arriving at a coherent and transparent system of applying and enforcing liberties and then, ultimately, at a new legal order taking care of this new society, are only further compounded in this scenario.

6. Conclusion

As on Earth, so in outer space ...

Most of those participating in humankind's venturing into outer space, in particular to the extent of establishing settlements there and thereby providing for a second home for humanity, would hope or even assume that humankind in so doing might be able to avoid the sometimes rather bloody developments that accompanied settlement in far-away territories on Earth. However, at least in the legal area, with a view to its role as providing not only justice and fairness but also predictability and foreseeability, logic dictates that terrestrial laws and experiences of these laws will certainly initially determine also the ways in, and extent to, which individual liberties, already addressed and acknowledged to such varying extents on Earth, will be applied and even enforced, or by contrast suppressed or extinguished, in outer space.

Given the major challenges of living in an environment which will remain extremely hostile at least for many decades or even centuries to come, the role of the law in balancing the overriding importance of collective public goods—access to oxygen and other life-or-death resources, settlement-encompassing safety, and avoidance of any armed conflicts—with the individual liberties that would in the last resort legitimize humankind's efforts to settle there in the first place is even more important than on Earth, and will require very thoughtful adaptation of legal concepts, principles, rules, rights, and obligations to that strange and dangerous environment.

Perhaps a variety of settlements, with a variety of legal approaches based on terrestrial background and experience, provides the best chance that ultimately, the legal regime which would be most beneficial and benevolent for humankind will triumph. At least in *that* respect current international space law, by way of its approaches to sovereignty and jurisdiction as illustrated above by the three scenarios, already provides an interesting and potentially very helpful point of departure.

Notes

1. At <https://en.wikipedia.org/wiki/Moon> accessed 11 October 2021.
2. Treaty on Principles Governing the Activities of States in the Exploration and Use of Outer Space, including the Moon and Other Celestial Bodies (hereafter Outer Space Treaty), London/Moscow/Washington, done 27 January 1967, entered into force 10 October 1967; 610 UNTS 205; TIAS 6347; 18 UST 2410; UKTS 1968 No. 10; Cmnd. 3198; ATS 1967 No. 24; 6 ILM 386 (1967).
3. Currently, it carries the ratification of 111 states (and the signature of 23 more); see A/AC.105/C.2/2021/CRP.10, of 31 May 2021.
4. The phrase 'the Moon and other celestial bodies' figures in almost every substantive (as opposed to procedural) clause of the Outer Space Treaty, see note 2.
5. Agreement Governing the Activities of States on the Moon and Other Celestial Bodies, New York, done 18 December 1979, entered into force 11 July 1984; 1363 UNTS 3; ATS 1986 No. 14; 18 ILM 1434 (1979).
6. Currently, it carries the ratification of 18 states, none of which—with the exception perhaps of Australia—could be viewed as leading spacefaring nations, and the signature of four more; see A/AC.105/C.2/2021/CRP.10, of 31 May 2021.
7. See on this in greater detail e.g. F. Tronchetti, *The Exploitation of Natural Resources of the Moon and Other Celestial Bodies* (2009), esp. 38–61, 118–23, 225–31.
8. At <https://en.wikipedia.org/wiki/Liberty> accessed 11 October 2021.
9. It may be noted that part of this sturdiness is also reflected by the fact that 'law' brings with it its own procedures for change, formalized and institutionalized so as to make sure it is not changed on a mere whim.
10. It may be added, that if the law lags behind *too* far in the view of (major elements of) society, this may result in such impatience also with the inherent procedures and formalities for changing it, which are then often seen as being part of the problem, as to incite what effectively amounts to revolution: a new start of the law including a new set of provisions on how it could be changed in the future.
11. Cf. already the discussion on the meaning and scope of 'space law' and 'space activities' at F. G. von der Dunk, 'Preface' in F. G. von der Dunk and F. Tronchetti (eds), *Handbook of Space Law* (Edward Elgar Publishing, 2015), xxvi, 29, and references.

12. This of course goes to the fundamental principle of sovereignty of a state over its own territory, i.e. to impose, adjudicate, and enforce its domestic laws at its sovereign discretion, including upon any foreigners happening to be present on that state's territory, unless international obligations binding upon that state would dictate otherwise.

13. Note that Article II, Outer Space Treaty (see note 2), precludes any appropriation by states of parts of the Moon (or indeed outer space as a whole), which is generally recognized as making the Moon a *res communis* or *terra communis*. See further e.g. S. R. Freeland and R. Jakhu, 'Article II' in S. Hobe, B. Schmidt-Tedd, and K. U. Schrogl (eds), *Cologne Commentary on Space Law*, Vol. I (Carl Heymanns Verlag, 2009) 44–63.

14. As per Article VIII, Outer Space Treaty (see note 2), and the Registration Convention (Convention on Registration of Objects Launched into Outer Space, New York, done 14 January 1975, entered into force 15 September 1976; 1023 UNTS 15; TIAS 8480; 28 UST 695; UKTS 1978 No. 70; Cmnd. 6256; ATS 1986 No. 5; 14 ILM 43 (1975)). See further e.g. B. Schmidt-Tedd and S. Mick, 'Article VIII' in *Cologne Commentary on Space Law* (see note 13) 146–68.

15. As enshrined in Article 1, Convention on International Civil Aviation, Chicago, done 7 December 1944, entered into force 4 April 1947; 15 UNTS 295; TIAS 1591; 61 Stat. 1180; Cmd. 6614; UKTS 1953 No. 8; ATS 1957 No. 5; ICAO Doc. 7300.

16. Note that, though as of yet no generally and authoritatively accepted lower boundary of outer space (as setting it off against the underlying airspace) could be established, some measure of convergence on an altitude of 100 km above sea level could be discerned. See further F. G. von der Dunk, 'International Space Law' in *Handbook of Space Law* (see note 11) 60–72; in exhaustive detail, T. Gangale, *How High The Sky?* (Brill Nijhoff, 2018).

17. As per Question of the peaceful use of outer space, UNGA Res. 1348 (XIII), of 13 December 1958; Resolutions adopted on the reports of the First Committee, General Assembly— Thirteenth Session, at 5. See further F. G. von der Dunk, 'The Undeniably Necessary Cradle—Out of Principle and Ultimately Out of Sense' in G. Lafferranderie (ed), *Outlook on Space Law over the Next 30 Years, Essays published for the 30th Anniversary of the Outer Space Treaty* (Kluwer Law International, 1997) 401–14.

18. Declaration of Legal Principles Governing the Activities of States in the Exploration and Use of Outer Space, UNGA Res. 1962(XVIII), of 13 December 1963; UN Doc. A/AC.105/572/Rev.1, at 37.

19. See note 2.

20. Emphasis added.

21. See further on Article I, Outer Space Treaty (see note 2), e.g. S. Hobe, 'Article I' in *Cologne Commentary on Space Law* (see note 13, 25–43).

22. *S.S. Lotus (Fr. v. Turk.)*, Judgment, 1927 P.C.I.J. (ser. A) No. 10, at 44 (7 September).

23. As per Article IV, Outer Space Treaty (see note 2).

24. As per Article IV, Outer Space Treaty (see note 2).

25. As per Article IX, Outer Space Treaty (see note 2).

26. Article VI, Outer Space Treaty (see note 2).

27. See further on Article VI, Outer Space Treaty (see note 2), e.g. M. Gerhard, 'Article VI' in *Cologne Commentary on Space Law* (see note 13) 103–25.

28. Charter of the United Nations, San Francisco, done 26 June 1945, entered into force 24 October 1945; USTS 993; 24 UST 2225; 59 Stat. 1031; 145 UKTS 805; UKTS 1946 No. 67; Cmd. 6666 & 6711.

29. Article III, Outer Space Treaty (see note 2). See further e.g.O. Ribbelink, 'Article III' in *Cologne Commentary on Space Law* (see note 13) 64–9.

30. Universal Declaration of Human Rights, Paris, UN GA Res. 217 A (III) of 10 December 1948; A/RES/217.

31. Emphasis added.
32. Cf. Articles 7–9, 12–13, 18–20, Universal Declaration of Human Rights (see note 30). See further for a broader perspective e.g. A. Cassese, *International Law* (Oxford University Press, 2001), esp. 351–69.
33. Emphasis added.
34. International Covenant on Civil and Political Rights, New York, done 19 December 1966, entered into force 23 March 1976; 999 UNTS 171; UKTS 1977 No. 6; Cm. 6702; 6 ILM 368 (1967). The Covenant currently counts 173 parties; <https://en.wikipedia.org/wiki/International_Covenant_on_Civil_and_Political_Rights> accessed 9 November 2021.
35. Article 3(a), International Covenant on Civil and Political Rights (see note 34).
36. Cf. e.g. Articles 6 ('inherent right to life'), 8 (prohibition of slavery and servitude), 9 ('right to liberty and security of person'), International Covenant on Civil and Political Rights (see note 34).
37. Article 4(1), International Covenant on Civil and Political Rights (see note 34).
38. International Covenant on Economic, Social and Cultural Rights, New York, done 19 December 1966, entered into force 3 January 1976; 993 UNTS 3; UKTS 1977 No. 6; Cm. 6702. The Covenant currently counts 171 parties; <https://en.wikipedia.org/wiki/International_Covenant_on_Economic,_Social_and_Cultural_Rights> accessed 9 November 2021.
39. Cf. e.g. Articles 1–5, International Covenant on Economic, Social and Cultural Rights, see note 38.
40. Cf. Article VIII, Outer Space Treaty, see note 2.

28

Welcoming disability as necessary in space travel

Sheri Wells-Jensen

1. Introduction

Before humans travelled to outer space, we spent generations dreaming about going there. We told ourselves astounding stories about courage and adventure and populated them with heroes who would live and work high above our heads, up beyond the clouds. They were strong, capable, fearless souls who emerged directly from our collective human myths: Buck Rogers, Flash Gordon, and John Carter of Mars.

And we loved them!

Early pulp science fiction was not always the finest literature but it did express some of our deepest dreams, and it set the stage and summoned the characters. No wonder, then, that when it came time to choose the real-life human beings who would begin to make these dreams into realities, there was little hesitation about which people we should send (Wolfe 1979; Hersch 2012; Cassutt 2018). We lifted our early astronauts and cosmonauts right off the pages of our favourite stories.

In the United States, the initial pool of applicants for these jobs were fighter pilots, and even these were scrupulously screened; anyone with any physical, mental, or social characteristics that could conceivably compromise either their abilities or their reputations was quickly eliminated from consideration. And that worked well.

Our rockets went up and the men aboard them accomplished their missions and returned to us. Regardless of what critiques we might have now from our current perspective, there is no doubt that we accomplished incredible things, including those exquisite moments when the first humans from Earth stepped onto the lunar surface.

Then, during the era of the US space shuttle and the International Space Station (ISS), things began to change. It is not that we ran out of steely-eyed missile men, it is just that our collective cultural appetite for them began to wane. There was initially the political (and socio-cultural) drive to supply historic firsts: the first woman (some 20 years after Major General Valentina Tereshkova), the first person of colour, the first openly gay astronaut, the first non-military scientist, and later the first citizen astronauts who flew to space because they had bought tickets themselves (Jaramillo 2020). Later, charitable organizations like Space for Humanity began giving away seats so that more and more regular people could experience the life-altering wonder of viewing the Earth from space. Space began to feel more and more like a place people could go—if they were very highly trained, or very, very rich—as long as they met the physical requirements (Ansari and Hickam 2010; Fisher and Garriot 2017).

It is this remaining prohibition—the imposition of strict (some might say excessive) physical requirements for traveling to space—that will occupy us in this chapter. It has effectively barred any disabled person from entering the realms occupied by our most exalted cultural heroes. Why was this prohibition imposed? Should it be lifted? If so, how? And what will happen to the human future in space if this barrier were to disappear?

2. Sickness vs Disability

In March 1970, as the Apollo 13 mission was poised to set off on its journey to the Moon, NASA discovered it had a problem. The crew (Commander Jim Lovell, Lunar Module Pilot Fred Haise, and Command Module Pilot Ken Mattingly) had been training together for years. They knew and trusted each other and had knitted themselves into a smoothly functioning team. Unfortunately, just days before launch, Ken Mattingly was exposed to German measles at a child's birthday party. Blood tests indicated that Lovell and Haise were immune, but there was a real chance that Ken Mattingly would begin to suffer from a high fever and exhaustion just as the crew was entering lunar orbit, impairing his abilities and putting all their lives in danger. As the mission could not be delayed, the only options available were to break up the primary team by replacing Mattingly with his backup, Jack Swigert, or to bench the entire primary crew, allowing Swigert and the rest of the backup team to take over as a unit (Lovell and Kluger 1994; Mars 2020). Aware that either decision would have consequences for the mission, Lovell weighed his options, then regretfully chose to replace Mattingly.

Although there was argument at the time about whether or not Mattingly would become ill, everybody agreed that it was undesirable to have a crew member become potentially seriously ill whilst orbiting the moon, 380,000 kilometres from home.

This is quite understandable, and was a necessary precaution to ensure the success of the mission. If Mattingly had become acutely ill he would not have been working at his peak ability. He might well be a liability, unable to perform his tasks in the expected way. His stamina, his perception, and even his judgement could be impaired, and the close-knit crew, who knew each other and could predict one another's responses, would be disrupted. The mission—even their lives—might be in danger. And so, if one accepted the premise that Mattingly could conceivably become seriously ill, he obviously could not fly. The dispute Lovell had with NASA was not that a sick crew member would be unfit to fly; that was understood. It was about the odds that he would get sick in the first place. Lovell might have been willing to throw the dice and take his chances, but NASA wasn't. The rules were clear: Sick people—even potentially sick people—should stay home.

Part of the prohibition against disabled people in space may have arisen from this same logic: if a person suffering from an illness is a liability, then a person with a permanent disability would be a veritable albatross.

I want to point out here that this comparison is profoundly flawed. A disabled person is in no way equivalent to a person who is temporarily side-lined by an illness or injury.

Take, for example, the (unfortunate) hypothetical astronaut who breaks a leg during an Extravehicular Activity (EVA). He is shocked, and in terrible pain. He has no idea how this will affect his mission. He doesn't know if the suit he is wearing, which has been protecting him from the vacuum around him, will support and protect his leg. Is he about to die? Further, he is unsure if he can trust his newly broken leg if he should need to push off from the surface of the ship towards the airlock. His body has become an unknown quantity. He has 1,000 unanticipated problems, and he and the crew around him are in high-alert emergency mode. All other activity stops as he quite reasonably becomes the sole focus of attention; any thoughts of completing the EVA are jettisoned in favour of his immediate health and safety.

On the other hand, a paraplegic scientist, comfortable in her adapted suit, floating in microgravity, has none of these problems. She is conversant with her equipment and aware of her body's capabilities. Her participation on the same EVA is nominal. She is a known quantity, and the crew who trained with her have no questions about her abilities. Her stamina, perceptions, and judgements remain constant, and she evokes (and needs) no emergency attention. She may actually have less physical capability than her companion with the broken leg, but she needs far less support. In fact, she may need no extra support at all, depending on how her environment has been constructed. She is a constant, known factor.

It is, of course, clear that some disabilities change over time. A crew member with retinitis pigmentosa, for example, will most likely lose significant vision during the course of a long (or permanent) stay in outer space. This, however, is also a known quantity, and unlike a temporary illness, the trajectory of the vision loss is predictable, and the crew member can (and should) incorporate training for eventual vision loss as part of her preparation for the mission. All crews must be ready to adapt to accidents or illnesses on a long mission, but one important difference between the progression of a known disability and illness is that the former is predictable and the latter is not.

Another difference is that whilst an illness or injury is unwelcome, disorienting, and debilitating, a disability is part of a person's known identity. If the disability imposes any difficulties, the disabled person has the training needed to adapt, and she will have, and know how to use, adaptive equipment.

In 2001, Canadian astronaut Chris Hadfield, whilst on his first EVA outside the ISS, was temporarily blinded by a wayward soap bubble inside his helmet. Although he was whirling around the Earth at 27,000 kilometres per hour and there were definitely things he should be doing, this brought him to a complete stop. He knew his only option was to stay put, stay calm, and radio back to Earth, hoping the collective knowledge of his support team would present him with a solution. There was no thought of his continuing to work whilst this was going on. He was completely, and terrifyingly, immobilized, hanging from his feet on the outside of the ISS, 300 kilometres above the Earth.

But Commander Hadfield's real problem was not that he could not see. It was that his training, his equipment, and the tasks before him were not designed to be completed successfully without looking. His blindness, which he managed superbly under the circumstances, was something that should not have happened. If, on the other hand, the Commander had been blind from birth or had acquired his blindness some years earlier, he would have had years of training, allowing him to understand his abilities and develop non-visual skills. And if blind astronauts were part of the NASA astronaut corps, his spacesuit would have been designed so that he could use these skills during his EVA. For example, significant attention would have been paid to ensuring the development of boots through which he could have felt his way along the exterior surface of the ISS, and that surface itself would have been designed with tactile indicators in place for anybody to use who could not see or who had to be looking elsewhere whilst moving about. It was his environment—not his blindness—that rendered him unable to move.

The distinction between illness and disability is important. An illness can be thought of as a change in body or mind that is problematic and that the person feels compelled to reverse if at all possible. A cold, a broken bone, a period of unexpected depression, and even a bad hangover fit this bill. It is understood that the person's ordinary abilities will be impaired until (or unless) her body naturally returns to its previous state or there is a medical intervention that returns her to her nominal starting point. She is not necessarily expected to function well whilst this is happening, and although her stiff upper lip will be appreciated, she can expect to be given some grace in the meantime. It is reasonable, then, to work hard to prevent this sort of problem, not only from a compassionate desire to ease suffering but from a pragmatic understanding that she obviously cannot be doing her best.

The disabled person, on the other hand, is neither suffering nor off his game. He can be a trusted member of the crew, and if the physical environment, equipment, and procedures are in place to allow him to do his job, no exceptions (compassionate or otherwise) are necessary.

Constructing an environment so that it is usable by both disabled and abled people alike is (quite suitably in our case) generally called 'universal design' (Steinfeld and Maisel 2012). This means it is clearly possible to work to prevent or mitigate the effects of illness whilst preparing to accept disability when that is the outcome.

3. Universal Design: Risk Assessment and Benefits

One might think, given NASA's response to Ken Mattingly's potential illness, that the rule for all space agencies, private and public, would be to take every single possible precaution to mitigate harm. And, yes, the physical systems aboard any vessel do have significant redundancies and backups built in. Procedures are also styled to force crews to be careful. The checklists attached to nearly every aspect of life in space create an environment in which every critical action is scheduled, acknowledged, performed, and confirmed. However, designers (and financiers) acknowledge that each mission is necessarily the result of compromises between safety, cost, and practical design considerations. And to date, this preference for redundancy does not extend to the creation of equipment and control systems that are accessible to disabled crew. It may be that, just as Commander Lovell felt that Mattingly's potential illness was unlikely, disability is treated as unlikely enough to be risked.

This is unwise.

During an extended mission, if no disability occurs, then arguably (all things being equal) any work done in advance to put accommodations in place is irrelevant, and the mission proceeds as planned. If, on the other hand, disabilities do occur, then the absence of such preparations could prove fatal. The brief table below lays out the scenarios in the simplest form: preparations are made in advance (yes or no) and significant disabilities occur (yes or no).

	Significant disability does not occur far from Earth	Significant disability occurs far from Earth
Preparations in advance	(No effect)	MISSION SAVED
No preparations in advance	(No effect)	MISSION IMPERILED

The overly simplistic analysis above does at least serve as a beginning point for conversation, but it has three obvious flaws. First, there is no way to accurately assign probabilities to any of the conditions above. Luckily, many more missions succeed than fail, and most failures are usually only partial. The Apollo 13 mission did not land on the moon, but the crew survived, and the changes made to the capsule afterward made subsequent flights safer. Second, these conditions are gradients rather than binary factors. All missions currently make some adjustments for individual crew members' bodies and abilities, from very basic accommodations such as providing clothing to fit each person and meeting individual dietary needs to major considerations such as, in the case of the Soyuz capsule, manufacturing launch couches moulded to fit a cosmonaut's body.

One could imagine instances where no adaptations are put in place for eyes-free access to systems used for docking or piloting, but where controls that must be activated manually, such as hatches or toilet controls, are modified for use by crew members with limited dexterity. More importantly, though, this analysis leaves out the most compelling argument for universal design, which is that the benefits are not limited to the nominal recipients. That is, an environment designed around universal principles is more comfortable, safer, and more efficient for everybody.

A classic Earthside example of universal design is curb cuts (also known as 'curb ramps' or 'dropped curbs'). These are graded concrete slopes which lead a pedestrian from the sidewalk down to street level and back up again. They are useful for people on bicycles, roller skates, or skate boards, and for people pushing strollers or pulling carts. They also eliminate the potential for a distracted pedestrian to twist an ankle stepping into the street. Designed initially for wheelchair users who could not easily navigate traditional curbs, they have become a mandated part of urban design in many countries, and millions of non-disabled people have benefitted. The same can be said of automatic doors, levers (rather than doorknobs), wide doorways, audible crossing signals, fire alarms which can be operated with one hand, captioning on movies, television shows, and newscasts, and myriad other environmental changes that happen to benefit disabled people. The case for accessible design in spacecraft is no different (Wells-Jensen 2018; Wells-Jensen, Miele, and Bohney 2019). A more accurate version of the table above would include not only the risk of not making spacecraft accessible but also the benefits of doing it.

	Significant disability does not occur far from Earth	Significant disability occurs far from Earth
Preparations in advance	Mission is safer and operation more convenient	Mission is safer and operation more convenient; MISSION SAVED
No preparations in advance	(No effect)	MISSION IMPERILED

Although, as stated earlier, it is impossible to assign percentages to any of these conditions, the accessible spacecraft not only makes mission failure less likely but also improves the chances for success.

4. Objections

Cost is the first objection that is generally made when extensive modifications to standard design are suggested, and it is certainly true that different designs come with different financial obligation. We are, however, at the beginning of a new era in space exploration when many different kinds of space vessels are being designed and tested, and it is always less expensive to plan for universal design initially than it is to renovate afterward. We might find that the resulting increase in comfort and safety make these modifications easy to approve, especially for long-range vessels and for environments which are intended for public use, such as permanent colonies or destinations directed at tourists who are more risk-averse and less willing to 'rough it'.

The second objection made to implementation of universal design in space is quite simply that we do not know, at this point, what exactly would be involved. What accommodations, for example, does the paraplegic astronaut actually need? Given that many astronauts currently anchor themselves whilst working by slipping a toe into a strap, what changes in the cabin would benefit her? What orientation aids might be useful in microgravity to assist a blind scientist in staying oriented whilst he works? How could we design boots that are thin enough to allow an astronaut to feel her way with her feet but still strong enough to protect her from the vacuum of space? What modifications to alert systems would be useful to the Deaf pilot? Are there ways in which the (sometimes noisy, sometimes chaotic) environment of a zero-gravity laboratory could be modified for the benefit of autistic payload specialists?

These questions must be answered before real design changes can be implemented; it does no good to advocate for change without data specifying what those changes might be.

5. The Quest for Answers

Incredibly, information about what is needed for disabled people to master space is forthcoming from at least three sources. First, there is the growing field of space tourism. Virgin Galactic and Blue Origin have both begun suborbital flights, giving passengers a few minutes of weightlessness and a peerless view out the window of Earth against a sea of stars, and although both companies are careful to protect the health and safety of their passengers, access to the ride does not depend on astronaut-level fitness. Neither 71-year-old Richard Branson (who is dyslexic), nor 90-year-old William Shatner (who has arthritis) would have passed muster if the old rules applied, but both went to space for a few precious minutes in 2021. Around the same time, 29-year-old physician's assistant Hayley Arceneaux, who has an artificial femur, spent almost three days in orbit aboard Inspiration IV, a flight launched by Space X. None of these civilian astronauts had any trouble at all fulfilling their (admittedly somewhat minimal) responsibilities whilst in space. This means that we now know for certain that these minimal disabilities represent no barrier at all to space, at least at this level. The door may not have been thrown wide, but it has certainly been strategically cracked open.

Also in 2021, the European Space Agency announced its Parastronaut program which is actively seeking a physically disabled candidate. This person will have a single leg amputation, have a significant difference in leg lengths, or be very short in stature. The intention is to eventually fly a disabled astronaut to the ISS, and to work in the meantime to clarify what questions need to be answered before this is feasible. ESA's understanding is that accommodations made for the nominal benefit of this candidate will benefit everyone who flies, and they have committed to doing the work to figure out how.

Finally, in an investigation of the feasibility of future astronauts who are blind, Deaf, or paraplegic, Mission: AstroAccess (MAA) sent 12 disabled scientists, artists, and athletes on a parabolic flight, during which they experienced Martian gravity, lunar gravity, and repeated periods of zero gravity whilst carrying out demonstrations and observations designed to begin systematically to address questions of disability in space. Findings from this inaugural flight are not yet available at this writing, but MAA is an ongoing project using parabolic flights to gather the data that will eventually answer the most important question, which is not *whether* disabled people will go to space, but *how* they will go.

6. Conclusions

NASA identifies five major sources of risk to humans in outer space (NASA 2019). These can be summarized with the acronym RIDGE, which stands for:

Radiation: exposure to radiation puts crew members at a dramatically increased risk for cancer, heart disease, cataracts, and other degenerative illnesses.
Isolation: physical, social, and psychological effects of prolonged isolation with a small group.
Distance from Earth: psychological risks associated with being far from Earth, including the inability to get help from home.

Gravity fields: prolonged exposure to microgravity causing muscle degradation and bone loss.

Enclosed environment: accumulation of health risks from living in an enclosed environment.

We will keep sending heroes into space. Sometimes these heroes really will look like figures from the early science fiction classics. But, perhaps most of the time, they will look like us: regular people embodying everything it is to be human, with our dazzling variety and complex set of abilities. It is up to us to set the stage for them. We cannot promise to keep these heroes from all harm but we can promise to design their environments so that, whatever happens, they can have the best chance of living well and comfortably as they go about their extraordinary business high above our heads, up beyond the clouds.

References

Ansari, A. and Hickam, H. *My Dream of Stars: From Daughter of Iran to Space Pioneer* (London: Palgrave Macmillan, 2010).

Cassutt, M. *The Astronaut Maker: How One Mysterious Engineer Ran Human Spaceflight for a Generation* (Chicago, IL: Chicago Review Press, 2018).

Fisher, D. and Garriot, R. *Explore/Create: My Life in Pursuit of New Frontiers, Hidden Worlds and the Creative Spark* (London: HarperCollins, 2017).

Hersch, M. *Inventing the American Astronaut* (London: Palgrave Macmillan, 2012).

Jaramillo, A. 'NASA Astronaut Corps More Diverse than Ever but Still Lacking' *Florida Today*, 14 February 2020.

Lovell, J. and Kluger, J. *Lost Moon: The Perilous Voyage of Apollo 13* (Boston, MA: Houghton Mifflin, 1994).

Mars, K. 'Fifty Years Ago, Apollo 13 and German Measles, or How a Three-Year-Old Boy Changed NASA's Plans for Apollo 13' (2020) *NASA History*.

NASA. 'Five Hazards of Spaceflight' <https://www.nasa.gov/hrp/5-hazards-of-human-spaceflight> accessed 8 July 2022.

Steinfeld, E. and Maisel, J. *Universal Design: Creating Inclusive Environments* (Oxford: Wiley, 2012).

Wells-Jensen, S. 'The Case for Disabled Astronauts' (30 May 2018) *Scientific American Observations*. https://blogs.scientificamerican.com/observations/the-case-for-disabled-astronauts/ (Accessed 24 August 2022).

Wells-Jensen, S., Miele, J. A., and Bohney, B. 'An Alternate Vision for Colonization' (2019) 110 *Futures* 50–3.

Wolfe, T. *The Right Stuff* (New York, NY: Farrar, Straus and Giroux, 1979).

29

Regulation—a restraint of liberty or an enabler?

Implementing sustainability guidelines for commercial space activities—normalising the regulatory 'race to the top' in an ESG world

Joanne Wheeler

1. Introduction

Whilst states have to date deferred and/or resisted the development of an international legally binding instrument on space debris mitigation, voluntary guidelines[1] and international soft-law instruments have been drafted and applied by some states in public procurement contracts and licence conditions. Such application is, however, inconsistent between states.

In the slowly burgeoning space industry over the last 50 years, one could argue that not enough attention was paid to the issue of the environmental sustainability of outer space and the commercial activities carried out there, particularly from the regulatory and legal perspective. Although the Scientific and Technical Subcommittee of the United National Committee on the Peaceful Uses of Outer Space (UNCOPUOS) has been discussing issues of space debris since 1994 (eventually leading to with the adoption of UNCOPUOS Space Debris Mitigation Guidelines in 2007),[2] the topic was not added to the agenda of the Legal Sub-Committee until 2009. The catalyst for the increasing attention to the issue of the sustainability of outer space is the advent of the large or 'mega' constellation operators, amplified perhaps by the celebrity of some of their CEOs. The issue of space debris, space situational awareness (SSA),[3] and space traffic management (STM)[4] has been a growing problem for many decades, one that if not addressed will potentially lead to some orbits being unusable, to the detriment of the value satellites can offer to us on Earth.

When the European Code of Conduct for Space Debris Mitigation[5] (the Code) was published in 2004, there was concern within the industry that its application by national regulators would unfairly prejudice operators whose licensing jurisdiction applied it because it would cause an unfair competitive advantage to operators established in, and licensing through, jurisdictions whose regulators did not apply the Code. There were also issues raised that industry had not been fully consulted on its content.

The risk was that application of the Code and similar space debris mitigation guidelines and practices might lead to regulatory 'forum shopping'; companies seeking to

Joanne Wheeler, *Regulation—a restraint of liberty or an enabler?*. In: *The Institutions of Extraterrestrial Liberty*. Edited by Charles S. Cockell, Oxford University Press. © Joanne Wheeler (2022). DOI: 10.1093/oso/9780192897985.003.0030 locate and establish in jurisdictions applying very light touch laws and regulations on space activities.

With careful contractual drafting, and corporate and financial structuring, companies can forum shop to avoid many regulatory requirements if they wish to. This is not always in their interests, however.

What has changed recently is the increasing move of some companies to accept, encourage, and seek out jurisdictions where regulators apply and implement space debris mitigation standards and sustainable best practices and other environmental, social and governance (ESG) requirements.

A market for licences applying space debris mitigation standards and sound SSA requirements has been created which offers companies granted such licences important benefits, including market access elsewhere in the world; better access to finance by demonstrably meeting investor ESG eligibility criteria; and enhanced supply chain credentials.

In effect, the effort of complying with tighter sustainability requirements and the perceived resulting restraint of a business' ambitions can offer not only a positive benefit to the outer space environment through compliance with sustainability objectives, but also increasingly important additional advantages through enhanced public and peer reputation, wider investment opportunity, and ease of market access. Each of these may in turn offer comparative advantage against non-compliant or less demonstrably compliant competitors.

The positive advantages to regulators and licensing authorities which require compliance with space debris mitigation guidelines and impose through licensing criteria can also be valuable in attracting foreign direct investment and in international regulatory diplomacy. Some regulators are looking to go one step further and develop the concept of some form of informal quality assurance 'kite mark' as a sign of good practice which can be granted, in addition to a launch and/or operations licence, to evidence that a licensee has met end-to-end sustainability criteria covering the design, manufacturing, launch, operations, and decommissioning of a satellite—and create a market around this.

This chapter will discuss the international legal framework governing activities in space and responsibility and liability for them, and the slow implementation of space debris mitigation standards and guidelines, the concept of forum shopping, and the regulatory 'race to the bottom'. This latter description will be balanced against the increasing approach of commercial companies to implement space debris mitigation guidelines and the resulting 'race to the top' philosophy incentivized by market and financial benefits, and thus how companies are willing to accept a restraint on some commercial liberty in order to gain market access and investment.

2. International Legal Framework Governing Space Activities

Outer space is an extension of the Earth environment to which international environmental law applies,[6] and we would do well to consider it as such rather than as a distinct domain for which we have no responsibility. The space environment presents an increasing number of issues to which international law must seek to respond.[7]

2.1 International legal framework; responsibility and liability

Satellites are by their nature extraterrestrial and extraterritorial.[8] Their use and the use and exploration of outer space is governed by an international legal framework, under the auspices of the United Nations (and implemented by UNCOPUOS), made up of treaties, agreements and conventions governed by international law, which may be implemented into national law.

The UN international space treaties[9] establish rights and obligations between states parties to the treaties which bear international responsibility for activities carried out by governmental entities (including national space agencies), commercial entities (such as launch service providers and satellite operators), and individuals conducting activities in outer space and increasingly on the Moon and other celestial bodies.

The primary treaty governing activities in outer space is the Treaty on Principles Governing the Activities of States in the Exploration and Use of Outer Space, including the Moon and Other Celestial Bodies 1967 (Outer Space Treaty). It is a universal treaty open for signature to all states.

The preamble to the Outer Space Treaty pronounces certain concerns underlying the decision to negotiate and draft it. These include, for example, the common interest in the progress of the exploration and use of outer space, its use for peaceful purposes,[10] and that the use should benefit 'all peoples irrespective of the degree of their economic or scientific development', the contribution to international cooperation in the scientific and legal aspects of the exploration and use of outer space and the need for mutual understanding. The concepts of cooperation, consultation, and due regard for the interests of other states permeate through the Outer Space Treaty, as does the application of international law.

The role of law in outer space is a basic principle contained in Article I and Article III of the Outer Space Treaty which inform us that the exploration and use of outer space is to be carried out in accordance with international law, which includes international terrestrial environmental law. Space is not free of legal constraint and restraint and there are limits to what one can lawfully do. This concept is generally held to form part of customary law and is therefore binding on states, even if they have not signed or ratified the Outer Space Treaty.

The catalysts for the practical implementation of the concepts of cooperation and consultation tend to be economic and financial interests rather than the bold aspirations in the Outer Space Treaty as one might expect. It is to such catalysts we would need to turn to inform decisions as to the implementation of space debris mitigation measures, as discussed later.

International responsibility for national activities in outer space is dealt with in Article VI of the Outer Space Treaty:

> States Parties to the Treaty shall bear international responsibility for national activities in outer space, including the Moon and other celestial bodies, whether such activities are carried on by government agencies or by non-governmental entities, and for assuring that national activities are carried out in conformity with the provisions set forth in the present Treaty. The activities of non-government

entities in outer space, including the Moon and other celestial bodies, shall require authorization and continuing supervision by the appropriate State Party to the Treaty.

States tend to exercise such authorisation and continuing supervision for national activities through legislation, regulation, and licensing conditions; implementing regulatory control of activities on, and from, their territory by their nationals through the licensing and authorisation of space activities.

Under Article VII of the Outer Space Treaty, if a state is the 'launching State' of a space object (if the state launches or procures the launch of an object into space or if its territory or facility is used to launch an object), it is internationally liable for damage to another state party to the Outer Space Treaty caused by that object. This is the case whether the object is an active satellite or can be termed to be space debris and whether the damage is caused by the object itself or its component parts on the Earth, in air space or outer space.

According to Article XIII of the Outer Space Treaty, if the state adds a space object to its national registry of objects, it retains jurisdiction and control over that object, whether the object remains in space, is space debris, or returns to Earth. This in practice means that the relevant state exercises 'ownership' over the space object.

2.2 Implementation at national level

The UN space treaties do not create obligations directly on non-governmental, commercial entities and it is at the discretion of states parties to the treaties to flow some of the obligations under the UN space treaties, such as responsibility and liability in case of damages, down to private commercial parties through the adoption of national space legislation or licensing regimes, or both.

In recent years, more states than ever before are implementing national laws and regulations to supervise the activities of their private commercial operators in space. The reasons for this include compliance with international legal requirements, to flow liability down to licensees (which can best take action to mitigate risk), but also to attract foreign direct investment and raise the international reputation of the state. There is a growing recognition of the value of space activities for economic growth.

There is also a growing awareness of the importance of space sustainability issues and the inclusion of mitigation, and even remediation, requirements in national legislation and licensing conditions in order to attract foreign direct investment and other investment.

2.3 Space debris mitigation guidelines and best practice

Space debris is not defined nor specifically dealt with under the Outer Space Treaty. At international level, however, various sets of voluntary guidelines and standards provide guidance on space debris mitigation. These have been issued by the Inter-Agency Space Debris Coordination Committee (IADC), the International Organization for Standardization (ISO), the United Nations and the International Telecommunication Union.[11]

The first international guidelines on space debris mitigation were the IADC Space Debris Mitigation Guidelines (IADC Guidelines),[12] published by the IADC in 2002. The IADC Guidelines were presented to UNCOPUOS in 2003 and have been revised three times, in 2007, 2020, and 2021.

In 2007, UNCOPUOS endorsed the UN's first 'Space Debris Mitigation Guidelines of the Committee on the Peaceful Uses of Outer Space' (UN Guidelines).[13] These voluntary UN Guidelines were modelled on the technical content and definitions in the IADC Guidelines but require a higher degree of compliance and risk management.

After the publication of the IADC Guidelines, a working group was established by the ISO to transform the IADC Guidelines into international standards. This resulted in the ISO's Space Debris Mitigation Standards 2010[14] with the aim of ensuring that space debris mitigation considerations are incorporated into the design, operation, and disposal of satellites. These standards adopt a hierarchical structure: at the top, ISO 24113:2010 Space systems—Space debris mitigation requirements (ISO 24113)[15] details the high-level debris mitigation requirements; below, ISO 24113, there are a series of lower-level standards and technical reports and specifications, outlining further, detailed requirements and the means of implementing ISO 24113. The ISO standards are continually revised, most recently in 2019.[16]

The ITU Recommendation S.1003.2 'Environmental protection of the geostationary-satellite orbit'[17] was also published in 2010. This voluntary ITU Recommendation outlines the guidance of the ITU-R Assembly for the member states of the ITU regarding the sustainable use of the geostationary-satellite orbit (GSO).[18]

The continued commitment by UNCOPUOS to the sustainable use of space was demonstrated by the development of the UN's Long-term Sustainability Guidelines 2019 (LTS Guidelines).[19] This was the result of a multi-year project to develop a set of international guidelines to recommend best practices that states could follow based on the efforts of the UNCOPUOS Working Group on the Long-term Sustainability of Outer Space Activities. The LTS Guidelines are intended to ensure the long-term operational stability and safe environment of outer space. The LTS Guidelines comprise 21 voluntary, non-binding guidelines that address a policy and regulatory framework for space activities; the safety of space operations; international cooperation, capacity building, and awareness; and scientific and technical research and development. States are encouraged by UNCOPUOS to take measures to implement the LTS Guidelines nationally and it is hoped that the LTS Guidelines will encourage some consistency in this regard.

2.4 Regulatory forum shopping and the 'race to the bottom'

States are obliged to exercise appropriate 'due diligence' as responsible states, under the Outer Space Treaty to ensure that an applicant for a licence can meet national regulatory and legal requirements, particularly in relation to safety, and ensure the safekeeping and sustainable use of the space environment. The extent of such due diligence is particular to the individual licensing state but usually includes an assessment of the corporate and financial standing of the applicant, the technology being licensed and the management of the applicant seeking the licence.

The regulator may also wish to have sight of the launch services agreement, if relevant, the satellite procurement contract, and any relevant satellite operations contracts to understand where liabilities and responsibilities lie (what the applicant is liable and responsible for) and which entity has direct and effective control over the spacecraft being licensed (which is usually a requirement for operations licensing). The regulator will also require evidence of adequate insurance cover, in particular third party liability cover, with the relevant government noted as a loss payee on the policy. In relation to the licensing of constellations, the regulator may also require the satellite operations centre, or SOC, to be based in its jurisdiction to ensure that continuing supervision and the 'direct and effective' control over the satellites can be exercised effectively.

Regulators use their authorisation and licensing conditions to balance issues such as government risk, safety, security, and the sustainable use and access to space; against the encouragement of commercialisation, innovation, and growth. The extent of a regulator's due diligence requirements therefore depends on its policy objectives and approach to achieving them.

Due diligence exercised in practice by regulators is not uniform or harmonised between states and often depends on national policy considerations, which leads to forum shopping for licensing states by commercial operators.

From the perspective of a commercial operator, the concept of regulatory forum shopping can be attractive; establishing businesses where the regulatory and business operational environment can be more enabling, cost-effective, straightforward, and time-efficient—and where certain requirements such as space debris mitigation requirements can be avoided. This can, however, also lead to opportunities for private entities to evade responsibility, as we saw in the case of the company Swarm,[20] and circumvent the need for national licensing and the requisite conditions of such licensing.

One issue is what measures can the international community and national regulators put in place to incentivise commercial companies to license through jurisdictions which require companies to meet sustainability guidelines?

To make matters more complicated, when establishing the responsibility of states, binding international laws would tend to be considered as part of responsible compliance, not voluntary and non-binding guidelines and standards, which are the only type of guidelines and standards which exist in relation to space debris mitigation and sustainability requirements.

3. Incentivising Compliance with Space Debris Mitigation Guidelines

3.1 Large constellations

Several large satellite communications operators, such as SpaceX, OneWeb, and Kuiper, are launching and propose to launch further LEO satellite constellations and other companies seek to enter the satellite communications market. Earth imaging satellite operators, such as Planet, Spire, and others, have also launched and propose to launch larger constellations of satellites. Several GSO operators such as

Inmarsat,[21] Telesat,[22] and EchoStar[23] are also looking to launch and operate LEO satellite constellations integrated with their existing GSO satellites.

LEO satellites orbit at an altitude of around 160 to 2,000 kilometres (approximately 100 to 1,200 miles) above the Earth's surface. Such satellites can provide continuous, global communications coverage as they orbit. A LEO satellite will typically orbit the Earth in less than two hours,[24] depending on its altitude. One satellite will, therefore, only be in view from the Earth for a few minutes. LEO satellite services must therefore be 'handed off' from satellite to satellite and between terrestrial 'cells' within the satellites' footprint to ensure the continuity of communications.

The advantage of LEO systems, compared to GSO networks, is that the satellites are smaller and cheaper and the satellites' proximity to the Earth enables them to communicate with minimal time delay (lower latency). An operator will, however, usually require a large number of satellites to achieve the required coverage, and many of the constellation's satellites, and related ground control equipment and network, must be in place before revenue can be derived for commercial systems.[25] National regulators and licensing agencies may therefore impose stricter financial standing obligations on licence applicants as part of their due diligence assessment in relation to the licensing of constellations. As mentioned, national governments are responsible internationally, and are potentially liable, for their national activities and are required to authorize and continually supervise them.

3.2 The need for space debris mitigation measures

At the G7 Leaders' Summit in Cornwall, UK in 2021,[26] delegates pledged to take action to tackle the growing hazard of space debris as Earth's orbit becomes increasingly crowded. This issue particularly relates to large constellations of satellites in LEO.

This pledge is not currently manifested through any binding space debris mitigation laws which would create a clear duty, and is instead a matter of voluntary activity. Whilst debris guidelines and standards exist,[27] they are voluntary and non-binding and applied inconsistently amongst states in national laws and regulations, if applied at all.

Individual national regulators, licensing authorities, and space agencies are increasingly implementing space debris mitigation guidelines[28] through launch and operations licensing conditions, procurement rules, and through the conditions of publicly funded missions.[29]

In June 2019, the United Kingdom (UK) encouraged and sponsored activities for the long-term sustainability in outer space at UN level by supporting LTS Guidelines, which at this time remain voluntary only. These provide an important regulatory framework for space activities, safety of space operations, international cooperation, capacity-building and awareness, and scientific and technical research and development.

The underlying fear of course is that space debris may accumulate to the point that it precludes access to space, and all the value and benefits such access provides to us on Earth, unless existing debris fields are reduced in size, removed, and remediated, and new debris creation is much curtailed.

The orbital population, particularly in LEO, is growing quickly and will increase even faster in the next years.[30] For a satellite constellation operator, the inter-constellation collision risk is increasing, as is the issue of deconflicting encounters with other operators. SpaceX seeks to 'passively deconflict' its own intra-constellation encounters through rigorous orbit design and orbit control, but nonetheless it still experiences close encounters within the constellation.

Operators face the need for more precise and robust orbit control capability, along with good tracking, to manage their constellations and the growing need to implement a large number of collision avoidance manoeuvres to ensure safety and sustainability.

There is always a certain level of 'acceptable' risk, and opportunities for error or a failure.[31] As constellation numbers rise, the law of very large numbers would indicate that an error or failure to manoeuvre will occur, almost no matter how much effort is put into trying to ensure the safety of the system, due to the large population of satellites and high number of encounters. One large risk is that of conjunction with lethal, non-trackable debris. This could result in loss of some functionality or loss of spacecraft, leading to increased collision risk and debris.

In short, space debris mitigation measures are now essential and the international community and national regulators should be considering the application of remediation measures in addition.

3.3 The increasing 'race to the top' philosophy of commercial operators

Certain freedoms are enshrined in the Outer Space Treaty such as the freedom of exploration and use of outer space for states[32] and free access to all areas of celestial bodies.[33] These freedoms are, however, limited. States must, for example, exercise these rights on the basis of equality,[34] in accordance with international law, without exercising national appropriation,[35] avoiding harmful contamination,[36] and 'with due regard to the corresponding interests of all other States Parties to the Treaty'.[37]

The drafters of the Outer Space Treaty did not envisage the increasing orbital populations we are witnessing. This raises several questions: should there be other restraints to these freedoms, such as complying with certain sustainability practices 'with due regard to the corresponding interests' of other states, and to what extent should national regulators be evolving the due diligence exercise they undertake in light of the increasing orbital populations and the raised probability of loss and damage being caused?

An effective, transparent, and proportionate regulatory framework is an enabler for raising investment. As part of such framework, regulators must assess how best to establish the financial standing of licence applicants, determine what insurance requirements to impose and conduct analysis of the technology to be launched. This needs to be done without implementing requirements which might place national entities at a disadvantage compared with entities licensing through another jurisdiction. If effectively structured and if meeting proportionate due diligence criteria, now also including space sustainability guidelines, a licence from a reputable licensing state can be a 'stamp of approval' for the licensed entity and its technology, one that investors and other regulators will recognize.

In an increasingly competitive market landscape we can look to market and investment incentives to assist to contribute to the solution. The suggestions below are not a panacea by any means but articulate two possible solutions which emanate from real and recent commercial experience. It will be some time before there is an international level playing field of national regulations or an internationally binding sustainability treaty.

3.4 Market incentives

There is an increasing correlation between:

1. the identity of the 'launching State' and the licensing state; and
2. the ease of obtaining market access in other jurisdictions.

For example, the US Federal Communications Commission (FCC) uses the following language (redacted to avoid commercial sensitivity) when granting US market access to entities licensed under the United Kingdom system:

> This grant is based on a finding that ... is and will be subject to direct and effective regulation by the United Kingdom concerning orbital debris mitigation.

> This grant will become effective and remain effective only to the extent that launch and space operations are authorized by the United Kingdom Space Agency under the United Kingdom Outer Space Act... . must file evidence in the public record of this procedure demonstrating grant of any such authorizations within five business days of action by the United Kingdom Space Agency.

One reason as to why the FCC calls out the United Kingdom expressly is that the United Kingdom applies the IADC Guidelines and ISO 24113 standards and exercises robust due diligence prior to the grant of a licence.

This extraterritorial jurisdiction request is an example of the growing link between the identity of a licensing state and its application of space debris mitigation guidelines, and the grant of market access in other jurisdictions due to the management by the licensing state and the licensed operator of the related risks.

Practically, by complying with space debris mitigation standards at international level and flowing them into national launch and operations licence conditions as mentioned above, the United Kingdom enables its licensees to obtain market access in the United States and elsewhere.

We are seeing companies looking to take, and be seen to take, a more 'responsible' attitude to compliance with sustainability criteria. Incentives such as the resulting enabling of market access encourages a 'race to the top' mentality. A licence which meets current sustainability guidelines acts as a 'stamp of approval' for market access and in raising investment.

Companies that previously migrated their licensing functions away from the United Kingdom to Germany and some African and Asian countries, for example, are now looking to seek Civil Aviation Authority launch and operations licences from the United Kingdom purely to gain the recognition of the United Kingdom licences as complying with space debris mitigation and sustainability guidelines in order to facilitate international market access.

There is also an increasing link between the reputation of a licensing state and compliance with ESG reporting standards when seeking to raise finance.

The widely hailed concept of ESG is now seen as being integral (at least in the West) to sound corporate and investment strategy, risk management, and the promotion of ethical, responsible and environmentally sustainable investment, including in the space and satellite industry.

The UN Environment Programme Finance Initiative recently stated that it is part of a company's fiduciary duty to integrate ESG objectives into its investment analysis.[38]

These ESG imperatives, which also increasingly drive investment decisions in the wider business environment, offer an opportunity for the commercial satellite industry to work closely with national regulators to make implementation of the international sustainability best practices and space debris mitigation guidelines a reality.

ESG, and the link to investment and even insurance decisions, could be the mechanism that forces the change in the approach of the space industry.

The LTS Guidelines state[39] that

> [i]In supervising space activities of non-governmental entities, States should ensure that entities under their jurisdiction and/or control that conduct outer space activities have the *appropriate structures and procedures* for planning and conducting space activities in a manner that *supports the objective of enhancing the long-term sustainability* of outer space activities, and that they have the *means to comply with relevant national and international regulatory frameworks*, requirements, policies and processes in this regard' (emphasis added).

The LTS Guidelines recognize the need for regulators to balance government risk, safety, security, and sustainable use and access to space against the aim of encouraging commercialization, innovation, and growth in the sector.

Opting to license operations through a jurisdiction which includes the recognized space debris mitigation and LTS Guidelines in its application assessment and conditions offers satellite operators a way to tangibly demonstrate their adherence to an evidenced ESG approach, which is increasingly critical for companies seeking to raise investment.

The United Kingdom Government's promotion of investment in the United Kingdom's UK's space economy and the United Kingdom's responsible approach to space evidenced by its support of the LTS Guidelines are well aligned and the investment needs of companies can therefore be a mutually supportive mechanism for driving the success of compliance with sustainability guidelines in the United Kingdom by reason of their fit with companies' investor ESG interests.

The synergy between the aims of the sustainability guidelines and ESG motivation for commercial companies and investors appears to be well-aligned. This can offer welcome momentum to encourage a 'race to the top' for regulators, commercial operators, and investors incentivizing sustainable behaviour.

National regulators competing at international level to attract high-tech satellite and space investment would be advised to consider the deployment of good forward-looking and enabling regulation for companies seeking favourable bases from which to license their operations.

Regulators successfully implementing space sustainability guidelines and best practices can engender a 'stamp of approval' reputation that effectively supports the commercial sector (even though it imposes greater restrictions on them) and aligns with its commercial aims by serving and supporting companies' and their investors' ESG reporting needs, and enabling market access across the globe.

This creates a new market in internationally recognized launch and operations licences. The overall ecosystem of international guidelines, national implementation, and commercial environmental, social, and corporate governance linked to investment and market access is a powerful one which should grow in importance.

National regulators, whose national companies compete at international level to attract high-tech satellite and space investment in the increasing forum shopping maze, should deploy best practice anticipatory and enabling regulation for companies seeking favourable bases from which to license their operations.

4. Conclusion

The dynamics, and challenges, of space activities are changing with aspiring space nations joining the international space community; new categories of non-state actors: large industrial players; start-ups; and universities; and the advent of large constellations, cubesats, robotics, and small launch facilities.

This growth in commercial activities and orbital populations has an increasing impact on the sustainability of long-term space activities. Therefore, whilst the international community and national regulators must enable the 'freedom' to access and use outer space, such access must be allowed in a way that ensures the safe and sustainable use of the space domain.

This can be achieved if both states and private entities accept greater responsibilities and private entities seeking licences accept the need to comply with sustainability requirements, which can be beneficial to them if seeking market access across the world and in raising finance.

Embracing a positively positioned 'race to the top' approach offers those regulators serious about confronting international regulatory forum shopping with a model to distinguish themselves to promote their national space economy and support investment and entrepreneurial vitality in the sector. Rather than being a negative attribute, seeking and obtaining a more robust and respected licence, which includes environmental sustainability criteria and restrictions as core elements will increasingly be regarded as a positive feature which enhances or is even necessary for achieving investment on attractive terms, market access and for public relations and acceptance.

Launch and satellite operators licensed through a regulator whose licence conditions include recognized space debris mitigation and sustainability criteria are likely, in general, to be a step ahead of those competitors who have opted for lighter touch licensing regimes in relation to demonstrating their good citizenship generally and more tangibly through what is now imperative ESG reporting, which if not required of the licensee itself by law and regulation, will almost certainly be required of investee companies by finance providers and insurers for their own reporting obligations.

Companies generally now need to be able to evidence that they are conscientious players with environmental planning and responsible local behaviour at the core of their business models—accepting certain restraints to allow greater liberty. This is necessary from a public relations perspective but also important from the perspective of key stakeholders in a business. For satellite operators (and especially constellation operators), being able to point to their regulatory compliance under a robust and well-regarded regulatory system which includes sustainability as a core component can go a long way to addressing third party interest from investors, customers, and suppliers in relation to ESG. If ESG motivations are now investment imperatives across all economic sectors, which they now seem to be, the 'race to the top' described above for the responsible choice of licensing states should not be a controversial strategy but rather a natural one for regulatory policy of companies and regulators alike to follow in the sector.

Notes

1. The Guidelines for the Long-term Sustainability of Outer Space Activities of the Committee on the Peaceful Uses of Outer Space 2018 <https://www.unoosa.org/res/oosadoc/data/documents/2018/aac_1052018crp/aac_1052018crp_20_0_html/AC105_2018_CRP20E.pdf> accessed 8 July 2022.
2. United Nations Office for Outer Space Affairs Space Debris Mitigation Guidelines of the Committee on the Peaceful Uses of Outer Space 2014 <https://www.unoosa.org/pdf/publications/st_space_49E.pdf> accessed 8 July 2022.
3. SSA refers to tracking objects in orbit, monitoring them, and predicting where they will be at any given time. <http://www.spacefoundation.org/space_brief/space-situational-awareness> accessed 8 July 2022.
4. STM is a framework for on-orbit coordination of activities to enhance the safety, stability, and sustainability of operations in the space environment. technology.nasa.gov/patent/TOP2-294
5. European Code of Conduct for Space Debris Mitigation, 28 June 2004 <https://www.unoosa.org/documents/pdf/spacelaw/sd/2004-B5-10.pdf> accessed 8 July 2022.
6. Treaty on Principles Governing the Activities of States in the Exploration and Use of Outer Space, including the Moon and Other Celestial Bodies 1967. 610 UNTS 205, entered into force 10 October 1967.
7. F. Lyall and P. Larson. *Space Law, A Treatise* (2nd edn, Abingdon: Routledge, 2018). 245.

8. J. Wheeler (ed). *The Space Law Review, The Law Reviews* (3rd edn, 2021) Chapter 1.
9. The five space treaties are as follows:

> *Treaty on Principles Governing the Activities of States in the Exploration and Use of Outer Space, including the Moon and Other Celestial Bodies 1967* (1968) 610 UNTS 205, in force 10 October 1967;
>
> *Agreement on the Rescue of Astronauts, the Return of Astronauts and the Return of Objects Launched into Outer Space 1968*, 672 UNTS 119, in force 3 December 1968;
>
> *Convention on International Liability for Damage Caused by Space Objects 1972*, 961 UNTS 187, in force 1 September 1972;
>
> *Convention on Registration of Objects Launched into Outer Space 1975*, 1023 UNTS 15, in force 15 September 1976; and
>
> *Agreement Governing the Activities of States on the Moon and Other Celestial Bodies*, UN Doc. A/34/664. Nov 1979, in force 11 July 1984.

Of most importance for, and the focus of, this chapter is the Outer Space Treaty 1969.

10. The preamble to the Outer Space Treaty calls out the United Nations General Assembly resolution 110(II) of 3 November 1947, 'which condemned propaganda designed or likely to provoke or encourage any threat to the peace, breach of the peace or act of aggression' and considered that this resolution applies to outer space.
11. J. Wheeler (ed) note 8.
12. <https://www.unoosa.org/documents/pdf/spacelaw/sd/IADC-2002-01-IADC-Space_Debris-GuidelinesRevision1.pdf> accessed 23 August 2021.
13. <https://www.unoosa.org/pdf/publications/st_space_49E.pdf> accessed 23 August 2021.
14. <https://www.unoosa.org/documents/pdf/spacelaw/sd/ISO20180921.pdf> accessed 23 August 2021.
15. <https://www.sis.se/api/document/preview/912414/> accessed 23 August 2021.
16. <https://www.sis.se/api/document/preview/80012832/> accessed 23 August 2021.
17. <https://www.unoosa.org/documents/pdf/spacelaw/sd/ITU-recommendation.pdf> accessed 23 August 2021.
18. The geostationary-satellite orbit, or GSO, is the circular orbit 35,785 km above the Earth's equator, used in particular for meteorological and communications satellites.
19. <https://www.unoosa.org/res/oosadoc/data/documents/2018/aac_1052018crp/aac_1052018crp_20_0_ html/AC105_2018_CRP20E.pdf> accessed 23 August 2021.
20. C. Henry. 'FCC Fines Swarm $900,000 for Unauthorized Smallsat Launch' *SpaceNews* 20 December 2018.
21. Inmarsat, 'Inmarsat Unveils the Communications Network of the Future' (Inmarsat, 29 July 2021) <https://www.inmarsat.com/en/news/latest-news/corporate/2021/inmarsat-announces-orchestra.html> accessed 23 August 2021.
22. <telesat.com/leo-satellites/> accessed 23 August 2021.
23. <Seradata.com/echostar-orders-two-small-satellites/> accessed 23 August 2021.
24. K. M. Peterson, 'Satellite Communications' *ScienceDirect* <https://www.sciencedirect.com/topics/engineering/low-earth-orbit> accessed 23 August 2021.
25. This is in contrast to geo-stationary (GEO) satellite networks, where a single satellite is capable of revenue generation.
26. T. Pultarova. 'The World Needs Space Junk Standards, G7 Nations Agree' (2021) Space.Com <https://www.space.com/g7-nations-commit-to-fight-space-debrishttps://www.gov.uk/government/news/g7-nations-commit-to-the-safe-and-sustainable-use-of-space> accessed 23 August 2021.
27. J. Wheeler (ed) note 8.

28. Inter-Agency Space Debris Coordination Committee, IADC Space Debris Mitigation Guidelines <https://www.unoosa.org/documents/pdf/spacelaw/sd/IADC-2002-01-IADC-Space_Debris-GuidelinesRevision1.pdf> accessed 23 August 2021. International Organization for Standardization, ISO 24,113 Space systems—Space debris mitigation requirements, <https://www.sis.se/api/document/preview/912414/> accessed 23 August 2021.

29. There are a number of international guidelines and best practice that the United Kingdom complies with including the IADC Guidelines, ISO Standards ISO 24113 and the LTS Guidelines. <https://www.caa.co.uk/space/guidance-and-resources/space-treaties/> accessed 23 August 2021.

30. If current LEO satellite constellation internet proposals become reality, about 50,000 active satellites will be in orbit within 10 years. Some authors predict that many more satellites will be launched in this time. C. Daehnick, I. Klinghoffer, B. Maritz, et al. 'Large LEO Satellite Constellations: Will It Be Different This Time?' (McKinsey & Company, 4 May 2020) <https://www.mckinsey.com/industries/aerospace-and-defense/our-insights/large-leo-satellite-constellations-will-it-be-different-this-time> accessed 23 August 2021. N. Scharping, 'The Future of Satellites Lies in the Constellations' (2021) 30 June *Astronomy* <https://astronomy.com/news/2021/06/the-future-of-satellites-lies-in-giant-constellations> accessed 23 August 2021.

31. <https://swfound.org/media/6575/swf_iridium_cosmos_collision_fact_sheet_updated_2012.pdf> accessed 8 July 2022.

32. Article I, paragraph 2 of the Outer Space Treaty.

33. Article I, paragraph 2 of the Outer Space Treaty.

34. Article I, paragraph 2 of the Outer Space Treaty.

35. Article II of the Outer Space Treaty.

36. Article IX of the Outer Space Treaty.

37. Article IX of the Outer Space Treaty.

38. <https://www.unepfi.org/fileadmin/documents/fiduciaryII.pdf> accessed 8 July 2022.

39. Guideline A.3, note 21.

30

The case for space is liberty

Robert Zubrin

I believe that the case for space is liberty. That's a very broad thesis. In this chapter, I'll try to back it up. But before I do, let's step back a little bit and talk about liberty itself.

The word 'liberty' has a Roman root originally, and I understand that it entered the English language through Norman French. There are two words for everything in English. There's the Norman word and the Saxon word. The Normans were the upper class after the conquest, so the Norman word is the polite word, whilst the Saxon word is the direct word. So, for instance, you can acquire or you can take, you can dine or you can eat. Someone could be deplorable or they can be bad. You can possess liberty or you can be free. Freedom is the direct word for liberty.

When politicians want to appear polished they employ the word liberty. But when they want people to really get what they mean, they use freedom. So, for example, during the Second World War, the Roosevelt administration wanted to try to define for Americans what they were fighting for; what it was about the war that was really worth risking fighting and even perhaps dying for; that the cause was not just to get back at the Japanese for their Pearl Harbor attack but that there were more fundamental principles at stake. So they set out such principles and enunciated them as 'the four freedoms', a piece of political science that everyday people in both uniform and out could truly understand. These four freedoms were freedom from want, freedom from fear, freedom of speech, and freedom of worship. This formulation was actually quite effective. I still have the wartime letters sent to my mother by my father and her brothers, all of whom were in the army, and they occasionally talk about the four freedoms.

Now the first two of the four freedoms, freedom from want and freedom from fear, are easy to understand. These are aspirations that are not even unique to humans; we share them with the other higher animals. They are certainly something that any Border Collie could comprehend. But as smart as they are, Border Collies might have difficulty with the next two. Freedom of speech they might get a narrow sense but not in the broader sense of which it's meant, which is political freedom. This is a unique to humans, on Earth anyway. It's something that we treasure, and it's something that people today, at least in the West, can still fully understand as something worth fighting and risking life for. The last, freedom of worship, is something of whose importance we have lost deep understanding because most people in our society—or any case those who read books of this kind—are secular. So it needs to be unpacked a bit as to why it was seen to be so critical that it was listed along with freedom from want, freedom from fear, and political freedom. I think this was so because to the people at that time this was really freedom to actualize your soul, freedom to be who

Robert Zubrin, *The case for space is liberty*. In: *The Institutions of Extraterrestrial Liberty*. Edited by Charles S. Cockell, Oxford University Press. © Robert Zubrin (2022). DOI: 10.1093/oso/9780192897985.003.0031 you really are. If it's understood that way, as a specific form of a broader human aspiration, a freedom to realize your life's true purpose in the universe, then I think it's something that we can grasp. I believe that space is fundamental to the future of all four of those freedoms.

But before we talk about the relationship between freedom and space, *per se*, let's go back in time just a little bit more, to 1876. That year, a remarkable speech—in my opinion one of the greatest speeches in American history—was given by Colonel Robert Ingersoll to an audience of civil war veterans. That speech can be found in many collections of great rhetoric, sometimes under the title of the 'Indianapolis Speech', or alternatively 'The Vision of War' (Ingersoll 1876). In it, Ingersoll tried to explain to veterans of the Union army what they had fought for. They had defeated slavery. So he defined the war in terms of the fight against slavery, and described the cruelty of slavery. But why were we able to get rid of slavery? How did we defeat slavery? Why don't we need slavery anymore? And he says, *we don't need human slavery anymore because the forces of nature are our slaves*. We make the lightning run our messages for us and we make the steam hammer and fashion what we need. The forces of nature are our slaves, he exclaims. 'They have no backs to be whipped, no hearts to be broken.' But how have we enslaved these forces, he asks. He answers, through our machines! But where did the machines come from? Where have all these marvellous machines that adorn our age, the steam engine, the railroad, the steamboat, the telegraph, and now in 1876, the telephone, come from? He answers, from free thought! Free thought, he says, has given us all the machines! This is the root of all of our progress, both material and moral. It has given us given us wealth, emancipation, and dignity by allowing us to take the stain of slavery off our flag.

So what does this have to do with space? Well, as Ingersoll said, it is free thought that gives us the machines. This is as true today as it was in 1876. The Space Age began as a contest between the free world and the unfree world, the communists. The race to the Moon would be the test of the two. This is a new ocean, John F. Kennedy said, and we are going to show that free men can sail it—and in fact, they can do it better than unfree men. Then, despite being behind initially, the United States caught up to the Soviet Union, then surpassed it, and was the first and, to date, the only country to land people on the Moon. The purpose of Apollo was not scientific. It was to astonish the world with what free people can do. I can tell you for a fact that it did just that, because when we landed on the Moon in July 1969, I was in Leningrad, playing chess. All the Russians I knew were amazed and thrilled. How did you do this? How is this possible? It made a point. This is something that free people can do.

The system that was set up to enable Apollo certainly exploited the creative potentials of a free country to solve all the manifold engineering challenges required to send people to the Moon in the 1960s. But it was imperfect, and it reached its limits. It's true that number of unfortunate political decisions in the early 1970s could have been made differently. If so, we might have progressed further, to establish a Moon base, or even to have reached Mars by the 1980s, as NASA planned to do at the time of Apollo landings. But ultimately, as a top-down enterprise, as soon as the political system failed to live up to its responsibilities, it would all come to a halt. This happened sooner than I would have preferred, but if it hadn't happened under President Nixon then it would have happened in the next administration or the one after that.

So, relatively speaking, after Apollo, our space programme stagnated. There was some progress, to be sure. In the following period, a number of important planetary probes were flown, but the stagnation of the human spaceflight programme was marked until a new force was activated. This was the potential of entrepreneurial creativity in enabling space activities. We are seeing this unfold now. The price of launch to orbit was static from 1970 to 2010, at $10,000 per kilogram (kg). For four decades this held, like a law of physics that was never going to change. But since 2010, the SpaceX company, led by Elon Musk, has cut the cost of space launch by a factor of five. It did this by first establishing a more rational business structure, eliminating the overheads caused by cost-plus contracting. But then the company started introducing partially reusable launch vehicles. It began with the Falcon 9, which reuses nine of its 10 engines, then followed with the Falcon Heavy, which reuses 27 of its 28 engines. As a result, they cut the cost of space launch from $10,000 per kilogram to $2,000 per kilogram, and if the company is successful in creating the starship, a fully reusable heavy lift launch system, the cost of space launch will be slashed to under $500 per kg. This is a remarkable thing. But the point here is not only has the SpaceX Company introduced some very useful, cost-effective launch systems, as well as Dragon capsules and so forth, but the company has proven a point: that it is possible for a well-led entrepreneurial team to do things that were previously thought needed to be done by the governments of superpowers, and not only that, but do it in one-third the time at one-tenth the cost or less than was previously deemed necessary. Not only that, this showed that such teams could do things that had been deemed entirely impossible altogether such as, for example, reusable launch vehicles that come back and land on the pad instead of being crashed into the ocean. This was something that was deemed more or less impossible by the space establishment, but now it has happened.

SpaceX has set off an entrepreneurial space race in space launch, in spacecraft, in spaceflight systems, all of which are going to serve to bring the cost of these systems down and increase their effectiveness. The cheaper space launches become, the more space launches there are going to be. That means satellites will get cheaper through mass production. The lower the launch cost, the less conservative spacecraft designers need to be. This will accelerate the progress of spacecraft technology. Taken together, this means that spacecraft will rapidly become much more cost-effective.

Furthermore, there is competition going on which will accelerate these trends. There is some free world competition from companies like Rocket Lab, Relativity Space, and Blue Origin. But there is also competition from China. There are at least five Chinese entrepreneurial launch companies that I'm aware of that have actually attracted investor funding, and who are currently trying to create systems that look a lot like Falcon 9's. The laws of physics are true for everyone. So it's a sure thing that these Chinese companies will be able to do it. Consequently, Musk is going to have to do something better, and, in fact, he is trying to do that right now. But if Starship is successful, his competitors will copy that too and he will have to innovate still further.

But if the Chinses or Russians ever really want to compete in space, they're going to need to invoke the forces of liberty. The reason why there is not a SpaceX in Russia right now is because Russia does not have liberty under law. There is certainly technical talent there, and there are people there with large amounts of capital who believe in the importance of human expansion into space. But nobody's willing to start a SpaceX in Russia because they could be expropriated instantly by the Kremlin kleptocrats if they were successful. Reason is a stick. If Russia wants to be truly competitive in this arena, it is going to have to create increased degrees of liberty. This will be true all around the world.

Liberty on Earth is something of great importance, but will there be liberty *in* space? Liberty will be necessary for us to settle space. We will need to create ever cheaper and more cost-effective launch systems, spacecraft, and space transportation systems, and this requires liberty. But why will liberty be necessary once we settle space, for example on Mars?

Some have speculated that there might not be liberty in extraterrestrial colonies because of the degree of individual dependence on their governments (Cockell 2013). The authorities in an extraterrestrial colony could turn your air off and kill you. Therefore, is it argued, these colonies will be totalitarian in their nature.

I believe that this thesis is false, and I'll give you a number of reasons why. First of all, if you look at the situation on Earth, the populations that are easiest to subject to tyranny are not highly urbanized, highly organized societies. They are peasant societies in which no single individual is particularly essential; that is, peasant populations and similar rural dispersed, nominally self-sufficient populations are the easiest ones for tyrants to exercise tyranny over. In contrast, as the medieval saying goes, 'city air makes a man free'. This is because in a highly organized society, individual people are more important *qua* individuals, they're more necessary to society and they can do it more damage, so they have to be respected more.

There's no-one who will need to be respected more than the inhabitant of a space colony because he or she could wreck the whole place if they were upset with the government. Dependence goes two ways. Individual dignity cannot be ignored in a highly organized society in which individual human skills and good will are at a premium. That's one point.

But then there's another point, which is: *why would anyone move to Mars if they knew they were going to move to a place that gave them less freedom?* They would not. There will almost certainly be many Mars colonies founded with different ideas behind them as to how they should be organized. But the ones that will grow are the ones that are most attractive to settlers, the ones that offer them the greatest opportunity to realize their human potential. That is why the 13 colonies and then the growing United States attracted so many more immigrants than Central or South America. They outgrew Latin America because they offered more liberty. There's nothing more valuable to people than liberty. People will cross oceans and take enormous risks, and leave their whole lives behind in order to obtain it. They will not do so in order to abandon it. So, from a Darwinian point of view, an extraterrestrial tyranny is an impossibility because that colony would not be able to grow, it would not be able to blossom. It would be outcompeted for immigrants by ones that offer greater liberty.

Furthermore, in order to grow not just in population but in prosperity, extraterrestrial settlements will have to offer liberty. This will be so because these extraterrestrial worlds will be frontier worlds. They will be places with a severe labour shortage, which means they're going to need immigrants. That's fundamental, but they're also going to need technology that is unique to their challenges. In colonial and nineteenth-century America, Americans became famous as gadgeteers, primarily in the area of labour-saving machinery, because an open frontier creates a severe labour shortage. America attempted to deal with that in two different ways. One was through slavery and the other was through invention. Invention won, hands down.

The frontier drove the creation of labour-saving machinery, but also social innovations, such as female schoolteachers, which then propagated east during the labour shortages induced by the Civil War. The frontier labour shortage which induced America's backyard inventor culture is going to be magnified a thousandfold on Mars. The Martians will be innovators not just in labour-saving machinery but also their more modern equivalents, which include automation, robotics, and artificial intelligence. Automation and robotics provide means to multiply the quantity of labour available in a population, whilst artificial intelligence is a way to multiply its diversity of skills. The Martian colonists are going to have to be inventors to create these technologies, which will be essential for their survival, and, once again, invention requires freedom. As Ingersoll said, free thought gave us all the machines. Furthermore, these inventions will not only meet the local needs of the colonists but will provide patents that can be licensed on Earth and create cash income to pay for necessary imports.

Another critical need on Mars will be energy. We need energy on Earth and we have lots of different ways to produce it. We can burn fossil fuels. There are tonnes of it here and an oxygen atmosphere to burn it. There are waterfalls, there's solar energy, there's wind. There are many options, so there is not a strong forcing function to create more. But now let's look at Mars. There is some solar energy on Mars, but it's only 40% as strong as the Earth. There won't be hydropower there until we terraform the planet, and the wind is too thin to be much good. You can make fossil fuels, but it requires energy to make them and the oxygen to burn them. So on Mars, fossil fuels are simply a way of storing energy, not a way of producing it. Nuclear fission could be used on Mars, but nuclear fission requires a massive industrial base to mine and refine fissile elements chemically, and then to perform isotopic separation, fuel element fabrication, and so forth. It would be very difficult to implement all that on an early Martian colony. However, Mars has deuterium.

Deuterium is six times as common on Mars as it is on Earth. Furthermore, all Martian life support systems involve electrolyzing water on a very large scale, proportional to population. Water electrolysis is the fundamental power requirement involved in separating heavy hydrogen from light hydrogen. So Mars settlements are going to have lots of deuterium available to use or export as a by-product of their life support system. Deuterium is the fuel for fusion power, so that will be a vital technology that the Martians will want very much to see developed. Therefore, they will do so.

Steam engines were invented in Britain but the steamboat was invented in the United States. This was because early America needed steamboats for transportation. The only highways young America had were rivers, and sail power would not do. For the same reason, Martians will be a driving force in creating practical fusion reactors. They will do so because they will need them. But to do so, they will need to be free. As Ingersoll said, it is free thought that gives us all the machines.

This is the driving force for extraterrestrial liberty. It is necessary for successful settlements. It is essential in order to create space launch and space transportation systems in the first place to get there, and it is essential to their growth. So, you might say from the point of view of natural selection, only free societies can become spacefaring civilizations. But one can go beyond this.

What *new* kinds of freedom can we gain by going into space? We will have the freedom to go to places where the rules haven't been written yet. That will provide an extraordinary opportunity. Writing at the time of the American Revolution, Thomas Paine said: 'We hold it in our power to begin the world anew.' Extraterrestrial colonists will be able to say the same thing. They will be able to write new rules, to try numerous noble experiments in many places on Mars and in the asteroid belt. Ultimately, they will be able to open many new worlds amongst the stars because Martian-developed fusion power will enable fusion rockets, and fusion rockets will give us interstellar capabilities.

If we do what we can do in our time, which is to establish the first human footholds on Mars to begin humanity's career as a spacefaring civilization, then 300 years from now, there will not only be a diverse set of new nations on Mars, and amongst the multitude of the asteroids and other bodies in the solar system, but on thousands of planets, orbiting thousands of stars in this realm of the galaxy. These will be and must be highly innovative, free cultures. That's because only free cultures *can* be highly innovative, and extraterrestrial settlers *must* be highly innovative if they are to be able to deal with the diverse novel environments that they will encounter. So there will be myriad new nations with new cultures, new languages, new literature's fantastical assortments of contributions to social thought, technology, invention, and histories of great deeds that will be used to inspire people to go further. That is something grand and wonderful, something that we, and all who participate in the great adventure in the future, can be responsible for developing and have the joy of creating.

The freedom to make a grand future is what we obtain by taking on the challenge of space. But there's more to it than that, because if we can even see that future, we can affect not just that far-off future but the near future, the one that we and our immediate descendants are going to have to deal with within the solar system and right here on Earth.

Let me explain what I mean. What is the great threat facing humanity today? Probably the most common answer to that question today is climate change. Climate change is real but I do not believe that it is an existential threat facing humanity in the century. Not at all. It is quite real. It is here. I'm not denying that at all, but it is not the immediate existential threat we face. Some people might say the threat is resource exhaustion. That's not real. In fact, our resources are growing exponentially because resources are a function of technology, which is a function of the number of inventors you have. The more inventors you have the more inventions you have, and thus the more resources. That is why our resources today are vastly more prolific than they were 100 years ago. But some people do believe in resource exhaustion anyway. A few might mention asteroid impacts but in fact that is only a remote probability. Furthermore, if we become spacefaring, we can certainly develop the capability to divert asteroids.

But there is an existential threat facing us. It is the same one that threatened us in the twentieth century This was not climate change, resource exhaustion, or asteroid impacts.

The catastrophes of the twentieth century were caused by something else entirely; bad ideas. In particular, they were caused by one bad idea that came in a multitude of forms. That idea was, and remains *that there isn't enough for everyone.* That being accepted, it follows that *we need to take what we need from other people, we need to kill them because otherwise they're going to take what they need from us.* This is the root idea that caused 1914 Europe, the most prosperous society that had ever existed in human history, to tear itself to pieces. It then caused it to happen again, in 1939. These were insane ideas, with no basis in reality, but they nevertheless had fatal effect. Take Hitler, claiming that 'the laws of existence require uninterrupted killing, so that the better may live', and 'Germany needs living space'. This was absolute nonsense. Germany didn't need living space. Germany today is smaller than the Third Reich. It has a larger population, yet it has a vastly higher standard of living. Why? Not because Germany succeeded invading other countries and killing people so they could steal their land and resources. Germany failed at that, and had it succeeded it would not have added anything to that nation's wealth. No, the reason why Germany today is vastly more prosperous than the Third Reich is because of the advance of science and technology, which has been a worldwide project accomplished by humanity at large. It's true Germans have contributed to this project in significant ways, but it has been much more broadly contributed to by people of many nations, including groups that the Nazis tried to exterminate. Had they succeeded in that Germany would be much poorer today, not richer. But nevertheless, this wrong idea propelled Germany to these catastrophic crimes. The same held true with Japan and China in the 1930s. The Japanese militarists reasoned that either Japan or China would eventually dominate Asia. We are going to fight for control of Asia sooner or later, they thought. Should it be sooner or later? They are weak now, so let's make it sooner, they thought.

If you come to the conclusion that you're going to have to fight someone sooner or later, it's always going to be the case for one side or the other that sooner is better than later. I happen to know for a fact, because I have spoken to them, that there are people in highly responsible positions in the US national security establishment who believe that war with China is inevitable. Why? Because there are 1.3 billion Chinese and if they all get cars and start driving around like Americans or Europeans, there won't be enough oil in the world for them and for us. You can bet your bottom dollar that these people have counterparts in Beijing who are looking at this issue from the opposite side of the chessboard, who think analogous thoughts, except the people to be crushed are us instead of them. If this kind of thinking is allowed to prevail, there's going to be war.

But this is a false idea. We are not in competition with the Chinese or anyone else for natural resources. In fact, there's no such thing as a natural resource. There are only natural raw materials. It is human inventiveness that creates the *technology that turns raw materials into resources.* Land was not a resource until people invented agriculture, and its value as a resource multiplied as agricultural technologies advanced. Oil was not a resource until we invented oil drilling and refining and machines that could run on the product. If Napoleon Bonaparte had gone to his general staff and asked them to list the natural resources of a country that they were contemplating invading, they wouldn't have listed oil, let alone aluminium or uranium. This would not have occurred to anyone. Uranium was not a resource until we invented nuclear power. It's not that we are going to get oil from Mars. There are no natural resources on Mars today. But there will be plenty of resources there once they are resourceful people there. That's why we need Mars: to refute this catastrophic concept that there's only so much to go around.

We must make the truth visible. The standard of living of people has gone up as the world's population has increased, so we are not competing for finite resources. The nations of the world are not, as Hitler claimed, in a struggle for existence over limited resources. Why is China progressing so rapidly today? It is because of inventions, ranging from electricity through to computers, that were created in the West. But the innovations of the West, made possible by the invention of paper and printing, were made possible only because these things were invented—by the Chinese. So, we are not enemies in a struggle for existence over limited resources. Rather, what we are as a family—a disorderly family, to be sure, but nevertheless a family of nations in a joint project to expand the human prospect through our collective creative capacities. If people can see this, they'll understand that there's no point going to war over provinces when working together, using the better side of our natures, we can open up planets.

Freedom will give us space. Space will give us freedom.

References

Cockell, C. S. *Extra-Terrestrial Liberty an Enquiry into the Nature and Causes of Tyrannical Government Beyond the Earth* (London: Shoving Leopard, 2013).

Ingersoll, R. G. 'Indianapolis Speech' (1876) <https://infidels.org/library/historical/robert-ingersoll-indianapolis-speech76/> accessed 27 November 2021.